ADVANCES IN ANALYSIS AND CONTROL OF TIME-DELAYED DYNAMICAL SYSTEMS

ADVANCES IN ANALYSIS AND CONTROL OF TIME-DELAYED DYNAMICAL SYSTEMS

Edited by

Jian-Qiao Sun

University of California-Merced, USA

Qian Ding

Tianjin University, China

高等教育出版社·北京

HIGHER EDUCATION PRESS BEIJING

World Scientific

NEW JERSEY • LONDON • SINGAPORE • BEIJING • SHANGHAI • HONG KONG • TAIPEI • CHENNAI

Editors
Jian-Qiao Sun
School of Engineering,
University of California-Merced
CA, USA

Qian Ding
Department of Mechanics
Tianjin University
Tianjin, China

Higher Education Press Limited Company
4 Dewai Dajie, Beijing 100120, P. R. China and
World Scientific Publishing Co Pte Ltd
5 Toh Tuck Link, Singapore 596224

ISBN 978-981-4522-02-1

Printed in P. R. China

Preface

Analysis and control of time-delayed systems have a large number of application areas including mechanical, control, transportation, economic, and biological systems. There has been a steady increase in the interest in time-delayed dynamic systems as demonstrated by a number of published articles on the subject shown in the figure below. This book presents a collection of papers presented at the International Workshop on Recent Advances in Analysis and Control of Time-delayed Dynamical Systems in Tianjin, China on July 2 to 4, 2012. The book takes a snap shot of this active research area, and contains the research work and survey articles from the world leading experts on analysis and control of dynamic systems with time delay.

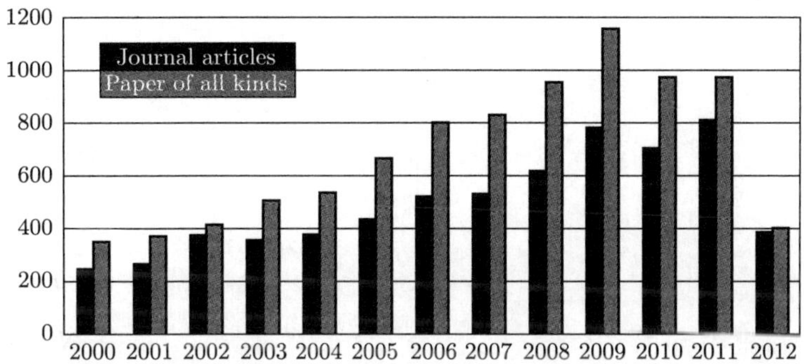

The trend of number of publications on dynamic systems with time delay in the past twelve years, based on a SCI search in June, 2012.

The chapters in the book cover a broad range of important topics. The complete quadratic Lyapunov stability conditions are discussed in Chapter 1, which offer necessary and sufficient conditions for stability of time-delayed dynamical systems. Chapter 2 describes recent advances in analysis and numerical methods of delayed differential equations (DDEs) with state-dependent time delay. State-dependent time delay can find important applications in complex dynamical systems, particularly in the systems with human interactions such as car following. Engineering applications of time-delayed dynamical systems are indeed plenty. Chapter 3 is focused on time-periodic time-delayed systems, and reports stimulating applications in machining process. Chapter 8, on the other hand, presents a methodology study of the Chebyshev spectral continuous time approximation for estimation and control of time-delayed dynamical systems. We have organized a few chapters on complex dynamics of time-delayed systems including synchronization in complex networks (Chapter 4), delay induced resonances (Chapter 6), Hopf bifurcation of time-delayed systems with complex coefficients (Chapter 7), and noise-induced dynamics of time-delayed systems (Chapter 9). Control problems of time-delayed systems have received a great deal of attention in the literature. This is reflected in the chapters of the book. Chapter 5 presents stochastic optimal control of quasi integrable Hamiltonian systems with delayed feedback control. Chapter 10 presents an overview of theoretical and experimental research on analysis, control, and time-delayed identification of structural systems. Chapter 11 deals with switching controls of uncertain linear and nonlinear dynamical systems with time delay, including supervisory control and sliding mode control.

The topics covered by the chapters are of broad interests to the research community of dynamics and control. The book can therefore be a good reference to the researchers in the community of dynamics and control including mechanical, civil, structural, aerospace, naval, and electrical engineers, and can serve as a study guide to graduate students pursuing research in the area of dynamics and control with a special interest in time-delayed dynamical systems.

Jian-Qiao Sun

Merced, California

Qian Ding

Tianjin, China

November, 2012

Contents

Chapter 1

Complete Quadratic Lyapunov-Krasovskii Functional: Limitations, Computational Efficiency, and Convergence

Keqin Gu

Department of Mechanical and Industrial Engineering
Southern Illinois University Edwardsville
Edwardsville, Illinois 62026-1805, *USA*
kgu@siue.edu

For linear time-delayed systems, the existence of a complete quadratic Lyapunov-Krasovskii functional is a necessary and sufficient condition for stability. A restriction of the kernel of integration to piecewise linear functions results in the discretized Lyapunov functional method. It is shown that the existence of such a piecewise linear Lyapunov-Krasovskii functional is also a necessary and sufficient condition for stability. Other issues related to complete quadratic Lyapunov-Krasovskii functional, its discretizations, acceleration of convergence, coupled differential-difference equations, and sum-of-square method are also discussed.

1. Introduction

Systems with time delays are used to model a variety of practical systems in engineering, life science, and economics. It has been very actively investigated in the control systems and dynamical systems circle. Basic theories of such systems may be found in Refs. 5, 11, 26, 31 and the references therein. The stability analysis of such systems may be divided into frequency domain methods and time-domain methods based on Lyapunov-Krasovskii functionals. This chapter is concentrated on time-domain methods. For frequency-domain methods, the readers are referred to Refs.19, 21, 62–65, 67 and the references therein.

The theoretical basis for extending the Lyapunov stability theory to time-delayed systems was established by Krasovskii[33]. To bring out the main spirit of such extension, consider the system described by

$$\dot{x}(t) = f(x(t), x(t-r), t), \tag{1}$$

where $x(t) \in \mathbb{R}^n$. Krasovskii showed that a traditional Lyapunov function that depends only on $x(t)$ and $x(t-r)$ is insufficient. In its place is a functional that depends on $x(t+\theta)$ for all $\theta \in [-r, 0]$, known now as a Lyapunov-Krasovskii functional. Let $\mathcal{C}$ represent the set of $\mathbb{R}^n$-valued continuous functions defined in the interval $[-r, 0]$, and let x_t represent a function in $\mathcal{C}$ obtained from x by shifting t

and restricting to the interval $[-r, 0]$, i.e.,

$$x_t(\theta) = x(t+\theta), \quad \theta \in [-r, 0]. \tag{2}$$

Then Krasovskii's theory indicates that a Lyapunov-Krasovskii functional must depend on x_t, as stated in the following theorem, which is a special case of Theorem 2.1 of Chapter 5 of Ref.26 or Theorem 1.3 of Ref.11.

Theorem 1. *Suppose the function f in (1) is such that*

$$\sup\{|f(x_0, x_r, t)| : x_0 \in X_0, x_r \in X_r, t \in \mathbb{R}\} < \infty$$

for any given bounded sets $X_0 \subset \mathbb{R}^n$ and $X_r \subset \mathbb{R}^n$. Suppose also that there exist functions $u, v, w : \bar{\mathbb{R}}_+ \to \bar{\mathbb{R}}_+$ that are continuous and nondecreasing, and the functions u and v also satisfy

$$u(0) = v(0) = 0,$$
$$u(s) > 0,\ v(s) > 0 \text{ for all } s > 0.$$

If there exists a continuous functional $V : \mathbb{R}\times\mathcal{C} \to \mathbb{R}$ that satisfies

$$u(|\phi(0)|) \le V(t, \phi) \le v(\|\phi\|) \tag{3}$$

and

$$\dot{V}(t, \phi) \le -w(|\phi(0)|), \tag{4}$$

then the trivial solution of (1) is uniformly stable. If $w(s) > 0$ for all $s > 0$, then it is uniformly asymptotically stable. If, in addition, $\lim_{s\to\infty} u(s) = \infty$, then it is globally uniformly asymptotically stable.

The notation used in the above theorem is defined below. $|\cdot|$ represents the norm of a vector in $\mathbb{R}^n$ for any finite n. For a function $\phi \in \mathcal{C}$,

$$\|\phi\| = \sup_{-r\le\theta\le 0} |\phi(\theta)|\,. \tag{5}$$

The notation $x(t; \tau, \phi)$ is used to represent the solution of (1) with the inital condition

$$x(\tau + \theta) = \phi(\theta), \ -r \le \theta \le 0, \tag{6}$$

and $x_t(\tau, \phi)$ is defined from $x(t; \tau, \phi)$ analogous to (2). For time-invariant systems, $x(t, \phi)$ and $x_t(\phi)$ are often used instead of $x(t+\tau; \tau, \phi)$ and $x_{t+\tau}(\tau, \phi)$, which are independent of τ. The notation $\dot{V}$ represents the derivative of the Lyapunov-Krasovskii functional V along the system trajectory

$$\begin{aligned}\dot{V}(\tau, \phi) &= \frac{d}{dt}V(t, x_t)\Big|_{t=\tau, x_t=\phi} \\ &= \limsup_{\Delta t\downarrow 0} \frac{1}{\Delta t}\left[V(\tau + \Delta t, x_{\tau+\Delta t}(\tau, \phi) - V(\tau, \phi)\right].\end{aligned} \tag{7}$$

While stronger statements on the general nonlinear systems, such as converse Lyapunov-like theorems, are available in Ref.33, they will not be discussed here. Rather, the focus will be on linear systems and their perturbations.

An alternative method that only uses $x(t)$ and $x(t-r)$, known as Razumikhim method, was developed later on Refs.55, 56. Direct application of Razumikhim method tends to be very conservative, although methods are available to reduce such conservatism, see, for example, Ref.26.

For a linear system, it was shown in Repin[57] that the system is exponentially stable if and only if there exists a quadratic Lyapunov-Krasovskii functional. The result was extended to the general systems of retarded type with multiple delays and distributed delays by Huang[28]. See Refs.29, 30 for additional results on quadratic Lyapunov-Krasovskii functional. Parallel results for coupled differential difference equations, which include systems of neutral type as a special case, are available in Refs.18, 20 for single delay case, and Ref.13 for multiple delay case. These quadratic Lyapunov-Krasovskii functionals are known as *complete quadratic* to distinguish them from the simple quadratic functionals such as discussed in Ref.68. Such a complete Lyapunov-Krasovskii functional is necessary for an accurate stability test for general linear system, in contrast to the case of simple quadratic Lyapunov-Krasovskii functionals that generally do not have the ability to approach analytical stability limit.

The first numerical implementation of stability conditions based on complete quadratic Lyapunov-Krasovskii functional was proposed in 1997 by Gu[17]. The method is based on piecewise linear discretization of the kernel of the complete quadratic Lyapunov-Krasovskii functional. For any given system that is exponentially stable, the stability conditions given in Ref.17 is satisfied for a sufficiently fine gridding. The resulting stability conditions are in the form of linear matrix inequalities (LMIs) for which efficient algorithms based on interior point method are available[2]. Of course, the sizes of the LMIs grow as the gridding becomes finer, and computational efficiency becomes the key to practical applications of the method. A drastic improvement on the convergence rate, and thus computational efficiency, was achieved in Ref.16 by introducing a number of new techniques, especially the application of the Jensen's inequality. The results have also been extended to more general cases, such as systems with multiple delays[15]. These methods are known as *discretized Lyapunov functional* (DLF) methods, and are summarized in the book[11]. Control design based on DLF method has also been developed[9].

Another interesting piecewise discretization of the quadratic Lyapunov-Krasovskii functional kernel was proposed by Ochoa and Kharitonov[41] in 2005. A number of extensions have also been pursued. See, for example, Ref.42. This method seeks to more closely approximate the "ideal" quadratic functional given in Ref.30. However, in practice, this method converges to the analytical stability limit much slower than the method given in Ref.16, and therefore, is much less efficient computationally.

Another numerically implementable and asymptotically accurate method, known as *sum-of-square* (SOS) method, was proposed by Peet et al.[46–49] in 1996. The method is based on approximating the kernel of the quadratic functional by positive polynomials. SOS method seems to be the only method currently available whose computational efficiency is comparable to that of DLF method.

It should be pointed out that not all the numerically implementable and asymptotically accurate stability analysis are based on a complete quadratic Lyapunov-Krasovskii functional. A notable example is the delay-decomposition (also known as delay-partitioning) method proposed by Gouaisbaut and Peaucelle[10] in 2006. It is even possible to use ordinary differential equations to approximate the differential-difference equations in stability analysis, see, for example, the works by Krasovskii[32], Repin[58], Ross and Flugge-Lotz[60], Ross[61], Delfour[4], and Halanay and Rasvan[24]. Again, these methods tend to be much slower in converging to the analytical stability methods, and are computationally not as efficient.

Another important development in the complete quadratic Lyapunov-Krasovskii functional method is the formulation of the coupled differential-difference equations. This formulation include systems of retarded and neutral type as special cases. While this model has been pursued as early as 1970s[52], most literature attempts to transform to the tradtitional form of neutral type differential-difference equations for analysis[25,36]. There is a resurgence of interest in direct analysis of such systems in recent years[50,51,54], and a recent planary session was devoted on this topic[53]. Relevant to the complete quadratic Lyapunov-Krasovskii functional are the formulation in Refs.18, 20 for systems with single delay, and Refs 13, 34 for systems with multiple delays. While the model has been used in the literature for some time, analyzed using frequency domain method[1] and known as the Roesser's model[35,59], its generality from the "pulling-out delay" perspective was discussed for the first time in Ref.13. In addition to the implication of covering a much larger class of systems, the formulation also has significant implication on computational efficiency for practical systems. Indeed, for both DLF method and SOS method, several order of magnitude of reduction of computational time was reported in typical systems[18,69]. The coupled differential-difference equation formulation also allows a more rigorous formulation of stability problems with uncertain time-varying delays[14,22].

The purpose of this chapter is to give a somewhat informal introduction of stability analysis based on complete Lyapunov-Krasovskii functional, and discuss a number of topics of practical importance that have been barely touched, or not touched at all in the literature. These include a rigorous proof of convergence, alternative formulations to improve computational efficiency, and theoretical rigor of some formulations in the literature. For those who are new to the topic, it is hoped that they can get a sense of the main spirit of the method before getting into the technical details. For those with some knowledge of the method, it is hoped that the author is able to share some perspectives that he gained from his exploration in the topic that is not easily presented in a more formal and rigorous fashion.

2. Complete Quadratic Lyapunov-Krasovskii Functional

Consider the linear time-invariant system with single delay described by the following equation

$$\dot{x}(t) = A_0 x(t) + A_1 x(t-r). \tag{8}$$

It has been well established[29, 30, 57] that a complete quadratic Lyapunov-Krasovskii functional is sufficient to study the stability of such a system. The presentation here follows Section 5.6 of Ref.11 where additional details may be found. A complete quadratic Lyapunov-Krasovskii functional can be motivated from obtaining a functional $v_W(\phi)$ whose derivative along the system trajectory satisfies

$$\dot{v}_W(x_t) = -x^T(t) W x(t) \tag{9}$$

for an arbitrarily given positive definite matrix W. For exponentially stable systems (which is equivalent to asymptotically stable systems in this case), it can be shown that

$$v_W(\phi) = \int_0^\infty x^T(t,\phi) W x(t,\phi) dt \tag{10}$$

satisfies this requirement. The expression above parallels the result for finite-dimensional system (see Section 5.6.1 in Ref.11).

The expression (10) is a quadratic expression of ϕ. Indeed, let $\Phi(t)$ be the fundamental solution of (8),

$$\Phi(t) = \mathcal{L}^{-1}\left[(sI - A_0 - e^{-rs}A_1)^{-1}\right], \tag{11}$$

where $\mathcal{L}^{-1}[\cdot]$ represents the inverse Laplace transform. Let

$$U_W(\tau) = \int_0^\infty \Phi^T(t) W \Phi(t+\tau) dt, \tag{12}$$

which is well defined when the system is exponentially stable. Then, it can be shown that

$$v_W(\phi) = \Phi^T(0) U_W(0) \phi(0) + 2\phi^T(0) \int_{-r}^0 U_W(-r-\theta) A_1 \phi(\theta) d\theta \tag{13}$$

$$+ \int_{-r}^0 \int_{-r}^0 \phi^T(\theta_1) A_1^T U_W(\theta_1 - \theta_2) A_1 \phi(\theta_2) d\theta_1 d\theta_2.$$

In the analysis later, it is useful to note that

$$U_W(-\tau) = U_W^T(\tau), \tag{14}$$

and $U_W(\tau)$ is continuous in $[-\tau, \tau]$, and twice differentiable in $[-\tau, 0)$ and $(0, \tau]$ with bounded second order derivative

$$\left| \frac{d^2 U_W(\tau)}{dt^2} \right| \leq M \text{ for some constant } M > 0, \tag{15}$$

which is a consequence of Theorem 5.18 in Ref.11.

It turns out that one may construct a complete quadratic Lyapunov-Krasovskii functional of the following form rather easily based on $v_W(\phi)$ in (13)

$$V(\phi) = \phi^T(0)P\phi(0) + 2\phi^T(0)\int_{-r}^{0} Q(\xi)\phi(\xi)d\xi \qquad (16)$$
$$+ \int_{-r}^{0}\int_{-r}^{0} \phi^T(\xi)R(\xi,\eta)\phi(\eta)d\eta + \int_{-r}^{0} \phi^T(\xi)S(\xi)\phi(\xi)d\xi,$$

where

$$P^T = P, \qquad (17)$$

$$R^T(\xi,\eta) = R(\eta,\xi), \qquad (18)$$

$$S^T(\xi) = S(\xi). \qquad (19)$$

It is not difficult to calculate that its derivative along the system trajectory can be obtained as

$$\dot{V}(\phi) = -\phi^T(0)[-PA_0 - A_0^T P - Q(0) - Q^T(0) - S(0)]\phi(0) \qquad (20)$$
$$-\phi^T(-r)S(-r)\phi(-r) - \int_{-r}^{0} \phi^T(\xi)\left[\frac{d}{d\xi}S(\xi)\right]\phi(\xi)d\xi$$
$$-\int_{-r}^{0} d\xi \int_{-r}^{0} \phi^T(\xi)\left[\frac{\partial}{\partial\xi}R(\xi,\eta) + \frac{\partial}{\partial\eta}R(\xi,\eta)\right]\phi(\eta)d\eta$$
$$+2\phi^T(0)[PA_1 - Q(-r)]\phi(-r)$$
$$+2\phi^T(0)\int_{-r}^{0}\left\{A_0^T Q(\xi) - \left[\frac{d}{d\xi}Q(\xi)\right] + R(0,\xi)\right\}\phi(\xi)d\xi$$
$$+2\phi^T(-r)\int_{-r}^{0}[A_1^T Q(\xi) - R(-r,\xi)]\phi(\xi)d\xi.$$

Indeed, it is not difficult to show that the system described by (8) is exponentially stable if and only if there exist a matrix P, matrix functions $Q(\xi)$, $R(\xi,\eta)$, and $S(\xi)$ that satisfy (17) to (19), such that

$$V(\phi) \geq \varepsilon\,|\phi(0)|^2, \qquad (21)$$

$$\dot{V}(\phi) \leq -\varepsilon\,|\phi(0)|^2 \qquad (22)$$

are satisfied for some $\varepsilon > 0$. In the next section, it will be shown that these matrix functions can be restricted to piecewise linear, which has an important implication in implementing such a Lyapunov-Krasovskii functional based stability test.

3. Discretized Lyapunov Functional Method

Checking the existence of P, $Q(\xi)$, $R(\xi,\eta)$, and $S(\xi)$ to satisfy conditions (21) and (22) is a formidable task in general as the space for matrix functions $Q(\xi)$, $R(\xi,\eta)$, and $S(\xi)$ are infinite-dimensional. To make the Lyapunov-Krasovskii functional based stability test practically implementable, restrict these matrix functions to be piecewise linear. To illustrate the basic idea without getting overly involved in technical detail, only the uniform gridding is considered here. For more general case, the readers are referred to Ref.16 or Section 5.6 of Ref.11.

Divide the interval $[-r, 0]$ to N smaller intervals of equal length

$$h = \frac{r}{N}, \tag{23}$$

and denote

$$\theta_p = -ph = -\frac{pr}{N}, \quad p = 0, 1, \ldots, N. \tag{24}$$

Then, the matrix functions $Q(\xi)$ and $S(\xi)$ are constrained to be linear within each interval $[\theta_p, \theta_{p-1}]$,

$$Q(\theta_p + \alpha h) = (1-\alpha)Q_p + \alpha Q_{p-1}, \tag{25}$$

$$S(\theta_p + \alpha h) = (1-\alpha)S_p + \alpha S_{p-1}, \tag{26}$$

and the matrix function $R(\xi, \eta)$ is constrained to be linear within small triangle regions

$$R(\theta_p+\alpha h, \theta_q+\beta h) = \begin{cases} (1-\alpha)R_{pq} + \beta R_{p-1,q-1} + (\alpha-\beta)R_{p-1,q}, & \text{if } \alpha \geq \beta, \\ (1-\beta)R_{pq} + \alpha R_{p-1,q-1} + (\beta-\alpha)R_{p,q-1}, & \text{if } \alpha < \beta, \end{cases} \tag{27}$$

for $\alpha, \beta \subset [0, 1]$. Through integration by parts and Jensen inequality, it is possible to write the conditions (16) and (20) in a form of linear matrix inequalities when the matrix functions are so constrained. The resulting stability conditions are known as *discretized Lyapunov funcitonal* (DLF) method.

While ample numerical evidence has been given about the fact that discretized Lyapunov functional method converges to the analytical stability limit, an accurate statement and rigorous proof have not been given yet in the literature. The remaining part of this section is devoted to this topic.

Lemma 1. *For any given exponentially stable system described by* (8)*, define*

$$Q_e(\xi) = Q(\xi) - U_W(-r-\xi)A_1,$$

$$R_e(\xi, \eta) = R(\xi, \eta) - A_1^T U_W(\xi - \eta)A_1.$$

Then there exist constants M_Q, M_R, K_Q, and K_R and piecewise linear matrix *functions* $Q(\xi)$ *and* $R(\xi,\eta)$ *as defined in* (25) *and* (27) *such that*

$$\max_{-r\leq\xi\leq 0}\|Q_e(\xi)\| \leq M_Q h^2, \tag{28}$$

$$\max_{\substack{-r\leq\xi\leq 0\\ -r\leq\eta\leq 0}}\|R_e(\xi,\eta)\| \leq M_R h^2, \tag{29}$$

$$\max_{-r\leq\xi\leq 0}\left\|\frac{dQ_e(\xi)}{d\xi}\right\| \leq K_Q h, \tag{30}$$

$$\max_{\substack{-r\leq\xi\leq 0\\ -r\leq\eta\leq 0}}\left\|\frac{\partial R_e(\xi,\eta)}{\partial\xi}\right\| \leq K_R h, \tag{31}$$

$$\max_{\substack{-r\leq\xi\leq 0\\ -r\leq\eta\leq 0}}\left\|\frac{\partial R_e(\xi,\eta)}{\partial\eta}\right\| \leq K_R h. \tag{32}$$

Proof. To show (28) and (30) choose

$$Q(\theta_p) = U_W(-r-\theta_p)A_1, \quad p = 0,1,\ldots,N. \tag{33}$$

Then,

$$Q_e(\theta_p) = 0, \quad p = 0,1,\ldots,N,$$

and

$$\left|\frac{d^2Q_e(\xi)}{d\xi^2}\right| = \left|-\frac{d^2}{d\xi^2}U_W(-r-\xi)A_1\right| \leq M\cdot|A_1|, \quad \xi\neq\theta_p, \quad p=0,1,\ldots,N. \tag{34}$$

For any $\xi\in[-r,0]$, let p be such that $\xi\in[\theta_p,\theta_{p-1}]$. Then for any $\theta\in[\theta_p,\theta_{p-1}]$, there exists an $\alpha\in[\theta_p,\theta_{p-1}]$ such that

$$Q_e(\theta) = Q_e(\xi) + \frac{dQ_e(\xi)}{d\xi}(\theta-\xi) + \frac{1}{2}\frac{d^2Q_e(\alpha)}{d\alpha^2}(\theta-\xi)^2.$$

Especially, for $\theta=\theta_p$ and θ_{p-1},

$$0 = Q_e(\xi) + \frac{dQ_e(\xi)}{d\xi}(\theta_p-\xi) + \frac{1}{2}\frac{d^2Q_e(\alpha_0)}{d\alpha_0^2}(\theta_p-\xi)^2,$$

$$0 = Q_e(\xi) + \frac{dQ_e(\xi)}{d\xi}(\theta_{p-1}-\xi) + \frac{1}{2}\frac{d^2Q_e(\alpha_1)}{d\alpha_1^2}(\theta_{p-1}-\xi)^2.$$

Solving for $Q_e(\xi)$ and $\dfrac{dQ_e(\xi)}{d\xi}$ in the above two equations yields

$$\begin{aligned} Q_e(\xi) &= -\frac{1}{2h}\frac{d^2Q_e(\alpha_0)}{d\alpha_0^2}(\theta_p-\xi)^2(\theta_{p-1}-\xi)\\ &\quad +\frac{1}{2h}\frac{d^2Q_e(\alpha_1)}{d\alpha_1^2}(\theta_{p-1}-\xi)^2(\theta_p-\xi),\\ \frac{dQ_e(\xi)}{d\xi} &= \frac{1}{2h}\frac{d^2Q_e(\alpha_0)}{d\alpha_0^2}(\theta_p-\xi)^2 - \frac{1}{2h}\frac{d^2Q_e(\alpha_1)}{d\alpha_1^2}(\theta_{p-1}-\xi)^2. \end{aligned}$$

From the above two expressions and (34), as well as

$$|\theta_p - \xi| \leq h,$$
$$|\theta_{p-1} - \xi| \leq h,$$

(28) and (30) can be concluded. The process of showing (29), (31), and (32) is very similar, but more tedious. Notice that the rectangular region $[-r, 0] \times [-r, 0]$ can be divided into $2N \times N$ triangular regions

$$\begin{aligned} \mathcal{T}_{pq}^{u} &= \{(\xi,\eta) \mid \theta_p \leq \xi \leq \theta_{p-1}, \theta_q \leq \eta \leq \theta_{q-1}, \xi - \theta_p \leq \eta - \theta_q\}, \\ \mathcal{T}_{pq}^{l} &= \{(\xi,\eta) \mid \theta_p \leq \xi \leq \theta_{p-1}, \theta_q \leq \eta \leq \theta_{q-1}, \xi - \theta_p \geq \eta - \theta_q\}, \\ & p = 1, 2, \ldots, N; \quad q = 1, 2, \ldots, N. \end{aligned}$$

Choose

$$R(\theta_p, \theta_q) = A_1^T U_W(\theta_p - \theta_q) A_1. \tag{35}$$

Then

$$R_e(\theta_p, \theta_q) = 0.$$

If $(\xi, \eta) \in \mathcal{T}_{pq}^{u}$, then

$$\begin{aligned} R_e(\theta,\tau) = R_e(\xi,\eta) &+ \frac{\partial R_e(\xi,\eta)}{\partial \xi}(\theta - \xi) + \frac{\partial R_e(\xi,\eta)}{\partial \eta}(\tau - \eta) \\ &+ \frac{1}{2}\frac{\partial^2 R_e(\alpha,\beta)}{\partial \xi^2}(\theta - \xi)^2 + \frac{\partial^2 R_e(\alpha,\beta)}{\partial \xi \partial \eta}(\theta - \xi)(\tau - \eta) \\ &+ \frac{1}{2}\frac{\partial^2 R_e(\alpha,\beta)}{\partial \eta^2}(\tau - \eta)^2, \end{aligned}$$

for some $(\alpha, \beta) \in \mathcal{T}_{pq}^{u}$. Apply the last expression to $(\theta, \tau) = (\theta_p, \theta_q)$, $(\theta_{p-1}, \theta_{q-1})$, and (θ_p, θ_{q-1}), and solve for $R_e(\xi, \eta)$, $\frac{\partial R_e(\xi,\eta)}{\partial \xi}$, and $\frac{\partial R_e(\xi,\eta)}{\partial \eta}$ in the resulting equations, (29), (31), and (32) can be obtained. The case for $(\xi, \eta) \in \mathcal{T}_{pq}^{l}$ is very similar, except $(\theta, \tau) = (\theta_p, \theta_q)$, $(\theta_{p-1}, \theta_{q-1})$, and (θ_{p-1}, θ_q) should be used instead. □

With the above lemma, the main result in this section can be shown.

Theorem 2. *The linear time-invariant system with single delay described by* (8) *is exponentially stable if and only if there exist a matrix P, matrix functions $Q(\xi)$, $R(\xi, \eta)$, and $S(\xi)$ that are piecewise linear as expressed in* (25) *to* (27) *such that the quadratic Lyapunov-Krasovskii functional $V(\phi)$ given in* (16) *satisfies* (21), *and its derivative $\dot{V}(\phi)$ given in* (20) *satisfies* (22) *for some $\varepsilon > 0$.*

Proof. The "if" part is a direct consequence of Theorem 1. To show the "only if" part, let the system be exponentially stable, and W be a symmetric positive definite matrix. We can define the matrix $U_W(\tau)$ accordingly. Let $\varepsilon_1 > 0$ sufficiently small, but independent of h. Choose piecewise linear matrix functions $Q(\xi)$, $R(\xi,\eta)$ according to (33) and (35), and

$$P = U_W(0) + \varepsilon_1 I, \tag{36}$$

$$S(\xi) = \frac{2r+\xi}{3r} W. \tag{37}$$

It is sufficient to show that $V(\phi)$ given in (16) satisfies (21) and the corresponding $\dot{V}(\phi)$ satisfies (22) when N is sufficiently large (and h is sufficiently small). To show (21), notice that

$$V(\phi) = v_W(\phi) + \varepsilon_1 |\phi(0)|^2 + \int_{-r}^{0} \phi^T(\xi)S(\xi)\phi(\xi)d\xi + V_e(\phi), \tag{38}$$

where

$$\begin{aligned} V_e(\phi) = {} & 2\phi^T(0)\int_{-r}^{0} Q_e(\xi)\phi(\xi)d\xi \\ & + \int_{-r}^{0}\int_{-r}^{0} \phi^T(\xi)R_e(\xi,\eta)\phi(\eta)d\xi d\eta. \end{aligned}$$

According to Lemma 1,

$$\begin{aligned} \phi^T(\xi)R_e(\xi,\eta)\phi(\eta) &\geq -|R_e(\xi,\eta)|\cdot|\phi(\xi)|\cdot|\phi(\eta)| \\ &\geq -M_R h^2 |\phi(\xi)|\cdot|\phi(\eta)|, \end{aligned}$$

and

$$\begin{aligned} \phi^T(0)Q_e(\xi)\phi(\xi) &\geq -|Q_e(\xi)|\cdot|\phi(0)|\cdot|\phi(\xi)| \\ &\geq -M_Q h^2 |\phi(0)|\cdot|\phi(\xi)|. \end{aligned}$$

Using the above and the Jensen inequality (see Ref.11), it can be concluded that

$$V_e(\phi) \geq -2M_Q h^2 |\phi(0)| \int_{-r}^{0} |\phi(\xi)|\, d\xi - M_R h^2 r \int_{-r}^{0} |\phi(\xi)|^2\, d\xi. \tag{39}$$

The expression (10) obviously implies

$$v_W(\phi) \geq 0. \tag{40}$$

Also

$$\phi^T(\xi)S(\xi)\phi(\xi) \geq \frac{1}{3}\lambda_m |\phi(\xi)|^2, \tag{41}$$

where $\lambda_m > 0$ is the smallest eigenvalue of W. Using (39), (40), and (41) in (38) and complete square, we arrive at

$$\begin{aligned} V(\phi) \geq {} & \varepsilon_1 |\phi(0)|^2 - 2M_Q h^2 |\phi(0)| \int_{-r}^{0} |\phi(\xi)|\, d\xi + \left(\frac{1}{3}\lambda_m - M_R h^2 r\right)\int_{-r}^{0} |\phi(\xi)|^2\, d\xi \\ = {} & \left(\frac{1}{3}\lambda_m - M_R h^2 r\right)\int_{-r}^{0}\left[|\phi(\xi)| - \frac{M_Q h^2}{\frac{1}{3}\lambda_m - M_R h^2 r}|\phi(0)|\right]^2 d\xi \\ & + \left[\varepsilon_1 - \frac{r\left(M_Q h^2\right)^2}{\frac{1}{3}\lambda_m - M_R h^2 r}\right]|\phi(0)|^2. \end{aligned}$$

When h is sufficiently small,

$$\varepsilon_1 - \frac{r\left(M_Q h^2\right)^2}{\frac{1}{3}\lambda_m - M_R h^2 r} > 0,$$

and (21) is satisfied for a sufficiently small $\varepsilon > 0$. To show (22), take derivative of (38) to obtain

$$\begin{aligned}\dot{V}(\phi) = &-\phi^T(0)W\phi(0) + 2\varepsilon_1\phi^T(0)\left[A_0\phi(0) + A_1\phi(-r)\right] \\ &+\phi^T(0)S(0)\phi(0) - \phi^T(-r)S(-r)\phi(-r) \\ &-\int_{-r}^{0}\phi^T(\xi)\left[\frac{d}{d\xi}S(\xi)\right]\phi(\xi)d\xi + \dot{V}_e(\phi),\end{aligned} \tag{42}$$

where

$$\begin{aligned}\dot{V}_e(\phi) = &-\int_{-r}^{0}\int_{-r}^{0}\phi^T(\xi)\left[\frac{\partial R_e(\xi,\eta)}{\partial\xi} + \frac{\partial R_e(\xi,\eta)}{\partial\eta}\right]\phi(\eta)d\xi d\eta \\ &+2\phi^T(0)\int_{-r}^{0}\left[A_0^T Q_e(\xi) - \frac{dQ_e(\xi)}{d\xi} + R_e(0,\xi)\right]\phi(\xi)d\xi \\ &+2\phi^T(-r)\int_{-r}^{0}\left[A_1^T Q_e(\xi) - R_e(-r,\xi)\right]\phi(\xi)d\xi.\end{aligned} \tag{43}$$

Using (37), we obtain

$$\begin{aligned}\dot{V}(\phi) = &-\frac{1}{3}\phi^T(0)W\phi(0) + 2\varepsilon_1\phi^T(0)\left[A_0\phi(0) + A_1\phi(-r)\right] \\ &-\frac{1}{3}\phi^T(-r)W\phi(-r) - \frac{1}{3r}\int_{-r}^{0}\phi^T(\xi)W\phi(\xi)d\xi + \dot{V}_e(\phi).\end{aligned} \tag{44}$$

Using the bound given in Lemma 1, an upper bound for $\dot{V}_e(\phi)$ may be obtained. A process of completing square very similar to the one in showing (21), albeit slightly more tedious, can also be used to show (22) from (44). □

It should be pointed out that Theorem 2 does not give any indication about the speed of convergence. Indeed, as the choice of matrix functions in the proof is included in the class of matrix functions permitted in Refs.16, 17, 41, all these discretization methods converge to the analytical stability limit. Practical computational experience indicates that the method given in Ref.16 converges much faster than those given in Refs.17, 41.

4. Coupled Differential-difference Equations

For linear systems with single delay, the coupled differential-difference equation model can be described by

$$\dot{x}(t) = Ax(t) + Bx(t-r), \tag{45}$$

$$y(t) = Cx(t) + Dx(t-r). \tag{46}$$

This model also includes time-delayed systems of neutral type when $D \neq 0$. While the above model is more direct in terms of reflecting the practical system[3], most early analysis of such systems was carried out by transforming it to the standard model of neutral type differential-difference equations[25, 36]. A notable exception is Ref.52, where direct analysis is carried out.

The stability conditions for such a system are remarkably similar to those for systems of retarded type given in (8). Indeed, it was shown in Refs.18, 20 that the system described by (45) and (46) is exponentially stable if and only if there exists a complete quadratic Lyapunov-Krasovskii functional of the form

$$V(\psi, \phi) = \psi^T P \psi + \psi^T \int_{-r}^{0} Q(\xi)\phi(\xi)d\xi \tag{47}$$
$$+ \int_{-r}^{0}\int_{-r}^{0} \phi^T(\xi)R(\xi,\eta)\phi(\eta)d\eta + \int_{-r}^{0} \phi^T(\xi)S(\xi)\phi(\xi)d\xi$$

such that

$$\varepsilon |\psi|^2 \leq V(\psi, \phi) \leq M \left\|(\psi, \phi)\right\|^2, \tag{48}$$
$$\dot{V}(\psi, \phi) \leq -\varepsilon \left\|\psi\right\|^2, \tag{49}$$

for some $\varepsilon > 0$. Again, the complete quadratic Lyapunov-Krasovskii functional may be restricted to be piecewise linear in order to render the condition to LMIs.

Even for systems of retarded type, using this model may have substantial advantage over the traditional model (8). Indeed, many practical delay elements are low-dimensional, often single-input–single-output. In such a case, the A_1 matrix in (8) has a low rank. This fact can be taken advantage of by writing

$$A_1 = BC,$$

where B is full column rank and C is full row rank. If the dimension of B is $n \times m$, and that of C is $m \times n$, then $m \ll n$. The resulting LMIs in stability analysis have a much lower dimension than those for the system (8). This idea may also be used for systems with multiple delays[13, 34]. For a typical system, several order of magnitude of reduction in computational time has been achieved by using this simple trick.

In the case of systems with multiple delays, the model can be derived by a process of "pulling-out delays"[13], which parallels that of "pulling-out uncertainties" discussed in Ref.6. It is interesting to observe that the resulting model has the form of "one-delay per channel"

$$\dot{x}(t) = Ax(t) + \sum_{k=1}^{K} B_k y_k(t - r_k), \tag{50}$$

$$y_i(t) = C_i x(t) + \sum_{k=1}^{K} D_{ik} y_k(t - r_k), \quad i = 1, 2, \ldots, K. \tag{51}$$

It was shown in Ref.13 that systems with multiple delays per channel, i.e., $y_k(t-r_j)$ with $k \neq j$ are also present in the model, may be transformed to the standard "one-delay per channel" model. Such a standard form permits a much simpler analysis. For example, the fundamental solution for (50) and (51) has a much simpler expression than those for the traditional model for multiple delays given in Ref.27.

For systems with multiple delays, an important property of time-delayed systems of neutral type, including coupled differential-difference equations, is the discontinuity of its spectrum as a function of delay ratios. The behavior is due to the discontinuity of the spectrum of the associated difference equations[27, 37]

$$y_i(t) = \sum_{k=1}^{K} D_{ik} y_k(t - r_k), \quad i = 1, 2, \ldots, K. \tag{52}$$

As was discussed in Ref.12, a sensible approach to consider such a system is to consider the *practical stability*, which was proposed in studying the sensitivity of Smith predictor under delay mismatch[43, 44]. An important application of practical stability formulation is to study the change of stability as distributed-delay feedback control is replaced by its discrete-delay approximation[12], which attracted substantial attention in the control systems community in recent years[7, 38–40, 66, 70, 71].

5. Miscellaneous Issues

In this section, a number of issues of theoretical and practical importance are discussed.

5.1. *Computational Efficiency*

In some literature, the number of decision variables in LMIs are used as the basis of comparison. The method with fewer parameters are often considered as better. See, for example, Ref.8. If everything else is equal, minimum number of decision variables indeed reduces the amount of computation in general. However, computational experience indicates that the dimensions of LMIs have much larger impact on computational efficiency[14]. For example, a recurring theme in time-delayed systems is the LMI of the form

$$\begin{bmatrix} P & Q_1 & Q_2 & \cdots & Q_K \\ Q_1^T & R_1 & 0 & \cdots & 0 \\ Q_2^T & 0 & R_2 & \cdots & 0 \\ \vdots & \vdots & \vdots & & \vdots \\ Q_K^T & 0 & 0 & \cdots & R_K \end{bmatrix} > 0, \tag{53}$$

where $P, Q_k, R_k, k = 1, 2, \ldots, K$ depend linearly on decision variables. According to Proposition 2 in Ref.16, the above is equivalent to the following $K+1$ LMIs

$$\begin{bmatrix} X_k & Q_k \\ Q_k^T & R_k \end{bmatrix} > 0, \quad k = 1, 2, \ldots, K, \tag{54}$$

$$P - \sum_{k=1}^{K} X_k > 0. \tag{55}$$

Even though (54) and (55) contain additional decision variables X_k, $k = 1, 2, \ldots, K$, they take much less time to solve than (53).

5.2. *Convergence Issue for Multiple Neutral Delays*

As was mentioned in Ref.34, when there are incommensurate neutral delays, i.e., those incommensurate delays appear in the associated difference equations, inability to converge to the analytical stability limit has been observed. This is due to the fact that the "theoretical" value of some matrix functions in the Lyapunov-Krasovskii functional may contain dense nonsmooth points, which makes it impossible for piecewise linear matrix functions to approximate these matrix functions with derivative errors arbitrarily small.

5.3. *Lyapunov-Krasovskii Functionals Containing State Derivatives*

Many Lyapunov-Krasovskii functional $V(x_t)$ proposed in the literature contains state derivatives $\dot{x}(t)$. On the other hand, the fundamental stability conditions given in Theorem 1 requires $V(x_t)$ to be upper-bounded by a continuous function of x_t. As $\dot{x}(t)$ is an unbounded function of x_t, this condition is not satisfied. For systems of retarded type, this does not pose any theoretical problem, as $\dot{x}(t)$ can be interpreted as the right hand side of the system equation (for example, $A_0 x(t) + A_1 x(t-r)$ for the system described in (8)). However, when the system is of neutral type, such as the case in Ref.8, this is no longer possible, and such formulation does not yet have a sound theoretical foundation as was commented in Ref.14.

6. SOS Method

An alternative method of reducing the infinite-dimensional space of matrix functions to a finite-dimensional space is to restrict the matrix functions $Q(\xi)$, $R(\xi, \eta)$, and $S(\xi)$ to be a special class of polynomial matrices. This formulation, known as sum-of-square (SOS) method, was proposed by Peet et al.[46–49]. The formulation can also be applied to the coupled differential-difference equation formulation. Further acceleration of convergence rate may be achieved by applying some equivalence relations for quadratic integral inequalities[23]. SOS is the only method in the current literature that has a comparable computational efficiency with the DLF method. Indeed, in some cases with multiple delays, a well-tuned method was found to be even more efficient than the DLF method[69], although part of the reason in this case is due to the fact that the code used to implement the DLF method still uses the less efficient formulation in the form of (53) rather than the more efficient formulation (54) and (55).

Unlike DLF method, the matrix function $R(\xi, \eta)$ used in SOS are smooth in the entire region $[-r, 0] \times [-r, 0]$. This is actually a disadvantage, because the "theoretical value" of R given in Theorem 2 is not smooth at the line $\xi = \eta$. Indeed, a strategy, known as "semi-separable" kernel, has been proposed to accommodate this[45].

For systems with multiple neutral delays, computational experience indicates that SOS has even more difficult time converging to the analytical solutions. For example, the current formulation cannot approach the analytical stability limit for the numerical example given in Ref.34.

7. Conclusions and Perspectives

For linear time-delayed systems, the existence of quadratic Lyapunov-Krasovskii functional with piecewise linear kernel is a necessary and sufficient condition for exponential stability. The remodeling of systems in the form of coupled differential-difference equations can often reduce the computational time by several orders of magnitude. Reducing a large LMI to several smaller LMIs and introducing additional decision parameters can often significantly reduce the computational time. Sum-of-square method provides a viable alternative to the discretized Lyapunov functional method for time-delayed systems of retarded type.

Lyapunov-Krasovskii functional method may provide solutions to many other problems in dynamical systems with delays. Such problems include, for example, systems with periodic coefficients. However, a blind use of the general theory may result in a method that requires excessive computation or is overly conservative as a method for stability test. Taking advantage of the special structure of the problem is often the key for obtaining a numerically implementable method.

References

1. P. Agathoklis and S. Foda, Stability and the matrix Lyapunov equation for delay differential systems, *Int. J. Control,* **49**(2), 417–432 (1989).
2. S. Boyd, L. El Ghaoui, E. Feron, and V. Balakrishnan, *Linear Matrix Inequalities in System and Control Theory* (SIAM, Philadelphia, 1994).
3. R. Brayton, Small signal stability criterion for electrical networks containing lossless transmission lines, *IBM J. Res. Dev.*, **12**, 431–440 (1968).
4. M. C. Delfour, The linear quadratic optimal control for hereditary differential systems: theory and numerical solution, *Appl. Math. and Optimization,* **3**(2–3), 101–162 (1976).
5. O. Diekmann, S. A. van Gils, S. M. Verduyn-Lunel, and H. -O. Walther, *Delay equations, Functional-, Complex- and Nonlinear Analysis* (Springer-Verlag, New York, 1995).

6. J. C. Doyle, J. Wall, and G. Stein, Performance and robustness analysis for structured uncertainty, *Proceedings of the* 20*th IEEE Conference on Decision and Control*, 629–636 (1982).
7. K. Engelborghs, M. Dambrine, and D. Roose, Limitations of a class of stabilization methods for delay systems, *IEEE Transactions on Autimatic Control*, **46**(2), 336–339 (2001).
8. E. Fridman and G. Tsodik, H_∞ control of distributed and discrete delay systems via discretized Lyapunov functional, *European Journal of Control,* **15**(1), 84–94 (2009).
9. E. Fridman, Descriptor discretized Lyapunov functional method: analysis and design, *IEEE Trans. on Automatic control*, **51**(5), 890–897 (2006).
10. F. Gouaisbaut and D. Peaucelle, Delay-dependent stability analysis of linear time delay systems, *The sixth IFAC workshop on time-delay systems*, L'Aquila, Italy (July 2006).
11. K. Gu, V. L. Kharitonov, and J. Chen, *Stability of Time-delay Systems* (Birkhäuser, Boston, 2003).
12. K. Gu, A review of some subtleties of practical relevance for time-delay systems of neutral type, *ISRN Applied Mathematics*, 725783 (2012).
13. K. Gu, Stability problem of systems with multiple delay channels, *Automatica*, **46**(4) 727–735 (2010).
14. K. Gu, Discussions on "H_∞ control of distributed and discrete delay systems via discretized Lyapunov functional", *European Journal of Control,* **15**(1), 95–96 (2009).
15. K. Gu, Refined discretized Lyapunov functional method for systems with multiple delays, *International Journal of Robust and Nonlinear Control*, **13**(11), 1017–1033 (2003).
16. K. Gu, A further refinement of discretized Lyapunov functional method for the stability of time-delay systems, *International Journal of Control*, **74**(10), 967–976 (2001).
17. K. Gu, Discretized LMI set in the stability problem of linear uncertain time-delay systems, *International Journal of Control*, **68**(4), 923–934 (1997).
18. K. Gu and Y. Liu, Lyapunov-Krasovskii functional for uniform stability of coupled differential-functional equations, *Automatica*, **45**(3), 798–804 (2009).
19. K. Gu, S. -I. Niculescu, and J. Chen, On stability of crossing curves for general systems with two delays, *J. Math. Anal. Appl.*, **311**, 231–253 (2005).
20. K. Gu. and S. -I. Niculescu, Stability analysis of time-delay systems: a Lyapunov approach, in A. Loría, F. Lamnabhi-Lagarrigue, and E. Panteley (Eds.), *Advanced Topics in Control Systems Theory*: *Lecture Notes from FAP* 2005, Lecture Notes in Control and Information Sciences, **328,** 139–170 (Springer-Verlag, Berlin, 2006).
21. K. Gu and M. Naghnaeian, Stability crossing set for systems with three delays, *IEEE Trans. Automat. Contr.*, **56**(1), 11–26 (2011).
22. K. Gu, Y. Zhang, and S. Xu, Small gain problem in coupled differential-difference equations, time-varying delays, and direct Lyapunov method, *International Journal of Robust and Nonlinear Control*, **21**(4), 429–451 (2011).
23. K. Gu, Y. Zhang, and M. M. Peet, Positivity of complete Lyapunov-Krasovskii functionals in time-delay systems, in R. Sipahi, T. Vyhlidal, S. -I. Niculescu, and P. Pepe, (Eds.), *Time Delay Systems*: *Methods, Applications and New Trends*, Lecture Notes in Control and Information Sciences, 35–47 (Springer, New York, 2012).
24. A. Halanay and V. Rasvan, Approximation of delays by ordinary differential equations, in R. Conti (Ed.), *Recent Advances in Differential Equations,* 155–198 (Academic Press, New York, 1981).
25. J. Hale and W. Huang, Variation of constants for hybrid systems of functional differential equations, *Proceedings of Royal Society of Edinburgh*, **125A**, 1–12 (1993).

26. J. K. Hale and S. M. Verduyn Lunel, *Introduction to Functional Differential Equations* (Springer-Verlag, New York, 1993).
27. D. Henry, Linear autonomous neutral functional differential equations, *J. Differential Eqns.*, **15**, 106–128 (1974).
28. W. Huang, Generalization of Liapunov's theorem in a linear delay system, *J. Math. Anal. Appl.*, **142**, 83–94 (1989).
29. E. F. Infante and W. V. Castelan, A Lyapunov functional for a matrix difference-differential equation, *J. Diff. Equations,* **29**, 439–451 (1978).
30. V. L. Kharitonov and A. P. Zhabko, Lyapunov-Krasovskii approach to the robust stability analysis of time-delay systems, *Automatica*, **39**, 15–20 (2003).
31. V. Kolmanovskii and A. Myshkis, *Introduction to the Theory and Applications of Functional Differential Equations* (Kluwer, Dordrecht, 1999).
32. N. N. Krasovskii, The approximation of a problem of analytic design of controls in a system with time lags, *J. Appl. Math. Mech.*, **28**, 876–885 (1964).
33. N. N. Krasovskii, *Stability of Motion* [Russian] (Moscow, 1959); [English translation] (Stanford University Press, Stanford, CA, 1963).
34. H. Li and K. Gu, Discretized Lyapunov-Krasovskii functional for coupled differential-difference equations with multiple delay channels, *Automatica*, **46**(5), 902–909 (2010).
35. J. J. Loiseau and D. Brethé, The use of 2-D systems theory for the control of time-delay systems, *JESA, European Journal of Automatic Systems*, **31**(6), 1043–1058 (1997).
36. P. Martinez-Amores, Periodic solutions of coupled systems of differential and difference equations, *Annali di matematica pura ed applicata,* **121**(1), 171–186 (1979).
37. W. R. Melvin, Stability properties of functional differential equations, *J. Math. Anal. Appl.*, **48**, 749–763 (1974).
38. W. Michiels, S. Mondié, D. Roose, and M. Dambrine, The effect of approximating distributed delay control laws on stability, in S. -I. Niculescu and K. Gu (Eds.), *Advances in Time-Delay Systems*, 207–222 (Springer, Berlin, 2004).
39. L. Mirkin, On the approximation of distributed-delay control laws, *Systems and Control Letters*, **51**, 331–342 (2004).
40. S. Mondié and W. Michiels, Finite spectrum assignment of unstable time-delay systems with a safe implementation, *IEEE Transactions on Automatic Control*, **48**(12), 2207–2212 (2003).
41. G. Ochoa and V. L. Kharitonov, Lyapunov matrices for neutral type of time delay systems, *The 2nd Int. Conf. on Electrical & Electronics engineering and 11th Conf. on Electrical Engineering* (Mexico City, September 7–9, 2005).
42. G. Ochoa and S. Mondie, Approximations of Lyapunov-Krasovskii functionals of complete type with given cross terms in the derivative for the stability of time-delay systems, *The 46th Conference on Decision and Control* (New Orleans, LA, December 12–14, 2007).
43. Z. J. Palmor, Time-delay compensation — Smith predictor and its modifications, in S. Levine (Ed.), *The Control Handbook*, 224–237 (CRC Press, Boca Raton, 1996).
44. Z. J. Palmor, Stability properties of Smith dead-time compensator controller, *International Journal of Control*, **32**(6), 937–949 (1980).
45. M. M. Peet and A. Papachristodoulou, Using polynomial semi-separable kernels to construct infinite-dimensional Lyapunov functions, in *Proceedings of the IEEE Conference on Decision and Control* (2008).
46. M. M. Peet, *Stability and control of functional differential equations*, Ph.D. dissertation (Stanford University, Stanford, CA, 2006).
47. M. M. Peet and A. Papachristodoulou, Positive forms and stability of linear time-delay systems, in *Proc. IEEE Conf. Decision Control,* 187–193 (San Diego, CA, USA, Dec.

2006).
48. M. M. Peet, A. Papachristodoulou, and S. Lall, On positive forms and the stability of linear time-delay systems, *The 45th Conference on Decision and Control*, 13–15 (San Diego, CA, Dec. 2006).
49. M. M. Peet, A. Papachristodoulou, and S. Lall, Positive forms and stability of linear time-delay systems, *SIAM J. Control Optim.*, **47**(6), 3237–3258 (2009).
50. P. Pepe and E. I. Verriest, On the stability of coupled delay differential and continuous time difference equations, *IEEE Trans. Automat. Control,* **48**(8), 1422–1427 (2003).
51. P. Pepe, Z. -P Jiang, and E. Fridman, A new Lyapunov-Krasovskii methodology for coupled delay differential and difference equations, *Int. J. Control,* **81**(1), 107–115 (2007).
52. V. Răsvan, Absolute stability of a class of control processes described by functional differential equations of neutral type, in P. Janssens, J. Mawhin, and N. Rouche (Eds.), *Equations Differentielles et Fonctionelles Nonlineaires* (Hermann, Paris, 1973).
53. V. Răsvan, Functional differential equations of lossless propagation and almost linear behavior, *The 6th IFAC Workshop on Time-Delay Systems,* Plenary Lecture. (L'Aquila, Italy, 2006).
54. V. Răsvan and S. -I. Niculescu, Oscillations in lossless propagation models: a Lyapunov-Krasovskii approach, *IMA J. Math. Control Inform.,* **19**(1–2), 157–172 (2002).
55. B. S. Razumikhin, On the stability of systems with a delay [Russian], *Prikl. Mat. Meh.,* **20**, 500–512 (1956).
56. B. S. Razumikhin, Application of Liapunov's method to problems in the stability of systems with a delay [Russian], *Automat. i Telemeh.*, **21**, 740–749 (1960).
57. Y. M. Repin, Quadratic Lyapunov functionals for systems with delay [Russian], *Prikl. Mat. Meh.,* **29**, 564–566 (1965).
58. Yu. M. Repin, On the approximate replacement of systems with lags by ordinary differential equations, *J. Appl. Math. Mech.*, **29**, 254–264 (1965).
59. J. P. Richard, Time-delay systems: an overview of some recent advances and open problems, *Automatica*, **39**(10), 1667–1694 (2003).
60. D. W. Ross and I. Flugge-Lotz, An optimal control problem for systems with differential difference equation dynamics, *SIAM J. Control,* **7**, 609–623 (1969).
61. D. W. Ross, Controller design for time lag systems with a quadratic criterion, *IEEE Trans. Aut. Control*, **16**, 664–672 (1971).
62. R. Sipahi, S. -I. Niculescu, C. T. Abdallah, W. Michiels, and K. Gu, Stability and stabilization of systems with time delay, *IEEE Control Systems Magazine*, **31**(1), 38–65 (2011).
63. R. Sipahi and N. Olgac, A unique methodology for the stability robustness of multiple time delay systems, *Systems and Control Letters*, **55**(10), 819–825 (2006).
64. R. Sipahi and N. Olgac, Complete stability analysis of neutral type first order — two time delay systems with cross-talking delays, *SIAM J. of Control and Optimization*, **45**(3), 957–971 (2006).
65. G. Stépán, *Retarded Dynamical Systems: Stability and Characteristic Function* (Longman Scientific, Essex, 1989).
66. V. Van Assche, M. Dambrine, J. -F. Lafay, and J. -P. Richard, Some problems arising in the implementation of distributed-delay control laws, *Proc. the 38th IEEE Conf. Decision and Control*, 4668–4672 (Pheonix, AZ, 1999).
67. K. Walton and J. E. Marshall, Direct method for TDS stability analysis, *IEEE Proc. part D*, **134**, 101–107 (1987).
68. M. Wu, Y. He, and J. -H. She, *Stability Analysis and Robust Control of Time-Delay*

Systems (Science Press, Beijing/Springer, Heidelberg, 2010).
69. Y. Zhang, M. M. Peet, and K. Gu, Reducing the complexity of the sum-of-squares test for stability of delayed linear systems, *IEEE Transactions on Automatic Control,* **56**(1), 229–234 (2011).
70. Q. -C. Zhong, On distributed delay in linear control laws—Part I: Discrete-delay implementations, *IEEE Transactions on Automatic Control,* **49**(11), 2074–2080 (2004).
71. Q. -C. Zhong, On distributed delay in linear control laws—Part II: Rational implementations inspired from the δ-operator, *IEEE Transactions on Automatic Control,* **50**(5), 729–734 (2005).

Chapter 2

Recent Approaches for the Numerical Solution of State-dependent Delay Differential Equations with Discontinuities

Alfredo Bellen

Department of Mathematics and Geosciences
University of Trieste
bellen@units.it

This chapter concerns with the termination and/or bifurcation of the solution of delay differential equations with state-dependent delays. Two introductory examples illustrate where the termination/bifurcation phenomenon comes from and how a generalized solution may be defined after termination. Subsequently, a set of possible "regularization" of the problem, developed in the last few years for the numerical approximation of the generalized solution of state dependent delay differential equations with neutral term are presented and compared.

1. Introduction

Let us consider the following Cauchy problem for State-Dependent Neutral Delay Differential Equations:

$$\text{(NP)} \qquad \begin{cases} \dot{y}(t) = f\Big(t, y, y\big(\alpha(t,y)\big), \dot{y}\big(\alpha(t,y)\big)\Big), & 0 \le t \le t_f, \\ y(0) = y_0, \\ y(t) = \phi(t), \quad t < 0, \\ \dot{y}(t) = \dot{\phi}(t), \quad t < 0, \end{cases}$$

where

1. $f(t, y, z, w) : [t_\ell, 0] \times \mathbb{R}^d \times \mathbb{R}^d \times \mathbb{R}^d \to \mathbb{R}^d$ is a smooth vector function;

2. the initial vector function $\phi : [t_\ell, 0] \to \mathbb{R}^d$ is piecewise continuous and piecewise continuously differentiable and has a left derivative $\dot{\phi}(0^-)$ and a left limit $\phi(0^-)$ at 0;

3. the deviating argument $\alpha : \mathbb{R} \times \mathbb{R}^d \to \mathbb{R}$ is a C^1 map that satisfies the condition (a posteriori)

$$t_\ell < \alpha\,(t, y(t)) < t, \quad \forall t \in [0, t_f].$$

It is typical of delay differential equations (DDE) with deviating argument satisfying 3 (non-vanishing delay) that possible jump discontinuities $\xi_i (i = 1, \dots)$ of the initial functions ϕ and $\dot{\phi}$, as well as inequalities $\phi(0) \neq y_0$ and $\dot{\phi}(0^-) \neq$

$f\big(0,\phi(0),\phi\left(\alpha(0,\phi(0))\right),\dot\phi\left(\alpha(0,\phi(0))\right)\big)$ at the initial point $\beta_0 = 0$, reflect in a set of subsequent jump discontinuity points $\beta_j > 0$ satisfying $\alpha(\beta_j, y(\beta_j)) = \beta_i$ for some $i < j$, where, in general, the vector field $\dot y$ is discontinuous. If this is the case, such point β_j gives in turn rise to additional points, the deviating argument $\alpha(t, y(t))$ maps into β_j, where $\dot y$ may be or may be not still discontinuous. Whereas in the non neutral case at such additional points $\dot y$ is smoothed out, for neutral equations $\dot y$ may be still discontinuous and, by iterating the process, an increasing number of discontinuity points could rise for $\dot y$ along the integration interval $[0, t_f]$. The set of these points β_i, called "breaking points", will be denoted by $\mathcal{B}$.

Since the derivative $\dot y$ is discontinuous at the breaking points, we must clarify what we mean by "solution" for the problem NP.

We say that a real valued vector function $y(t)$ is a solution of the problem NP if

(i) it is continuous on $[0, t_f]$;
(ii) it is continuously differentiable in $[0, t_f] \setminus \mathcal{B}$;
(iii) it satisfies NP in $[0, t_f] \setminus \mathcal{B}$;
(iv) at those points $\{\beta_i\} \in \mathcal{B}$ where NP is not satisfied we have that

$$\lim_{t \searrow \beta_i} \dot y(t) = f\big(\beta_i, y(\beta_i), y\big(\alpha(\beta_i, y(\beta_i))\big), z\big(\alpha(\beta_i, y(\beta_i))\big)\big)$$

where

$$z\big(\alpha(\beta_i, y(\beta_i))\big) = \lim_{t \searrow \beta_i} \dot y\big(\alpha(t, y(t))\big).$$

A peculiarity of DDE with state-dependent deviating argument is that appearance of a breaking points is an "event" that can not be determined "a priori" and require the use of suitable intergrators for its accurate computation.

In order to stress the effect of the delay when it depends on $y(t)$, observe that for a non state dependent delays $\alpha(t)$, conditions $1, 2, 3$ are sufficient for the existence and uniqueness of the solution of the problem NP. On the contrary, for equations with state dependent delay, possible appearance of breaking points in $\dot y$ originated by some discontinuity in the initial data, may cause the termination or bifurcation of the solution.

The following example of non neutral equation, going back to El'sgol'ts and Norkin[5], illustrates how the presence of a jump discontinuity point $\xi < 0$ in the initial function ϕ may cause the termination of the sought solution at any breaking point $\beta > 0$ such that $\alpha(\beta, y(\beta)) = \xi$. The equation, in the simplified form

$$\dot y(t) = f\big(y(\alpha(t, y(t)))\big),$$

is

$$\begin{cases} \dot y(t) = -y\big(t - 2 - y(t)^2\big) + 5, & t \geq 0, \\ y(t) = \phi(t), & t \leq 0, \end{cases} \tag{1}$$

where the initial function

$$\phi(t) = \begin{cases} \dfrac{9}{2}, & t < -1, \\ \\ -\dfrac{1}{2}, & -1 \le t \le 0 \end{cases} \tag{2}$$

has a jump discontinuity at the point $\xi = -1$. By direct computation, one easily see that the solution in $\left[0, \dfrac{125}{121}\right]$ is given by

$$y(t) = \begin{cases} \dfrac{1}{2}(t-1), & 0 \le t \le 1, \\ \\ \dfrac{11}{2}(t-1), & 1 \le t \le \dfrac{125}{121}. \end{cases} \tag{3}$$

We observe, a posteriori, that the point $t = 1$ is a breaking point where the derivative has a jump discontinuity. Nevertheless, the solution exists to the right of this point. On the other hand, it is not difficult to see that the solution cannot be continued beyond the point $t = \dfrac{125}{121}$ where $y\left(\dfrac{125}{121}\right) = \dfrac{2}{11}$. In fact, at $t = \dfrac{125}{121}$ the deviated argument $\alpha\,(t, y(t)) = t - 2 - y(t)^2$ is equal to -1 and therefore, in a right neighborhood of such a point, $y(t - 2 - y(t)^2)$ is given by one of the two values of $\phi(t)$. Thus the solutions of (1) should take the form

$$y(t) = c\Big(t - \frac{125}{121}\Big) + \frac{2}{11},$$

with

$$c = \frac{1}{2}, \quad \text{if } t - 2 - y(t)^2 < -1$$

and

$$c = \frac{11}{2}, \quad \text{if } t - 2 - y(t)^2 \ge -1.$$

Now, each choice of c leads to a solution $y(t)$ that contradicts the assumption made on $t - 2 - y(t)^2$ and hence the solution does not exist for $t > \dfrac{125}{121}$.

The termination is essentially due to the fact that in a neighborhood of any breaking point $t > 0$ where the deviated argument $\alpha(t, y(t))$ crosses the discontinuity point $t = -1$, the vector field f undergoes a discontinuity and the problem (1) become an ordinary differential equation with discontinuous right-hand side for which the solution may continue to exist (as at $t = 1$), or cease to exist $\left(\text{as at } t = \dfrac{125}{121}\right)$.

It is worth remarking that, from the numerical point of view, integration after a breaking point is a very delicate issue that must be carefully analysed. In fact,

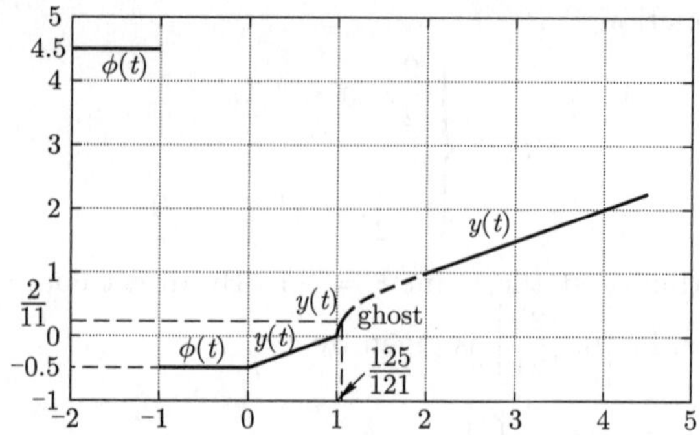

Fig. 1. Solutions (solid) and ghost solution (dashed) of (1).

the implementation of the numerical method may lead to surprising and misleading results. For example, in a right neighborhood of the termination point $t_N = \frac{125}{121}$, where $y_N \approx \frac{2}{11}$, the *Forward Euler* (FE) method reads

$$y_{n+1} = y_n + h_{n+1}\left(-\frac{9}{2} + 5\right), \quad \text{if } t_n - 2 - y_n^2 < -1$$

and

$$y_{n+1} = y_n + h_{n+1}\left(\frac{1}{2} + 5\right), \quad \text{if } t_n - 2 - y_n^2 \geq -1,$$

and nothing prevents to implement it for $n \geq N$. The resulting approximation is plotted in Fig.1 where, for $t \geq \frac{125}{121}$, a *ghost solution* (see Bellen and Zennaro[3]) appears that approximates the function $\sqrt{t-1}$. Such a function is not a solution of (1) but stays close to the manifold $\mathcal{M} = \{y : \alpha(t, y(t)) = -1\}$. In fact, at each step after t_N, the FE method provides a solution that stays on one side of the manifold where the vector field push it towards the opposite side. Actually, the numerical solution is forced to attain values y_n such that the deviated arguments $t_n - 2 - y_n^2$ oscillate around the discontinuity point -1. Such a situation corresponds, in control theory, to the so called "sliding mode" where, according to the theory of Filippov[6] on discontinuous ordinary differential equations , after the termination point $\beta = \frac{125}{121}$ a "weak solution" $y(t)$ exists which stays on the manifold $\mathcal{M}$ and $y(\alpha(t, y(t)))$ (which then equals $\phi(-1)$) is allowed to be multi-valued. In other words the weak solution stays on the manifold $\mathcal{M}$ and, simultaneusly, rather than satisfying the equation (1), satisfies the inclusion

$$\dot{y}(t) \in f\big(y(\alpha(t, y(t)))\big)$$

being the right hand side any value of the function f between $f(\phi^-(-1))$ and $f(\phi^+(-1))$ (in our example between $\frac{11}{2}$ and $\frac{1}{2}$ respectively).

The Filippov constrains for the inclusion above is then

$$\dot{y}(t) = \lambda(t)\, \frac{1}{2} + (1 - \lambda(t))\, \frac{11}{2},$$

$\lambda(t) \in [0,1]$. Actually, the function $y(t) = \sqrt{t-1}$ is a solution of the following differential algebraic equation of index two

$$\begin{cases} \dot{y}(t) = \lambda(t)\, \frac{1}{2} + (1 - \lambda(t))\, \frac{11}{2}, \\ \alpha(t, y(t)) - 1 = 0, \end{cases} \tag{4}$$

for $\lambda(t) = -\frac{1}{10\sqrt{t-1}} + \frac{11}{10} \in [0,1]$.

So, we can conclude that the "naive" implementation of the Forward Euler method has provided, for our equation, an approximation of the weak solution in the sense of Filippov.

Contrary to the FE method, the *Backward Euler* (BE) method reads

$$y_{n+1} = y_n + h_{n+1}\left(-\frac{9}{2} + 5\right), \quad \text{if } t_{n+1} - 2 - y_{n+1}^2 < -1$$

and

$$y_{n+1} = y_n + h_{n+1}\left(\frac{1}{2} + 5\right), \quad \text{if } t_{n+1} - 2 - y_{n+1}^2 \geq -1.$$

It is not difficult to see that, for small values of h_{N+1}, no solution y_{N+1} exists. This agrees perfectly with the expected termination of the solution, but may result in a very large number of rejected steps in the root-finding mechanism before the overall procedure stops.

Going back to the FE method , once the ghost (weak) solution has arrived at the new breaking point $t = 2$, with a value $y(2) \approx 1$, the integration goes on and the method provides the approximation of a classical solution that to the right of $t = 2$ exists again and is given by $y(t) = t/2$. Solutions which are defined on a disconnected set of values t are called *lacunary solutions.*

Summarizing, for the problem (1) a continuous *generalized solution* exists for any $t > 0$. It is given by the classical solution outside and by the ghost (week) solution inside the interval $\left[\frac{125}{121}, 2\right]$.

Beside termination and lacunary solutions, other interesting examples of singularities such as coalescing, bifurcation and continual bifurcation of the solutions, emerging from discontinuities in the initial data are presented in El'sgol'ts and Norkin[5] for neutral and non neutral equations.

Despite the first example of termination and bifurcation for state-dependent neutral DDEs was given by Driver[4] in the far 1961, a sistematic analysis for the numerical integration of such equations has been faced only recently in Bellen and Guglielmi[1], Guglielmi and Hairer[9,10] (see also Bellen, Guglielmi, Maset, and Zennaro[2] and the bibliography therein).

Before presenting some example of neutral equations with singularities, let us illustrate a general theory for detecting possible termination and/or bifurcation for the solution of NP after a breaking point. As shown by the BE method in the introductory example, checking numerically whether or not at some breaking point the solution ceases to exist is an important issue that prevents the codes to reduce the step size up to a minimal value before stopping integrating.

Once we have approached a breaking point ζ such that $\alpha(\zeta, y(\zeta))-\xi = 0,\ \ \xi \in \mathcal{B}$, we can (locally) write NP as an ordinary differential equation with discontinuous right-hand side

$$\dot{y}(t) = \begin{cases} h_1(t,y), & \text{if } \alpha(t,y(t)) - \xi \geq 0, \\ h_2(t,y), & \text{if } \alpha(t,y(t)) - \xi < 0, \end{cases} \tag{5}$$

where

$$h_1(t,y) = f\Big(t, y(t), x^+\big(\alpha(t,y(t))\big), \dot{x}^+\big(\alpha(t,y(t))\big)\Big),$$
$$h_2(t,y) = f\Big(t, y(t), x^-\big(\alpha(t,y(t))\big), \dot{x}^-\big(\alpha(t,y(t))\big)\Big),$$

where x^+ and x^- are smooth prolongations of the right and the left branch of $y(t)$ around the point $t = \xi$ where $\dot{y}$ is discontinuous.

Since we have removed the discontinuity, both equations (5) admit one and only one solution $y(t)$ in $[\zeta, \zeta + \epsilon]$, for which we have the following four possibilities.

$$\dot{y}(t) = \begin{cases} h_1(t,y) \Rightarrow \alpha(t,y(t)) - \xi \geq 0, & \text{solution of (5)} \\ h_2(t,y) \Rightarrow \alpha(t,y(t)) - \xi \geq 0, & \text{not solution of (5)} \end{cases} \quad \textbf{uniqueness}$$

$$\dot{y}(t) = \begin{cases} h_1(t,y) \Rightarrow \alpha(t,y(t)) - \xi < 0, & \text{not solution of (5)} \\ h_2(t,y) \Rightarrow \alpha(t,y(t)) - \xi < 0, & \text{solution of (5)} \end{cases} \quad \textbf{uniqueness}$$

$$\dot{y}(t) = \begin{cases} h_1(t,y) \Rightarrow \alpha(t,y(t)) - \xi \geq 0, & \text{solution of (5)} \\ h_2(t,y) \Rightarrow \alpha(t,y(t)) - \xi < 0, & \text{solution of (5)} \end{cases} \quad \textbf{bifurcation}$$

$$\dot{y}(t) = \begin{cases} h_1(t,y) \Rightarrow \alpha(t,y(t)) - \xi < 0, & \text{not solution of (5)} \\ h_2(t,y) \Rightarrow \alpha(t,y(t)) - \xi \geq 0, & \text{not solution of (5)} \end{cases} \quad \textbf{termination}$$

being in any case $\alpha(\zeta, y(\zeta)) - \xi = 0$.

A different and more illuminating approach for detecting uniqueness, bifurcation, and termination is the following. Consider the picture in Fig.2 where the manifold $\mathcal{M} = \{y : \alpha(t, y(t)) - \xi = 0\}$ of codimension 1 separates the two regions S^+, S^- where the vector field of the equation (5) is smooth , and let $\bar{t}$ be a breaking point such that $\bar{y} = y(\bar{t}) \in \mathcal{M}$.

Moreover, let

$$\mathbf{n}(\bar{y}) = \alpha'(\bar{y})/\|\alpha'(\bar{y})\|$$

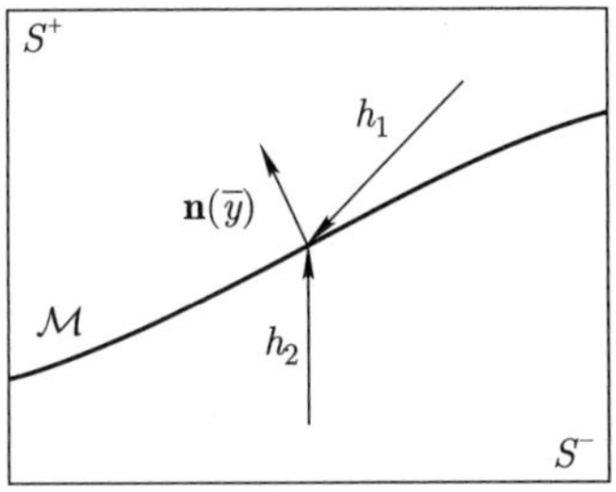

Fig. 2. The vector field around the manifold $\mathcal{M}$ at $\bar{y}(\bar{t})$ where the solution ceases to exist.

be the normal to $\mathcal{M}$ at $\bar{y}$ and let h_1 and h_2 be instances of the vector fields in the regions S^+ (identified by $\mathbf{n}$) and S^-. The solution ceases to exist at $\bar{y} \in \mathcal{M}$ if

$$\begin{cases} \langle h_1(\bar{t}, \bar{y}), \mathbf{n}(\bar{y}) \rangle < 0, \\ \langle h_2(\bar{t}, \bar{y}), \mathbf{n}(\bar{y}) \rangle > 0 \end{cases} \tag{6}$$

as drawn in the picture. In fact, in this case, both flows h_1 and h_2 push the solution towards the maniford $\mathcal{M}$ where it remains trapped and ceases to exist.

On the contrary, in the two cases when

$$\langle h_1(\bar{t}, \bar{y}), \mathbf{n}(\bar{y}) \rangle \cdot \langle h_2(\bar{t}, \bar{y}), \mathbf{n}(\bar{y}) \rangle > 0 \tag{7}$$

a unique solution exists in a right neighborhood of $\bar{t}$ that crosses traversally the manifold $\mathcal{M}$.

More delicate are the remaining cases. The first one is

$$\langle h_1(\bar{t}, \bar{y}), \mathbf{n}(\bar{y}) \rangle \cdot \langle h_2(\bar{t}, \bar{y}), \mathbf{n}(\bar{y}) \rangle < 0$$

with

$$\begin{cases} \langle h_1(\bar{t}, \bar{y}), \mathbf{n}(\bar{y}) \rangle > 0, \\ \langle h_2(\bar{t}, \bar{y}), \mathbf{n}(\bar{y}) \rangle < 0 \end{cases}$$

where both flows push the solution away form the manifold and, in principle, two solutions could leave the manifold at $\bar{y}$. On the other hand, it is evident that such an occurence cannot take place simultaneously. In fact, by the continuity of the vector field in S_1 and S_2, if both flows push the solution away from the manifold it could not have reached the manifold itself at $\bar{t}$. So, the significant remaining cases are

$$\langle h_1(\bar{t}, \bar{y}), \mathbf{n}(\bar{y}) \rangle \cdot \langle h_2(\bar{t}, \bar{y}), \mathbf{n}(\bar{y}) \rangle = 0$$

where having assumed $\bar{t}$ to be a breaking point,

$$\text{either} \quad \langle h_1(\bar{t}, \bar{y}), \mathbf{n}(\bar{y}) \rangle = 0 \quad \text{or} \quad \langle h_2(\bar{t}, \bar{y}), \mathbf{n}(\bar{y}) \rangle = 0$$

that is, either h_1 or h_2 is tangent to the manifold at $(\bar{t}, \bar{y})$.

$$\begin{cases} \langle h_1(\bar{t},\bar{y}), \mathbf{n}(\bar{y})\rangle = 0 \ \text{ and } \langle h_1(t,y(t)), \mathbf{n}(y)\rangle \geq 0 \\ \qquad\qquad\qquad\qquad\qquad \text{for } t \text{ in a right neighborhood of } \bar{t}, \\ \langle h_2(\bar{t},\bar{y}), \mathbf{n}(\bar{y})\rangle < 0 \end{cases} \tag{8}$$

and

$$\begin{cases} \langle h_1(\bar{t},\bar{y}), \mathbf{n}(\bar{y})\rangle > 0, \\ \langle h_2(\bar{t},\bar{y}), \mathbf{n}(\bar{y})\rangle = 0 \ \text{ and } \langle h_2(t,y(t)), \mathbf{n}(y)\rangle \leq 0 \\ \qquad\qquad\qquad\qquad\qquad \text{for } t \text{ in a right neighborhood of } \bar{t} \end{cases} \tag{9}$$

for both of which one solution is admissible and leaves the manifold tangentially.

Remark that if

$$\langle h_1(\bar{t},\bar{y}), \mathbf{n}(\bar{y})\rangle = 0, \quad \langle h_2(\bar{t},\bar{y}), \mathbf{n}(\bar{y})\rangle = 0$$

then two solutions are admissible (bifurcation) and leave the manifold tangentially.

2. Weak Solutions

Once the termination has been detected at some breaking point ζ generated by ξ, such that $\alpha(\zeta, y(\zeta)) - \xi = 0$, the problem is how to define and then to compute numerically a weak solution after ζ. The most natural Filippov-like approach consist in forcing the weak solution to stay on the manifold $\mathcal{M}$, that is $\alpha(t, y(t)) - \xi = 0$ and $\dot{y}$ to belong to the convex hull

$$\dot{y}(t) = \lambda f\Big(t, y(t), y(\xi), \dot{y}^+(\xi)\Big) + (1-\lambda) f\Big(t, y(t), y(\xi), \dot{y}^-(\xi)\Big),$$

$\lambda \in [0,1]$. This reflects in the following differential algebraic problem of index two

$$\begin{cases} \dot{y}(t) = \lambda(t) f\Big(t, y(t), y(\xi), \dot{y}^+(\xi)\Big) + (1-\lambda(t)) f\Big(t, y(t), y(\xi), \dot{y}^-(\xi)\Big), \\ \alpha(t, y(t)) - \xi = 0, \end{cases} \tag{10}$$

for $t \geq \zeta$.

A similar approach, due to Utkin[13], sets the vector field along the manifold as

$$\dot{y}(\alpha(y(t))) = \mu(t)\dot{y}^+(\xi) + (1-\mu(t))\dot{y}^-(\xi)$$

and hence

$$\begin{cases} \dot{y}(t) = f\Big(t, y(t), y(\xi), \mu(t)\dot{y}^+(\xi) + (1-\mu(t))\dot{y}^-(\xi)\Big), \\ \alpha(t, y(t)) - \xi = 0, \end{cases} \tag{11}$$

which coincide with the Filippov approach if $f\Big(t, y(t), y(\xi), z\Big)$ is linear w.r.t. z.

A possible alternative approach consists in solving the following delay differential algebraic problem of index two

$$\begin{cases} \dot{y}(t) = f\Big(t, y(t), y(\alpha(t, y(t))), \mu(t)\Big), \\ \alpha(t, y(t)) - \xi = 0. \end{cases} \tag{12}$$

For all approaches, after each step of the numerical integration the code must check if a classical solution revivals and, in this case, switch to the original problem.

Application of the previous analysis is illustrated by the following example (see Bellen and Guglielmi[1]) with lacunary solution and bifurcation.

$$\dot{y}(t) = -a(t)\dot{y}\Big(y(t) - 2\Big), \quad t \geq 0, \tag{13}$$

with initial function

$$\phi(t) = 1 - t, \quad \text{for } t < 0$$

where

$$a(t) = \begin{cases} 1, & t < 1, \\ 2 - t, & t \geq 1. \end{cases}$$

Equation (13) has the solution $y(t) = 1 + t$ on the interval $(0, 1)$, and a breaking point at $\zeta = 1$, created by the discontinuity at $\xi = 0$, where, according to (6), the solution ceases to exist. In fact in this simple case we have $\alpha'(y(\zeta)) = 1$ and

$$h_1(\zeta, y(\zeta)) = f(\zeta, y(\zeta), \dot{x}^+(\xi)) = -1, \qquad h_2(\zeta, y(\zeta)) = f(\zeta, y(\zeta), \dot{x}^-(\xi)) = 1.$$

A deeper analysis aimed to detect numerically possible termination or bifurcation of the solution in more general cases has been satisfactorily carried out in Bellen and Guglielmi[1] and Guglielmi and Hairer[9]. We refer the interested reader to the reference above.

Being the problem of dimension 1, the weak solution after $\zeta = 1$ is trivially given by $y(t) = 2$. As for its interpretation in the form (10), we have

$$\begin{cases} \dot{y}(t) = \lambda(t)\ 2a(t)\ - a(t), \\ y(t) - 2 = 0, \end{cases}$$

and hence $\lambda(t) = \dfrac{1}{2}$. In other words, along the manifold the derivative of the weak solution is constantly equal to the average of the left $(f(\zeta, y(\zeta), \dot{x}^+(\xi)) = -1)$ and right $(f(\zeta, y(\zeta), \dot{x}^-(\xi)) = 1)$ values of the vector field f at the breaking point $t = \zeta$, that is $\dot{y}(t) = 0$.

Similarly, in the form (12) the weak solution solves the system

$$\begin{cases} \dot{y}(t) = -a(t)\mu(t), \\ y(t) - 2 = 0, \end{cases}$$

which gives $\mu(t) = 0$. Now the derivative of the weak solution on the manyfold is given by $f(\zeta, y(\zeta), 0)$ and hence again $\dot{y}(t) = 0$.

After $t = \zeta$ the weak solution stays on the manifold up to $t = 2$ where, being $a(2) = 0$, the vector field is continuous,

$$f(2, y(2), \dot{x}^-(\alpha(2, y(2)))) = f(2, y(2), \dot{x}^+(\alpha(2, y(2)))) = 0$$

and both conditions (8) and (9) hold. Then the solution revivals and bifurcates. Summing up, the generalized solutions of (13) are given by

$$y(t) = \begin{cases} 1 + t, & 0 \le t < 1, \\ 2, & 1 \le t \le 2, \\ 2 \pm \frac{1}{2}(t-2)^2, & 2 < t \le 2 + \sqrt{2}, \end{cases}$$

and one branch is depicted in Fig.3.

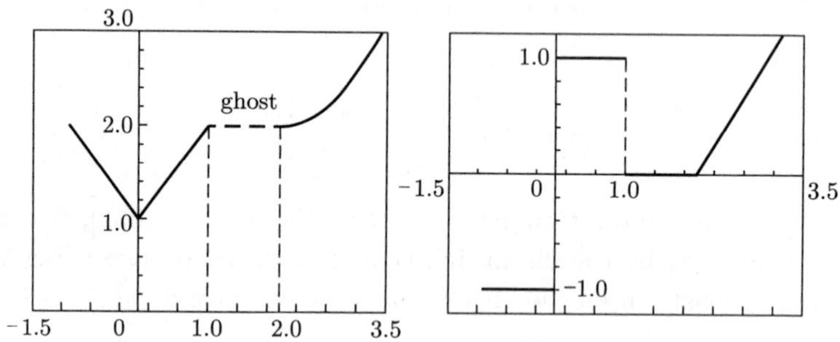

Fig. 3. Generalized solution of (13) (left) and its derivative (right).

An useful application of the generalized solution occurs in the search of periodic orbits of neutral equations by numerical integrazion (see Bellen and Guglielmi[1]). Actually, if the initial function $\phi(t)$ is not close enough to the periodic orbit, the numerical solution might cease to exist at some breaking point before approaching it. The generalized solution avoid this inconvenience and eventually approaches the periodic orbit.

Remark that for equations with more than one delay, each delay provides a manifold of codimension 1, where a weak solution evolves, and the intersection of two of them gives rise to a manifold of codimension 2 where the dynamics of the system gets more and more complicated. This case has been recently approached in Guglielmi and Hairer[11].

3. Regularization Techniques

The previous two examples have outlined the following issues to be taken into account in the analytical and numerical analysis of DDEs with state dependent delays.

1. The presence of breaking points originated by various discontinuities in the data of the problem NP may lead to various singularities in the solution, in particular termination, bifurcation, and lacunary solutions. It is worth remarking that, for neutral equations, even if $y(0) = \phi(0)$ and all the functions f, ϕ and α are continuous and smooth, in order to avoid the appearance of breaking points we must also assume the *splicing condition*

$$\dot{\phi}(0^-) = f\big(0, \phi(0), \phi\left(\alpha(0, \phi(0))\right), \dot{\phi}\left(\alpha(0, \phi(0))\right)\big).$$

If this is not the case, termination of the solution is possible at some breaking point.

2. Being the delay dependent on the solution, breaking points can not be computed a priori and a mechanism suitable to compute them during the integration is needed (see Bellen, Guglielmi, Maset, and Zennaro[2]).

3. Once a breaking point has been captured, termination and existence of one or more classical solution beyond this point must be checked before continuing the integration.

4. If the breaking point is a termination point for the classical solution a weak solution (in some sense to be defined) must be computed beyond this point.

5. During the computation of the weak solution a test must be carried out, step be step, in order to verify whether a classical solution revivals.

Codes taking into account items 2–5 for the computation of generalized solution are complicated, often not efficient and, in any case, should require important modifications of the existing codes for delay differential equations. Recently, different approaches have been carried out which are based on various regularizations of NP involving small parameters ϵ aimed to remove the discontinuities in the derivative of the solution. The underlying idea is that, for any ϵ, the regularized equation admit a unique "regularized" solutions $y_\epsilon(t)$ in $[0, t_f]$ which is computable by existing codes disregarding the location of possible breaking points and hence ignoring possible termination or bifurcation of the classical solutions $y(t)$. Once the regularized solution has been computed, the problem consists in understanding if the regularized solution converges , as $\epsilon \to 0$, to the classical solution, where it does exist, and what is the meaning of the limit where the classical solution does not. In other words, is the limit a weak solution in some known sense or is itself a new kind of weak solution? Moreover, which solution will be tracked by the regularized equation when the classical solution bifurcates? Also, when the generalized solution bifurcates in a classical and a weak solution, does the regularized equation tracks the classical one? And other more...

Assuming that the splicing condition does not hold

$$\dot{\phi}(0^-) \neq f\big(0, \phi(0), \phi\left(\alpha(0, \phi(0))\right), \dot{\phi}\left(\alpha(0, \phi(0))\right)\big),$$

various kind of regularization for equations with neutral term are presented together with the answers, if available, to the questions stated above.

Even confining to the case of equation with one single delay, the analytical and numerical analysis requires sophisticated tools and, in some particular case, the resulting dynamic turns out to be very intriguing and difficult to explain completely.

- **Regularization by the FE Method**

The first regularization of the problem NP consists in substituting the equation

$$\dot{y}(t) = f\big(t, y, y\big(\alpha(t,y)\big), \dot{y}\big(\alpha(t,y)\big)\big)$$

by the difference equation

$$y_{n+1} = y_n + \epsilon f\Big(t_n, y_n, p_\epsilon\big(\alpha(t_n, y_n)\big), \dot{p}_\epsilon\big(\alpha(t_n, y_n)\big)\Big)$$

where $y_0 = \phi(0)$, $p_\epsilon(t) = \phi(t)$ for $t \le 0$ while, for $t > 0$ it is a suitable piecewise approximation of order 1 of the discretized function $\{y_i\}_{i\in N}$, in general linear interpolation between the nodes to the left and to the right of $\alpha(t_n, y_n)$ (supposed $< t_n$). Finally, the regularization parameter ϵ is the step size $t_{n+1} - t_n$.

The piecewise approximation $p_\epsilon(t)$ is the regularized solution and the $\lim\limits_{\epsilon \downarrow 0} p_\epsilon(t)$ is the generalized solution. The convergence of the regularized solution to the classical solution, when it exists, and to some weak solution of the problem NP when it does not, has been shown in the introductory example of non neutral equation.

- **Regularization by Time-averaging**

In order to simplify the notation we can consider, with no loose of generality, the problem

$$\dot{y}(t) = f(y(t), \dot{y}(\alpha(y(t)))), \quad t \in [0, t_f] \tag{14}$$

with $y(t) = \phi(t)$ $(\dot{y}(t) = \dot{\phi}(t))$, $t \le 0$.

Being the splicing condition not assumed, let $\zeta > 0$ be any breaking point where the deviating argument $\alpha(y(\zeta)) = 0$ and the vector field of (14) has a discontinuity.

For $t \ge 0$, (14) can be written as an ordinary differential equations with discontinuous r.h.s.

$$\dot{y}(t) = \begin{cases} h_1(y(t)), & \text{if } \alpha(y(t)) \ge 0, \\ h_2(y(t)), & \text{if } \alpha(y(t)) < 0, \end{cases} \tag{15}$$

where

$$\begin{aligned} h_1(y(t)) &= f\Big(y(t), \dot{y}^+\big(\alpha(y(t))\big)\Big), \quad \text{if} \quad \alpha(y(t)) \ge 0, \\ h_2(y(t)) &= f\Big(y(t), \dot{y}^-\big(\alpha(y(t))\big)\Big), \quad \text{if} \quad \alpha(y(t)) < 0, \end{aligned} \tag{16}$$

with $\dot{y}^+$ and $\dot{y}^-$ the derivatives of the solution to the right and to the left of 0 respectively.

By introducing the average of the r.h.s.

$$\dot{y}(t) = \frac{1}{\varepsilon} \int_{t-\varepsilon}^{t} f(y(s), \dot{y}(\alpha(y(s))))ds. \tag{17}$$

Fusco and Guglielmi[7] regularized the equation (14) as

$$\dot{y}_\epsilon(t) = \frac{1}{\varepsilon}\int_{t-\varepsilon}^{t}\left[H\Big(\alpha(y_\epsilon(s))\Big)h_1\left(y_\epsilon(s)\right) + H\Big(-\alpha(y_\epsilon(s))\Big)h_2\left(y_\epsilon(s)\right)\right]ds \qquad (18)$$

where $H(\cdot)$ is the Heviside step function. A solution $y_\epsilon(t)$ of class C^1 exists and GF proved that it consists in a smooth function $x(t)$ and an highly oscillatory function $\epsilon\, v\,(t/\epsilon)$. As for the convergence of the regularized solution, the following theorem holds.

Theorem 1. *There exists $\varepsilon_0 > 0$ such that $\forall\varepsilon \in (0, \varepsilon_0)$ there exists a $T > 0$ such that the problem (18) has a C^1 solution $y_\varepsilon : [\zeta, \zeta + T] \to \mathbb{R}^d$*

- $|\alpha\left(y_\varepsilon(t)\right)| \le C\varepsilon,\ \forall t \in [\zeta, \zeta + T]$.

Moreover there is a C^1 function y^0 such that

- $\alpha\left(y^0(t)\right) \equiv 0$;
- $\lim\limits_{\varepsilon\to 0} \|y_\varepsilon - y^0\|_{C^0[\zeta,\zeta+T]} = 0$.
- *y^0 solves (according to Filippov theory) the following Cauchy problem:*

$$\begin{cases} \dot{y}(t) = \mu(t)f^-(y(t)) + (1-\mu(t))f^+(y(t)), \\ 0 = \alpha\left(y(t)\right), \end{cases}$$

where $f^-(y(t)) = f\left(y(t), \dot{\phi}^-(0)\right)$ and $f^+(y(t)) = f\left(y(t), \dot{y}^+(0)\right)$, and $\mu(t) \in [0,1]$ is given by

$$\mu(t) = \frac{\langle f^+(y(t), \mathbf{n}(y(t))\rangle}{\langle f^+(y(t)) - f^-(y(t)), \mathbf{n}(y(t))\rangle}.$$

Remark that the Filippov vector field turns out to be tangent to the manifold $\mathcal{M}$ the weak solution y^0 belongs to.

Let us consider the following system presented in Ref.7:

$$\begin{cases} \dot{y}_1(t) = y_2(y_1(t)^2 - y_2(t)) - \dot{y}_1(y_1(t)^2 - y_2(t)), \\ \dot{y}_2(t) = y_1(y_1(t)^2 - y_2(t)) - y_2(y_1(t)^2 - y_2(t)) \end{cases} \qquad (19)$$

for $t \ge 0$, with initial conditions

$$\begin{cases} \phi(t) = t^2 - t, \\ \dot{\phi}(t) = 1/2 \end{cases}$$

for $t \le 0$. Being $t = 0$ a discontinuity point for the vector field, the manifold $\mathcal{M}$ is given by the functions $y(t) = (y_1(t), y_2(t))$ such that the deviating argument

$\alpha(y) = y_1^2 - y_2$ satisfies $y_1^2 - y_2 = 0$. For any $\bar{y} = (\bar{y}_1, \bar{y}_2) \in \mathcal{M}$, direct computations of the functions (16) provides, for equation (19),

$$h_1(\bar{y}(t)) = \begin{pmatrix} -2 \\ -1/2 \end{pmatrix}, \quad h_2(\bar{y}(t)) = \begin{pmatrix} 3/2 \\ -1/2 \end{pmatrix},$$

and hence

$$\begin{cases} \langle h_1(\bar{y}), \mathbf{n}(\bar{y}) \rangle = \dfrac{1}{\sqrt{4\bar{y}_1^2 + 1}}(1/2 - 4\bar{y}_1), \\ \langle h_2(\bar{y}), \mathbf{n}(\bar{y}) \rangle = \dfrac{1}{\sqrt{4\bar{y}_1^2 + 1}}(1/2 + 3\bar{y}_1). \end{cases} \tag{20}$$

So, by (7) and (6), we can conclude that for $\bar{y}_1 \in (-1/6, 1/8)$ a classical solution exists and, for $\bar{y}_1 > 1/8$, the classical solution ceases to exist.

The regularized solution $(y_{\epsilon 1}, y_{\epsilon 2})$ computed by (18) is depicted in Fig.4. A breaking point appears at $t \approx 0.32$, with $y_{\epsilon 1} \approx 0.7$ and $y_{\epsilon 2} \approx 0.49$, where the classical solution ceases to exist. After then, a weak solution continues (with highly oscillatory behaviour) up to a second braking point $t \approx 1.28$, with $y_{\epsilon 1} \approx 1/8$ and $y_{\epsilon 2} \approx 1/64$ where it leaves the manifold and a classical solution appears again.

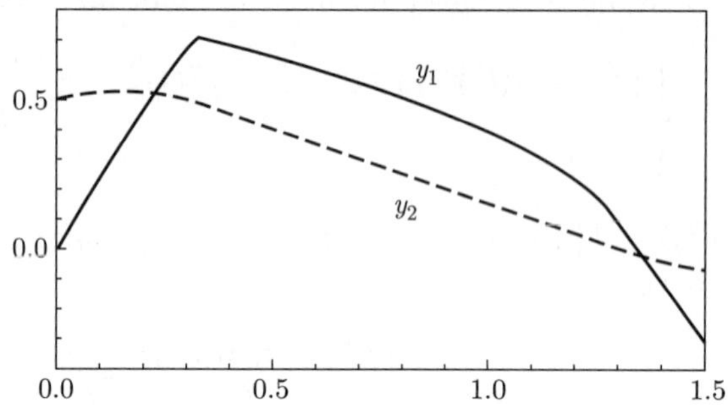

Fig. 4. Solution of (19) regularized by time-averaging with $\epsilon = 10^{-2}$. After the breaking point $t \approx 0.32$ the regularized solution y_ϵ approximates the weak solution and then, after $t \approx 1.28$, the classical solution again.

• Regularization by Singular Perturbation

Let's consider again the problem

$$\dot{y}(t) = f(y(t), \dot{y}(\alpha(y(t)))), \quad t \in [0, t_f] \tag{21}$$

with $y(t) = \phi(t)$ $(\dot{y}(t) = \dot{\phi}(t))$, $t \leq 0$.

The approach, introduced experimentally in Bellen and Guglielmi[1] and put on a firm mathematical basis in Guglielmi and Hairer[9] consists in writing the problem as an equivalent non-neutral delay differential algebraic system of index one

$$\begin{cases} \dot{y}(t) = z(t), \\ 0 = f\big(y(t), z(\alpha(y(t)))\big) - z(t), \end{cases} \quad t \in [0, t_f] \tag{22}$$

with initial data: $y(t) = \phi(t)$, $\dot{z}(t) = \dot{\phi}(t)$ for $t \leq 0$, and introducing a singular perturbation in the algebraic equation. This leads to the following singularly perturbed regularization.

$$\begin{cases} \dot{y}_\epsilon(t) &= z_\epsilon(t), \\ \epsilon \dot{z}_\epsilon(t) &= f\big(y_\epsilon(t), z_\epsilon(\alpha(y_\epsilon(t)))\big) - z_\epsilon(t), \end{cases} \qquad t \in [0, t_f] \tag{23}$$

with $y_\epsilon(t) = \phi(t)$, $\dot{z}_\epsilon(t) = \dot{\phi}(t)$ for $t \leq 0$. Remark that beyond the first breaking point ζ_0, as well as other breaking points originated by the discontinuity at $t = 0$, the regularized solution $y_\epsilon(t)$ of (23) exists and is uniquely defined. The approach overcomes the difficulties outlined at the beginning of this section and any code designed for stiff, state-dependent delay equation, such as RADAR5[a] , can solve the system (23).

On the other hand, after the breaking point ζ_0 the equation (22) may have a classical or a weak solution and even both. The problem is to determine which is the case and which solution $y(t)$ is approximated by $y_\epsilon(t)$ as $\epsilon \to 0$.

In order to answer the questions stated above, Guglielmi and Hairer[9] introduced the following scalar function

$$g(\theta) = \begin{cases} \alpha'(y(\zeta_0)) f\Big(y(\zeta_0), \theta \dot{y}^-(0) + (1-\theta)\dot{y}^+(0)\Big), & \text{for } \theta \leq 1, \\ \alpha'(y(\zeta_0)) f\Big(y(\zeta_0), \dot{y}^-(0)\Big), & \text{for } \theta \geq 1, \end{cases} \tag{24}$$

that, being ζ_0 the first breaking point where we assume that $y(t)$ enters traversally the manifold, satisfies $g(1) > 0$. According to the conditions (6), (7) of the previous section we immediately have:

- $g(0) > 0$ implies the existence of a classical solution,
- $g(0) < 0$ implies the termination of the classical solution.

They also proved that

- the existence of $\theta_0 \in (0,1)$ satisfying $g(\theta_0) = 0$ and $g'(\theta_0) \neq 0$ implies the presence of a weak solution in the sense of Utkin (11), that is satisfying

$$\begin{cases} \dot{y}(t) = f\Big(y(t), \mu(t)\dot{y}^-(0) + (1-\mu(t))\dot{y}^+(0)\Big), \\ \alpha(t, y(t)) = 0, \end{cases} \tag{25}$$

with $\dot{y}^+(\zeta_0) = f\Big(y(\zeta_0), \theta_0 \dot{y}^-(0) + (1-\theta_0)\dot{y}^+(0)\Big)$.

The last item says that a weak solution may exist together with a classical one. As for the limit of the regularized solution $y_\epsilon(t)$ the analysis is carried out by a

[a]The code is based on the adaptation of the 3-stage Radau IIA method (see Guglielmi and Hairer[8]) and is presently being distributed at the websites http://univaq.it/≈guglielm and http://www.unige.ch/≈hairer/software.html.

sophisticated generalization of the asymptotic expansion of the solution for singularly perturbed ordinary differential equation developed in Hairer and Wanner[12]. Here, after the first breaking point the righ-hand side of the system (23) contains the parameter ϵ and hence a more general asymptotic expansion is proposed in the form

$$y(\zeta_0(\epsilon)+t) = \sum_{j\geq 0} \epsilon^j\, y_j(t) + \epsilon \sum_{j\geq 0} \epsilon^j\, \eta_j(t/\epsilon),$$

$$z(\zeta_0(\epsilon)+t) = \sum_{j\geq 0} \epsilon^j\, z_j(t) + \sum_{j\geq 0} \epsilon^j\, \zeta_j(t/\epsilon),$$

where $\eta_j(\tau)$ and $\zeta_j(\tau)$ decay exponentially and $\zeta_0(\epsilon)$ is the approximation of ζ_0 satisfying $\alpha(y_\epsilon(\zeta_0(\epsilon))) = 0$.

In order to find the leading terms η and ζ of the expansion in the two cases of weak and classical solution the following dynamical system, valid in both cases, must be solved

$$\begin{cases} \dot\theta = -\theta\zeta, & \theta(0) = 1, \\ \dot\zeta = -\zeta + g(\theta), & \zeta(0) = g(1), \end{cases} \tag{26}$$

where $\theta(\tau) = e^{-\eta(\tau)}$.

The system has two stationary points. The first, $(0, g(0))$, is attractive for $g(0) > 0$, and the second, $(\theta_0, 0)$ with $g(\theta_0) = 0$, $\theta_0 \in (0,1)$, is attractive for $g'(\theta_0) > 0$. The remarkable result is that the asymptotic behaviour of the solution of (26) determines the dynamic of the regularized solution y_ϵ as well as of the exact solution $y(t)$ beyond the breaking point ζ_0. More precisely,

- if the solution of (26) converges to $(0, g(0))$ with $g(0) > 0$ then $y_\epsilon(t)$ converges to a classical solution of (21) as an $\mathcal{O}(\epsilon)$;
- if the solution of (26) converges to $(\theta_0, 0)$ with $g(\theta_0) = 0$ and $g'(\theta_0) > 0$ then $y_\epsilon(t)$ converges to a weak solution of (21), in the sense of Utkin with $\dot y^+(\zeta_0) = f\big(y(\zeta_0), \theta_0\dot y^-(0) + (1-\theta_0)\dot y^+(0)\big)$, as an $\mathcal{O}(\epsilon)$.

The results are illustrated by the following examples:

Example 1

$$\begin{cases} \dot y(t) = 1 - \dfrac{1}{2}\dot y\big(y(t) - 1\big), & t \geq 0, \\ y(t) = 0, \quad t \leq 0. \end{cases} \tag{27}$$

The solution for $t > 0$ is $y(t) = t$ up to the breaking point $\zeta_0 = 1$. By (24) we have $g(\theta) = (\theta + 1)/2$ for which $g(0) > 0$ and hence the classical solution crosses the manifold at the breaking point ζ_0 and continue to exist for $t > 1$ as $y = (t+1)/2$. The regularized solution with $\epsilon = 10^{-2}$ is depicted, around the breaking point, in Fig.5.

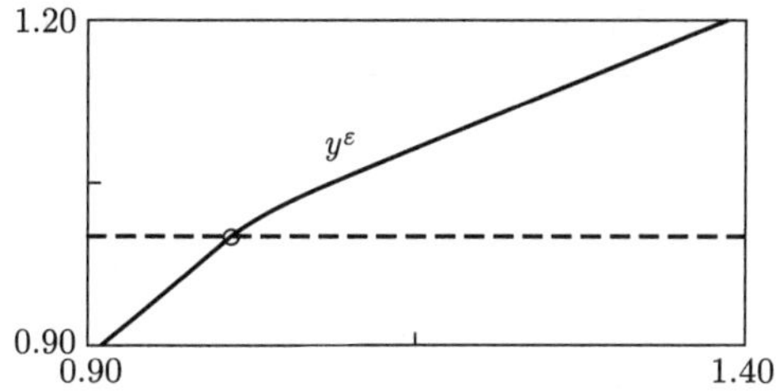

Fig. 5. Solution of (27) regularized by singular perturation with $\epsilon = 10^{-2}$. The regularized solution y_ϵ converges to the classical solution $y = (t+1)/2$.

Example 2

$$\begin{cases} \dot{y}(t) = 1 - 2\dot{y}\big(y(t) - 1\big), & t \geq 0, \\ y(t) = 0, & t \leq 0. \end{cases} \tag{28}$$

The solution for $t > 0$ is again $y(t) = t$ up to the breaking point $\zeta_0 = 1$. Now $g(\theta) = -1 + 2\theta$ and, being $g(0) < 0$, the solution ceases to exist. On the other hand $g(1/2) = 0$ and $g'(1/2) = 2$ and thus beyond ζ_0 the regularized solution, depicted in Fig.6, converges to the weak solution $y(t) = 1$.

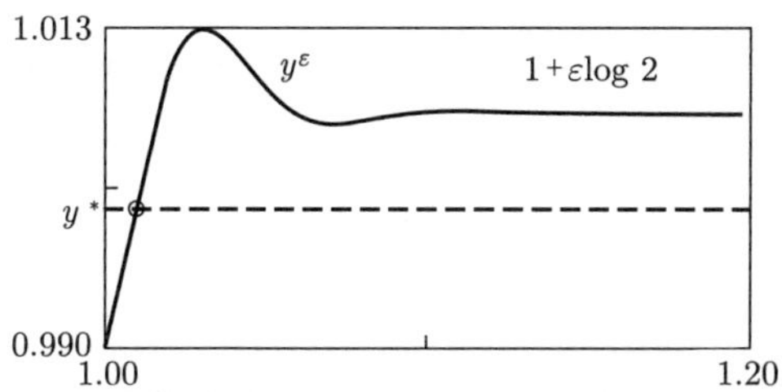

Fig. 6. Solution of (28) regularized by singular perturation with $\epsilon = 10^{-2}$. The regularized solution y_ϵ remains ϵ-close to the weak solution $y(t) = 1$.

Example 3

$$\begin{cases} \dot{y}(t) = -13\Big(\dot{y}(y(t)-1) - 0.15\Big)\Big(\dot{y}(y(t)-1) - 0.5\Big)\Big(\dot{y}(y(t)-1) - 0.85\Big), & t \geq 0, \\ y(t) = 0, \quad t \leq 0. \end{cases} \tag{29}$$

Direct computation provides, for $t > 0$, the solution $y(t) = At$ with $A = 0.82875$ and hence the breaking point $\zeta_0 = 1/A$. The function g takes the form

$$g(\theta) = -13\Big((1-\theta)A - 0.15\Big)\Big((1-\theta)A - 0.5\Big)\Big((1-\theta)A - 0.85\Big)$$

and thus $g(0) > 0$. This guaranties the continuation of a classical solution beyond ζ_0. On the other hand the equation $g(\theta) = 0$ has three roots in $(0, 1)$ where $g' \neq 0$ and hence weak solutions exist beyond ζ_0. Only the central root θ_0 fulfils $g'(\theta_0) > 0$

and this explains the surprising result that the regularized solution y_ϵ converges to the weak solution $y(t) = 1$ rather than to the classical one (Fig.7).

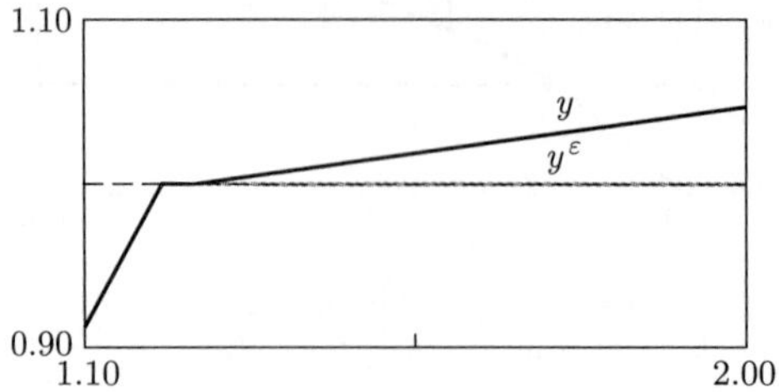

Fig. 7. Solution of (29) regularized by singular perturation with $\epsilon = 10^{-2}$. After the breaking point the regularized solution y_ϵ tracks the weak solution $y(t) = 1$ rather than the classical one.

• **Regularization by Smoothing the Data**

Let's consider again the problem (21) in the equivalent non-neutral delay differential algebraic system of index one form (22) with initial data: $y(t) = \phi(t)$, $\dot{z}(t) = \dot{\phi}(t)$ for $t \leq 0$. Being the variables y and z independent on each other and being the raise of breaking points caused by the discontinuity of $z(t)$ at the initial point $t = 0$,

$$\dot{\phi}(0) \neq \dot{y}_0^+, \qquad \dot{y}_0^+ = f\left(\phi(0), \dot{\phi}(\alpha(\phi(0)))\right),$$

it is natural to remove it by regularizing the initial condition in a small interval $-\epsilon \leq t \leq 0$ by setting

$$z(t) = \dot{\phi}(-\epsilon) + \chi\left(\frac{t}{\epsilon}\right)(\dot{y}_0^+ - \dot{\phi}(-\epsilon)), \tag{30}$$

where $\chi(s)$, $s \in [0,1]$, is any smooth increasing function satisfying $\chi(-1) = 0$, $\chi(0) = 1$, i.e., linear interpolation. Remark that, contrary to the singular perturbation regularization where the equation changes and the initial data remain fixed, here the equation is fixed and the initial data, in particular $\dot{\phi}$, changes. For any ϵ the resulting regularized problem admits a classical solution $y_\epsilon(t)$ which, for $\epsilon \to 0$, is expected to converge either to the classical solution or, after its termination, to the weak solution of (21).

The approach has been deeply investigated by Guglielmi and Hairer[10] where the existence and convergence results are still based on suitable asymptotic expansion of the form

$$y(t_0(\epsilon) + t) = \sum_{j\geq 0} \epsilon^j\, y_j(t) + \epsilon \sum_{j\geq 0} \epsilon^j\, \eta_j(t/\epsilon), \tag{31}$$

where, in this case, the asymptotic expansion of the solution $y_\epsilon(t)$ must be considered to the right of the *artificial* breaking point $t_0(\epsilon)$ such that $\alpha(y_\epsilon(t_0(\epsilon))) = -\epsilon$, as long as $-\epsilon \leq \alpha(y_\epsilon(t_0(\epsilon))) \leq 0$ and then to the right of the breaking point $\zeta_0(\epsilon)$ such that $\alpha(y(\zeta_0(\epsilon))) = 0$.

Giving up to go into details, the main results may be summarized as follows:

- After a breaking point we can have a unique weak solution which has the form (31), or a unique classical solution which has the form

$$y(t_0(\epsilon) + t) = y_0(t) + \epsilon \sum_{j \geq 0} \epsilon^j \, \eta_j(t/\epsilon)$$

  or both.
- If the unique solution is classical or weak, the regularized solution converge to it.
- If classical and weak solutions co-exist, the regularized solution always converge to the weak one.

4. Comparing Regularizations

In comparing the numerical efficiency of the algorithms based on the regularizations above, the general situation is the following. Due to the high oscillations in the time average-based regularization, the standard numerical integration is characterized by a severe stepsize restriction throughout the interval $[0, t_f]$. On the contrary, in the singular perturbation-based regularization stepsize restrictions occur only close to the breaking points where, in general, the solution shows damped high oscillations. On the other hand, possible but well understood pitfalls may take place in the choice weak/classical solution as in the *Example* 3. As for the regularization by smoothing the data, the significant difference with singular perturbation-based regularization is the unavoidable lost of the classical solution if a weak solution co-exists.

Acknowledgments. The author is grateful to Nicola Guglielmi for the useful discussion during the draft of the paper and for providing some of the included pictures.

References

1. A. Bellen and N. Guglielmi, Soving neutral delay differential equations with state-dependent delays. *J. Comp. Appl. Math.*, **229**, 350–362 (2009).
2. A. Bellen, N. Guglielmi, S. Maset, and M. Zennaro, Recent trends in numerical solution of retarded differential equations, *Acta Numerica*, 1–110 (2009).
3. A. Bellen and M. Zennaro, *Numerical Methods for Delay Differential Equations* (Oxford University Press, Oxford, 2003).
4. R. D. Driver, A functional-differential system of neutral type arising in a two-body problem of classical electrodynamics, *Int. Symp. Nonlinear Diff. Equations and Nonlinear Mech. Colorado Springs* (1961), 474–484 (Acad. Press, New York-London, 1963).
5. L.E. El'sgol'ts and S. B. Norkin, Introduction to the theory and application of differential equations with deviating arguments, *Mathematics in Science and Engineering*, **105**, (1973).

6. A. F. Filippov, Differential equations with discontinuous right-hand sides, *Trans. Amer. Mah. Soc.*, **42**, 199–231 (1964).
7. G. Fusco and N. Guglielmi, A regularization for discontinuous differential equations with application to state-dependent delay differential equations of neutral type, *J. Diff. Eq.*, **250**, 3230–3279 (2011).
8. N. Guglielmi and E. Hairer, Implementin Radau IIA methods for stiff delay differential equations, *Computing*, **67**, 1–12 (2001).
9. N. Guglielmi and E. Hairer, Asymptotic expansion for regularized state-dependent neutral delay equations, *SIAM J. Math. Anal.*, **44**(4), 2428–2458 (2012).
10. N. Guglielmi and E. Hairer, Recent approaches for state-dependent neutral delay equations with discontinuities, *Mathematics and Computers in Simulation* (in press, 2011).
11. N. Guglielmi and E. Hairer, Regularization of neutral delay differential equations with several delays, *J. Dynamics and Differential Equations* (in press, 2012).
12. E. Hairer and G. Wanner, Solving Ordinary Differential Equations II. Stiff and Differential Algebraic Problems, *Springer Series in Computational Mathematics*, **14**, 2nd edition (Springer-Verlag, Berlin, 1996).
13. V. I. Utkin, Sliding Modes in Control and Optimization, *Comm. Control Engrg. Ser.* (Springer-Verlag, Berlin, 1992).

Chapter 3

Engineering Applications of Time-periodic Time-delayed Systems

Gábor Stépán

Department of Applied Mechanics
Budapest University of Technology and Economics
Muegyetem rkp 3, Budapest H-1111, Hungary
stepan@mm.bme.hu

Time delays often occur in engineering systems related to mechanical contact tasks and/or control tasks with relevant information processing and transmission time. Parametric excitation is also a typical vibration phenomenon in many well-known engineering applications. There are not many models, however, that combine delay and time-periodic coefficients, in spite of the fact that machine tool vibrations and many control problems, especially digital control tasks, may often lead to time-periodic delay differential equations with discrete delays or even with distributed ones. A paradigm for such delayed time-periodic systems is the delayed Mathieu equation, the stability analysis and stability charts of which will be presented. Then the numerical method called semi-discretization is introduced as a time-periodic perturbation of the time delay. Among the applications, the milling process and turning with spindle-speed variation is mentioned, while the so-called "act-and-wait" control is presented for robotic force control problems.

1. Introduction

Mathematical modeling of engineering systems with delay-differential equations has become increasingly efficient and popular as the mathematical theory developed during the recent decades. This process started with the proper formulation of the infinite dimensional state spaces of these systems by Myshkis[1], continued with the generalization of Floquet Theory to delay systems by Halanay[2], then followed by the construction of a unified theory for functional differential equations by Hale[3], then by Hale and Lunel[4] and many others.

However, engineering problems leading to delay-differential equation (DDE) models appeared much earlier, and they provided a driving force for the development of the mathematical theory. The famous population dynamics models of Volterra[5] in the 1920's already included time delays and produced oscillatory behavior observed in nature even for first order scalar systems. From mechanical engineering view-point, however, the second order models are relevant since many of them are based to Newton's Laws where the acceleration appears as second derivative of the state variable. The corresponding models are also called *delayed oscillators*. These are oscillatory even without time delay, so the presence of delay makes their dynamics quite intricate.

The classical self-excited vibrations in the delayed oscillators are described by autonomous DDEs. Parametric excitation, however, is also an important source of oscillations, and the corresponding delayed oscillators are often governed by time-periodic non-autonomous DDEs. In the subsequent section of this review paper, the delayed Mathieu equation paradigm is discussed leading to the presentation of the corresponding stability chart. In Section 3, the basic idea of the semi-discretization method is summarized for the numerical stability analysis of large systems. Then two time-periodic DDE models are derived for milling processes: the first involves time-periodic coefficients; the second involves time-periodic time delays. Finally, the act-and-wait control concept is presented for a force-control problem with delay, which is an application of time-periodic gains for controlling delayed oscillators.

2. Delayed Mathieu Equation

The linear non-autonomous DDEs are considered in the general form

$$\dot{\mathbf{y}}(t) = \int_{-\tau_{\max}}^{0} \mathbf{y}(t+\vartheta) d_{\vartheta}\boldsymbol{\eta}(t,\vartheta) \,, \tag{1}$$

which is the representation of a linear non-autonomous functional differential equation $\dot{\mathbf{y}}(t) = L\,\mathbf{y}_t$ by means of a Stieltjes-integral in accordance with the Riesz Representation Theorem (see Hale[3]). Here $\mathbf{y}_t(\vartheta) = \mathbf{y}(t+\vartheta)$, $\vartheta \in [-\tau_{\max}, 0]$ is a function in the space $C^0_{[-\tau_{\max},0]}$ of continuous functions mapping the given time interval of length $\tau_{\max}$ into $\mathbb{R}^n$, and the $n \times n$ matrix $\boldsymbol{\eta}$ is a function of bounded variation in its second variable. If $\boldsymbol{\eta}$ is also time-periodic in its first variable, that is, $\boldsymbol{\eta}(t+T, \cdot) = \boldsymbol{\eta}(t, \cdot)$, then a time-periodic DDE is presented in (1).

It is well-known that all high order differential equations can be transformed into first order high dimensional differential equations. Since the Newtonian systems involve second time derivatives, the simplest delayed oscillator can be represented in the form of a 2-dimensional system. As the basic example for a delayed and parametrically excited oscillator, let us choose the measure $\boldsymbol{\eta}$ in the following simple form:

$$\boldsymbol{\eta}(t,\vartheta) = \begin{cases} \begin{bmatrix} 0 & 0 \\ 0 & 0 \end{bmatrix}, & \vartheta \in [-4\pi, -2\pi), \\ \begin{bmatrix} 0 & 0 \\ b & 0 \end{bmatrix}, & \vartheta \in [-2\pi, 0), \\ \begin{bmatrix} 0 & 1 \\ b-(\delta+\varepsilon\cos t) & 0 \end{bmatrix}, & \vartheta = 0, \end{cases} \qquad t \in [t_0 - 2\pi, \infty), \tag{2a}$$

and substitute it into (1) by using Stieltjes calculus $d_\vartheta \boldsymbol{\eta}(\,\cdot\,,\vartheta) = \frac{d}{d\vartheta}\boldsymbol{\eta}(\,\cdot\,,\vartheta)$ where $\boldsymbol{\eta}$ is differentiable, and $d_\vartheta \boldsymbol{\eta}(\,\cdot\,,\vartheta) = \boldsymbol{\eta}(\,\cdot\,,\vartheta) - \boldsymbol{\eta}(\,\cdot\,,\vartheta - 0)$ where it is discontinuous. The resulting periodic DDE will have the form

$$\begin{bmatrix} \dot{y}_1(t) \\ \dot{y}_2(t) \end{bmatrix} = \begin{bmatrix} 0 & 1 \\ -(\delta + \varepsilon \cos t) & 0 \end{bmatrix} \begin{bmatrix} y_1(t) \\ y_2(t) \end{bmatrix} + \begin{bmatrix} 0 & 0 \\ b & 0 \end{bmatrix} \begin{bmatrix} y_1(t-2\pi) \\ y_2(t-2\pi) \end{bmatrix} . \tag{2b}$$

After the introduction of the notation $x := y_1$, this can be transformed to the scalar second order periodic DDE

$$\ddot{x}(t) + (\delta + \varepsilon \cos t\,)x(t) = bx(t - 2\pi)\,, \tag{3}$$

which is also called *delayed Mathieu equation*, since it gives the Mathieu[6] equation for $b = 0$, while it describes the delayed oscillator (see Bhatt and Hsu[7]) for $\varepsilon = 0$. This equation is special in the sense that the time-periodicity and the time delay are both equal to 2π. Still, this special case is important since a closed form analytical stability chart can be constructed, and also some mechanical models like those of milling operations have the same property, the delay and the time periodicity are just equal to each other.

The stability charts present those parameter domains where the trivial solution $x(t) \equiv 0$ is stable in Lyapunov sense. For the special case $b = 0$, the chart in the (δ, ε) parameter plane was constructed by van der Pol and Strutt[8] and can also be found in the book of Ince[9], while the chart in the (δ, b) parameter plane for $\varepsilon = 0$ is presented, for example, in the books of Stepan[10], Hu and Wang[11] or Michiels and Niculescu[12].

The stability chart in the 3-dimensional (δ, b, ε) parameter space can be determined by the application of the infinite dimensional extension of the Floquet theory and the analytical calculations can be carried out by means of the generalization of Hill's infinite dimensional determinant method. The details of this calculation together with its proof are presented in Insperger and Stepan[13] and the resulting stability chart is shown in Fig.1. Clearly, it is a kind of direct product of the two special cases discussed above, that is, the stability boundaries can be obtained by

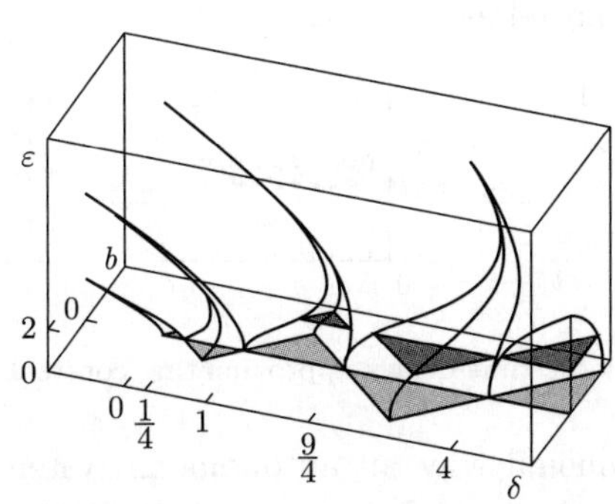

Fig. 1. Stability chart of the delayed Mathieu equation (3). Sections of stable regions are shaded.

shifting the tongues of the Strutt-Van der Pol stability chart along the triangles of the Bhatt-Hsu stability chart.

Along the stability surfaces, all the three kinds of possible bifurcations in periodic systems will appear: Neimark-Sacker (or secondary Hopf), flip (or period doubling), and fold (or secondary saddle-node) as explained by Insperger and Stepan[13].

3. Semi-discretization Method for Periodic DDEs

The stability of parametrically excited delayed oscillators can rarely be studied analytically. The delayed Mathieu equation paradigm discussed in the previous section is one of these exceptions that serves also as a test example for checking the numerical methods developed during the last decade, like the semi-discretization method of Insperger and Stepan[14], its version improved by Sheng, Elbeyly, and Sun[15], the time finite element method of Bayly, Halley, Mann, and Davies[16], the Chebishev polynomial method of Butcher et al.[17], the pseudo-spectral method of Breda, Maset, and Vermiglio[18]. In this section, the semi-discretization method is discussed as a numerical method based on the introduction of a special time-periodic time delay, that is, the introduction of a special parametric excitation for the delay itself.

Consider the delayed oscillator with constant parameters $c_{0,1}$ in the form of the scalar second order autonomous DDE

$$\ddot{x}(t) + c_0 x(t) = c_1\, x(t - 2\pi)\,, \tag{4}$$

and also an approximating delayed oscillator

$$\ddot{x}(t) + c_0 x(t) = c_1 x(t - \tau(t)) \tag{5}$$

with the time-periodic delay

$$\tau(t) = t + (m - \text{int}\,(t/\Delta t))\Delta t\,, \tag{6}$$

which is periodic with the time-step $\Delta t = 2\pi/(m+1/2)$ where m is an appropriately chosen integer, also called approximation number. As Fig.2 shows, the constant time delay 2π can be approximated as $m \to \infty$.

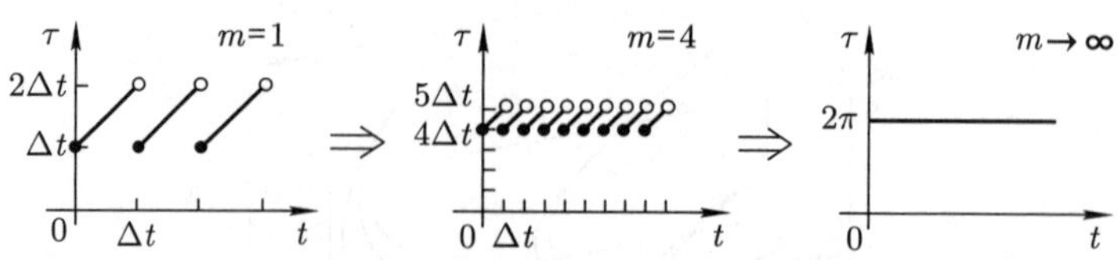

Fig. 2. Periodic time delays approximating constant time delay.

First, it may look unusual why an autonomous system is approximated by a non-autonomous one, but the special time periodicity of the delay results in a piece-wise constant term in the right-hand-side of (4) for the intervals $t \in [t_i, t_{i+1}) =$

$[i\Delta t, (i+1)\Delta t)$, $i = 0, 1, 2, \ldots$ since

$$x(t - \tau(t)) \equiv x((i-m)\Delta t) =: x_{i-m}. \tag{7}$$

This way, an infinite series of non-homogeneous ordinary differential equations (ODE) has to be solved with concatenated initial conditions:

$$\ddot{x}(t) + c_0 x(t) = c_1 x_{i-m}, \quad x(t_i) = x_i, \; \dot{x}(t_i) = \dot{x}_i. \tag{8}$$

By means of the initial conditions, the coefficients K_{1i}, K_{2i} in the general solution

$$x(t) = K_{1i}\cos(\sqrt{c_0}t) + K_{2i}\sin(\sqrt{c_0}t) + \frac{c_1}{c_0}x_{i-m}, \quad t \in [t_i, t_{i+1}) \tag{9}$$

can be expressed by means of x_i, $\dot{x}_i$, x_{i-m}. Consequently, when $x_{i+1} := x(t_{i+1}) = x(t_i + \Delta t)$ is calculated in (9), then x_{i+1} can be expressed as a linear combination of x_i, $\dot{x}_i$, x_{i-m}. The same calculation can be repeated for $\dot{x}_{i+1}$. If all these values are collected in the vector $\mathbf{y}_i = \text{col}\,(\dot{x}_i \; x_i \; x_{i-1} \; \cdots \; x_{i-m}) \in \mathbb{R}^{m+2}$, then a linear map is compiled and the so-called characteristic multipliers are calculated

$$\mathbf{y}_{i+1} = \mathbf{A}\mathbf{y}_i \quad \Rightarrow \quad \det(\mu\mathbf{I} - \mathbf{A}) = 0 \quad \Rightarrow \quad \mu_{1,2,\ldots,m+2}. \tag{10}$$

The coefficient matrix $\mathbf{A}$ is a constant and all its elements are expressed by the coefficients $c_{0,1}$ of the original delayed oscillator (4) and the approximation parameter m. The stable parameter regions in the parameter plane are determined by the condition $|\mu_{1,2,\ldots,m+2}| < 1$. The larger m is, the better the stability chart is approximated, as it is shown in Fig.3. For $m = 10$, the exact triangle-shaped stability chart (see the shaded regions for $\varepsilon = 0$ in Fig.1.) is already well approached.

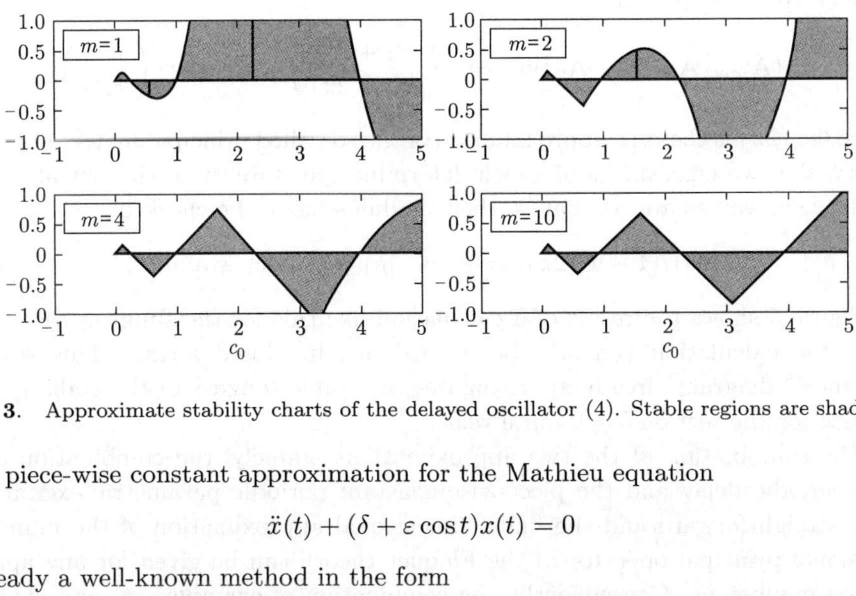

Fig. 3. Approximate stability charts of the delayed oscillator (4). Stable regions are shaded.

A piece-wise constant approximation for the Mathieu equation

$$\ddot{x}(t) + (\delta + \varepsilon\cos t)x(t) = 0 \tag{11}$$

is already a well-known method in the form

$$\ddot{x}(t) + c_{0i}x(t) = 0, \quad x(t_i) = x_i, \; \dot{x}(t_i) = \dot{x}_i \tag{12}$$

for $t \in [t_i, t_{i+1}) = [\,i\Delta t\,, (i+1)\Delta t\,)$, $\Delta t = 2\pi/m$ with $c_{0i} = \delta + \varepsilon \cos t_i$ (see also Fig.4).

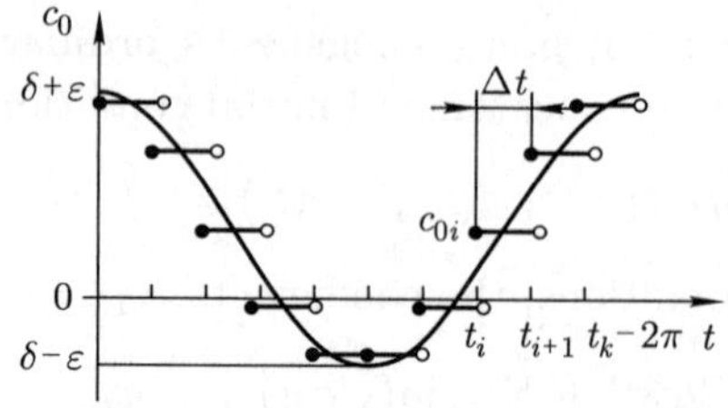

Fig. 4. Approximation of harmonic parameter excitation in the Mathieu equation.

The piece-wise specific solution of (12) is

$$x(t) = x_i \cos\left(\sqrt{c_{0i}}(t-t_i)\right) + \frac{\dot{x}_i}{\sqrt{c_{0i}}} \sin\left(\sqrt{c_{0i}}(t-t_i)\right)\,, \qquad t \in [t_i, t_{i+1})\,, \tag{13}$$

which means that again, the calculation of $x_{i+1} := x(t_{i+1}) = x(t_i + \Delta t)$ leads to the construction of a discrete map for the vector $\mathbf{y}_i = \mathrm{col}(x_i\ \dot{x}_i) \in \mathbb{R}^2$ in the form

$$\mathbf{y}_{i+1} = \mathbf{A}\mathbf{y}_i, \quad \mathbf{A}_i = \begin{pmatrix} \cos\left(\Delta t\sqrt{c_{0i}}\right) & \dfrac{\sin\left(\Delta t\sqrt{c_{0i}}\right)}{\sqrt{c_{0i}}} \\ -\sqrt{c_{0i}}\sin\left(\Delta t\sqrt{c_{0i}}\right) & \cos\left(\Delta t\sqrt{c_{0i}}\right) \end{pmatrix}. \tag{14}$$

Then the subsequent application of this iteration provides the mapping of the vector between two time periods:

$$\mathbf{y}_m = (\mathbf{A}_{m-1}\mathbf{A}_{m-2}\cdots\mathbf{A}_0)\mathbf{y}_0 \;\Leftrightarrow\; \begin{pmatrix} x(t+2\pi) \\ \dot{x}(t+2\pi) \end{pmatrix} = \mathbf{\Phi}_m(2\pi)\begin{pmatrix} x(t) \\ \dot{x}(t) \end{pmatrix}, \tag{15}$$

where $\mathbf{\Phi}_m(2\pi)$ is the mth approximation of the so-called principal matrix of Floquet theory, the two eigenvalues of which determine the stability of the trivial solution in the same way as above, that is, their modulus has to be checked:

$$\det(\mu\mathbf{I} - \mathbf{\Phi}\,(2\pi)) = 0 \;\Rightarrow\; |\mu_{1,2}| \le 1 \Leftrightarrow \text{stability.} \tag{16}$$

Figure 5 shows the result of a calculation like this for the simplest case $m = 2$ when the calculation can still be carried out in closed form. This so-called Meissner[19] diagram already approximates the stable tongues of the stability chart in Fig.1 for the section $b = 0$ quite well.

The combination of the two approximations, namely, the combination of the time-periodic delay and the piece-wise constant periodic parametric excitation is quite straightforward and the finite dimensional approximation of the infinite dimensional principal operator of the Floquet theory can be given for any approximation number m. Consequently, the combination of examples (4) and (11) with increasing m converges to the stability regions of the delayed Mathieu equation

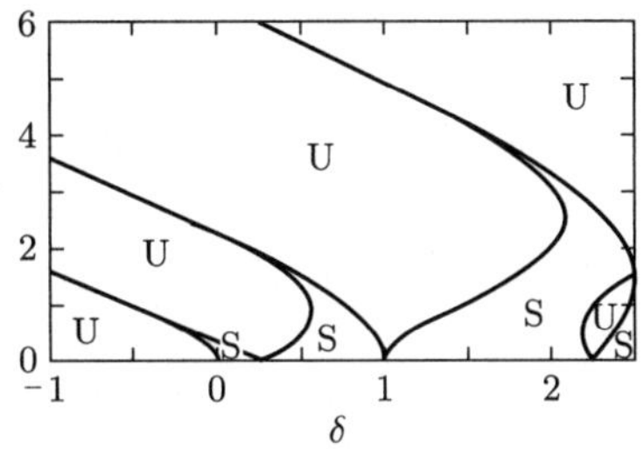

Fig. 5. Stable regions (S) of approximated harmonic parametric excitation equation (12) for $m = 2$.

in Fig.1. The method can be generalized to any linear time-periodic functional differential equation of the form (1) as proved by Insperger, Stepan, and Turi[20].

As an example, the stability chart of the damped delayed Mathieu equation

$$\ddot{x}(t) + \kappa\dot{x}(t) + (\delta + \varepsilon\cos t\,)x(t) = bx(t - 2\pi) \tag{17}$$

is presented in Fig.6 for $\varepsilon = 1$ and different values of the damping parameter κ.

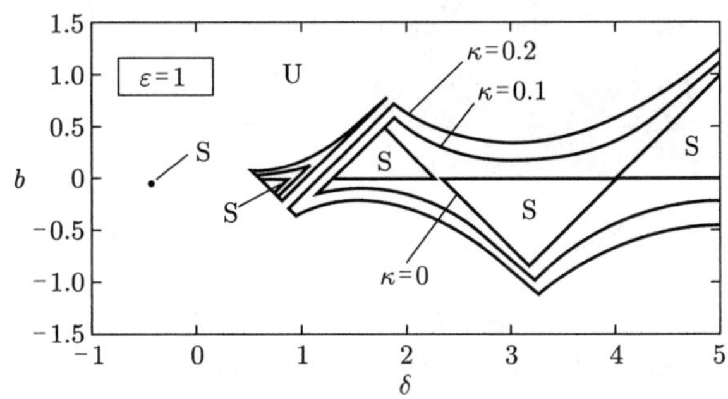

Fig. 6. Stable regions (S) of damped delayed Mathieu equation (17) by semi-discretization.

4. Engineering Applications

In the subsequent subsections, three engineering applications are presented where the mechanical models lead to the parametric excitation of delayed oscillatory systems and the stability analyses of the trivial solutions provide relevant information for the design and development of these structures.

4.1. *Modeling and Stability of Milling Operations*

Machine tools often have a characteristic lowest vibration mode. If this mode is approximated by a single degree-of-freedom oscillator of mass m, stiffness k, and (small) damping b, a parametrically excited delayed oscillator model can be derived to describe the most critical vibrations of machine tools called chatter or regenerative vibrations.

The cutting force F_j acting on the jth cutting edge of the finger-like milling tool depends on the actual chip thickness $h(t)$ at the time instant t, and its variation can be approximated as a linear function of the chip thickness variation, which is proportional to the difference of two cutting edge positions: one is the past position of the preceding cutting edge at the time instant when it passed the same angular position, the other is the present position of the cutting edge. The resultant cutting force is the sum of the cutting force vectors acting at those cutting edges that are in contact with the workpiece. This leads to the mathematical model

$$m\ddot{x}(t) + b\dot{x}(t) + kx(t) = k_1(t)\left(x(t-\tau) - x(t)\right), \quad k_1(t+\tau) = k_1(t). \tag{18}$$

Fig. 7. Single degree-of-freedom mechanical model of milling.

The so-called cutting coefficient $k_1(t)$ is time-periodic with the time delay due to the periodic entering and leaving of the cutting edges into and from the material of the workpiece with the angular velocity Ω, that is, the period and the delay are both $\tau = 2\pi/\Omega$ (see details in Insperger, Stepan, Bayly, and Mann[21]).

The governing equation (17) is a generalized damped and delayed Mathieu equation with time-periodic parameters both in the stiffness and the gain parameter. The corresponding stability chart in Fig.8 is constructed by means of the semi-discretization method.

As predicted by the analytical stability chart of the delayed Mathieu equation, both Neimark-Sacker (that is, secondary Hopf) and flip (that is, period dobling) bifurcations occur at the stability limits. Figure 8 also shows the corresponding vibration frequencies at the stability limits and a typical time-variation of the cutting coefficient. All the parameters are dimensionless in the diagram, tilde refers to time-average value, $N = 4$ is the number of cutting edges, $f_n = \omega_n/(2\pi) = \sqrt{k/m}/(2\pi)$ is the natural frequency of the machine tool structure.

The stability charts, like the one in Fig.8, help engineers to design optimal cutting parameters where the material removal rate (MRR) is the largest, that is, the average cutting coefficient is maximized within the stability limit since it is

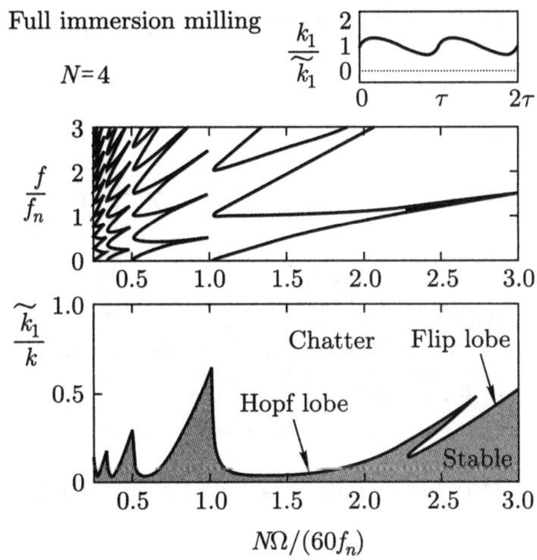

Fig. 8. Typical stability chart of full immersion milling operation and frequencies of self-excited vibrations at the stability limits also called lobes.

linearly proportional to the average chip width.

4.2. *Cutting with Varying Spindle Speed*

When turning processes are modeled as shown in Fig.9, the cutting coefficient is constant since there is always one cutting edge in contact with the workpiece. This means that there is no time-periodic stiffness or gain parameter in the governing equation, consequently, no flip bifurcations appear in the stability charts, only Hopf bifurcation occurs with a single frequency in it.

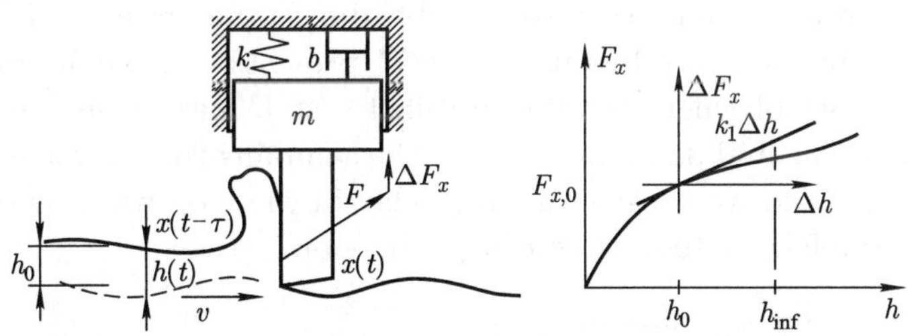

Fig. 9. Mechanical model of turning processes and the definition of the cutting coefficient k_1.

Still, the equation of motion includes time delay that is inversely proportional to the cutting speed v in the same way as explained above for milling. Parametric excitation can also be introduced by means of the periodic variation of the cutting speed, which results in a time-periodic time delay in the equation of motion:

$$m\ddot{x}(t) + b\dot{x}(t) + kx(t) = k_1\left(x(t-\tau(t)) - x(t)\right), \quad \tau = \tau_0 + \tau_1\cos(\omega_m t). \qquad (19)$$

Now, the time period $2\pi/\omega_m$ of the time delay is, of course, different from the time delay itself, but it can also be independent of the average value τ_0 of the delay. The stability chart of (19) is presented in Fig.10.

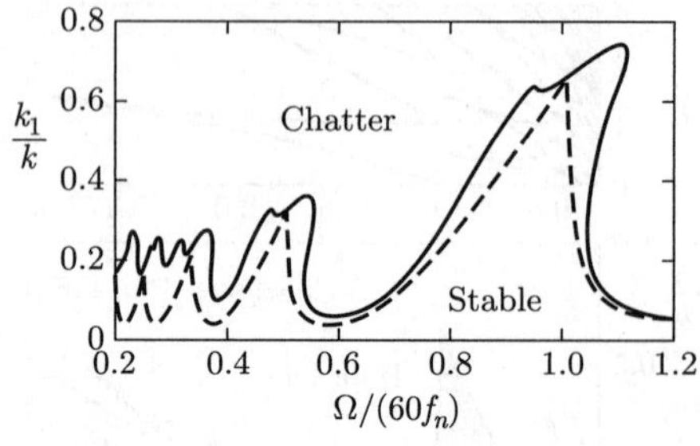

Fig. 10. Stability chart of (19) for spindle speed variation (continuous line) and for constant cutting speed (dashed line).

The stability chart of Fig.10 is presented with continuous lines for $T_m/\tau_0 = 2\pi/(\omega_m \tau_0) = 2$ and $\tau_1/\tau_0 = 0.1$ in case of spindle speed variation, while the stability boundaries are also presented with dashed lines for constant spindle speed when $\tau_1/\tau_0 = 0$. The calculations were carried out by the semi-discretization method in Insperger and Stepan[22]. The improvement in the stability of the turning process may be relevant for low cutting speeds when large periodic perturbation is introduced at the spindle speed. Similar results were obtained by Faassen, van de Wouw, Nijmeijer, and Oosterling[23].

4.3. *Act-and-wait Control of Force Controlled Robots*

Force control of robots often becomes unstable due to the presence of digital effects and time delays in the control loop. Figure 11 presents a single degree-of-freedom model of force control where the uncontrolled system has an undamped angular natural frequency ω_n and damping ratio ζ. The sampling time in the digital control is denoted by Δt, and we consider a time delay in the loop which is characterized by the integer r referring to a delay $r\Delta t$ in the loop.

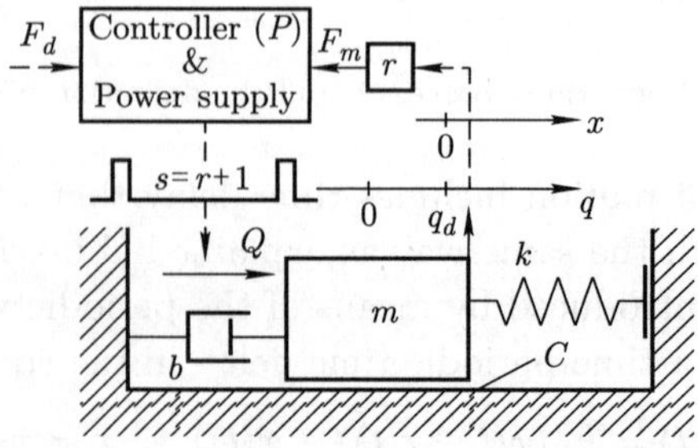

Fig. 11. Mechanical model of force control with act-and-wait feedback.

The desired contact force is $F_d = kq_d$ where q denotes the position of the robotic arm that touches the environment via the spring of stiffness k, while the measured or sensed force is $F_m = kq$. The traditional proportional control applies the control force

$$Q(t) = F_d - P(F_m(t_{j-r}) - F_d), \quad t \in [t_j, t_{j+1}), \quad t_j = j\Delta t, \quad j = 0, 1, 2, \ldots, \tag{20}$$

where P is a dimensionless proportional gain and t_j is the jth sampling instant of the digital control. In the presence of a Coulomb friction force C, this results in a static force error $F_e = C/(1+P)$ around the desired force F_d, which can be decreased by the increase of the gain P. However, there is a maximum value for the gain P above which the desired position of the system becomes unstable and self excited vibrations appear. The larger the sampling time Δt and/or the time delay $r\Delta t$ are, the smaller this maximal gain is, and consequently, the larger the static error of the system is.

The application of a time-periodic gain in the control loop gives the possibility to increase the maximum values of the gain and so to decrease the static force error. As it is explained in the introduction of the so-called act-and-wait control strategy by Stepan and Insperger[24], a quasi optimal time-periodicity is achieved when the proportional control is switched on and off, as represented in Fig.11 by the $s\Delta t$ periodicity of the control force in Fig.11:

$$Q_{a\&w}(t) = \begin{cases} F_d - P(F_m(t_{j-r}) - F_d), & \text{if} \quad t \in [t_{hs}, t_{hs+1}), \quad h \in \mathbb{Z}, \\ F_d, & \text{otherwise} \end{cases} \tag{21}$$

with subscript $a\&w$ referring to act-and-wait.

If the small perturbation x is introduced by $q(t) = q_d + x(t)$ around the desired position of the system, the equation of motion assumes the form

$$\ddot{x}(t) + 2\zeta\omega_n\dot{x}(t) + \omega_n^2 x(t) = -g_j\omega_n^2 P x(t_{j-r}), \qquad t \in [t_j, t_{j+1}), \tag{22}$$

where a switching function g_j is introduced by

$$g_j = \begin{cases} 1, & \text{if} \quad j = hs, \quad h \in \mathbb{Z}, \\ 0, & \text{otherwise} \end{cases} \tag{23}$$

corresponding to the act-and-wait control force (21). The actual interpretation of (22) is that it is a time periodic DDE with a piece-wise constant periodic parametric excitation at the gain and a piece-wise linear parametric excitation at the time delay. The stability of the trivial solution can be analyzed directly in the discrete state space representation with $\mathbf{y} = \mathrm{col}(x\ \dot{x}) \in \mathbb{R}^2$:

$$\begin{aligned} \mathbf{y}(j+1) &= \mathbf{Q}\mathbf{y}(j) + \mathbf{R}\mathbf{w}(j-r), \\ \mathbf{w}(j) &= g_j\mathbf{H}\mathbf{x}(j), \end{aligned} \tag{24}$$

where the matrices are defined as

$$\mathbf{Q} = \exp(\mathbf{A}\Delta t), \quad \mathbf{R} = (\exp(\mathbf{A}\Delta t) - \mathbf{I})\mathbf{A}^{-1}\mathbf{B}, \quad \mathbf{H} = [-\omega_n^2 P, \ 0], \tag{25}$$

and the matrices $\mathbf{A}$ and $\mathbf{B}$ are given by the DDE (22) in the form

$$\mathbf{A} = \begin{bmatrix} 0 & 1 \\ -\omega_n^2 & -2\zeta\omega_n \end{bmatrix}, \quad \mathbf{B} = \begin{bmatrix} 0 \\ 1 \end{bmatrix}. \tag{26a}$$

In the same way as the semi-discretization method was introduced, a discrete map

$$\underbrace{\begin{bmatrix} \mathbf{y}(j+1) \\ \mathbf{w}(j) \\ \mathbf{w}(j-1) \\ \vdots \\ \mathbf{w}(j-r+1) \end{bmatrix}}_{\mathbf{z}_{j+1}} = \underbrace{\begin{bmatrix} \mathbf{Q} & \mathbf{0} & \cdots & \mathbf{0} & \mathbf{R} \\ g_j\mathbf{H} & \mathbf{0} & \cdots & \mathbf{0} & \mathbf{0} \\ \mathbf{0} & \mathbf{I} & \cdots & \mathbf{0} & \mathbf{0} \\ \vdots & \vdots & & \vdots & \vdots \\ \mathbf{0} & \mathbf{0} & \cdots & \mathbf{I} & \mathbf{0} \end{bmatrix}}_{\mathbf{G}_j} \underbrace{\begin{bmatrix} \mathbf{y}(j) \\ \mathbf{w}(j-1) \\ \mathbf{w}(j-2) \\ \vdots \\ \mathbf{w}(j-r) \end{bmatrix}}_{\mathbf{z}_j} \tag{26b}$$

can be constructed where the coefficient matrix varies with periodicity s in accordance with the definition of the switching function g_j in (23). If the discrete maps are concatenated, the linear discrete map has the form

$$\mathbf{z}_s = \mathbf{G}_{s-1} \cdots \mathbf{G}_1\mathbf{G}_0\mathbf{z}_0 \;\Rightarrow\; \mathbf{z}_s = \mathbf{G}_1^{s-1}\mathbf{G}_0\mathbf{z}_0 \;\Rightarrow\; \mathbf{z}_s = \mathbf{\Phi}_{r,s}\mathbf{z}_0\,, \tag{27}$$

where the matrix $\mathbf{\Phi}_{r,s}$ is the actual form of the principal operator of the periodic DDE (23) with periodicity s and delay r.

It is shown by Insperger, Wahi, Colombo, Stepan, di Bernardo, Hogan[25] that the largest stable parameter domains can be achieved when the period s is just larger than the delay r, so when $s = r + 1$, as it is also represented in Fig.11. The stability chart in Fig.12 is constructed for the modal parameters of a Hirata robot used in laboratory experiments. The stability limits and the frequencies of the emerging self-excited vibrations are presented by grey lines for constant proportional gains and by black lines for periodically varying gains with $r = 1$, $s = 2$. Apart of the Hopf bifurcation, flip bifurcations can occur in case of the act-and-wait control, while the maximal gains within the stability limits are doubled in this case, which means that the static force error is reduced by a factor of 2.

These theoretical predictions were also confirmed experimentally in Insperger, Kovacs, Galambos, and Stepan[26] where the force control of a Hirata robot was established: the robotic arm pushed a helical spring with a prescribed (or desired) contact force against a fixed wall. After the accurate identification of the modal parameters, the above calculations were repeated and the theoretical and the experimental results are compared in Fig.12.

5. Conclusions

Parametric excitation may cause vibration phenomena that are often difficult to handle in certain engineering tasks. Time delays may also cause unexpected self-excited vibrations even for very small delays. When parametric excitation appears in combination with time delays, the emerging vibrations are often intricate, quasi

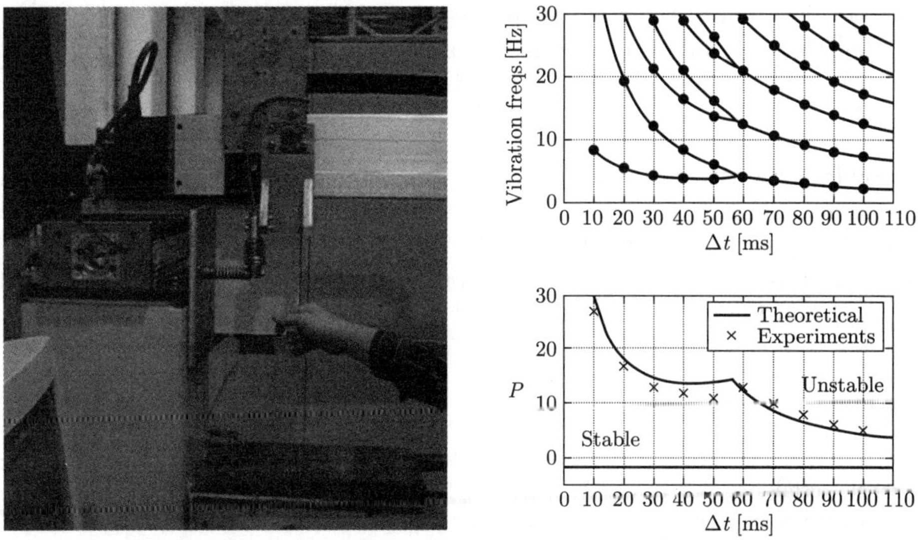

Fig. 12. Force controlled Hirata robot in laboratory experiments and experimental stability chart with measured and theoretically predicted vibration frequencies of self-excited vibrations originated in Hopf and period doubling bifurcations.

periodic or even chaotic, and it is difficult to find where and how to tune the system parameters to eliminate them.

The mathematical models of parametrically excited delayed oscillators are time-periodic delay-differential equations. Their linearization and stability analysis requires the application of an infinite dimensional Floquet theory. The application of the theory for engineering problems is difficult due to the lack of analytically proven reference examples and efficient numerical methods.

The delayed Mathieu equation paradigm is an essential example for a delayed oscillator subjected to harmonic parametric excitation at the stiffness. The stability chart was constructed analytically, which means that it can serve as a reference example for testing numerical methods. Among the numerical methods, the basic idea of the semi-discretization method was described (see Insperger and Stepan[27]) as a time-periodic perturbation of the time delay.

The efficiency of the introduced method was presented in case of three relevant engineering problems. In case of cutting operations, the time delay occurs due to the contact of the tool and the workpiece. Milling operations involve time periodic parametric excitation with time period equal to the time delay. In case of turning, this kind of parametric excitation cannot occur, but the time-periodic spindle speed variation results parametric excitation at the delay. This can be used to improve the stability properties of turning processes. The third example discussed in this study is the act-and-wait force control, where time periodic gains are used to improve the stability properties and to decrease the static force error of force controlled robots.

The exploration of the stability properties of time-periodic delayed oscillators partly helps understanding the peculiar vibration properties of these systems, partly it helps to explore new parameter domains where those delayed systems can be stabilized with parametric excitation, which would be unstable otherwise.

Acknowledgments. The research work was supported by the Hungarian National Science Foundation under grant no. OTKA K101714 and it is also connected to the scientific program of the "Development of quality-oriented and harmonized R+D+I strategy and functional model at BME" project. This project is supported by the New Szechenyi Plan, Project ID: TAMOP-4.2.1/B-09/1/KMR-2010-0002.

References

1. A. D. Myshkis, *Lineare Differentialgleichungen mit Nacheilendem Argument* (Deutscher Verlag der Wissenschaften, Berlin, 1955).
2. A. Halanay, Stability theory of linear periodic systems with delay (in Russian), *Rev Roum Math Pure A*, **6**, 633–653 (1961).
3. J. K. Hale, *Theory of Functional Differential Equations* (Springer-Verlag, New York, 1977).
4. J. K. Hale and S. M. V. Lunel, *Introduction to Functional Differential Equations* (Springer-Verlag, New York, 1993).
5. V. Volterra, Sur la theorie mathematique des phenomenes hereditaires, *J. Math. Pure Appl.*, **7**, 149–192 (1928).
6. E. Mathieu, Memoire sur le mouvement vibratoire d'une membrane de forme elliptique, *J. Math. Pure Appl.*, **13**, 137–203 (1868).
7. S. J. Bhatt and C. S. Hsu, Stability criteria for second-order dynamical systems with time lag. *J. Appl. Mech-T ASME*, **33**, 113–118 (1966).
8. F. van der Pol and M. J. O. Strutt, On the stability of the solutions of Mathieu's equation, *Philosophical Magazine and Journal of Science*, **5**, 18–38 (1928).
9. E. L. Ince, *Ordinary Differential Equations* (Longmans, Green and Co., London, 1926).
10. G. Stepan, *Retarded Dynamical Systems* (Longman, Harlow, 1989).
11. H. Hu and Z. Wang, *Dynamics of Controlled Mechanical Systems with Delayed Feedback* (Springer, Berlin, 2002).
12. W. Michiels and S. I. Niculescu, *Stability and Stabilization of Time-delay Systems: an Eigenvalue-based Approach* (SIAM Publications, Philadelphia, 2007).
13. T. Insperger and G. Stepan, Stability chart for the delayed Mathieu equation, *Proc. R. Soc. Lond A-Math Phy.*, **458**, 1989–1998 (2002).
14. T. Insperger and G. Stepan, Semi-discretization method for delayed systems. *Int. J. Numer. Meth. Eng.*, **55**, 503–518 (2002).
15. J. Sheng, O. Elbeyli, and J. Q. Sun, Stability and optimal feedback controls for time-delayed linear periodic systems, *AIAA J.*, **42**, 908–911 (2004).
16. P. V. Bayly, J. E. Halley, B. P. Mann, and M. A. Davies, Stability of interrupted cutting by temporal finite element analysis, *J. Manuf. Sci. E–T ASME*, **125**, 220–225 (2003).
17. E. A. Butcher, H. Ma, E. Bueler, V. Averina, and Z. Szabo, Stability of linear time-periodic delay-differential equations via Chebyshev polynomials, *Int. J. Numer. Meth. Eng.*, **59**, 895–922 (2004).

18. D. Breda, S. Maset, and R. Vermiglio, Pseudospectral differencing methods for characteristic roots of delay differential equations, *SIAM J. Sci. Comput.*, **27**, 482–495 (2005).
19. R. Zurmuhl, *Numerische Behandlung von Schwingungsaufgaben mittels Übertragungsmatrizen* (Dusseldorf, VDI, 1959).
20. T. Insperger, G. Stepan, and J. Turi, On the higher-order semi-discretizations for periodic delayed systems, *J. Sound Vib.*, **313**, 334–341 (2008).
21. T. Insperger, G. Stepan, P. V. Bayly, and B. P. Mann, Multiple chatter frequencies in milling processes, *Journal of Sound and Vibration*, **262**, 333–345 (2003).
22. T. Insperger and G. Stepan, Stability analysis of turning with periodic spindle speed modulation via semi-discretization, *J. Vib. Control*, **10**, 1835–1855 (2004).
23. R. P. H. Faassen, N. van de Wouw, H. Nijmeijer, and J. A. J. Oosterling, An improved tool path model including periodic delay for chatter prediction in milling, *J. Comput. Nonlin. Dyn–T ASME*, **2**, 167–179 (2007).
24. G. Stepan and T. Insperger, Stability of time-periodic and delayed systems — a route to act-and-wait control, *IFAC Annual Reviews in Control*, **30**, 159–168 (2006).
25. T. Insperger, P. Wahi, A. Colombo, G. Stepan, M. di Bernardo, and J. S. Hogan, Full characterization of act-and-wait control for first order unstable lag processes, *Journal of Vibration and Control*, **16**, 1209–1233 (2010).
26. T. Insperger, L. L. Kovacs, P. Galambos, and G. Stepan, Increasing the accuracy of digital force control process using the act-and-wait concept, *IEEE-ASME T Mech.*, **15**, 291–298 (2010).
27. T. Insperger and G. Stepan, *Semi-discretization for Time-delay Systems* (Springer, New York, 2011).

Chapter 4

Synchronization in Delay-coupled Complex Networks

Eckehard Schöll
Institute of Theoretical Physics, TU Berlin
Hardenbergstr 36, 10623 Berlin, Germany
schoell@physik.tu-berlin.de

We review synchronization in networks of oscillators with time-delayed coupling. The stability of synchronization generally depends in a complicated way on the local node dynamics and the coupling topology. However, for large coupling delays synchronizability relates in a simple way to the spectral properties of the network topology, i.e., the eigenvalue spectrum of the coupling matrix, allowing for a universal classification of networks. For smaller coupling delays the synchronization properties depend in a more subtle way upon the local dynamics and the details of the network topology. Various cluster synchronization states, where certain clusters inside the network show isochronous synchronization, can be realized by tuning the coupling parameters such as the coupling phase, coupling strength, and delay time. Adaptive synchronization is possible by applying the speed-gradient method from control theory. We consider different generic local dynamics like oscillatory (Stuart-Landau), excitable type I (Saddle-node bifurcation on invariant cycle), and type II (FitzHugh-Nagumo), and various coupling topologies like rings, small-world, and random networks with excitatory and inhibitory links. Transitions between synchronization and desynchronization can be induced by changing the balance between excitatory and inhibitory links.

1. Introduction

The control of nonlinear dynamical systems has evolved into a wide interdisciplinary area of research over the past decades[1]. In particular, control of dynamics on complex networks has recently gained much interest. Synchronization phenomena in networks are of great importance[2] in many areas ranging from physics and chemistry to biology and engineering. Chaos synchronization of lasers, for instance, may lead to new secure communication schemes[3–5]. The synchronization of neurons is believed to play a crucial role in the brain under normal conditions, for instance in the context of cognition and learning[6], and under pathological conditions such as Parkinson's disease[7]. Time delay effects are a key issue in realistic networks. For example, the finite propagation time of light between coupled semiconductor lasers[8–12] significantly influences the dynamics. Similar effects occur in neuronal[13,14] and biological[15] networks.

To determine the stability of a synchronized state in a network of identical units, a powerful method has been developed[16], i.e., the master stability function (MSF). This approach has been extended to networks with coupling delays[17–22], where the MSF depends non-trivially on delay times. There exist different forms of synchronization, i.e., complete or isochronous (zero-lag) synchronization, generalized synchronization, cluster or group synchronization, and many other forms. Chimera states, where a network of identical oscillators splits into distinct coexisting domains of coherent (phase-locked) and incoherent (desynchronized) behaviour, have gained much attention recently[23–30].

In this chapter we review recent work on the stability of synchronization in delay-coupled networks of oscillators. In general, the stability depends in a complicated way on the local dynamics of the nodes and the coupling topology. However, for large coupling delays synchronizability relates in a simple way to the spectral properties of the network topology, characterized by the eigenvalue spectrum of the coupling matrix. The master stability function used to determine the stability of synchronous solutions has a universal structure in the limit of large delay: it is rotationally symmetric and increases monotonically with the radius in the complex plane. This allows for a universal classification of networks with respect to synchronization properties[20]. For smaller coupling delays the synchronization properties depend in a more subtle way upon the local dynamics, and the details of the network topology. Various cluster synchronization states, where certain clusters inside the network show isochronous synchronization, can be realized by tuning the coupling parameters such as the coupling phase, coupling strength, and delay time[18,22]. To find appropriate values of these control parameters, the speed-gradient method from control theory can be applied to achieve a desired state of generalized synchrony (*adaptive synchronization*)[31].

We focus on different generic local dynamics of the nodes, and various coupling topologies like rings, small-world, and random networks with excitatory and inhibitory links. In particular, we study oscillatory local dynamics (Stuart-Landau oscillator, i.e., generic expansion near a Hopf bifurcation), type I excitable dynamics (near a saddle-node bifurcation on an invariant cycle, or *saddle-node infinite period* (SNIPER) bifurcation), and type II excitable dynamics (FitzHugh-Nagumo model). Transitions between synchronization and desynchronization can be induced by introducing inhibitory links into an excitable regular network with a probability p in a small-world like fashion, and changing the balance between excitatory and inhibitory links[32,33].

2. Stability of Synchronization for Large Delay

First, we show that in the limit of large coupling delays the MSF has a very simple structure[20,21,34,35]. This solves the problem of complete zero-lag synchronization for networks with large coupling delay. After briefly introducing the notion of the MSF, we demonstrate the implications for large coupling delays based on a scaling theory[36–41]. This allows us to describe the synchronizability of networks with strongly delayed coupling depending on the type of node dynamics and spectral properties of the network topology. For example, as recently conjectured[19], networks for which the trajectory of an uncoupled unit is also a solution of the network cannot exhibit chaos synchronization for large coupling delay. The results presented here confirm and generalize these previous findings.

Consider a system of N identical units connected in a network with a coupling delay τ ($x^i \in \mathbb{R}^n$, $i = 1, \ldots, N$)

$$\frac{d}{dt}x^i(t) = f\left[x^i(t)\right] + \sum_{j=1}^{N} g_{ij} h\left[x^j(t-\tau)\right]. \tag{1}$$

Here, $g_{ij} \in \mathbb{R}$ is the coupling matrix determining the coupling topology and the strength of each link in the network, f is the (non-linear) function describing the dynamics of an isolated unit, and h is a possibly non-linear coupling function. A synchronized solution can only exist if the row sum of the matrix is the same for each row, i. e., $\sigma = \sum_{j=1}^{N} g_{ij}$ independent of i. In this case if the systems start in a synchronized state $x^1 = x^2 = \cdots = x^N = \overline{x}(t)$, they will remain synchronized since the feedback term will be equal for all x^i. The synchronized solution $\overline{x}(t)$ is then determined by

$$\frac{d}{dt}\overline{x}(t) = f\left[\overline{x}(t)\right] + \sigma h\left[\overline{x}(t-\tau)\right]. \tag{2}$$

To calculate the stability of this synchronized solution, we consider the evolution of small perturbations $\xi^i(t)$

$$x^i(t) = \overline{x}(t) + \xi^i(t).$$

Inserting this ansatz into Eq.(1) and linearizing in ξ^i we find

$$\frac{d}{dt}\xi^i(t) = Df\left[\overline{x}(t)\right]\xi^i(t) + \sum_{j=1}^{N} g_{ij} Dh\left[\overline{x}(t-\tau)\right]\xi^i(t-\tau), \tag{3}$$

where Df and Dh are Jacobians. Using the vector

$$\Xi(t) = (\xi^1(t), \xi^2(t), \ldots, \xi^N(t))$$

the system of linear equations (3) can be written as

$$\frac{d}{dt}\Xi(t) = I_N \otimes Df\left[\overline{x}(t)\right]\Xi(t) + g \otimes Dh\left[\overline{x}(t-\tau)\right]\Xi(t-\tau), \tag{4}$$

where I_N denotes the N-dimensional identity matrix and $g = (g_{ij})$ is the coupling matrix. We assume that the coupling matrix g is diagonalizable, i.e., there exists a unitary transformation U such that

$$UgU^{-1} = \mathrm{diag}(\sigma, \gamma_1, \gamma_2, \ldots, \gamma_{N-1}).$$

Here, σ is the row sum of g, which is always an eigenvalue of g corresponding to the eigenvector (1, 1, ..., 1). We call this the *longitudinal* eigenvalue of g. The other eigenvalues γ_k we then call the *transversal* eigenvalues of g.

Diagonalizing g in Eq.(4) with the transformation U does not affect the term $I_N \otimes Df[\overline{x}(t)]$ in Eq.(4), since this transformation acts only on the matrix I_N. Thus after the diagonalization equation (4) is transformed into N equations

$$\frac{d}{dt}\xi(t) = Df\left[\overline{x}(t)\right]\xi(t) + \sigma\, Dh\left[\overline{x}(t-\tau)\right]\xi(t-\tau), \tag{5}$$

$$\frac{d}{dt}\xi(t) = Df\left[\overline{x}(t)\right]\xi(t) + \gamma_k\, Dh\left[\overline{x}(t-\tau)\right]\xi(t-\tau) \tag{6}$$

with $k = 1, \ldots, N-1$. The first equation corresponds to perturbations in the direction of the vector (1, 1, ..., 1), which act equally on each individual system and thus do not cause desynchronization. A growing perturbation in this direction indicates that the synchronized solution of the network is chaotic.

The $N-1$ other equations in Eq.(6) describe perturbations transversal to the synchronization manifold. The synchronized solution is stable if and only if these perturbations decay, i. e., if the maximum Lyapunov exponent arising from the variational equation (6) is negative for all transversal eigenvalues γ_k. The Lyapunov exponents λ measure the rate of separation of infinitesimally close trajectories. If the initial separation is δx_0, then in the long-time limit (provided the linearization is valid) $|\delta x(t)| \approx e^{\lambda t}|\delta x_0|$.

The master stability function (MSF)[16] is defined as a function $\lambda_{\max} : \mathbb{C} \rightarrow \mathbb{R}$, which maps a complex number $re^{i\phi}$ to the maximum Lyapunov exponent arising from the variational equation

$$\frac{d}{dt}\xi(t) = Df\left[\overline{x}(t)\right]\xi(t) + re^{i\phi}\, Dh\left[\overline{x}(t-\tau)\right]\xi(t-\tau). \tag{7}$$

Once this function is known on a sufficiently large domain in $\mathbb{C}$, one can immediately decide for any network structure whether synchronization will be stable or not. One only needs to evaluate the MSF at the transversal eigenvalues γ_k of the particular network's coupling matrix. Thus the condition for stable synchronization is then

$$\lambda_{\max}(\gamma_k) < 0, \quad k = 1, \ldots, N-1. \tag{8}$$

This way the problem has been separated into one part that only depends on the dynamics of the individual system, and another part that only depends on the coupling topology.

The Lyapunov exponents arising from Eq.(7) can be analyzed in the limit of large[20] τ. For a detailed derivation for maps see Ref 35. The analysis proceeds by first studying the two simpler cases, where the dynamics in the synchronization manifold is a fixed point or a periodic orbits, and then extending the results to chaotic trajectories in the synchronization manifold by using the fact that unstable periodic orbits are dense in a chaotic attractor.

For fixed points of delay differential equations there exists a scaling theory for the eigenvalues of the fixed point in the limit of large delay[37,39,40,42]. Recently this theory has been generalized to the scaling of Floquet exponents of periodic orbits[41]. In both cases the eigenvalues or Floquet spectrum consist of two parts: a strongly unstable part arising from unstable eigenvalues of the system without delay and a pseudo-continuous spectrum, for which the real part of the eigenvalues approach zero in the limit of large delay. For chaotic attractors, averaging over the unstable periodic orbits embedded in the attractor allows us to calculate the natural measure of the chaotic attractor, which is concentrated at the unstable periodic orbits and can in fact be expressed in terms of the orbit's Floquet multipliers[43,44].

From the scaling theory it follows that for large τ the MSF is rotationally symmetric in the complex plane about the origin. If without feedback ($r = 0$) the MSF is positive, then it is constant in the limit of large delay. Otherwise it is a monotonically increasing function of r and there is a critical radius r_0 where it changes sign.

Let us now discuss what this structure of the MSF means for the synchronizability of networks. We can classify networks into three types depending on the magnitude of the largest transversal eigenvalue $\gamma_{\max}$ in relation to the magnitude of the row sum σ: (A) the largest transversal eigenvalue is strictly smaller than the magnitude of the row sum ($|\gamma_{\max}| < |\sigma|$), (B) the largest transversal eigenvalue has the same magnitude as the row sum ($|\gamma_{\max}| = |\sigma|$), and (C) the largest transversal eigenvalue has a larger magnitude than the row sum ($|\gamma_{\max}| > |\sigma|$).

At $r = |\sigma|$ the MSF is positive ($r_0 < |\sigma|$) for chaotic dynamics in the synchronization manifold and negative ($|\sigma| < r_0$) for dynamics on a stable periodic orbit or a fixed point. This gives us a lower or an upper bound on r_0 and we can thus give the classification as shown in Table 1.

Table 1. Stability of chaotic and non-chaotic synchronized solutions for the three types of networks.

		chaotic dynamics in the SM ($r_0 < \|\sigma\|$)	Periodic orbit or fixed point in the SM ($\|\sigma\| < r_0$)
(A)	$\|\gamma_{\max}\| < \|\sigma\|$	synchr. stable if $\|\gamma_{\max}\| < r_0$	synchr. stable
(B)	$\|\gamma_{\max}\| = \|\sigma\|$	synchr. unstable	synchr. stable
(C)	$\|\gamma_{\max}\| > \|\sigma\|$	synchr. unstable	synchr. stable if $\|\gamma_{\max}\| < r_0$

In networks of type (A) and (B) synchronization on a fixed point or a periodic orbit, which is stable within the synchronization manifold, is always stable. For type (C) this dynamics may be stable or not depending on the particular network (value of $|\gamma_{\max}|$) and the dynamics in the synchronization manifold (value of r_0). On the other hand chaos synchronization is always unstable in networks of type (B) and (C) and it may be stable or not in networks of type (A) again depending on the particular network and the dynamics.

Note that for autonomous flows with a stable periodic orbit in the synchronization manifold we always have $r_0 = |\sigma|$, due to the Goldstone mode. Thus for this case we cannot decide whether synchronization for type (B) networks will be stable or not. This depends on whether the $\lambda_{\max} = 0$ contour line of the MSF approaches the circle with radius $r_0 = |\sigma|$, locally, at the transversal eigenvalues with $|\gamma_k| = |\sigma|$, from the outside (stable) or from the inside (unstable).

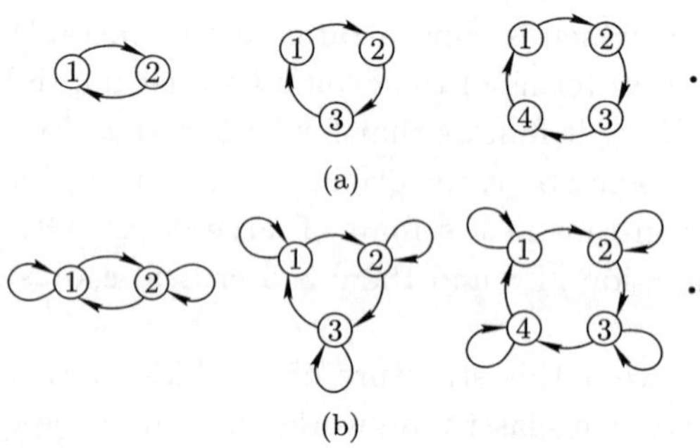

Fig. 1. (a) Unidirectionally coupled rings without feedback: class (B) or (C) (no chaos synchronization), (b) unidirectionally coupled rings with feedback: class (A) (chaos synchronization possible).

We now list some examples for the three types of networks (Fig.1). The classification follows from the eigenvalue structure (spectral radius) for the corresponding matrices, which can, for instance, be derived using Gerschgorin's circle theorem[45].

- Mean field coupled systems (all-to-all coupling) have $\gamma_k = 0$ for all k and are thus of type (A).
- Networks with only inhibitory connections (negative entries) or only excitatory connections (positive entries) are up to the row sum factor stochastic matrices, i. e., the coupling matrix G can be written as

$$G = \sigma P,$$

  where P is a stochastic matrix (positive entries and unity row sum). For stochastic matrices it is well-known that the spectral radius is one, i. e., all eigenvalue have magnitude smaller than or equal to one. The proof utilizes Gerschgorin's theorem[32]. Thus it follows for G that no eigenvalue has magnitude larger than $|\sigma|$ and these networks are of type (A) or (B).
- Any network with zero row sum ($\sigma = 0$) is of type (B) (trivial case) or (C).
- Rings of uni-directionally coupled elements and two bidirectionally coupled elements are of type (B)[18].
- Unidirectionally coupled rings with self-feedback are of type (A).
- Bidirectionally coupled rings with even number of elements without self-feedback are of type (B).

- Bidirectionally coupled rings with odd number of elements are of type (A).

In the literature there is a great amount of material on the relation of the spectral radius and the row sum for certain types of matrices. These results are immediately applicable to our classification. For a concrete network topology the classification is of course very simple.

Networks with $\sigma = 0$ belong to class (B) (trivial case) or to class (C). This confirms the conjecture stated in Ref.19: Networks for which the trajectory of an uncoupled unit is also a solution of the network ($\sigma = 0$) cannot exhibit chaos synchronization for large coupling delay.

For the chaotic case there may exist another radius r_b, with $0 \leq r_b \leq r_0$, where the first periodic orbit in the attractor loses its transverse stability and the attractor undergoes a bubbling bifurcation[46–48]. Then any network with $r_b < |\gamma_{\max}| < r_0$ will exhibit bubbling in the presence of noise (or parameter mismatch), while any network with $|\gamma_{\max}| < r_b$ will show stable synchronization, even in the presence of noise.

For chaotic synchronization in networks with time-delayed coupling the notion of strong and weak chaos has been introduced[21]. It is distinguished by the scaling properties of the maximum Lyapunov exponent within the synchronization manifold for large delay times, and is related to the condition for stable or unstable chaotic synchronization, respectively. Transitions from weak to strong and back to weak chaos can be induced by monotonically increasing the coupling strength. Strong chaos is characterized by a Lyapunov exponent that stays constant with increasing delay time. As recently shown[21] chaos synchronization is not possible in the large delay limit for *strong chaos*. The other case of *weak chaos* occurs when the largest Lyapunov exponent scales as $1/\tau$ for $\tau \to \infty$. In this case chaos synchronization is possible and the critical radius is determined by

$$r_0 = |\sigma| e^{-\lambda_m \tau} , \tag{9}$$

where λ_m is the maximum Lyapunov exponent of the system[21].

3. Cluster Synchronization

Next we shall study more general types of synchronization than complete or isochronous (zero-lag, in-phase) synchronization. More complicated synchronization patterns may be observed in the form of cluster, group, and sublattice synchronization[18,30,49–54,73]. Cluster synchronization in delay-coupled networks, where certain clusters inside the network show isochronous synchronization, will be investigated in this section. These results can be extended to group synchronization, i.e., a generalization of cluster synchronization where the local dynamics of the nodes in each group differs. The details of this formalism are described elsewhere[22].

Using a master stability approach, one observes that the master stability function shows a discrete rotational symmetry depending on the number of groups[22,49].

The coupling matrices that permit solutions on group or cluster synchronization manifolds show a very similar symmetry in their eigenvalue spectrum, which helps to simplify the evaluation of the master stability function. Our approach allows for the characterization of stability of different patterns of synchronized dynamics in networks with multiple delay times, multiple coupling functions, but also with multiple kinds of local dynamics in the nodes of the network[22]. This method can also be applied to very complex topologies, for instance, hierarchical network structures. A hierarchical network usually consists of topological clusters that are densely coupled inside, while links to other such topological clusters are sparse. The hierarchy is then built by larger topological clusters that contain the smaller ones[55,56]. This procedure can be continued over many levels of hierarchy. It has been shown that each cluster can exhibit isochronous synchronization under certain conditions[22].

As a simple example for cluster synchronization we consider a ring network of N delay-coupled Stuart-Landau oscillators

$$\dot{z}_j(t) = f[z_j(t)] + Ke^{i\beta}\sum_{n=1}^{N} a_{jn}[z_n(t-\tau) - z_j(t)] \tag{10}$$

with $z_j = r_j e^{i\varphi_j} \in \mathbb{C}$, $j = 1, \ldots, N$. The coupling matrix $A = \{a_{ij}\}_{i,j=1}^{N}$ determines the topology of the network. The local dynamics of each element is given by the normal form of a supercritical Hopf bifurcation, also known as Stuart-Landau oscillator,

$$f(z_j) = [\lambda + i\omega - (1 + i\gamma)|z_j|^2]z_j \tag{11}$$

with real constants $\lambda, \omega \neq 0$, and γ. In Eq.(10), τ is the delay time. K and β denote the amplitude and phase of the complex coupling constant, respectively. Such kinds of networks are used in different areas of nonlinear dynamics, e.g., to describe neural activities[57].

Synchronous in-phase, cluster, and splay states are possible solutions of Eqs. (10) and (11) as shown schematically in Fig.2. They exhibit a common amplitude $r_j \equiv r_{0,m}$ and phases given by $\varphi_j = \Omega_m t + j\Delta\varphi_m$ with a phase shift $\Delta\varphi_m = 2\pi m/N$ and collective frequency Ω_m. The integer m determines the specific state: in-phase oscillations correspond to $m = 0$, while splay and cluster states correspond to $m = 1, \ldots, N-1$. The cluster number d, which determines how many clusters of

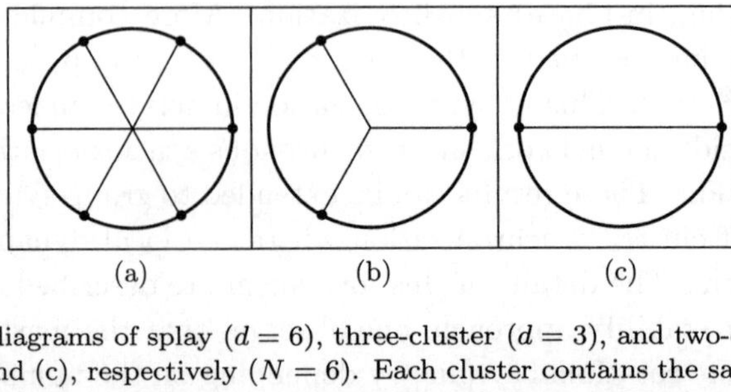

Fig. 2. Schematic diagrams of splay ($d = 6$), three-cluster ($d = 3$), and two-cluster ($d = 2$) states in panels (a), (b), and (c), respectively ($N = 6$). Each cluster contains the same number of nodes.

oscillators exist, is given by the least common multiple of m and N divided by m, and $d = N$ (e.g., $m = 1$), corresponds to a splay state.

The stability of synchronized oscillations in networks can be determined numerically by the master stability function. In the case of the Stuart-Landau oscillators it is possible to obtain the Floquet exponents of different cluster states analytically with this technique[18]. By these means it has been demonstrated that the unidirectional ring configuration of Stuart-Landau oscillators exhibits in-phase synchrony, splay states, and clustering, respectively, depending on the choice of the control parameter β. For $\beta = 0$, there exists multistability of the possible synchronous states in a large parameter range. However, when tuning the coupling phase to an optimal value $\beta = \Omega_m\tau - 2\pi m/N$ according to a particular state m, this synchronous state is monostable for any values of the coupling strength K and the time delay[18] τ.

4. Adaptive Synchronization

In this section, an adaptive algorithm will be introduced to find optimal values of the control parameters K, β, and τ to achieve cluster synchronization in the Stuart-Landau oscillator network studied in the previous section[31,58].

4.1. *Speed-gradient Method*

First we briefly review an adaptive control scheme based on optimization of cost or goal functions called *speed-gradient* (SG) *method*[59–61]. Consider a general nonlinear dynamical system

$$\dot{x} = F(x, u, t) \tag{12}$$

with state vector $x \in \mathbb{C}^n$, input (control) variables $u \in \mathbb{C}^m$, and nonlinear function F. Define a control goal

$$\lim_{t\to\infty} Q(x(t), t) = 0, \tag{13}$$

where $Q(x, t) \geq 0$ is a smooth scalar goal function.

In order to design a control algorithm, the scalar function $\dot{Q} = \omega(x, u, t)$ is calculated, that is, the speed (rate) at which $Q(x(t), t)$ is changing along trajectories of Eq. (12):

$$\omega(x, u, t) = \frac{\partial Q(x, t)}{\partial t} + [\nabla_x Q(x, t)]^T F(x, u, t). \tag{14}$$

Then we evaluate the gradient of $\omega(x, u, t)$ with respect to the control variables:

$$\nabla_u \omega(x, u, t) = \nabla_u [\nabla_x Q(x, t)]^T F(x, u, t).$$

Finally, we set up a differential equation for the input variables u

$$\frac{du}{dt} = -\Gamma \nabla_u \omega(x, u, t), \tag{15}$$

where $\Gamma = \Gamma^T > 0$ is a positive definite gain matrix. The algorithm (15) is called *speed-gradient* (SG) *algorithm*, since it suggests to change u proportionally to the gradient of the speed of changing Q.

The idea of this algorithm is the following. The term $-\nabla_u \omega(x,u,t)$ points to the direction in which the value of $\dot{Q}$ decreases with the highest speed. Therefore, if one forces the control signal to "follow" this direction, the value of $\dot{Q}$ will decrease and finally be negative. When $\dot{Q} < 0$, then Q will decrease and, eventually, tend to zero.

We shall now apply the speed-gradient method to networks of Stuart-Landau oscillators. Since the coupling phase β is the crucial parameter that determines stability of the possible in-phase, cluster, and splay states, we use this control parameter as the input variable u. Setting $u = \beta$, $x = (z_1, \ldots, z_N)$ and $\Gamma = \Gamma_\beta$, equation (10) takes the form of Eq.(12) with state vector $x \in \mathbb{C}^N$ and input variable $\beta \in \mathbb{R}$, and nonlinear function $F(x,\beta,t) = [f(z_1), \ldots, f(z_N)] + Ke^{i\beta}[Ax(t-\tau) - x(t)]$.

The SG control equation (15) for the input variable β then becomes

$$\frac{d\beta}{dt} = -\Gamma_\beta \frac{\partial}{\partial \beta}\omega(x,\beta,t) = -\Gamma_\beta \left(\frac{\partial F}{\partial \beta}\right)^T \nabla_x Q(x,t), \tag{16}$$

where $\Gamma_\beta > 0$ is now a scalar.

4.2. *Zero-lag Synchronization*

To apply the SG method for the selection of in-phase (zero-lag) synchronization we need to find an appropriate goal function Q. It should satisfy the following conditions: the goal function must be zero for an in-phase synchronous state and larger than zero for other states. Hence, a simple goal function can be introduced by considering a function based on the order parameter

$$R_1 = \frac{1}{N}\left|\sum_{j=1}^{N} e^{i\varphi_j}\right|. \tag{17}$$

This global parameter can conveniently be measured in real-world systems, e.g., lasers. It is obvious that $R_1 = 1$ if and only if the state is in-phase synchronized. For other cases we have $R_1 < 1$. Using this observation we can introduce the following goal function

$$Q_0 = 1 - \frac{1}{N^2}\sum_{j=1}^{N} e^{i\varphi_j}\sum_{k=1}^{N} e^{-i\varphi_k}. \tag{18}$$

From $\dot{\beta} = -\Gamma_\beta \frac{\partial}{\partial \beta}\dot{Q}_0$ we derive an adaptive law:

$$\dot{\beta} = \Gamma_\beta \frac{2K}{N^2}\sum_{k=1}^{N}\sum_{j=1}^{N}\sin(\varphi_k - \varphi_j)\sum_{n=1}^{N} a_{jn}\left(\frac{r_{n,\tau}}{r_j}\cos(\beta + \varphi_{n,\tau} - \varphi_j) - \cos\beta\right). \tag{19}$$

Figure 3 shows the results of a numerical simulation for an Erdős-Rényi random network with $N = 6$ nodes and row sum normalized to unity. Unless otherwise stated, we use $\Gamma_\beta = 1$. According to the numerical simulations decreasing Γ_β will yield a decrease of the speed of convergence. On the other hand, if Γ_β is too big, undesirable oscillations appear. The model parameters are chosen as in Ref.18. The amplitude and phase approach appropriate values that lead to in-phase synchronization. Note that the obtained value of β does not converge to the one for which the analytical approach[18] has established stability of the in-phase oscillation (dotted line), but to another limit value. This can be explained as follows: There exists a whole interval of acceptable values of β around the value of the coupling phase for which an analytical treatment is possible, such that for any value from this interval an in-phase state is stable. Our SG algorithm finds one of them, depending upon initial conditions.

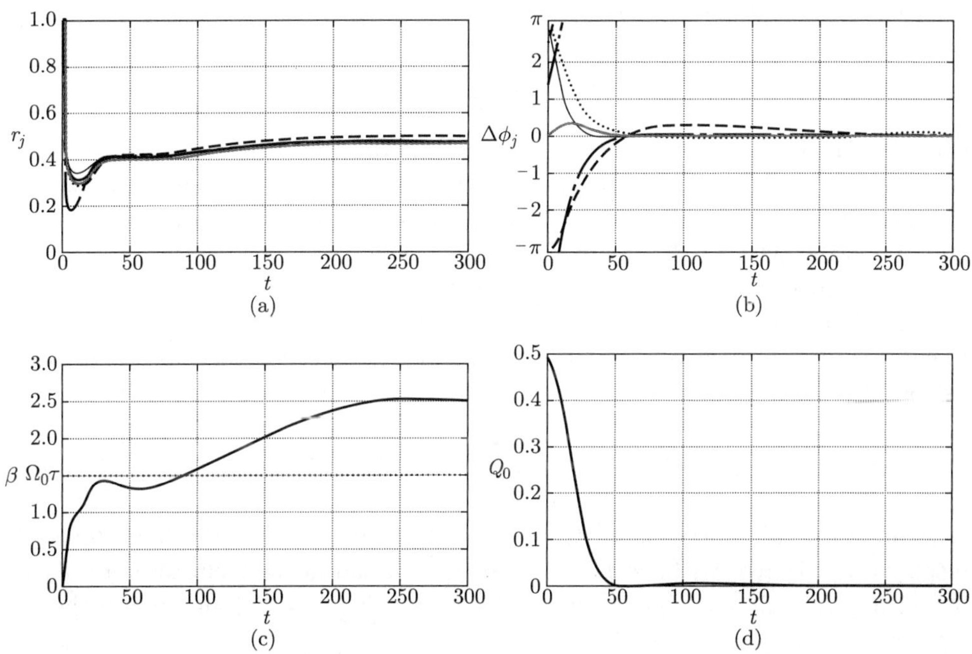

Fig. 3. Adaptive control of in-phase oscillations with goal function Eq. (18). (a) absolute values $r_j = |z_j|$ for $j = 1, \ldots, 6$; (b) phase differences $\Delta\phi_j = \varphi_j - \varphi_{j+1}$ for $j = 1, \ldots, 5$; (c) temporal evolution of β, dotted line: reference value for $\Omega_0 = 0.92$; (d) goal function. Parameters: $\lambda = 0.1, \omega = 1, \gamma = 0$, $K = 0.08$, $\tau = 0.52\pi$, $N = 6$. Initial conditions for r_j and φ_j are chosen randomly from $[0, 4]$ and $[0, 2\pi]$, respectively. The initial condition for β is zero.[58]

4.3. *Splay State and Cluster Synchronization*

Now we will consider unidirectionally coupled rings with $N = 6$ nodes. That is, the coupling matrix has the following form:

$$A = \begin{bmatrix} 0 & 1 & 0 & \cdots & 0 \\ 0 & 0 & 1 & \cdots & 0 \\ \vdots & \vdots & \vdots & & \vdots \\ 0 & 0 & 0 & \cdots & 1 \\ 1 & 0 & 0 & \cdots & 0 \end{bmatrix}.$$

Let $1 \leq m \leq N - 1$. Then $d = \mathrm{LCM}(m, N)/m$, where LCM denotes the least common multiple, is the number of different clusters of a synchronized solution. A splay state corresponds to $d = N$ while cluster states yield $d < N$. In order to extend the goal function Eq. (18) such that we can stabilize splay and cluster states, we define a generalized order parameter

$$R_d = \frac{1}{N} \left| \sum_{k=1}^{N} e^{di\varphi_k} \right| \tag{20}$$

with $d \in \mathbb{N}$. However, if we derive a goal function from this order parameter in an analogous way as in Eq. (18), this function will not have a unique minimum at the d-cluster state because $R_d = 1$ holds also for the in-phase state and for other p-cluster states where p are divisors of d.

For example, suppose that the system has six nodes. Then states for which equation (20) yields $R_d = 1$ (for $d = 6$, 3, and 2) hold are schematically depicted in Fig.2(a),(b),(c), respectively. In order to distinguish between these three cases, let us consider the functions

$$f_p(\varphi) = \frac{1}{N^2} \sum_{j=1}^{N} e^{pi\varphi_j} \sum_{k=1}^{N} e^{-pi\varphi_k}. \tag{21}$$

A splay state (Fig.2(a)) yields $f_1 = f_2 = f_3 = 0$, while in the 3-cluster state displayed in Fig.2(b) we have $f_1 = f_2 = 0$, $f_3 = 1$, and in the 2-cluster-state shown in Fig.2(c) $f_1 = f_3 = 0$, $f_2 = 1$. Hence, we obtain $\sum_p f_p = 0$ if and only if there is a state with d clusters, where the sum is taken over all divisors of d.

Combining all previous results we adopt the following goal function:

$$Q_d = 1 - f_d(\varphi) + \frac{N^2}{2} \sum_{p|d, 1 \leq p < d} f_p(\varphi), \tag{22}$$

where $p|d$ means that p is a factor of d. This goal function contains f_d as the primary contribution for the d-cluster state, but also a sum of penalty terms that counteract reaching other cluster states in which f_d is also unity. Whenever one of those unwanted cluster states is approached, the penalty term will lead to a gradient away from it. The prefactor $N^2/2$ is chosen for convenience to secure faster convergence of the algorithm. From $\dot{\beta} = -\Gamma_\beta \frac{\partial}{\partial \beta}\dot{Q}_d$ one can derive the adaptation law

$$\dot{\beta} = -\Gamma_\beta K \sum_{j=1}^{N}\sum_{k=1}^{N}\left\{ \sum_{p|d,1\le p<d} p\sin[p(\varphi_k-\varphi_j)] - \frac{2d}{N^2}\sin[d(\varphi_k-\varphi_j)]\right\} \quad (23)$$
$$\cdot \sum_{n=1}^{N} a_{jn}\left[\frac{r_{n,\tau}}{r_j}\cos(\beta+\varphi_{n,\tau}-\varphi_j)-\cos\beta\right].$$

In Fig.4 we show the results of a numerical simulation for splay state stabilization ($d = N = 6$, $m = 1$). The phase differences are $\Delta\phi_j = \varphi_j - \varphi_{j+1} = 2\pi - 2\pi/N$, which corresponds to the splay state. In Fig.4(c) one can see that the adaptively obtained value of β converges to that for which stability was shown analytically in Ref.18 (dotted line).

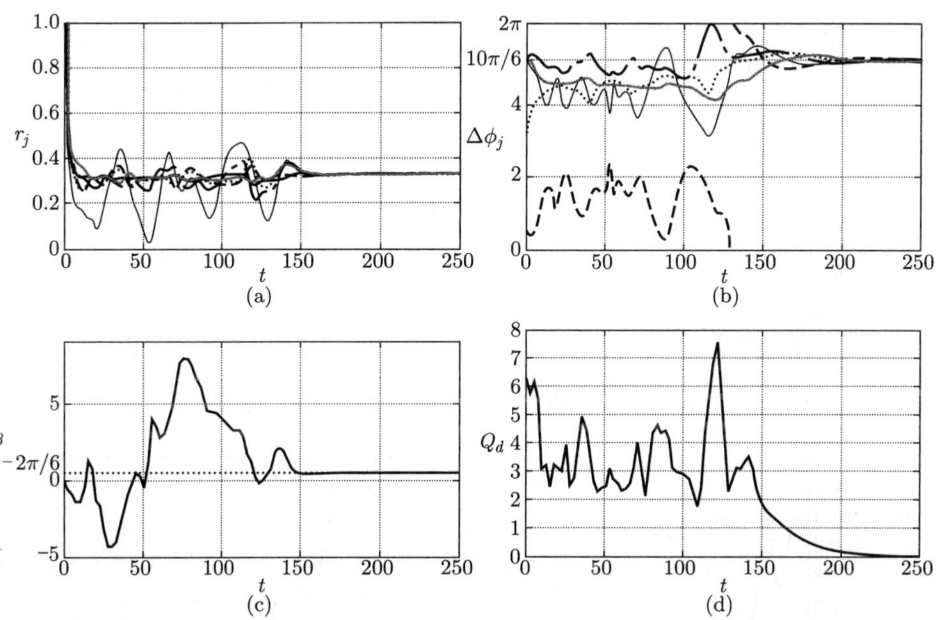

Fig. 4. Adaptive control of splay state with goal function Eq. (22). (a) absolute values $r_j = |z_j|$; (b) phase differences $\Delta\phi_j = \varphi_j - \varphi_{j+1}$; (c) temporal evolution of β, dotted line: reference value for $\Omega_1 = 0.96$; (d) goal function. Other parameters as in Fig.3.[58]

Figures 5 and 6 depict the results of numerical simulations for two clusters ($d = 2$, $m = 3$) and three clusters ($d = 3$, $m = 2, 4$), respectively. Again we note that the obtained value of β comes close to the one for which stability was shown analytically in Ref.18.

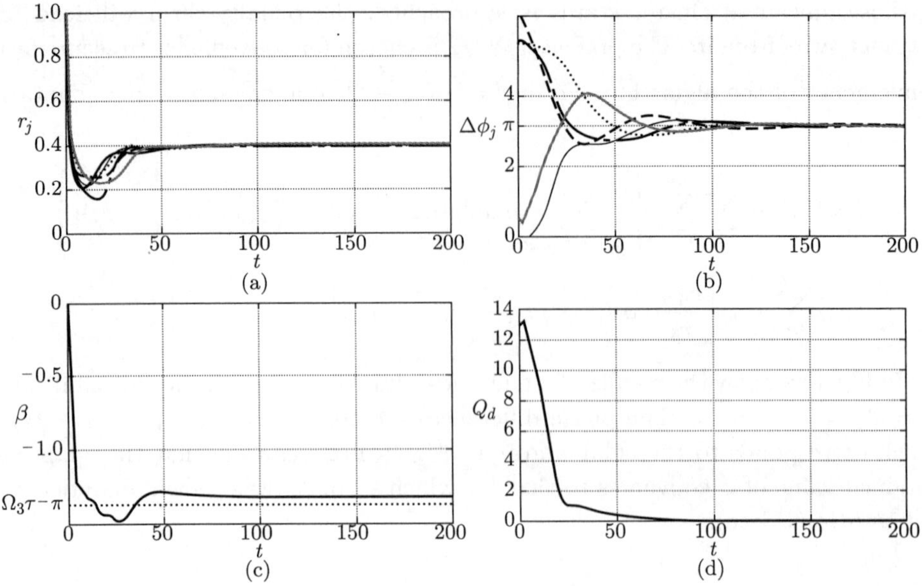

Fig. 5. Adaptive control of 2-cluster state ($m = 3$) with goal function Eq. (22). (a) absolute values $r_j = |z_j|$; (b) phase differences $\Delta\phi_j = \varphi_j - \varphi_{j+1}$; (c) temporal evolution of β, dotted line: reference value for $\Omega_3 = 1.08$; (d) goal function. Other parameters as in Fig.3.[58]

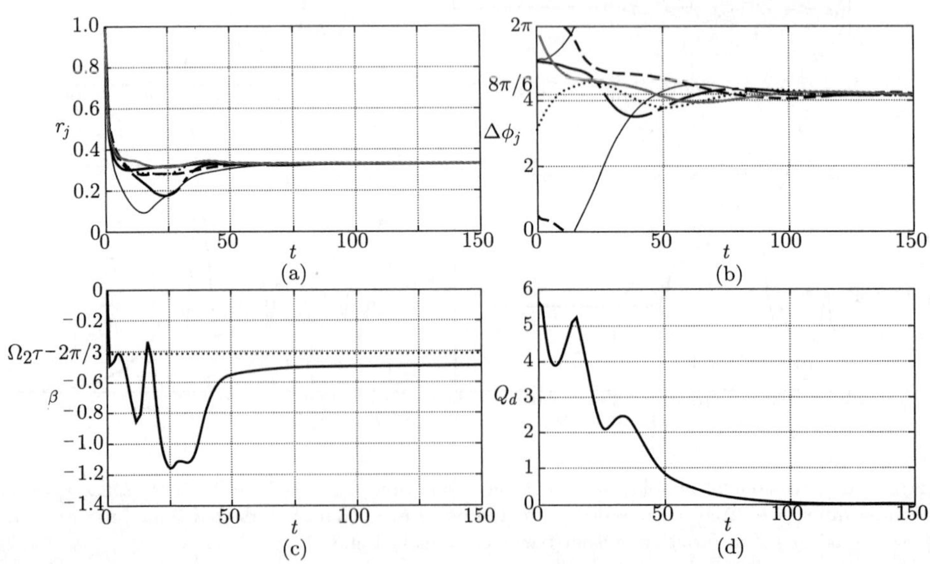

Fig. 6. Adaptive control of 3-cluster state ($m = 2, 4$) with goal function Eq.(22). (a) absolute values $r_j = |z_j|$; (b) phase differences $\Delta\phi_j = \varphi_j - \varphi_{j+1}$; (c) temporal evolution of β, dotted line: reference value for $\Omega_2 = 1.03$; (d) goal function. Other parameters as in Fig.3.[58]

The above results indicate that the speed-gradient method is able to drive the network dynamics into the desired cluster or splay state by adaptively adjusting the coupling phase, where the goal function is chosen according to the corresponding target state. Control is possible in a wide range of the coupling parameters[31] K and τ. When using too large delay times, however, overshoots of the control lead to a failure.

4.4. *Controlling Several Parameters Simultaneously*

The general form of the SG method as given in Eq.(15) is also suitable for controlling more than one parameter. Here this is demonstrated by controlling β, K, and τ simultaneously. The vector u in Eq.(15) is then given by $u = (\beta, K, \tau)$. We choose Γ as a diagonal matrix with the diagonal elements $\Gamma_{11} \equiv \Gamma_\beta$, $\Gamma_{22} \equiv \Gamma_K$, and $\Gamma_{33} \equiv \Gamma_\tau$. Using the goal function Q_d of Eq.(22) we obtain for β the adaptive algorithm given by Eq.(23). For $\dot{K} = -\Gamma_K \frac{\partial}{\partial K} \dot{Q}_d$ we obtain

$$\dot{K} = -\Gamma_K \sum_{j=1}^{N} \sum_{k=1}^{N} \left\{ \sum_{p|d, 1 \le p < d} p \sin[p(\varphi_k - \varphi_j)] - \frac{2d}{N^2} \sin[d(\varphi_k - \varphi_j)] \right\} \tag{24}$$

$$\cdot \sum_{n=1}^{N} a_{jn} \left[\frac{r_{n,\tau}}{r_j} \sin(\beta + \varphi_{n,\tau} - \varphi_j) - \sin(\beta) \right],$$

and for $\dot{\tau} = -\Gamma_\tau \frac{\partial}{\partial \tau} \dot{Q}_d$

$$\dot{\tau} = -\Gamma_\tau \sum_{j=1}^{N} \sum_{k=1}^{N} \left\{ \sum_{p|d, 1 \le p < d} p \sin[p(\varphi_k - \varphi_j)] - \frac{2d}{N^2} \sin[d(\varphi_k - \varphi_j)] \right\} \tag{25}$$

$$\cdot \sum_{n=1}^{N} a_{jn} \left[-\frac{\dot{r}_{n,\tau}}{r_j} \sin(\beta + \varphi_{n,\tau} - \varphi_j) - \dot{\varphi}_{n,\tau} \frac{r_{n,\tau}}{r_j} \cos(\beta + \varphi_{n,\tau} - \varphi_j) \right].$$

Figure 7 shows the successful control of a 3-cluster state in a network consisting of 6 nodes where appropriate values of β, K, and τ are found adaptively.

5. Transitions between Synchronization and Desychronization

In this section we focus on excitable local dynamics, and consider two paradigmatic models which are of importance for applications to neural networks: the FitzHugh-Nagumo (FHN) model[62,63] and the model of a saddle-node infinite period (SNIPER) bifurcation[64]. An excitable system rests in a stable fixed point, but a sufficiently strong perturbation across a threshold can excite a single pulse (spike), after which the system returns to the rest state. Such behavior typically

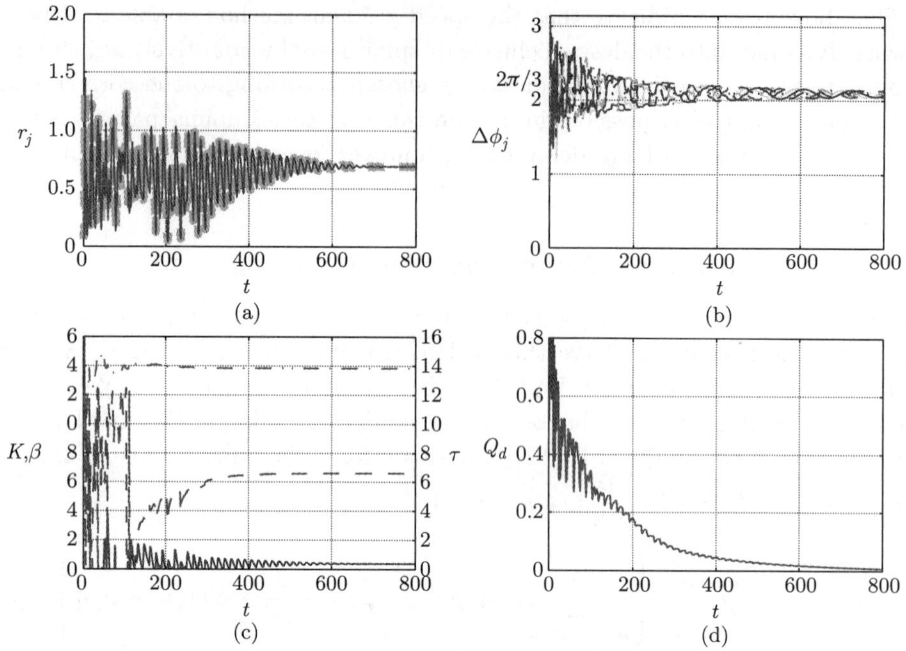

Fig. 7. Adaptive control of 3-cluster state in a network of 6 nodes by simultaneously tuning K, β, and τ according to Eqs. (23), (25), and (24), respectively. (a) absolute values $r_j = |z_j|$; (b) phase differences $\Delta\phi_j = \varphi_j - \varphi_{j+1}$; (c) temporal evolution of control parameters β (red dashed line), K (blue solid line), τ (green dashed-dotted line); (d) goal function. $\Gamma_\beta = \Gamma_K = 10$, $\Gamma_\tau = 0.05$. Other parameters as in Fig.3.[58] (color figure at the end of the book)

occurs below a global SNIPER bifurcation, also called saddle-node bifurcation on invariant cycle (*excitability of type* I), or a Hopf bifurcation (*excitability of type* II).

In neural networks, the type of coupling, namely with positive sign (excitatory or attractive), or with negative sign (inhibitory or repulsive), is also an important aspect. Inhibition plays a prominent role in many neural processes[65]. Without an inhibitory mechanism, excitation in a compound system would not decay, but spread through the whole network, finally leading to persistent spiking of all neurons. Thus, encoding and processing of information would be impossible. There is usually a delicate balance between excitatory and inhibitory couplings[66].

We investigate the stability of synchronization in networks of delay-coupled excitable neural oscillators. With the help of the master stability function formalism we demonstrate that synchronization is always stable for excitatory coupling independently of the delay and coupling strength. Superimposing inhibitory links randomly on top of a regular ring of excitatory coupling, which yields a small-world-like network topology, we find a phase transition to desynchronization as the probability of inhibitory links exceeds a critical value[32]. For certain types of local dynamics and network topologies, multiple synchronization/desynchronization transitions are possible[33]. There is evidence that neuronal networks exhibit the properties of a small-world (SW) network[67]: The high clustering coefficient enhances local communication efficiency, while the small shortest path length enables efficient global communication. Thus, the SW architecture is optimal for processing and transmission of signals within and between brain areas.

In this section, we combine both fundamental aspects — inhibition and SW property — in order to emphasize the important interplay of excitation and inhibition in synchronizing and desynchronizing complex networks.

5.1. *Excitability of Type* II

First, we consider a network of N delay-coupled FHN oscillators. The network dynamics is described by

$$\epsilon \dot{u}_i = u_i - \frac{u_i^3}{3} - v_i + \sigma \sum_{j=1}^{N} G_{ij}[u_j(t-\tau) - u_i], \tag{26a}$$

$$\dot{v}_i = u_i + a, \tag{26b}$$

where u_i and v_i denote the activator and inhibitor variables of the nodes $i = 1, \ldots, N$, respectively. The parameter a determines the threshold of excitability. A single FHN oscillator is excitable for $a > 1$ and exhibits self-sustained periodic firing for $a < 1$, i.e., beyond the Hopf bifurcation at $a = 1$. Here, we will focus on the excitable regime with $a = 1.3$. The time-scale parameter ϵ is chosen as $\epsilon = 0.01$. σ is the coupling strength. $\mathbf{G} = \{G_{ij}\}$, $i, j = 1, \ldots, N$, denotes the coupling matrix that determines the topology of the network. In the following we will assume unity row sum of $\mathbf{G}$. This ensures that each neuron receives the same input if the network is synchronized. The delay time τ takes into account the finite propagation speed of an action potential. We investigate complete (isochronous) synchronization with $(u_i(t), v_i(t)) = (u_s(t), v_s(t)) \equiv x_s(t)$ for $i = 1, \ldots, N$, and consider σ and τ sufficiently large such that the time-delayed coupling induces periodic oscillations[68].

In the following we address the question whether the oscillatory solution on the synchronization manifold is transversely stable.

Figure 8 depicts the MSF for the network of FHN oscillators given by Eqs. (26) as a function of the complex parameter $re^{i\phi} = \alpha + i\beta$. Blue colors mark the stable region. As an illustration the rescaled eigenvalues $\gamma_i = \{\sigma\nu_i\}$ (where ν_i, $i = 1, \ldots, N$, are the eigenvalues of $\mathbf{G}$) of a bidirectionally coupled ring ($N = 8$) are shown as red symbols. The corresponding coupling matrix is given by $G_{i,i+1 \bmod N} = G_{i,i-1 \bmod N} = 1$ ($i = 1, \ldots, N$) and zero otherwise. The rescaled longitudinal eigenvalue $\sigma\nu_N = \sigma$ is depicted by a red square. All rescaled transversal eigenvalues (red circles) lie inside the stable region indicating that the synchronization of the bidirectionally coupled ring is stable.

The MSF must be calculated for each combination of σ and τ. Although different σ and τ lead to quantitatively different maximum Lyapunov exponents $\lambda_{\max}$, the shape of the stable regions remains qualitatively very similar. In particular, it is in

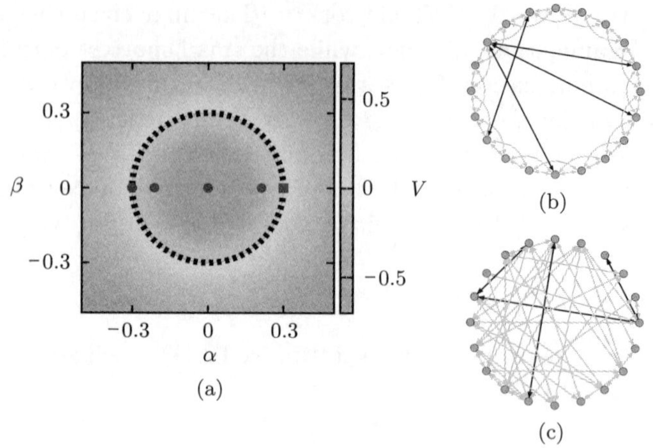

Fig. 8. (a) Master stability function for a network of FHN systems given by Eqs. (26). Dotted curve: $S((0,0),\sigma)$. Red circles (square): rescaled transversal (longitudinal) eigenvalues $\gamma_i = \sigma\nu_i$ of a bidirectionally coupled ring with $N = 8$ nodes. Parameters: $a = 1.3$, $\epsilon = 0.01$, $\sigma = 0.3$, $\tau = 1$. (b) Scheme of a bidirectional regular network ($N = 20$, $k = 2$), and (c) a random network ($N = 20$, fixed number of links kN) with excitatory coupling (green arrows) on which inhibitory links (red arrows) are superimposed.[32] (color figure at the end of the book)

very good approximation given by the circle $S((0,0),\sigma)$ with center at the origin and radius σ (dotted circle in Fig.8) independent of the specific values of σ and τ. While this is generally expected for large τ, as shown above[20], for the FHN model this rotational symmetry is retained approximately even for small[32] τ. The positive α-axis is always intersected at $\alpha = \sigma$, which corresponds to $\lambda_{\max}(\sigma) = 0$ as discussed above. For any choice of τ and σ that leads to periodic dynamics on the SM, the circle $S((0,0),\sigma)$ serves as a lower bound for the stability boundary. We conclude that the stability of the synchronized periodic dynamics, if such a solution exists, depends only on the topology and neither on the coupling strength nor on the delay time.

For excitatory coupling, i.e., $G_{ij} \geq 0$, all eigenvalues of $\mathbf{G}$ are located inside the stable region. Using Gerschgorin's circle theorem[69], which gives an upper bound of the eigenvalues, and the constant row sum assumption, all Gerschgorin circles ($i = 1, \ldots, N$), centered at G_{ii} with radius $\sum_{j \neq i} G_{ij} = 1 - G_{ii}$ because of the unity row sum, lie inside the unit circle. Thus all rescaled eigenvalues $\{\sigma\nu_i\}$, $i = 1, \ldots, N$, are located inside $S((0,0),\sigma)$, i.e., inside the stable region. Networks with purely excitatory coupling will always exhibit stable synchronization.

As a consequence of this result, desynchronization can only be achieved by introducing negative entries in the coupling matrix $\mathbf{G}$, i.e., inhibitory coupling between neurons. This inhibition is a crucial feature in neural processes, e.g., to overcome unwanted synchronization associated with pathological states. In particular, we consider the following variation of the Watts-Strogatz SW network[67,70]: (i) Start with a one-dimensional ring of N nodes, where every node is connected by excitatory links to its k neighbors on either side. (ii) For each of the kN links of the network add an inhibitory link with probability p connecting two randomly chosen nodes. (iii) Do not allow self-coupling or more than one link between any pair of nodes. (iv) Normalize the entries of the coupling matrix $\mathbf{G}$ by dividing each row by the absolute value of its row sum. In the case that the row sum of the ith row is negative we set $G_{ii} = 2$ to ensure unity row sum. Figure 8(b) illustrates such a SW network for $N = 20$ and $k = 2$, where green and red arrows indicate excitatory and inhibitory coupling, respectively. For each realization of such a network, we determine the stability of synchronization by checking whether the full eigenspectrum $\sigma\nu_i$ of the coupling matrix is contained in the stable region $S((0,0),\sigma)$. Hereby we compute the fraction f of desynchronized networks. Figure 9 shows f as a function of p for different coupling ranges k. For fixed k a sharp transition between synchronization and desynchronization takes place as p approaches a critical value[32] p_c. The transition becomes increasingly sharp as N increases.

To verify that this transition is common in networks with excitatory and inhibitory links, we replace the regular excitatory network by a random network with fixed number of excitatory links kN equivalently to the regular network used before, as shown in Fig.8(c). We only consider realizations where this underlying excitatory network is fully connected. The construction of the inhibitory links then follows steps (ii–iv) as above. We find that a desynchronization transition to still occurs at a critical probability of inhibitory links p_c. We observe, however, that the values of p_c are higher, i.e., the random network can tolerate more inhibitory links than the SW network before desynchronizing. Furthermore, for random networks with constant k, p_c is independent of N for sufficiently large N, while for SW networks it approaches zero. Recall that p_c is the mean value of the ratio of inhibitory to excitatory links. Thus, we conclude that in SW networks with increasing network size N but same local structure (constant k) only an infinitesimally small ratio of inhibition to excitation is needed to prevent synchronization, while in a random network even for very large networks only a non-vanishing ratio impairs synchronization. This difference to SW networks can be understood in an intuitive way: in a SW, any added inhibitory link is part of a shortest path for many pairs of nodes, as it shortens the mean path length considerably with respect to the underlying regular ring. In the random network, however, where the mean path length is relatively low even without added shortcuts, only few node pairs will gain shorter paths by adding inhibitory links. Considering the dynamics on a network, perturbations from the synchronized state spread along the shortest paths first, changing the response of the receiving node, and information flow along longer paths will reach the receiving node only at a later time and will not influence the change of the initial response. In conclusion, if a large fraction of the inhibitory links is part of the shortest paths, like in the small-world topology superimposed to a regular ring, these inhibitory shortcuts become dominant.

5.2. *Excitability of Type* I

Next, we consider as local dynamics a generic normal-form of a SNIPER bifurcation[64,71]:

$$\dot{r} = r\left(1 - r^2\right), \tag{27}$$

$$\dot{\varphi} = b - r\cos\varphi, \tag{28}$$

where r, φ are polar coordinates, and $b > 0$ is the bifurcation parameter. For $b < 1$ there exists an unstable focus at the origin, as well as a saddle point and a stable node situated on the unit circle at $(x_i, y_i) = (b, \sqrt{1-b^2})$ and $(b, -\sqrt{1-b^2})$, respectively. At $b = 1$ the saddle point and stable node collide, so that for $b > 1$ a limit cycle exists along the unit circle. In the case of $b < 1$ (excitable regime), a perturbation which pushes the system from the stable node beyond the saddle point can result in a single oscillations along the unit circle. Delayed coupling can induce a homoclinic bifurcation, such that a limit cycle is produced that bypasses the saddle point and stable node[71,72]. Here, we will consider the excitable regime with $b = 0.95$ and investigate for which topologies the synchronized dynamics of complex networks is stable[33].

In the following, the coupling function h is taken to be the 2×2 identity matrix in cartesian coordinates. According to Ref.20, for τ in the order of the system's characteristic time scale and above, the MSF will always tend towards a rotational symmetry. As in the case of the FitzHugh-Nagumo system, for a network with purely excitatory coupling, synchronization will then always be stable. Similarly, if additional inhibitory links are introduced with probability p to construct a small-world network as described above, transitions from stable to unstable synchronization are found with increasing probability of inhibition p.

However, if the delay time is not large enough, the rotational symmetry of the MSF no longer holds, in contrast to the FHN model. In fact, in the regime of small delay, small changes in τ can have a great impact on the stability. This becomes clear by plotting the MSF versus the real part α of the eigenvalue (with $\beta = 0$) for varying τ. Taking this one slice of the eigenvalue plane gives a good indication of the growth and decay of the stable region in the MSF while changing τ. Figure 10 shows the MSF as a function of the real part α ($\beta = 0$) and the delay time τ, with fixed coupling strength of 0.30. This produces an MSF on an (α, τ)-plane, which can be used for network topologies with a coupling matrix that yields real eigenvalues, e.g., symmetric matrices for undirected (bidirectionally coupled) networks. In the following we restrict ourselves to undirected networks.

Note that the stability depends on both σ and τ. In Fig. 9 the vertical boundary line at $\alpha = \sigma$, corresponding to the longitudinal eigenvalue (i.e., $\nu_0 = 1$, where $\lambda_{\max} = 0$), can be easily identified. It separates regimes of stable and unstable synchronization. Another common characteristic is that the τ-dependent MSFs have a critical delay time τ_c at which the stable region splits into two separate, disconnected regions. For values above τ_c the stable region is found to the left of the longitudinal eigenvalue; whereas for values below τ_c there may be one stable region to the right of the longitudinal eigenvalue and one to the left. Hence τ_c marks the most significant τ-dependent qualitative change in the MSF. Ultimately, the MSF can be divided into three regimes of τ: (i) two or more smaller unconnected regions of stability exist when $\tau < \tau_c$; (ii) one large region of stability exists when $\tau > \tau_c$; (iii) the MSF has one rotationally symmetric region of stability for large τ.

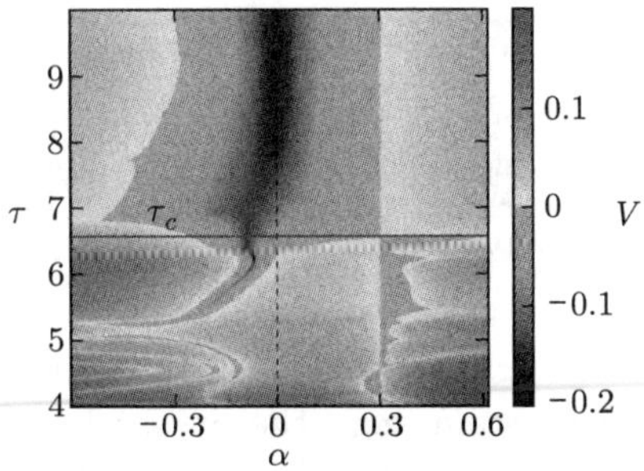

Fig. 9. Master stability function $\lambda_{\max}$ for a fixed coupling strength $\sigma = 0.3$ in the plane of the real part α ($\beta = 0$) and the delay time τ. The horizontal red line shows the position of the critical delay time τ_c. Parameter: $b = 0.95$.[33] (color figure at the end of the book)

A consequence is that networks with purely excitatory coupling, which are always stable for large delay times, may not show stable synchronization for $\tau < \tau_c$. For small-world networks with inhibitory links of probability p desynchronization can occur, as discussed above for excitability of type II, because a larger probability of inhibition can push a part of the eigenvalue spectrum of any network beyond the longitudinal eigenvalue at $\alpha = \sigma$ (because Gerschgorin's circle theorem no longer holds) and into the unstable region of the MSF. Now, however, for smaller τ, there may be a pocket of stability to the right of the longitudinal eigenvalue, so that increasing inhibition can make the otherwise unstable synchronization of a network stable. This means that the transitions between stable and unstable synchronization as a function of the probability of inhibition p are now sensitive to the coupling parameters, not just the network parameters N and k. Due to the multiple regions of stability, eigenvalues may wander in and out of stable regions, while increasing the probability of inhibition p. For large τ, in contrast, increasing p in the SW network model only results in one transition where the fraction of desynchronizing networks $f(p)$ switches from 0 to 1 (cf. Fig.10). For small τ, it is possible that $f(p)$ jumps back to 0, before increasing again to 1. This will occur if there is a separate region of stability to the right side in the MSF that is large enough that all the eigenvalues lying over there can fit inside. For a small-world network the eigenvalue spectrum may have gaps, which must be matched with the gaps of instability in the master stability function in a subtle interplay. Thus multiple transitions between synchronization and desynchronization are possible, and can be explained by the change in the eigenvalue spectrum with increasing[33] p. For random Erdős-Rényi networks the eigenvalue spectrum has no large gaps, and hence multiple transitions will not occur.

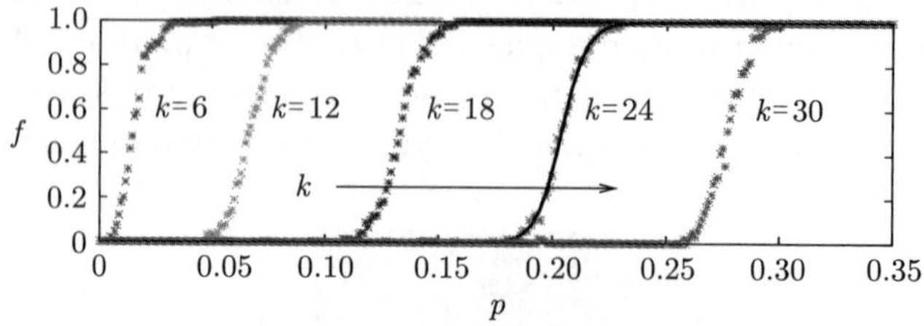

Fig. 10. Fraction of desynchronized networks f vs the probability of additional inhibitory links p for $N = 100$. k varies from 6 to 30. Number of realizations: 500 for each value of k. Parameters as in Fig.8.[32] (color figure at the end of the book)

It should be noted that such unconnected stable regions may also appear in the MSF of a purely oscillatory system, e.g., the Stuart-Landau oscillator, and hence multiple transitions in small-world networks can also be controlled by the balance of excitatory and inhibitory couplings[33].

6. Conclusion and Outlook

The master stability function of delay-coupled networks has a simple universal structure in the limit of large delay: it is rotationally symmetric around the origin and either positive and constant (if it is positive at the origin), or monotonically increasing and becomes positive at a minimum radius r_0. As a result, network structures can be classified into three types depending on the magnitude of the maximum transversal eigenvalue of the coupling matrix in relation to the magnitude of the row sum, resulting in distinct synchronization properties[20]. The rotational symmetry of the master stability function in delay-coupled networks of FitzHugh-Nagumo excitable systems has been established numerically and analytically even if the delay time τ is not large[32]. So the symmetry properties of the master stability function seem to be valid in even more general cases.

Furthermore, the master stability function approach can be extended to more general states of synchrony, i.e., cluster synchronization, and these can be stabilized by tuning the coupling parameters such as the coupling phase, coupling strength, and delay time[18,22]. Adaptive control of these various synchronous states can be achieved by the speed gradient method. Choosing an appropriate goal function, a desired state of generalized synchrony can be selected by the self-adaptive automatic adjustment of a control parameter, i.e., the coupling phase. This goal function, which is based on a generalization of the Kuramoto order parameter, vanishes for the desired state, e.g., in-phase, splay, or cluster states, irrespectively of the ordering of the nodes[31]. The speed gradient method can also be applied to simultaneously tune the coupling phase, strength, and the time delay. In this way control of cluster and splay synchronization is possible without any a priori knowledge of the coupling parameters. Given the paradigmatic nature of the Stuart-Landau oscillator as a generic model, one may expect broad applicability, for instance to synchronization of networks in medicine, chemistry, or mechanical engineering.

We have shown how the interplay of excitatory and inhibitory couplings leads to transitions between synchronization and desynchronization in networks of neural oscillators[32,33]. The desynchronization is achieved via a phase transition from a completely synchronized state when the fraction of inhibitory links is increased. In addition to the FitzHugh-Nagumo model, which is paradigmatic for type II excitability, we have considered the normal form of a saddle-node bifurcation on an invariant cycle as a generic model of type I excitability. For sufficiently large delay times and coupling strength the desynchronization transition is similar in type I and type II systems, but for small delay in type I systems, in contrast to the FitzHugh-Nagumo model, multiple transitions between synchronization and desynchronization can be induced by increasing the probability of inhibitory shortcuts in a small-world topology. This shows that the balance between excitatory and inhibitory couplings controls synchronization in a subtle way.

Acknowledgments. This work was supported by DFG in the framework of Sfb 910. I am indebted to stimulating collaboration and discussion with A. Amann, K. Blyuss, C.-U. Choe, T. Dahms, B. Fiedler, I. Fischer, V. Flunkert, A. Fradkov, S. Heiligenthal, P. Hövel, T. Isele, J. Jiang, T. Jüngling, W. Just, I. Kanter, A. Keane, W. Kinzel, Y. Kyrychko, J. Lehnert, K. Lüdge, Y. Maistrenko, I. Omelchenko, A. Panchuk, R. Roy, A. S. Sachin, A. Selivanov, S. Yanchuk, and A. Zakharova.

References

1. E. Schöll and H. G. Schuster (Eds.), *Handbook of Chaos Control* (Wiley-VCH, Weinheim, 2008). Second completely revised and enlarged edition.
2. A. S. Pikovsky, M. G. Rosenblum, and J. Kurths, *Synchronization, A Universal Concept in Nonlinear Sciences* (Cambridge University Press, Cambridge, 2001).
3. K. M. Cuomo and A. V. Oppenheim, Circuit implementation of synchronized chaos with applications to communications, *Phys. Rev. Lett.*, **71**(1), 65–68 (1993).
4. S. Boccaletti, J. Kurths, G. Osipov, D. L. Valladares, and C. S. Zhou, The synchronization of chaotic systems, *Phys. Rep.*, **366**, 1–101 (2002).
5. I. Kanter, E. Kopelowitz, and W. Kinzel, Public channel cryptography: chaos synchronization and Hilbert's tenth problem, *Phys. Rev. Lett.*, **101**(8), 084102 (2008).
6. W. Singer, Neuronal synchrony: a versatile code review for the definition of relations, *Neuron.*, **24**, 49–65 (1999).
7. P. A. Tass, M. G. Rosenblum, J. Weule, J. Kurths, A. S. Pikovsky, J. Volkmann, A. Schnitzler, and H. J. Freund, Detection of n:m phase locking from noisy data: application to magnetoencephalography, *Phys. Rev. Lett.*, **81**, 3291 (1998).
8. H. J. Wünsche, S. Bauer, J. Kreissl, O. Ushakov, N. Korneyev, F. Henneberger, E. Wille, H. Erzgräber, M. Peil, W. Elsäßer, and I. Fischer, Synchronization of delay-coupled oscillators: a study of semiconductor lasers, *Phys. Rev. Lett.*, **94**, 163901 (2005).

9. T. W. Carr, I. B. Schwartz, M. Y. Kim, and R. Roy, Delayed-mutual coupling dynamics of lasers: scaling laws and resonances, *SIAM J. Appl. Dyn. Syst.*, **5**(4), 699–725 (2006). doi: 10.1137/050647918.
10. H. Erzgräber, B. Krauskopf, and D. Lenstra, Compound laser modes of mutually delay-coupled lasers, *SIAM J. Appl. Dyn. Syst.*, **5**(1), 30–65 (2006). doi: 10.1137/040619958.
11. I. Fischer, R. Vicente, J. M. Buldú, M. Peil, C. R. Mirasso, M. C. Torrent, and J. García-Ojalvo, Zero-lag long-range synchronization via dynamical relaying, *Phys. Rev. Lett.*, **97**(12), 123902 (2006). doi: 10.1103/physrevlett.97.123902.
12. O. D'Huys, R. Vicente, T. Erneux, J. Danckaert, and I. Fischer, Synchronization properties of network motifs: influence of coupling delay and symmetry, *Chaos.*, **18**(3), 037116 (2008). doi: 10.1063/1.2953582.
13. E. Rossoni, Y. Chen, M. Ding, and J. Feng, Stability of synchronous oscillations in a system of Hodgkin-Huxley neurons with delayed diffusive and pulsed coupling, *Phys. Rev. E.*, **71**(6), 061904 (2005). doi: 10.1103/physreve.71.061904.
14. C. Masoller, M. C. Torrent, and J. García-Ojalvo, Interplay of subthreshold activity, time-delayed feedback, and noise on neuronal firing patterns, *Phys. Rev. E.*, **78**, 041907 (2008). doi: 10.1103/physreve.78.041907.
15. A. Takamatsu, R. Tanaka, H. Yamada, T. Nakagaki, T. Fujii, and I. Endo, Spatiotemporal symmetry in rings of coupled biological oscillators of physarum plasmodial slime mold, *Phys. Rev. Lett.*, **87**, 078102 (2001). doi: 10.1103/physrevlett.87.078102.
16. L. M. Pecora and T. L. Carroll, Master stability functions for synchronized coupled systems, *Phys. Rev. Lett.*, **80**(10), 2109–2112 (1998). doi: 10.1103/physrevlett.80.2109.
17. M. Dhamala, V. K. Jirsa, and M. Ding, Enhancement of neural synchrony by time delay, *Phys. Rev. Lett.*, **92**(7), 074104 (2004). doi: 10.1103/physrevlett.92.074104.
18. C. U. Choe, T. Dahms, P. Hövel, and E. Schöll, Controlling synchrony by delay coupling in networks: from in-phase to splay and cluster states, *Phys. Rev. E.*, **81**(2), 025205 (2010). doi: 10.1103/physreve.81.025205.
19. W. Kinzel, A. Englert, G. Reents, M. Zigzag, and I. Kanter, Synchronization of networks of chaotic units with time-delayed couplings, *Phys. Rev. E.*, **79**(5), 056207 (2009). doi: 10.1103/physreve.79.056207.
20. V. Flunkert, S. Yanchuk, T. Dahms, and E. Schöll, Synchronizing distant nodes: a universal classification of networks, *Phys. Rev. Lett.*, **105**, 254101 (2010). doi: 10.1103/physrevlett.105.254101.
21. S. Heiligenthal, T. Dahms, S. Yanchuk, T. Jüngling, V. Flunkert, I. Kanter, E. Schöll, and W. Kinzel, Strong and weak chaos in nonlinear networks with time-delayed couplings, *Phys. Rev. Lett.*, **107**, 234102 (2011). doi: 10.1103/physrevlett.107.234102.
22. T. Dahms, J. Lehnert, and E. Schöll, Cluster and group synchronization in delay-coupled networks, *Phys. Rev. E.*, **86**(1), 016202 (2012). doi: 10.1103/physreve.86.016202.
23. Y. Kuramoto and D. Battogtokh, Coexistence of coherence and incoherence in nonlocally coupled phase oscillators., *Nonlin. Phen. in Complex Sys.*, **5**(4), 380–385 (2002).
24. D. M. Abrams and S. H. Strogatz, Chimera states for coupled oscillators, *Phys. Rev. Lett.*, **93**(17), 174102 (2004). doi: 10.1103/physrevlett.93.174102.
25. E. A. Martens, C. R. Laing, and S. H. Strogatz, Solvable model of spiral wave chimeras, *Phys. Rev. Lett.*, **104**(4), 044101 (2010). doi: 10.1103/physrevlett.104.044101.
26. A. E. Motter, Nonlinear dynamics: spontaneous synchrony breaking, *Nature Physics.*, **6**(3), 164–165 (2010).
27. I. Omelchenko, Y. L. Maistrenko, P. Hövel, and E. Schöll, Loss of coherence in dy-

namical networks: spatial chaos and chimera states, *Phys. Rev. Lett.*, **106**, 234102 (2011). doi: 10.1103/physrevlett.106.234102.
28. I. Omelchenko, B. Riemenschneider, P. Hövel, Y. L. Maistrenko, and E. Schöll, Transition from spatial coherence to incoherence in coupled chaotic systems, *Phys. Rev. E.*, **85**, 026212 (2012). doi: 10.1103/physreve.85.026212.
29. A. Hagerstrom, T. E. Murphy, R. Roy, P. Hövel, I. Omelchenko, and E. Schöll, Experimental observation of chimeras in coupled-map lattices, *Nature Physics.*, **8**, 658–661 (2012). doi: 10.1038/nphys2372.
30. M. R. Tinsley, S. Nkomo, and K. Showalter, Chimera and phase cluster states in populations of coupled chemical oscillators, *Nature Physics.*, **8**, 662–665 (2012). doi: 10.1038/nphys2371.
31. A. A. Selivanov, J. Lehnert, T. Dahms, P. Hövel, A. L. Fradkov, and E. Schöll, Adaptive synchronization in delay-coupled networks of Stuart-Landau oscillators, *Phys. Rev. E.*, **85**, 016201 (2012). doi: 10.1103/physreve.85.016201.
32. J. Lehnert, T. Dahms, P. Hövel, and E. Schöll, Loss of synchronization in complex neural networks with delay, *Europhys. Lett.*, **96**, 60013 (2011). doi: 10.1209/0295-5075/96/60013.
33. A. Keane, T. Dahms, J. Lehnert, S. A. Suryanarayana, P. Hövel, and E. Schöll, Synchronisation in networks of delay-coupled type-I excitable systems, *Eur. Phys. J. B.*, **85**, 407 (2012).
34. V. Flunkert and E. Schöll, Chaos synchronization in networks of delay-coupled lasers: role of the coupling phases, *New. J. Phys.*, **14**, 033039 (2012). doi: 10.1088/1367-2630/14/3/033039.
35. V. Flunkert, S. Yanchuk, T. Dahms, and E. Schöll, Synchronizability of networks with strongly delayed links: a universal classification, *Contemporary Mathematics. Fundamental Directions*, **48**, (June 2013).
36. J. D. Farmer, Chaotic attractors of an infinite-dimensional dynamical system, *Physica D.*, **4**, 366 (1982).
37. G. Giacomelli and A. Politi, Relationship between delayed and spatially extended dynamical systems, *Phys. Rev. Lett.*, **76**, 2686 (1996).
38. B. Mensour and A. Longtin, Chaos control in multistable delay-differential equations and their singular limit maps, *Phys. Rev. E.*, **58**, 410 (1998).
39. M. Wolfrum and S. Yanchuk, Eckhaus instability in systems with large delay, *Phys. Rev. Lett.*, **96**, 220201 (2006).
40. S. Yanchuk, M. Wolfrum, P. Hövel, and E. Schöll, Control of unstable steady states by long delay feedback, *Phys. Rev. E.*, **74**, 026201 (2006).
41. S. Yanchuk and P. Perlikowski, Delay and periodicity, *Phys. Rev. E.*, **79**(4), 046221 (2009). doi: 10.1103/physreve.79.046221.
42. M. Lichtner, M. Wolfrum, and S. Yanchuk, The spectrum of delay differential equations with large delay, *SIAM J. Math. Anal.*, **43**, 788–802 (2011).
43. C. Grebogi, E. Ott, and J. A. Yorke, Unstable periodic orbits and the dimensions of multifractal chaotic attractors, *Phys. Rev. A.*, **37**, 1711–1724 (1988). doi: 10.1103/physreva.37.1711.
44. Y. C. Lai, Y. Nagai, and C. Grebogi, Characterization of the natural measure by unstable periodic orbits in chaotic attractors, *Phys. Rev. Lett.*, **79**, 649–652 (1997).
45. S. A. Gerschgorin, Über die Abgrenzung der Eigenwerte einer Matrix, *Izv. Akad. Nauk. SSSR*, **7**, 749–754 (1931).
46. E. Ott and J. C. Sommerer, Blowout bifurcations: the occurrence of riddled basins and on-off intermittency, *Phys. Lett. A.*, **188**(1), 39–47 (1994).
47. P. Ashwin, J. Buescu, and I. Stewart, From attractor to chaotic saddle: a tale of

transverse instability, *Nonlinearity*, **9**(3), 703–737 (1996).
48. V. Flunkert, O. D'Huys, J. Danckaert, I. Fischer, and E. Schöll, Bubbling in delay-coupled lasers, *Phys. Rev. E.*, **79**, 065201 (2009). doi: 10.1103/physreve.79.065201.
49. F. Sorrentino and E. Ott, Network synchronization of groups, *Phys. Rev. E.*, **76**(5), 056114 (2007). doi: 10.1103/physreve.76.056114.
50. J. Kestler, W. Kinzel, and I. Kanter, Sublattice synchronization of chaotic networks with delayed couplings, *Phys. Rev. E.*, **76**(3), 035202 (2007). doi: 10.1103/physreve.76.035202.
51. J. Kestler, E. Kopelowitz, I. Kanter, and W. Kinzel, Patterns of chaos synchronization, *Phys. Rev. E.*, **77**(4), 046209 (2008). doi: 10.1103/physreve.77.046209.
52. I. Kanter, M. Zigzag, A. Englert, F. Geissler, and W. Kinzel, Synchronization of unidirectional time delay chaotic networks and the greatest common divisor, *Europhys. Lett.*, **93**(6), 60003 (2011).
53. I. Kanter, E. Kopelowitz, R. Vardi, M. Zigzag, W. Kinzel, M. Abeles, and D. Cohen, Nonlocal mechanism for cluster synchronization in neural circuits, *Europhys. Lett.*, **93**(6), 66001 (2011).
54. Y. Aviad, I. Reidler, M. Zigzag, M. Rosenbluh, and I. Kanter, Synchronization in small networks of time-delay coupled chaotic diode lasers, *Opt. Express.*, **20**(4), 4352–4359 (2012). doi: 10.1364/oe.20.004352.
55. C. Zhou, L. Zemanova, G. Zamora, C. C. Hilgetag, and J. Kurths, Hierarchical organization unveiled by functional connectivity in complex brain networks, *Phys. Rev. Lett.*, **97**(23), 238103 (2006). doi: 10.1103/physrevlett.97.238103.
56. C. Zhou and J. Kurths, Hierarchical synchronization in complex networks with heterogeneous degrees, *Chaos*, **16**(1), 015104 (2006). doi: 10.1063/1.2150381.
57. C. Hauptmann, O. E. Omel'chenko, O. V. Popovych, Y. L. Maistrenko, and P. A. Tass, Control of spatially patterned synchrony with multisite delayed feedback, *Phys. Rev. E.*, **76**, 066209 (2007). doi: 10.1103/physreve.76.066209.
58. E. Schöll, A. A. Selivanov, J. Lehnert, T. Dahms, P. Hövel, and A. L. Fradkov, Control of synchronization in delay-coupled networks, *Int. J. Mod. Phys. B.*, **26**(25), 1246007 (2012). doi: 10.1142/s0217979212460071.
59. A. L. Fradkov, Speed-gradient scheme and its application in adaptive control problems, *Autom. Remote Control*, **40**, 1333–1342 (1979).
60. A. L. Fradkov, Application of cybernetic methods in physics, *Physics-Uspekhi.*, **48**(2), 103–127 (2005). doi: 10.1070/pu2005v048n02abeh002047.
61. A. L. Fradkov, *Cybernetical Physics: From Control of Chaos to Quantum Control.* (Springer, Heidelberg, Germany, 2007).
62. R. FitzHugh, Impulses and physiological states in theoretical models of nerve membrane, *Biophys. J.*, **1**, 445–466 (1961).
63. J. Nagumo, S. Arimoto, and S. Yoshizawa., An active pulse transmission line simulating nerve axon., *Proc. IRE*, **50**, 2061–2070 (1962).
64. G. Hu, T. Ditzinger, C. Z. Ning, and H. Haken, Stochastic resonance without external periodic force, *Phys. Rev. Lett.*, **71**, 807 (1993).
65. B. Haider, A. Duque, A. R. Hasenstaub, and D. A. McCormick, Neocortical network activity in vivo is generated through a dynamic balance of excitation and inhibition, *J. Neurosci.*, **26**(17), 4535–4545 (2006). doi: 10.1523/jneurosci.5297-05.2006.
66. T. P. Vogels, H. Sprekeler, F. Zenke, C. Clopath, and W. Gerstner, Inhibitory plasticity balances excitation and inhibition in sensory pathways and memory networks, *Science*, **334**(6062), 1569–1573 (2011). doi: 10.1126/science.1211095.
67. D. J. Watts and S. H. Strogatz, Collective dynamics of "small-world" networks, *Nature*, **393**, 440–442 (1998).

68. E. Schöll, G. Hiller, P. Hövel, and M. A. Dahlem, Time-delayed feedback in neurosystems, *Phil. Trans. R. Soc. A.*, **367**, 1079–1096 (2009). doi: 10.1098/rsta.2008.0258.
69. M. G. Earl and S. H. Strogatz, Synchronization in oscillator networks with delayed coupling: a stability criterion, *Phys. Rev. E.*, **67**(3), 036204 (2003). doi: 10.1103/physreve.67.036204.
70. M. E. J. Newman and D. J. Watts, Renormalization group analysis of the small-world network model, *Phys. Lett. A.*, **263**(4–6), 341–346 (1999). doi: 10.1016/s0375-9601(99)00757-4.
71. J. Hizanidis, R. Aust, and E. Schöll, Delay-induced multistability near a global bifurcation, *Int. J. Bifur. Chaos*, **18**(6), 1759–1765 (2008). doi: 10.1142/s0218127408021348.
72. R. Aust, P. Hövel, J. Hizanidis, and E. Schöll, Delay control of coherence resonance in type-I excitable dynamics, *Eur. Phys. J. ST*, **187**, 77–85 (2010). doi: 10.1140/epjst/e2010-01272-5.
73. C. R. S. Williams, T. E. Murphy, R. Roy, F. Sorrentino, T. Dahms, and E. Scholl, Experimental observations of group synchrony in a system of chaotic optoelectronic oscillators, *Phys. Rev. Lett.*, **110**, 064104 (2013).

Chapter 5

Stochastic Dynamics and Optimal Control of Quasi Integrable Hamiltonian Systems with Time-delayed Feedback Control

Weiqiu Zhu
Department of Mechanics, Zhejiang University
Hangzhou, China
wqzhu@yahoo.com

Zhonghua Liu
Department of Civil Engineering, Xiamen University
Xiamen, China
zhliuzju@yahoo.com

Some results on the dynamics and optimal control of quasi integrable Hamiltonian systems with time-delayed feedback control under various stochastic excitations obtained by the authors and their coworkers are presented. First, the stochastic averaging method is briefly introduced for quasi integrable Hamiltonian systems with time-delayed feedback control under excitations of Gaussian white noise, wide-band random process, narrow-band bounded noise, and both harmonic function and one of above random processes. Then, the response, stochastic stability, stochastic bifurcation and first passage failure of such systems predicted by using the stochastic averaging method are presented. Finally, the optimal control of such systems investigated by using the stochastic averaging method and the stochastic dynamical programming principle is reviewed. The focus of the present study is the effects of the time delay in feedback control on the dynamics and control performance of the system.

1. Introduction

In the implementation of feedback control of a dynamical system, time delay is usually unavoidable due to the time spent in measuring and estimating the system state, calculating and executing the control forces, etc. This time delay causes unsynchronized application of the control. This unsynchronization may not only deteriorate the control performance, but also cause instability of the system. Thus time delay problem has drawn much attention of the control community.

Systems with time delay under deterministic excitation have been studied extensively[1–7]. Time-delayed systems under stochastic excitation have attracted many researches recently. The multi-scale analysis has been adopted to study the effect of noise near critical delay in stochastic delay differential equations by Klosek and Kuske[8]. The center manifold reduction of delay differential equations was used by Fofana[9] to deal with machine-tool chatter problem. The linearly controlled system with deterministic and random time delays excited by Gaussian white noise has been treated by Grigoriu[10] and the stability of such a system has been investigated by means of Lyapunov exponent. The effects of time delay on the controlled linear systems under Gaussian random excitation have been studied by Di Paola and Pirrotta[11] using Taylor expansion of the control force.

In the present chapter, some results obtained by the present authors and their coworkers on the dynamics and optimal control of quasi integrable Hamiltonian systems with time-delayed feedback control under various stochastic excitations are presented. First, the stochastic averaging method is briefly introduced for quasi integrable Hamiltonian systems with time-delayed feedback control under excitations of Gaussian white noise, wide-band random process, narrow-band bounded noise, or both harmonic function and one of above random processes[12–14]. Then, the response[12–17], stochastic stability[18–22], stochastic bifurcation[23,24], and first passage failure[25–28] of such systems predicted by using the stochastic averaging method are presented. Finally, the optimal control[29,30] of such systems investigated by using the stochastic averaging method and the stochastic dynamical programming principle is reviewed. Some examples are given to illustrate the application and effectiveness of the proposed procedure and the effect of the time delay in feedback control on the dynamics and control of the system.

2. Stochastic Averaging Method for Quasi Integrable Hamiltonian Systems with Time-delayed Feedback Control

Consider an n-degree-of-freedom (nDOF) quasi Hamiltonian system with delayed feedback control governed by the following equations:

$$\begin{aligned} \dot{Q}_i &= \frac{\partial H'}{\partial P_i}, \\ \dot{P}_i &= -\frac{\partial H'}{\partial Q_i} - \varepsilon c'_{ij}\frac{\partial H'}{\partial P_j} + \varepsilon u_i(\mathbf{Q}_\tau, \mathbf{P}_\tau) + \varepsilon^{1/2} f_{ik}\xi_k(t), \\ & i, j = 1, 2, \ldots, n; \quad k = 1, 2, \ldots, m \end{aligned} \tag{1}$$

where Q_i and P_i are generalized displacements and momenta, respectively; $H' = H'(\mathbf{Q}, \mathbf{P})$ is twice differentiable Hamiltonian; ε is a small positive parameter; $\varepsilon c'_{ij} = \varepsilon c'_{ij}(\mathbf{Q}, \mathbf{P})$ represent the coefficients of lightly quasi linear dampings; $\varepsilon^{1/2} f_{ik} = \varepsilon^{1/2} f_{ik}(\mathbf{Q}, \mathbf{P})$ represent the amplitudes of weakly stochastic excitations; $\varepsilon u_i(\mathbf{Q}_\tau, \mathbf{P}_\tau)$ with $\mathbf{Q}_\tau = [\mathbf{Q}(t-\tau_1), \mathbf{Q}(t-\tau_2), \ldots, \mathbf{Q}(t-\tau_r)]$ and $\mathbf{P}_\tau = [\mathbf{P}(t-\tau_1), \mathbf{P}(t-\tau_2), \ldots, \mathbf{P}(t-\tau_r)]$ denote weakly multi-time-delayed feedback control forces, where τ_i $(r = 1, 2, \ldots, r)$ are i time delays and $\varepsilon u_i(\mathbf{Q}_\tau, \mathbf{P}_\tau) = 0$ when $t \in [0, \tau]$; $\xi_k(t)$ are stochastic excitations.

When $\varepsilon = 0$, system (1) is reduced to nDOF Hamiltonian system. It is called integrable or completely integrable if there are n independent integrals of motion, which are in involution. In this case, a canonical transformation

$$I_i = I_i(\mathbf{q}, \mathbf{p}), \quad \theta_i = \theta_i(\mathbf{q}, \mathbf{p}), \quad i = 1, 2, \ldots, n \tag{2}$$

may be introduced so that the Hamiltonian equations of an integrable Hamiltonian system are of the form

$$\dot{I}_i = -\frac{\partial}{\partial \theta_i} H'(\mathbf{I}) = 0, \quad \dot{\theta}_i = \frac{\partial}{\partial I_i} H'(\mathbf{I}) = \omega_i(\mathbf{I}) \tag{3}$$

where I_i and θ_i are action-angle variables and $\omega_i(\mathbf{I})$ are the frequencies of the system. Equation (3) can be easily solved to yield

$$I_i = const, \quad \theta_i = \omega_i(\mathbf{I})t + \delta_i, \quad i = 1, 2, \ldots, n \tag{4}$$

where δ_i are constants. It is seen from Eq.(4) that the motion of an integrable Hamiltonian system is almost periodic or periodic depending upon the number of the internal resonant relations of the form

$$k_i^u \omega_i = 0, \quad u = 1, 2, \ldots, \alpha; \quad i = 1, 2, \ldots, n \tag{5}$$

where k_i^u are integers and α is the number of resonant relations. If there is no resonant relation, then the Hamiltonian system is called non-resonant. The motion of non-resonant integrable Hamiltonian system is almost periodic and a single orbit covers an n-dimensional tori uniformly. If there are $n - 1$ resonant relations, then the system is called completely resonant and the motion of the system is periodic. If the number of resonant relations is between 1 and $n - 1$, then the system is called partially resonant and the motion of the system is periodic or almost periodic depending on the specific resonant relations.

It is noted that n action variable I_i can be regarded as n independent integrals of motion in involution, satisfying Eq.(3), and the frequencies of truly nonlinear Hamiltonian systems are functions of integrals of motion or action variables.

If the Hamiltonian system associated with Eq.(1) is integrable, then system (1) is called quasi integrable Hamiltonian system. The quasi integrable Hamiltonian system with multi-time-delayed feedback control is studied in the following.

Assume that the Hamiltonian H' associated with system (1) is of the form

$$H' = \sum_{i=1}^{n} H_i'(q_i, p_i), \quad H_i' = \frac{1}{2} p_i^2 + G(q_i) \tag{6}$$

where $G(q_i) \geq 0$ is symmetric with respect to the $q_i = 0$, and with minimum at $q_i = 0$. Then the associated Hamiltonian system has a family of periodic solutions and the quasi integrable Hamiltonian system has randomly periodic solution of the form[12]

$$Q_i(t) = A_i \cos \Phi_i(t), \quad P_i(t) = -A_i \frac{d\Theta_i}{dt} \sin \Phi_i(t), \quad \Phi_i(t) = \Theta_i(t) + \Gamma_i(t) \tag{7}$$

where $\cos\Phi(t)$ and $\sin\Phi(t)$ are called generalized harmonic functions, $A_i(t)$ and $\Gamma_i(t)$ are slowly varying processes and the average value of the instantaneous frequency $d\Theta_i/dt$ is equal to[12] $\omega_i(A_i)$. If $A_i(t-\tau_r)$and $\Gamma_i(t-\tau_r)$ are approximated by$A_i(t)$ and $\Gamma_i(t)$, respectively, and $\Theta_i(t-\tau_r)$is approximated by $\Theta_i(t)-\omega_i\tau_r$, then we have the following approximate expressions:

$$\begin{aligned}
Q_i(t-\tau_r) &= A_i(t-\tau_r)\cos\Phi_i(t-\tau_r) \doteq A_i(t)\cos[\omega_i(t-\tau_r)+\Gamma_i(t)] \\
&= A_i(t)\{\cos[\omega_i t+\Gamma_i(t)]\cos\omega_i\tau_r + \sin[\omega_i t+\Gamma_i(t)]\sin\omega_i\tau_r\} \\
&= Q_i(t)\cos\omega_i\tau_r - \frac{P_i}{\omega_i}\sin\omega_i\tau_r, \\
P_i(t-\tau_r) &= -A_i(t-\tau_r)\frac{d\Theta_i(t-\tau_r)}{dt}\sin\Phi_i(t-\tau_r) \\
&\doteq -A_i(t)\omega_i\sin[\omega_i(t-\tau_r)+\Gamma_i(t)] \\
&= -A_i(t)\omega_i\{\sin[\omega_i t+\Gamma_i(t)]\cos\omega_i\tau_r + \cos[\omega_i t+\Gamma_i(t)]\sin\omega_i\tau_r\} \\
&= P_i\cos\omega_i\tau_r + Q_i(t)\omega_i\sin\omega_i\tau_r.
\end{aligned} \tag{8}$$

In the feedback bang-bang control of quasi integrable Hamiltonian systems, the optimal control forces are of the form of

$$u_i = u_i(P_i) = -b_i\text{sgn}\,(P_i)\,, \quad i=1,2,\ldots,n. \tag{9}$$

The delayed control forces $u_i(P_{i\tau_i})$ can be equivalently substituted by $K_i u_i(P_i)$ in the sense of averaging, i.e.,

$$\int_0^{2\pi/\omega_i} u_i\left(P_{i\tau_i}\right)P_i(t)dt = \int_0^{2\pi/\omega_i} K_i u_i\left(P_i(t)\right)P_i(t)dt. \tag{10}$$

Substituting the approximate relation for $P_i(t)$ and $P_i(t-\tau_i)$ in Eq.(8) to Eq.(10) yields

$$K_i = \cos\omega_i\tau_i. \tag{11}$$

So the delayed feedback bang-bang control forces are of the form[13]

$$u_i\left(P_{i\tau}\right) = u_i\left(P_i(t-\tau_i)\right) = u_i\left(P_i(t)\right)\cos\omega_i\tau_i = -b_i\cos\omega_i\tau_i\text{sgn}\left(P_i(t)\right). \tag{12}$$

The numerical results for the examples described in the following sections will show that equation (12) is acceptable even for some large delay time τ_i.

After substituting Eq.(8) or (12) into Eq.(1), the time-delayed control forces of system (1) can be split into two parts: one has the effect of modifying the conservative forces and the other modifying the damping forces. The first part can be combined with $-\partial H'/\partial Q_i$ to form an overall effective conservative force $-\partial H/\partial Q_i$ with a new Hamiltonian $H=H(\mathbf{Q},\mathbf{P};\tau)$ and with $\partial H/\partial P_i = \partial H'/\partial P_i$. The second part may be combined $-\varepsilon c'_{ij}\partial H'/\partial P_j$ to constitute to an effective damping force $-\varepsilon c_{ij}\partial H/\partial P_i$ with $c_{ij}=c_{ij}(\mathbf{Q},\mathbf{P};\tau)$. With these accomplished, equation (1) can be rewritten as

$$\begin{aligned}\dot{Q}_i &= \frac{\partial H}{\partial P_i},\\ \dot{P}_i &= -\left(\frac{\partial H}{\partial Q_i} + \varepsilon c_{ij}\frac{\partial H}{\partial P_j}\right) + \varepsilon^{1/2} f_{ik}\xi_k(t),\\ i,j &= 1,2,\ldots,n; \quad k = 1,2,\ldots,m\end{aligned} \tag{13}$$

where $H = H(\mathbf{Q}, \mathbf{P}; \tau)$, $c_{ij} = c_{ij}(\mathbf{Q}, \mathbf{P}; \tau)$. Suppose that the modified Hamiltonian is still separable, i.e.,

$$H = \sum_{i=1}^{n} H_i(q_i, p_i; \tau_i), \quad H_i = p_i^2/2 + G_i(q_i; \tau_i). \tag{14}$$

Equation (13) describes a quasi integrable Hamiltonian system without time delay.

There are several versions of stochastic averaging method for the quasi integrable Hamiltonian systems under different stochastic excitations.

2.1. *Gaussian White Noise Excitations*[12,13]

Suppose that the stochastic excitations $\xi_k(t)$ are Gaussian white noises. After adding the Wong-Zakai correction terms, then the Itô stochastic differential equations for quasi integrable Hamiltonian systems are of the form

$$\begin{aligned}dQ_i &= \frac{\partial H}{\partial P_i}dt,\\ dP_i &= -\left(\frac{\partial H}{\partial Q_i} + \varepsilon m_{ij}(\mathbf{Q}, \mathbf{P}; \tau)\frac{\partial H}{\partial P_j}\right)dt + \varepsilon^{1/2}\sigma_{ik} dB_k(t),\\ i,j &= 1,2,\ldots,n; \quad k = 1,2,\ldots,m.\end{aligned} \tag{15}$$

Note that here H represents the Hamiltonian and m_{ij} the daming coefficients possibley modified by Wong-Zakai corrections. The stochastic averaging method for quasi integrable Hamiltonian systems has been well developed[31]. The dimension and form of the averaged Itô and FPK equations depend upon the resonance of the associated Hamiltonian system with modified Hamiltonian H.

2.1.1. *Non-resonant Case*

In this case, the averaged Itô equations are of the form

$$dI_r = \varepsilon U_r(\mathbf{I}; \tau)dt + \varepsilon^{1/2} V_{rk}(\mathbf{I}) dB_k(t), \quad r = 1,2,\ldots,n; \quad k = 1,2,\ldots,m \tag{16}$$

and the averaged FPK equation is of the form

$$\frac{\partial p}{\partial t} = \varepsilon\left\{-\frac{\partial}{\partial I_r}[a_r(\mathbf{I}; \tau)p] + \frac{1}{2}\frac{\partial^2}{\partial I_r \partial I_s}[b_{rs}(\mathbf{I})p]\right\}. \tag{17}$$

In Eqs.(16) and (17),

$$\begin{aligned} a_r(\mathbf{I};\tau) &= U_r(\mathbf{I};\tau) = \frac{1}{(2\pi)^n}\int_0^{2\pi}\left(-m_{ij}\frac{\partial H}{\partial p_j}\frac{\partial I_r}{\partial p_i} + D_{kl}f_{ik}f_{jl}\frac{\partial^2 I_r}{\partial p_i \partial p_j}\right)d\boldsymbol{\theta}, \\ b_{rs}(\mathbf{I}) &= [\mathbf{V}\mathbf{V}^T]_{rs} = \frac{1}{(2\pi)^n}\int_0^{2\pi}\left(2D_{kl}f_{ik}f_{jl}\frac{\partial I_r}{\partial p_i}\frac{\partial I_s}{\partial p_j}\right)d\boldsymbol{\theta}, \\ r,s,i,j &= 1,2,\ldots,n; \quad k,l = 1,2,\ldots,m \end{aligned} \tag{18}$$

in which $\boldsymbol{\theta} = [\theta_1, \theta_2, \ldots, \theta_n]^T$, $\mathbf{V} = [V_{rk}]$, and $\int_0^{2\pi}[\cdot]d\boldsymbol{\theta}$ denotes an n-fold integral.

If the action-angle vectors $\mathbf{I}$, $\boldsymbol{\theta}$ for Hamiltonian system with Hamiltonian H cannot be obtained, then the averaged Itô equations for independent integrals of motion may be derived. In this case, the averaged Itô equations are of the form

$$dH_r = \varepsilon\bar{m}_r(\mathbf{H};\tau)dt + \varepsilon^{1/2}\bar{\sigma}_{rk}(\mathbf{H})dB_k(t), \quad r = 1,2,\cdots,n; \quad k = 1,2,\ldots,m \tag{19}$$

and the averaged FPK equation is of the form

$$\frac{\partial p}{\partial t} = \varepsilon\left\{-\frac{\partial}{\partial H_r}\left[a_r(\mathbf{H};\tau)p\right] + \frac{1}{2}\frac{\partial^2}{\partial H_r \partial H_s}\left[b_{rs}(\mathbf{H})p\right]\right\} \tag{20}$$

where the averaged drift and diffusion coefficients are

$$\begin{aligned} a_r(\mathbf{H};\tau) &= \bar{m}_r(\mathbf{H};\tau) = \frac{1}{T}\oint\left(-m_{ij}\frac{\partial H}{\partial p_j}\frac{\partial H_r}{\partial p_i} + \frac{1}{2}\sigma_{ik}\sigma_{jk}\frac{\partial^2 H_r}{\partial p_i \partial p_j}\right)\prod_{u=1}^{n}\left(1\Big/\frac{\partial H_u}{\partial p_u}\right)dq_u, \\ b_{rs}(\mathbf{H}) &= \bar{\sigma}_{rk}\bar{\sigma}_{sk}(\mathbf{H}) = \frac{1}{T}\oint\prod_{u=1}^{n}\left(\sigma_{ik}\sigma_{jk}\frac{\partial II_r}{\partial p_i}\frac{\partial H_r}{\partial p_j}\Big/\frac{\partial H_u}{\partial p_u}\right)dq_u \end{aligned} \tag{21}$$

in which

$$T = T(\mathbf{H}) = \prod_{u=1}^{n} T_u(H_u) = \prod_{u=1}^{n}\oint\left(1/\frac{\partial H_u}{\partial p_u}\right)dq_u. \tag{22}$$

2.1.2. *Resonant Case*

Suppose that the modified Hamiltonian system with Hamiltonian H is integrable and resonant with α weak resonant relations of the form

$$k_r^u\omega_r = 0(\varepsilon), \quad u = 1,2,\ldots,\alpha; \quad r = 1,2,\ldots,n. \tag{23}$$

Then the averaged Itô equations are of the form

$$\begin{aligned} dI_r &= \varepsilon\bar{m}_r(\mathbf{I},\mathbf{\Psi};\tau)dt + \varepsilon^{1/2}\bar{\sigma}_{rk}(\mathbf{I},\mathbf{\Psi})dB_k(t), \\ d\Psi_u &= \varepsilon\bar{m}_{n+u}(\mathbf{I},\mathbf{\Psi};\tau)dt + \varepsilon^{1/2}\bar{\sigma}_{n+u,k}(\mathbf{I},\mathbf{\Psi})dB_k(t), \\ r &= 1,2,\ldots,n; \quad u = 1,2,\ldots,\alpha; \quad k = 1,2,\ldots,m \end{aligned} \tag{24}$$

and the averaged FPK equation is of the form

$$\frac{\partial p}{\partial t} = \varepsilon\left[-\frac{\partial}{\partial I_r}(a_r p) - \frac{\partial}{\partial \psi_u}(a_{n+u}p) + \frac{1}{2}\frac{\partial^2}{\partial I_r \partial I_s}(b_{rs}p) + \frac{\partial^2}{\partial I_r \partial \psi_u}(b_{r,n+u}p) + \frac{1}{2}\frac{\partial^2}{\partial \psi_u \partial \psi_v}(b_{n+u,n+v}p)\right] \tag{25}$$

where $\mathbf{\Psi} = [\Psi_1, \Psi_2, \ldots, \Psi_\alpha]^T$, $\psi_u = k_r^u \Theta_r$, $k = 1, 2, \ldots, \alpha$ and

$$\bar{m}_r(\mathbf{I}, \mathbf{\Psi}; \tau) = \frac{1}{(2\pi)^{n-\alpha}} \int_0^{2\pi} \left(-m_{ij}\frac{\partial H}{\partial p_j}\frac{\partial I_r}{\partial p_i} + \frac{1}{2}\sigma_{ik}\sigma_{jk}\frac{\partial^2 I_r}{\partial p_i \partial p_j}\right) d\boldsymbol{\theta}_1,$$

$$\bar{m}_{n+u}(\mathbf{I}, \mathbf{\Psi}; \tau) = \frac{1}{(2\pi)^{n-\alpha}} \int_0^{2\pi} \left(O_u(\varepsilon) - m_{ij}\frac{\partial H}{\partial p_j}\frac{\partial \Psi_r}{\partial p_i} + \frac{1}{2}\sigma_{ik}\sigma_{jk}\frac{\partial^2 \Psi_r}{\partial p_i \partial p_j}\right) d\boldsymbol{\theta}_1,$$

$$\bar{\sigma}_{ik}\bar{\sigma}_{sk}(\mathbf{I}, \mathbf{\Psi}) = \frac{1}{(2\pi)^{n-\alpha}} \int_0^{2\pi} \sigma_{ik}\sigma_{jk}\frac{\partial I_r}{\partial p_i}\frac{\partial I_s}{\partial p_j} d\boldsymbol{\theta}_1,$$

$$\bar{\sigma}_{rk}\bar{\sigma}_{n+u,k}(\mathbf{I}, \mathbf{\Psi}) = \frac{1}{(2\pi)^{n-\alpha}} \int_0^{2\pi} \sigma_{ik}\sigma_{jk}\frac{\partial I_r}{\partial p_i}\frac{\partial \Psi_u}{\partial p_j} d\boldsymbol{\theta}_1,$$

$$\begin{aligned} \bar{\sigma}_{n+u,k}\bar{\sigma}_{n+v,k}(\mathbf{I}, \mathbf{\Psi}) &= \frac{1}{(2\pi)^{n-\alpha}} \int_0^{2\pi} \sigma_{ik}\sigma_{jk}\frac{\partial \Psi_u}{\partial p_i}\frac{\partial \Psi_v}{\partial p_j} d\boldsymbol{\theta}_1, \\ a_r &= a_r(\mathbf{I}, \boldsymbol{\psi}; \tau) = \bar{m}_r(\mathbf{I}, \mathbf{\Psi}; \tau), \\ a_{n+u} &= a_{n+u}(\mathbf{I}, \boldsymbol{\psi}; \tau) = \bar{m}_{n+u}(\mathbf{I}, \mathbf{\Psi}; \tau), \\ b_{rs} &= b_{rs}(\mathbf{I}, \boldsymbol{\psi}) = \bar{\sigma}_{rk}\bar{\sigma}_{sk}(\mathbf{I}, \mathbf{\Psi}), \\ b_{r,n+u} &= b_{r,n+u}(\mathbf{I}, \boldsymbol{\psi}) = \bar{\sigma}_{rk}\bar{\sigma}_{n+u,k}(\mathbf{I}, \mathbf{\Psi}), \\ b_{n+u,n+v} &= b_{n+u,n+v}(\mathbf{I}, \boldsymbol{\psi}) = \bar{\sigma}_{n+u,k}\bar{\sigma}_{n+v,k}(\mathbf{I}, \mathbf{\Psi}), \\ r, s, i, j &= 1, 2, \ldots, n; \quad u, v = 1, 2, \ldots, \alpha; \quad k, l = 1, 2, \ldots, m \end{aligned} \tag{26}$$

in which $\boldsymbol{\theta}_1 = [\theta_1, \theta_2, \ldots, \theta_{n-r}]^T$.

For quasi integrable Hamiltonian systems with time delay, the stochastic averaging method described in this section has been extended to the case of nonwhite random excitations, including wide-band random excitation, combined harmonic and Gaussian white noise excitations, narrow-band bounded noise and pure harmonic excitation. In the case of harmonic excitation, combined harmonic and white noise excitations or bounded noise excitation, possible external resonance should be considered except possible internal resonance considered in Section 2.1.2.

2.2. *Wide-band Random Excitations*

In this case, the stochastic excitations $\xi_k(t)$ in Eq.(1) are independent wide-band random processes with the following power spectral densities:

$$S_k(\omega) = \sigma_k^2 \sum_{u=0}^{r-1} b_{ku}\omega^{2u} \Big/ \sum_{u=0}^{r} a_{ku}\omega^{2u}, \quad k = 1, 2, \dots, m \tag{27}$$

where σ_k, a_{ku}, b_{ku} are constants.

Assume that the Hamiltonian is the form (14) with $g_i(q_i) = dG_i(q_i)/dq_i$. Introducing a similar transformation as that in Eq.(7) to Eq.(13) leads to the following stochastic differential equations for A_i and Γ_i:

$$\begin{aligned} \dot{A}_i &= \varepsilon F_{1i}(\mathbf{A}, \mathbf{\Gamma}; \tau) + \varepsilon^{1/2} G_{1ik}(\mathbf{A}, \mathbf{\Gamma})\xi_k(t), \\ \dot{\Gamma}_i &= \varepsilon F_{2i}(\mathbf{A}, \mathbf{\Gamma}; \tau) + \varepsilon^{1/2} G_{2ik}(\mathbf{A}, \mathbf{\Gamma})\xi_k(t), \\ i &= 1, 2, \dots, n; \quad k = 1, 2, \dots, m \end{aligned} \tag{28}$$

where

$$\begin{aligned} F_{1i}(\mathbf{A}, \mathbf{\Gamma}; \tau) &= -\frac{A_i\omega_i(A_i, \Phi_i)\sin\Phi_i}{g_i(A_i)} c_{ij}(\mathbf{A}, \mathbf{\Phi}; \tau) A_j \omega_j(A_j, \Phi_j)\sin\Phi_j, \\ F_{2i}(\mathbf{A}, \mathbf{\Gamma}; \tau) &= -\frac{\omega_i(A_i, \Phi_i)\cos\Phi_i}{g_i(A_i)} c_{ij}(\mathbf{A}, \mathbf{\Phi}; \tau) A_j \omega_j(A_j, \Phi_j)\sin\Phi_j, \\ G_{1ik}(\mathbf{A}, \mathbf{\Gamma}) &= -\frac{A_i\omega_i(A_i, \Phi_i)\sin\Phi_i}{g_i(A_i)} f_{ik}(\mathbf{A}, \mathbf{\Phi}), \\ G_{2ik}(\mathbf{A}, \mathbf{\Gamma}) &= -\frac{\omega_i(A_i, \Phi_i)\sin\Phi_i}{g_i(A_i)} f_{ik}(\mathbf{A}, \mathbf{\Phi}). \end{aligned} \tag{29}$$

It is seen from Eq.(28) that $A_i(t)$, $\Gamma_i(t)$are really slowly varying processes when ε is small.

2.2.1. *Non-resonant Case*

In non-resonant case, the averaged Itô equations are of the form

$$dA_r = \varepsilon m_r(\mathbf{A}; \tau)dt + \varepsilon^{1/2}\sigma_{rk}(\mathbf{A})dB_k(t), \quad r = 1, 2, \dots, n; \quad k = 1, 2, \dots, m \tag{30}$$

where the drift and diffusion coefficients are determined by the following equations:

$$\begin{aligned} m_i(\mathbf{A}; \tau) &= \varepsilon\left\langle F_{1i} + \int_{-\infty}^{0} \left(\frac{\partial G_{1ik}}{\partial A_j}\bigg|_t G_{1jl}\big|_{t+\delta} + \frac{\partial G_{1ik}}{\partial \Gamma_j}\bigg|_t G_{2jl}\big|_{t+\delta} \right) R_{kl}(\delta)d\delta \right\rangle_{\mathbf{\Phi}}, \\ \sigma_{ik}(\mathbf{A})\sigma_{jk}(\mathbf{A}) &= \varepsilon\left\langle F_{1i} + \int_{-\infty}^{\infty} \left(G_{1ik}\big|_t + G_{1jl}\big|_{t+\delta} \right) R_{kl}(\delta)d\delta \right\rangle_{\mathbf{\Phi}} \end{aligned} \tag{31}$$

in which $\langle\cdot\rangle_{\mathbf{\Phi}}$ denote the deterministic averaging with respect to $\mathbf{\Phi}$. Following the Itô differential rule, the averaged Itô differential equations for sub-Hamiltonian H_i

can be derived from the averaged equation (30) for amplitudes A_i as follows[14]:

$$\begin{aligned} dH_i &= \bar{m}_i(\mathbf{H};\tau)dt + \bar{\sigma}_{ik}(\mathbf{H})dB_k(t), \\ \bar{m}_i(\mathbf{H};\tau) &= \{g_i(A_i)m_i(\mathbf{A}) + [g_i(A_i)]'\sigma_{ik}\sigma_{ik}(\mathbf{A})/2\}|_{A_i=G_i^{-1}(H_i)}, \\ \bar{\sigma}_{ik}\bar{\sigma}_{jk}(\mathbf{H}) &= g_i(A_i)g_j(A_j)\sigma_{ik}\sigma_{jk}(\mathbf{A})|_{A_i=G_i^{-1}(H_i)}, \\ i &= 1,2,\ldots,n; \quad k = 1,2,\ldots,m. \end{aligned} \tag{32}$$

2.2.2. *Resonant Case*

In the case of internal resonance with α weak resonant relations, introduce new variables $\Psi_u = k_i^u \Phi_i$. The equations for Ψ_u can be obtained from combination of the second part of Eq.(28). Then equation (28) is replaced by the following equations:

$$\begin{aligned} \dot{A}_i &= \varepsilon \bar{F}_{1i}(\mathbf{A},\mathbf{\Gamma}_1,\mathbf{\Psi};\tau) + \varepsilon^{1/2}\bar{G}_{1ik}(\mathbf{A},\mathbf{\Gamma}_1,\mathbf{\Psi})\xi_k(t), \\ \dot{\Gamma}_\rho &= \varepsilon \bar{F}_{2\rho}(\mathbf{A},\mathbf{\Gamma}_1,\mathbf{\Psi};\tau) + \varepsilon^{1/2}\bar{G}_{2\rho k}(\mathbf{A},\mathbf{\Gamma}_1,\mathbf{\Psi})\xi_k(t), \\ \dot{\Psi}_u &= \varepsilon \bar{F}_{n+u}(\mathbf{A},\mathbf{\Gamma}_1,\mathbf{\Psi};\tau) + \varepsilon^{1/2}G_{n+u,k}(\mathbf{A},\mathbf{\Gamma}_1,\mathbf{\Psi})\xi_k(t), \\ i &= 1,2,\ldots,n; \ \rho = 1,2,\ldots,n-\alpha; \ u = 1,2,\ldots,\alpha; \ k = 1,2,\ldots,m \end{aligned} \tag{33}$$

where the drift coefficients and diffusion coefficients can be obtained from Eq.(28).

After averaging Eq.(34), the Itô equations are of the form[14]

$$\begin{aligned} dA_i &= \varepsilon m_i(\mathbf{A},\mathbf{\Psi};\tau)dt + \varepsilon^{1/2}\sigma_{ik}(\mathbf{A},\mathbf{\Psi})dB_k(t), \\ d\Psi_u &= \varepsilon m_{n+u}(\mathbf{A},\mathbf{\Psi};\tau)dt + \varepsilon^{1/2}\sigma_{n+u,k}(\mathbf{A},\mathbf{\Psi})dB_k(t), \\ i &= 1,2,\ldots,n; \ u = 1,2,\ldots,\alpha; \ k = 1,2,\ldots,m. \end{aligned} \tag{34}$$

2.3. *Narrow-band Bounded Noise Excitation*

In this case, the stochastic excitations $\xi_k(t)$ in Eq.(1) are independent bounded noises of the form

$$\xi_k(t) = \cos(\Omega_k t + \sigma_k B_k(t) + \Delta_k), \quad k = 1,2,\ldots,l \tag{35}$$

where Ω_k and σ_k^2 are constants representing the center frequency and intensity of frequency stochastic perturbation, respectively; $B_k(t)$ are standard Wiener processes; Δ_k are independent random phases uniformly distributed in $[0, 2\pi]$. It can be shown that $\xi_k(t)$ are stationary processes in wide sense with spectral densities

$$S_k(\omega) = \frac{\sigma_k^2}{4\pi} \frac{\omega^2 + \Omega_k^2 + \sigma_k^4/4}{\left(\omega^2 - \Omega_k^2 - \sigma_k^4/4\right)^2 + \sigma_k^4\omega^2} \tag{36}$$

and auto-correlation functions

$$R_k(\tau) = \frac{1}{2}\exp\left(-\frac{\sigma_k^2}{2}|\tau|\right)\cos\Omega_k\tau. \tag{37}$$

The bandwidths of $\xi_k(t)$ depend mainly on parameters σ_k. They are narrow-band when σ_k are small and wide-band when σ_k are large. Here we assume that σ_k are small and $\xi_k(t)$ are narrow-band processes.

The differential equations for the action-angle variables of the system (13) are of the form

$$\begin{aligned}\dot{I}_r &= -\varepsilon c_{ij}\frac{\partial H}{\partial P_j}\frac{\partial I_r}{\partial P_i} + \varepsilon\frac{\partial I_r}{\partial P_i}h_{ik}\xi_k(t),\\ \dot{\theta}_r &= \omega_r - \varepsilon c_{ij}\frac{\partial H}{\partial P_j}\frac{\partial \theta_r}{\partial P_i} + \varepsilon\frac{\partial \theta_r}{\partial P_i}h_{ik}\xi_k(t),\\ r,i,j &= 1,2,\ldots,n; \quad k = 1,2,\ldots,l.\end{aligned} \tag{38}$$

The form and dimension of averaged equations depend on the internal and external resonances of system (38). Two cases are considered here.

2.3.1. *External Resonance Only*

Consider the case where there are external resonances but no internal resonance. Suppose that there are β external primary resonant relations between the first β oscillators and the first β bounded noise excitations, i.e.,

$$M_v\Omega_v + L_v\omega_v = \varepsilon\delta_v, \quad v = 1,\ldots,\beta \tag{39}$$

where M_v and L_v are positive or negative integers and there is no summation over subscript v. Introduce β new variables

$$\psi_v = M_v(\Omega_v t + \sigma_v B_v(t) + \Delta_v) + L_v\theta_v. \tag{40}$$

The following differential equations for $\mathbf{I}, \boldsymbol{\psi}, \boldsymbol{\theta}$ can be obtained from Eq.(38):

$$\begin{aligned}\frac{dI_r}{dt} &= \varepsilon\left[-c_{ij}\frac{\partial H}{\partial P_j}\frac{\partial I_r}{\partial P_i} + h_{ik_1}\frac{\partial I_r}{\partial P_i}\cos\left(\frac{1}{M_{k_1}}\psi_{k_1} - \frac{L_{k_1}}{M_{k_1}}\theta_{k_1}\right) + h_{ik_2}\frac{\partial I_r}{\partial P_i}\xi_{k_2}(t)\right],\\ \frac{d\psi_v}{dt} &= \varepsilon\left[\delta_v + L_r c_{ij}\frac{\partial H}{\partial P_j}\frac{\partial \theta_v}{\partial P_i} + L_v h_{ik_1}\frac{\partial \theta_v}{\partial P_i}\cos\left(\frac{1}{M_{k_1}}\psi_{k_1} - \frac{L_{k_1}}{M_{k_1}}\theta_{k_1}\right)\right.\\ &\quad\left. + L_v h_{ik_2}\frac{\partial \theta_v}{\partial P_i}\xi_{k_2}(t)\right] + M_v\sigma_v\frac{dB_v(t)}{dt},\\ \frac{d\theta_r}{dt} &= \omega_r - \varepsilon c_{ij}\frac{\partial H}{\partial P_j}\frac{\partial \theta_r}{\partial P_i} + \varepsilon h_{ik}\xi_k\frac{\partial \theta_r}{\partial P_i},\\ r,i,j &= 1,\ldots,n; \quad k_1, v = 1,\ldots,\beta; \quad k_2 = \beta+1,\ldots,l; \quad k = 1,\ldots,l\end{aligned} \tag{41}$$

where $\mathbf{I} = [I_1,\ldots,I_n]^T$, $\boldsymbol{\psi} = [\psi_1,\ldots,\psi_\beta]^T$, and $\boldsymbol{\theta} = [\theta_1,\ldots,\theta_n]^T$. Applying deterministic averaging to Eq.(41) with respect to $\theta_1,\ldots,\theta_n$, yields the following averaged Itô equations[15]

$$\begin{aligned}&dI_r = \varepsilon\bar{U}_r(\mathbf{I},\boldsymbol{\psi};\tau)dt, \quad d\psi_v = \varepsilon\bar{V}_v(\mathbf{I},\boldsymbol{\psi};\tau)dt + M_v\sigma_v dB_v(t),\\ &r = 1,\ldots,n; \quad v = 1,\ldots,\beta.\end{aligned} \tag{42}$$

2.3.2. *Both Internal and External Resonances*

Consider the case where there are β external resonant relations and α internal resonant relations, i.e.,

$$M_v\Omega_v + L_v\omega_v = \varepsilon\delta_v, \quad \sum_{i=1}^{n} N_i^u\omega_i = \varepsilon\sigma_v, \quad v = 1,\ldots,\beta;\ u = 1,\ldots,\alpha. \tag{43}$$

where M_v and L_v are positive or negative integers and there is no summation over subscript v. N_i^u are integers and not all zero for a given u. Introduce new variables

$$\begin{aligned} \psi_v &= M_v(\Omega_v t + \sigma_v B_v(t) + \Delta_v) + L_v\theta_v, \\ \Phi_u &= \sum_{i=1}^{n} N_i^u\theta_i, \quad v = 1,\ldots,\beta; \quad u = 1,\ldots,\alpha. \end{aligned} \tag{44}$$

The differential equations in this case are of the form

$$\begin{aligned} \frac{dI_r}{dt} &= \varepsilon\left[-c_{ij}\frac{\partial H}{\partial P_j}\frac{\partial I_r}{\partial P_i} + h_{ik_1}\frac{\partial I_r}{\partial P_i}\cos\left(\frac{\psi_{k_1}}{M_{k_1}} - \frac{L_{k_1}}{M_{k_1}}\theta_{k_1}\right) + h_{ik_2}\frac{\partial I_r}{\partial P_i}\xi_{k_2}(t)\right], \\ \frac{d\psi_v}{dt} &= \varepsilon\left[\delta_v + L_r c_{ij}\frac{\partial H}{\partial P_j}\frac{\partial \theta_v}{\partial P_i} + L_v h_{ik_1}\frac{\partial \theta_v}{\partial P_i}\cos\left(\frac{\psi_{k_1}}{M_{k_1}} - \frac{L_{k_1}}{M_{k_1}}\theta_{k_1}\right)\right. \\ &\quad \left. + L_v h_{ik_2}\frac{\partial \theta_v}{\partial P_i}\xi_{k_2}(t)\right] + M_v\sigma_v\frac{dB_v(t)}{dt}, \\ \frac{d\Phi_u}{dt} &= \varepsilon\left[\sigma_u - N_{u_1}^u c_{ij}\frac{\partial H}{\partial P_j}\frac{\partial \theta_{u_1}}{\partial P_i} + N_{u_1}^u h_{ik_1}\frac{\partial \theta_{u_1}}{\partial P_i}\cos\left(\frac{\psi_{k_1}}{M_{k_1}} - \frac{L_{k_1}}{M_{k_1}}\theta_{k_1}\right)\right. \\ &\quad \left. + N_{u_1}^u h_{ik_2}\frac{\partial \theta_{u_1}}{\partial P_i}\xi_{k_2}(t)\right], \\ \frac{d\theta_s}{dt} &= \omega_s - \varepsilon c_{ij}\frac{\partial H}{\partial P_j}\frac{\partial \theta_s}{\partial P_i} + \varepsilon h_{ik}\xi_k\frac{\partial \theta_s}{\partial P_i}, \\ r,i,j,u_1 &= 1,\ldots,n; \quad k_1, u = 1,\ldots,\alpha; \quad v = 1,\ldots,\beta; \quad k_2 = \beta+1,\ldots,l; \\ k &= 1,\ldots,l; \quad s = \alpha+1,\ldots,n. \end{aligned} \tag{45}$$

Applying deterministic averaging to Eq.(45) with respect to $\theta_{\alpha+1},\ldots,\theta_n$, yields the following averaged Itô equations[15]

$$\begin{aligned} dI_r &= \varepsilon\tilde{U}_r(\mathbf{I}, \boldsymbol{\psi}, \boldsymbol{\Phi}; \tau)dt, \quad d\psi_v = \varepsilon\tilde{V}_v(\mathbf{I}, \boldsymbol{\psi}, \boldsymbol{\Phi}; \tau)dt + M_v\sigma_v dB_v(t), \\ d\Phi_u &= \varepsilon\tilde{W}_u(\mathbf{I}, \boldsymbol{\psi}, \boldsymbol{\Phi}; \tau)dt, \quad r = 1,\ldots,n; \quad v = 1,\ldots,\beta; \quad u = 1,\ldots,\alpha. \end{aligned} \tag{46}$$

The drift and diffusion coefficients can be obtained, but are omitted here due to space limit.

2.4. *Combined Excitations of Harmonic Function and One Kind of above Random Processes*

In this case, the stochastic excitations $\xi_k(t)$ in Eq.(1) contain both harmonic functions and random noises, i.e.,

$$\varepsilon^{1/2} f_{ik}\xi_k(t) = \varepsilon h_{ik}\cos(\Omega_k t + \gamma_k) + \varepsilon^{1/2} g_{is} W_s(t), \quad k = 1, \ldots, l; \quad s = 1, \ldots, m \tag{47}$$

where h_{ik} are amplitudes of harmonic excitations; g_{is} are amplitudes of random excitations; $W_s(t)$ are random processes. For simplicity, here $W_s(t)$ are assumed to be independent Gaussian white noises in the sense of Stratonovich with intensities $2D_s$. In this case, the differential equations for the action-angle variables of system (13) are

$$\begin{aligned} dI_r &= \varepsilon\left[-c_{ij}\frac{\partial H}{\partial P_j}\frac{\partial I_r}{\partial P_i} + \frac{\partial I_r}{\partial P_i}h_{ik}\cos\Gamma_k(t) + D_{\bar{u}\bar{v}}g_{i\bar{u}}D_{j\bar{v}}\frac{\partial^2 I_r}{\partial P_i \partial P_j}\right]dt \\ &\quad + \varepsilon^{1/2}\frac{\partial I_r}{\partial P_i}\sigma_{is}dB_s(t) \\ &= \varepsilon U_u(I,\theta,t;\tau)dt + \varepsilon^{1/2}\sigma_{rs}^{(1)}(I,\theta)dB_s(t), \\ d\theta_r &= \left[\omega_r - \varepsilon c_{ij}\frac{\partial H}{\partial P_j}\frac{\partial \theta_r}{\partial P_i} + \varepsilon\frac{\partial \theta_r}{\partial P_i}h_{ik}\cos\Gamma_k(t) + \varepsilon D_{\bar{u}\bar{v}}g_{i\bar{u}}D_{j\bar{v}}\frac{\partial^2 \theta_r}{\partial P_i \partial P_j}\right]dt \\ &\quad + \varepsilon^{1/2}\frac{\partial \theta_r}{\partial P_i}\sigma_{is}dB_s(t) \\ &= [\omega_r + \varepsilon V_u(I,\theta,t;\tau)]dt + \varepsilon^{1/2}\sigma_{rs}^{(2)}(I,\theta)dB_s(t), \\ r,i,j &= 1,2,\ldots,n; \quad k = 1,2,\ldots,l; \quad s,\bar{u},\bar{v} = 1,\ldots,m. \end{aligned} \tag{48}$$

The form and dimension of averaged equations depend on the internal and external resonances of the system associated with Eq.(48). Three cases are considered here.

2.4.1. *Internal Resonance Only*

Consider the case where there is no external resonance but are α internal resonant relations of the form

$$L_i^u \omega_i = o(\varepsilon), \quad u = 1, \ldots, \alpha \tag{49}$$

where L_i^u are positive or negative integers and not all zeros for a fixed u. In this case, the terms containing $\cos\Gamma_k(t)$ in Eq.(48) can be neglected in the first approximation. The averaged Itô differential equations are essentially the same as those in Section 2.1.2.

2.4.2. *External Resonance Only*

Consider the case where there is no internal resonance but are β external primary resonant relations of the form

$$M_k^v \Omega_k + L_r^v \omega_r = \varepsilon \delta_v, \quad r = 1, \ldots, n; \quad k = 1, \ldots, l; \quad v = 1, \ldots, \beta \tag{50}$$

where M_k^v and L_r^v are positive or negative integers and not all zero for a given v. Introducing β combinations ψ_v of angle variables and phases angles of harmonic excitations

$$\psi_v = L_r^v\theta_r + M_k^v\Gamma_k(t), \quad r = 1, \ldots, n; \quad k = 1, \ldots, l; \quad v = 1, \ldots, \beta. \tag{51}$$

The Itô differential equations are obtained from Eq.(48) by using Itô differential rule as follows:

$$\begin{aligned} dI_r &= \varepsilon U_r(\mathbf{I}, \boldsymbol{\psi}, \boldsymbol{\theta}, \boldsymbol{\Gamma}; \tau)dt + \varepsilon^{1/2}\sigma_{rs}^{(1)}(\mathbf{I}, \boldsymbol{\psi}, \boldsymbol{\theta})dB_s(t), \\ d\psi_v &= \varepsilon[\delta_v + L_r^v V_r(\mathbf{I}, \boldsymbol{\psi}, \boldsymbol{\theta}, \boldsymbol{\Gamma}; \tau)]dt + \varepsilon^{1/2}L_r^v\sigma_{rs}^{(2)}(\mathbf{I}, \boldsymbol{\psi}, \boldsymbol{\theta})dB_s(t), \\ d\theta_r &= [\omega_r + \varepsilon V_r(\mathbf{I}, \boldsymbol{\psi}, \boldsymbol{\theta}, \boldsymbol{\Gamma}; \tau)]dt + \varepsilon^{1/2}\sigma_{rs}^{(2)}(\mathbf{I}, \boldsymbol{\psi}, \boldsymbol{\theta})dB_s(t), \\ & r = 1, \ldots, n; \quad v = 1, \ldots, \beta; \quad s = 1, \ldots, m. \end{aligned} \tag{52}$$

Applying deterministic averaging to Eq.(52) with respect to $\theta_1, \ldots, \theta_n$, yields the following averaged equations[16]:

$$\begin{aligned} dI_r &= \varepsilon \bar{U}_r(\mathbf{I}, \boldsymbol{\psi}; \tau)dt + \varepsilon^{1/2}\bar{\sigma}_{rs}^{(1)}(\mathbf{I}, \boldsymbol{\psi})dB_s(t), \\ d\psi_v &= \varepsilon \bar{V}_v(\mathbf{I}, \boldsymbol{\psi}; \tau)dt + \varepsilon^{1/2}\bar{\sigma}_{vs}^{(2)}(\mathbf{I}, \boldsymbol{\psi})dB_s(t). \end{aligned} \tag{53}$$

2.4.3. *Both Internal and External Resonances*

Consider the case where there are α internal resonant relations and β external primary resonant relations, i.e.,

$$\begin{aligned} & L_r^u\omega_r = \varepsilon\sigma_u, \quad M_k^v\Omega_k + L_r^v\omega_r = \varepsilon\delta_v, \\ & u = 1, \ldots, \alpha; \quad v = 1, \ldots, \beta; \quad r = 1, \ldots, n; \quad k = 1, \ldots, m \end{aligned} \tag{54}$$

where L_r^u, L_r^v, and M_k^v are integers and not all zero for a given u or v. Introducing α combinations Φ_u of angles variables and β combinations ψ_v of angle variables and excitation phase angles of harmonic excitations

$$\begin{aligned} & \Phi_u = L_r^u\theta_r, \quad \psi_v = L_r^v\theta_r + M_k^v\Gamma_k(t), \\ & r = 1, \ldots, n; \quad k = 1, \ldots, l; \quad u = 1, \ldots, \alpha; \quad v = 1, \ldots, \beta. \end{aligned} \tag{55}$$

The Itô differential equations are obtained from Eq.(48) by using Itô differential rule as follows:

$$\begin{aligned} dI_r &= \varepsilon U_r(\mathbf{I}, \boldsymbol{\Phi}, \boldsymbol{\psi}, \boldsymbol{\theta}_1, \boldsymbol{\Gamma}; \tau)dt + \varepsilon^{1/2}\sigma_{rs}^{(1)}(\mathbf{I}, \boldsymbol{\Phi}, \boldsymbol{\psi}, \boldsymbol{\theta}_1)dB_s(t), \\ d\Phi_u &= \varepsilon[\sigma_u + L_r^u V_r(\mathbf{I}, \boldsymbol{\Phi}, \boldsymbol{\psi}, \boldsymbol{\theta}_1, \boldsymbol{\Gamma}; \tau)]dt + \varepsilon^{1/2}L_r^u\sigma_{rs}^{(2)}(\mathbf{I}, \boldsymbol{\Phi}, \boldsymbol{\psi}, \boldsymbol{\theta}_1)dB_s(t), \\ d\psi_v &= \varepsilon[\delta_v + L_r^v V_r(\mathbf{I}, \boldsymbol{\Phi}, \boldsymbol{\psi}, \boldsymbol{\theta}_1, \boldsymbol{\Gamma}; \tau)]dt + \varepsilon^{1/2}L_r^v\sigma_{rs}^{(2)}(\mathbf{I}, \boldsymbol{\Phi}, \boldsymbol{\psi}, \boldsymbol{\theta}_1)dB_s(t), \\ d\theta_{u'} &= [\omega_{u'} + \varepsilon V_{u'}(\mathbf{I}, \boldsymbol{\Phi}, \boldsymbol{\psi}, \boldsymbol{\theta}_1, \boldsymbol{\Gamma}; \tau)]dt + \varepsilon^{1/2}L_r^v\sigma_{u's}^{(2)}(\mathbf{I}, \boldsymbol{\Phi}, \boldsymbol{\psi}, \boldsymbol{\theta}_1)dB_s(t), \\ & r = 1, \ldots, n; \quad u = 1, \ldots, \alpha; \quad v = 1, \ldots, \beta; \quad u' = \alpha + 1, \ldots, n; \quad s = 1, \ldots, m \end{aligned} \tag{56}$$

where $\mathbf{\Phi} = \{\Phi_1, \ldots, \Phi_\alpha\}$, $\boldsymbol{\theta}_1 = \{\theta_{\alpha+1}, \ldots, \theta_n\}$. Applying deterministic averaging to Eq.(56) with respect to $\boldsymbol{\theta}_1$, yields the following averaged equations[16]:

$$\begin{aligned} dI_r &= \varepsilon\tilde{U}_r(\mathbf{I}, \mathbf{\Phi}, \boldsymbol{\psi}; \tau)dt + \varepsilon^{1/2}\tilde{\sigma}^{(1)}_{rs}(\mathbf{I}, \mathbf{\Phi}, \boldsymbol{\psi})dB_s(t), \\ d\Phi_u &= \varepsilon\tilde{V}_u(\mathbf{I}, \mathbf{\Phi}, \boldsymbol{\psi}; \tau)dt + \varepsilon^{1/2}\tilde{\sigma}^{(2)}_{us}(\mathbf{I}, \mathbf{\Phi}, \boldsymbol{\psi})dB_s(t), \\ d\psi_v &= \varepsilon\tilde{W}_v(\mathbf{I}, \mathbf{\Phi}, \boldsymbol{\psi}; \tau)dt + \varepsilon^{1/2}\tilde{\sigma}^{(2)}_{vs}(\mathbf{I}, \mathbf{\Phi}, \boldsymbol{\psi})dB_s(t). \end{aligned} \tag{57}$$

The expressions for the coefficients in Eqs.(53) and (57) are omitted due to space limit.

3. Stochastic Dynamics of Quasi Integrable Hamiltonian Systems with Time-delayed Feedback Control

By using the averaged Itô or FPK equations, stochastic dynamics of quasi integrable Hamiltonian systems with time-delayed feedback control can be investigated.

3.1. *Response*

The response of the system can be studied by using the averaged FPK equation. The exact stationary solution can be obtained by using stationary potential technique. For example, in the case of Gaussian white noise excitations and no resonance, the exact stationary solution of FPK equation (17) with vanished probability potential flow at infinite boundary is of the form

$$p(\mathbf{I}) = C\exp[-\lambda(\mathbf{I})] \tag{58}$$

where C is a normalization constant and $\lambda(\mathbf{I})$ is the so-called probability potential which is governed by equations

$$b_{rs}\frac{\partial\lambda}{\partial I_s} = \frac{\partial b_{rs}}{\partial I_s} - 2a_r, \quad r, s = 1, 2, \ldots, n. \tag{59}$$

If diffusion matrix $\mathbf{B} = [b_{rs}]$ is not singular, i.e., its inverse matrix $\mathbf{B}^{-1} = \mathbf{G} = [g_{rs}]$ exists, then equation (59) can be converted into

$$\frac{\partial\lambda}{\partial I_i} = g_{ir}\left(\frac{\partial b_{rs}}{\partial I_s} - 2a_r\right). \tag{60}$$

Furthermore, if the following compatibility conditions

$$\frac{\partial}{\partial I_j}g_{ir}\left(\frac{\partial b_{rs}}{\partial I_s} - 2a_r\right) = \frac{\partial}{\partial I_i}g_{jr}\left(\frac{\partial b_{rs}}{\partial I_s} - 2a_r\right) \tag{61}$$

are satisfied, then the probability potential is

$$\lambda(\mathbf{I}) = \lambda_0 + \int_0^{\mathbf{I}} \frac{\partial\lambda}{\partial I_s}dI_s \tag{62}$$

where $\lambda_0 = \lambda(0)$ and the second term is a summation of line integrals over $s = 1, 2, \ldots, n$. The exact stationary solution $p(\mathbf{I})$ of averaged FPK equation (17) is obtained by substituting Eq.(62) into Eq.(58). The approximate stationary probability density of system (1) is then

$$p(\mathbf{q},\mathbf{p}) = p(\mathbf{I},\boldsymbol{\theta})\left|\frac{\partial(\mathbf{I},\boldsymbol{\theta})}{\partial(\mathbf{q},\mathbf{p})}\right| = p(\boldsymbol{\theta}|\mathbf{I})p(\mathbf{I})\left|\frac{\partial(\mathbf{I},\boldsymbol{\theta})}{\partial(\mathbf{q},\mathbf{p})}\right| = \frac{1}{(2\pi)^n}p(\mathbf{I}) \tag{63}$$

where $|\partial(\mathbf{I},\boldsymbol{\theta})/\partial(\mathbf{q},\mathbf{p})|$ is the absolute value of the Jacobian determinant of the canonical transformations from $\mathbf{q},\mathbf{p}$ to $\mathbf{I},\boldsymbol{\theta}$, which is always equal to unity.

The exact stationary solution $p(\mathbf{H})$ to averaged FPK equation (20) can be obtained similarly and the approximate stationary probability density of system (1) is then

$$p(\mathbf{q},\mathbf{p}) = p(\mathbf{H})/T(\mathbf{H}). \tag{64}$$

In resonant case, the exact stationary solution to averaged FPK equation (25) is of the form

$$p(\mathbf{I},\boldsymbol{\psi}) = C\exp[-\lambda(\mathbf{I},\boldsymbol{\psi})] \tag{65}$$

where $\lambda(\mathbf{I},\boldsymbol{\psi})$ can be obtained by expanding $\lambda(\mathbf{I},\boldsymbol{\psi})$, the averaged drift and diffusion coefficients into α-fold Fourier expansions of $\boldsymbol{\psi}$, substituting them into averaged FPK equation (25) with $\partial p/\partial t = 0$ and obtaining the Fourier coefficients of $\lambda(\mathbf{I},\boldsymbol{\psi})$. The approximate stationary solution of system (1) is then

$$\begin{aligned} p(\mathbf{q},\mathbf{p}) &= p(\mathbf{I},\boldsymbol{\psi},\boldsymbol{\theta}_1)\left|\frac{\partial(\mathbf{I},\boldsymbol{\psi},\boldsymbol{\theta}_1)}{\partial(\mathbf{q},\mathbf{p})}\right| = p(\boldsymbol{\theta}_1|\mathbf{I},\boldsymbol{\psi})p(\mathbf{I},\boldsymbol{\psi})\left|\frac{\partial(\mathbf{I},\boldsymbol{\psi},\boldsymbol{\theta}_1)}{\partial(\mathbf{q},\mathbf{p})}\right| \\ &= \frac{1}{(2\pi)^{(n-\alpha)}}p(\mathbf{I},\boldsymbol{\psi})\left|\frac{\partial(\mathbf{I},\boldsymbol{\psi},\boldsymbol{\theta}_1)}{\partial(\mathbf{q},\mathbf{p})}\right| \end{aligned} \tag{66}$$

where $|\partial(\mathbf{I},\boldsymbol{\psi},\boldsymbol{\theta}_1)/\partial(\mathbf{q},\mathbf{p})|$is the absolute value of the Jacobian determinant for the transformation from $\mathbf{q}$, $\mathbf{p}$ to $\mathbf{I}$, $\boldsymbol{\psi}$, $\boldsymbol{\theta}_1$, which is an integer.

Example 1[12]

Consider two linear oscillators coupled by linear and polynomial type nonlinear dampings subject to external excitations of two uncorrelated Gaussian white noises and delayed velocity feedback control. The equations of the system are of the form

$$\begin{aligned} &\ddot{X}_1 + \varepsilon[\alpha'_{11}\dot{X}_1 + \alpha_{12}\dot{X}_2 + \beta_1\dot{X}_1(X_1^2 + X_2^2)] + \omega_1'^2 X_1 = -\varepsilon\eta_1\dot{X}_{1\tau} + \varepsilon^{1/2}W_1(t), \\ &\ddot{X}_2 + \varepsilon[\alpha_{21}\dot{X}_1 + \alpha'_{22}\dot{X}_2 + \beta_2\dot{X}_2(X_1^2 + X_2^2)] + \omega_2'^2 X_2 = -\varepsilon\eta_2\dot{X}_{2\tau} + \varepsilon^{1/2}W_2(t) \end{aligned} \tag{67}$$

where $\alpha'_{ii}, \alpha_{ij}, \beta_i, \eta_i, \omega'_i$ $(i,j = 1,2)$ are constants; $\varepsilon\eta_i$ $(i = 1,2)$ are the feedback gain and $\eta_i\dot{X}_{i\tau} = \eta_i\dot{X}_i(t-\tau)$ are time-delayed feedback control forces; $W_i(t)(i = 1,2)$ are uncorrelated Gaussian white noises with intensities $2D_{ii}$.

The Hamiltonian system associated with Eq.(67) is linear and the Hamiltonian can be expressed in terms of action-angle variables as

$$H' = \sum_{i=1}^{2}\omega'_i I'_i, \quad I'_i = \frac{1}{2\omega'_i}(\dot{X}_i^2 + \omega_i'^2 X_i^2), \quad \theta'_i = -\arctan\left(\frac{\dot{X}_i}{\omega'_i X_i}\right). \tag{68}$$

The time-delayed system state in system (67) can be approximately converted into that without time delay as

$$\dot{X}_{i\tau} = \dot{X}_i \cos \omega_i' \tau + X_i \omega_i' \sin \omega_i' \tau. \tag{69}$$

The modified Hamiltonian is of the form $H = \omega_1 I_1 + \omega_2 I_2$, and $\omega_i^2 = \omega_i'^2 + \varepsilon \eta_i \omega_i' \sin \omega' \tau$. I_i and θ_i are of the same form as I_i' and θ_i' in Eq.(68) with ω_i' replaced by ω_i. Also, the damping coefficients α_{ii}' become $\alpha_{ii} = \alpha_{ii}' + \varepsilon \eta_i \omega_i \sin \omega_i \tau$. Equation (67) can be rewritten as the following Itô stochastic differential equations:

$$\begin{aligned} dI_i &= \varepsilon \left\{ -[\alpha_{i1}\dot{X}_1 + \alpha_{i2}\dot{X}_2 + \beta_i \dot{X}_i (X_1^2 + X_2^2)] \frac{\dot{X}_i}{\omega_i} + \frac{D_{ii}}{\omega_i} \right\} dt + \varepsilon^{1/2} \frac{\dot{X}_i}{\omega_i} dB_t(t), \\ d\theta_i &= \omega_i + \varepsilon \left\{ [\alpha_{i1}\dot{X}_1 + \alpha_{i2}\dot{X}_2 + \beta_i \dot{X}_i (X_1^2 + X_2^2)] \frac{\omega_i X_i}{\omega_i^2 X_i^2 + \dot{X}_i^2} \right. \\ &\quad \left. + D_{ii} \frac{2\omega_i X_i \dot{X}_i}{(\omega_i^2 X_i^2 + \dot{X}_i^2)^2} \right\} dt - \varepsilon^{1/2} \frac{\omega_i X_i}{\omega_i^2 X_i^2 + \dot{X}_i^2} dB_i(t). \end{aligned} \tag{70}$$

Note that the repeated subscripts in Eq.(70) do not imply a summation. Two cases are considered in the following.

Non-resonant Case. $r\omega_1 + s\omega_2 \neq o(\varepsilon), r, s$ are integers. In this case the averaged FPK equation is of the form of Eq.(17) with the following drift and diffusion coefficients

$$\begin{aligned} a_1 &= -\alpha_{11} I_1 - \frac{\beta_1}{2\omega_1} I_1^2 - \frac{\beta_1}{\omega_2} I_1 I_2 + \frac{D_{11}}{\omega_1}, \\ a_2 &= -\alpha_{22} I_2 - \frac{\beta_2}{2\omega_2} I_2^2 - \frac{\beta_2}{\omega_1} I_1 I_2 + \frac{D_{22}}{\omega_2}, \\ b_{11} &= \frac{2}{\omega_1} D_{11} I_1, \quad b_{22} = \frac{2}{\omega_2} D_{22} I_2, \quad b_{12} = b_{21} = 0. \end{aligned} \tag{71}$$

The stationary solution of the averaged FPK equation is of the form of Eq. (58), where $\partial\lambda/\partial I_s$ satisfy the following equations:

$$\begin{aligned} \frac{2D_{11}I_1}{\omega_1} \frac{\partial \lambda}{\partial I_1} &= \frac{2D_{11}}{\omega_1} - 2\left(-\alpha_{11} I_1 - \frac{\beta_1}{2\omega_1} I_1^2 - \frac{\beta_1}{\omega_2} I_1 I_2 + \frac{D_{11}}{\omega_1} \right), \\ \frac{2D_{22}I_2}{\omega_2} \frac{\partial \lambda}{\partial I_2} &= \frac{2D_{22}}{\omega_2} - 2\left(-\alpha_{22} I_2 - \frac{\beta_2}{2\omega_2} I_2^2 - \frac{\beta_2}{\omega_1} I_1 I_2 + \frac{D_{22}}{\omega_2} \right). \end{aligned} \tag{72}$$

If $(\beta_1/D_{11})(\omega_1/\omega_2) = (\beta_2/D_{22})(\omega_2/\omega_1) = \gamma$, the averaged FPK equation has an exact stationary solution

$$p(I_1, I_2) = C \exp[-\lambda(I_1, I_2)] \tag{73}$$

where

$$\lambda(I_1, I_2) = \frac{1}{D_{11}} \left(\alpha_{11}\omega_1 I_1 + \frac{\beta_1}{4} I_1^2 \right) + \frac{1}{D_{22}} \left(\alpha_{22}\omega_2 I_2 + \frac{\beta_2}{4} I_2^2 \right) + \gamma I_1 I_2. \tag{74}$$

The approximate stationary probability density of the displacements and velocities of the original system (67) is then

$$p(x_1, x_2, \dot{x}_1, \dot{x}_2) = \frac{1}{4\pi^2} p(I_1, I_2)\bigg|_{I_i=(\dot{x}_i^2+\omega_i^2 x_i^2)/2\omega_i}. \tag{75}$$

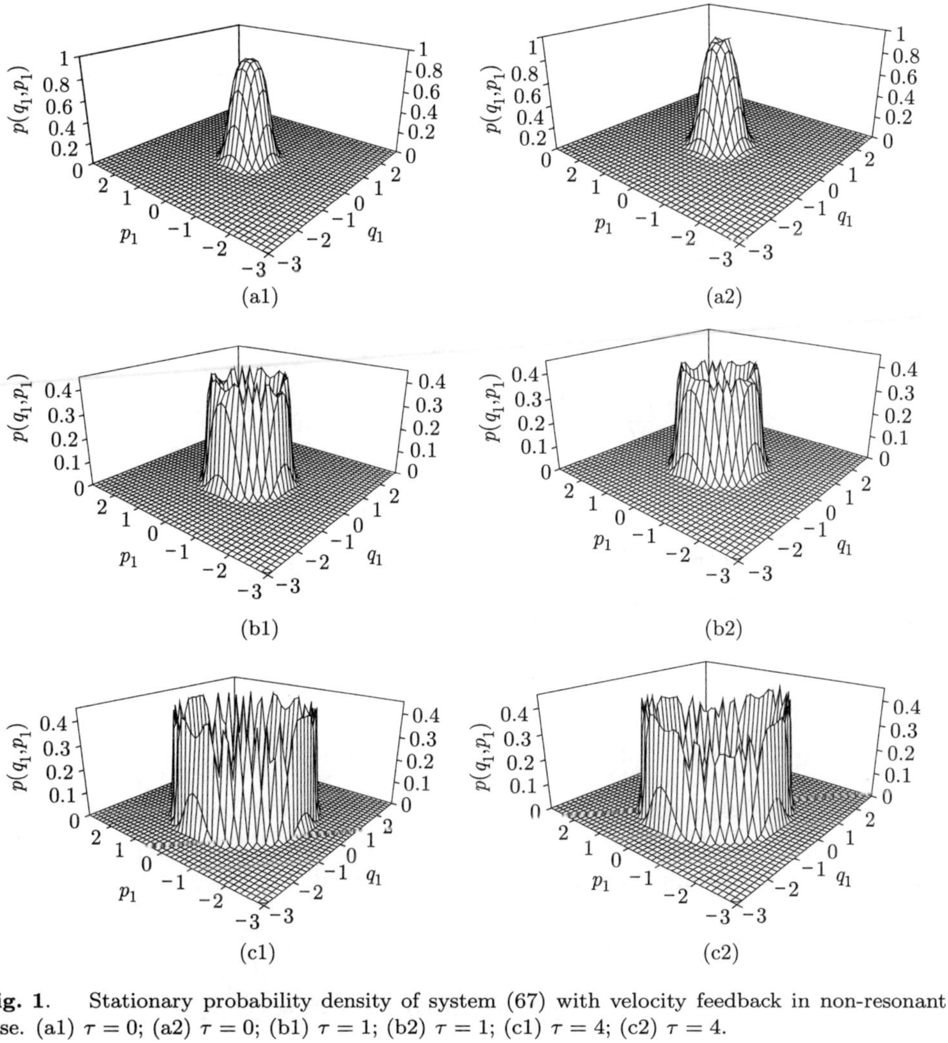

Fig. 1. Stationary probability density of system (67) with velocity feedback in non-resonant case. (a1) $\tau = 0$; (a2) $\tau = 0$; (b1) $\tau = 1$; (b2) $\tau = 1$; (c1) $\tau = 4$; (c2) $\tau = 4$.

Primary Resonant Case. $\omega_1 = \omega_2 = \omega$. Let $\theta_1 - \theta_2 = \psi$. The averaged FPK equation in this case is of the form of Eq.(25) with the following drift and diffusion coefficients:

$$a_1 = -\alpha_{11} I_1 - \alpha_{12}\sqrt{I_1 I_2}\cos\psi - \frac{\beta_1}{2\omega} I_1^2 - \frac{\beta_1}{\omega} I_1 I_2 \left(1 - \frac{1}{2}\cos 2\psi\right) + \frac{D_{11}}{\omega},$$

$$a_2 = -\alpha_{22} I_2 - \alpha_{21}\sqrt{I_1 I_2}\cos\psi - \frac{\beta_2}{2\omega} I_2^2 - \frac{\beta_2}{\omega} I_1 I_2 \left(1 - \frac{1}{2}\cos 2\psi\right) + \frac{D_{22}}{\omega},$$

$$a_3 = \frac{1}{2}\left(\alpha_{12}\sqrt{\frac{I_2}{I_1}} + \alpha_{21}\sqrt{\frac{I_1}{I_2}}\right)\sin\psi - \frac{1}{4\omega}(\beta_1 I_2 + \beta_2 I_1)\sin 2\psi, \tag{76}$$

$$b_{11} = \frac{2}{\omega}D_{11}I_1, \quad b_{22} = \frac{2}{\omega}D_{22}I_2, \quad b_{33} = \frac{1}{2\omega}\left(\frac{D_{11}}{I_1} + \frac{D_{22}}{I_2}\right),$$

$$b_{12} = b_{21} = b_{13} = b_{31} = b_{23} = b_{32} = 0.$$

The stationary solution of the averaged FPK equation is of the form

$$p(I_1, I_2, \psi) = C\exp[-\lambda(I_1, I_2, \psi)] \tag{77}$$

where $\lambda(I_1, I_2, \psi)$ satisfies the following partial differential equations:

$$\begin{aligned} \frac{2D_{11}I_1}{\omega}\frac{\partial\lambda}{\partial I_1} &= \frac{2D_{11}}{\omega} - 2\left[-\alpha_{11}I_1 - \alpha_{12}\sqrt{I_1 I_2}\cos\psi \right.\\ &\quad \left. -\frac{\beta_1}{2\omega}I_1^2 - \frac{\beta_1}{\omega}I_1 I_2\left(1 - \frac{1}{2}\cos 2\psi\right) + \frac{D_{11}}{\omega}\right],\\ \frac{2D_{22}I_2}{\omega}\frac{\partial\lambda}{\partial I_2} &= \frac{2D_{22}}{\omega} - 2\left[-\alpha_{22}I_2 - \alpha_{21}\sqrt{I_1 I_2}\cos\psi \right.\\ &\quad \left. -\frac{\beta_2}{2\omega}I_2^2 - \frac{\beta_2}{\omega}I_1 I_2\left(1 - \frac{1}{2}\cos 2\psi\right) + \frac{D_{22}}{\omega}\right],\\ \left(\frac{D_{11}}{2\omega I_1} + \frac{D_{22}}{2\omega I_2}\right)\frac{\partial\lambda}{\partial\psi} &= -\left(\alpha_{12}\sqrt{\frac{I_2}{I_1}} + \alpha_{21}\sqrt{\frac{I_1}{I_2}}\right)\sin\psi\\ &\quad + \frac{1}{2\omega}(\beta_1 I_2 + \beta_2 I_1)\sin\psi. \end{aligned} \tag{78}$$

Let

$$\lambda(I_1, I_2, \psi) = \lambda_0(I_1, I_2) + \lambda_1(I_1, I_2)\cos\psi + \lambda_2(I_1, I_2)\cos 2\psi. \tag{79}$$

Substituting Eq.(79) into Eq.(78), we obtain three sets of partial differential equations for λ_0, λ_1, and λ_2. In the case that $\beta_1/D_{11} = \beta_2/D_{22} = \gamma_1$, $\alpha_{12}/D_{11} = \alpha_{21}/D_{22} = \gamma_2$, we obtain the exact stationary solution (77) with

$$\begin{aligned} \lambda(I_1, I_2, \psi) &= \frac{\alpha_{11}\omega}{D_{11}}I_1 + \frac{\alpha_{22}\omega}{D_{22}}I_2 + \frac{\beta_1}{4D_{11}}I_1^2 + \frac{\beta_2}{4D_{22}}I_2^2\\ &\quad + \gamma_1 I_1 I_2 - \frac{\gamma_1}{2}I_1 I_2\cos 2\psi + 2\gamma_2\omega\sqrt{I_1 I_2}\cos\psi. \end{aligned} \tag{80}$$

The approximate stationary probability density of the displacements and velocities of the original system (67) is then

$$p(x_1, \dot{x}_1, x_2, \dot{x}_2) = \frac{1}{2\pi}p(I_1, I_2, \psi) \tag{81}$$

where $\psi = \theta_1 - \theta_2$; I_i and θ_i are functions of x_i and $\dot{x}_i$.

Letq_1, p_1 represent the displacement and velocity of the first oscillator respectively. Some numerical results for stationary probability density $p(q_1, p_1)$ obtained by using the proposed stochastic averaging method ((a1), (b1), (c1)) and from digital simulation ((a2), (b2), (c2)) are shown in Fig.1 for non-resonant case and in Fig.2 for primary resonant case. It is seen that the proposed method yields very good prediction even the time delay approaches to one period. For both non-resonant and resonant cases, the results for several τ values are given to illustrate the effect of time delay in control forces on the response of the system. It is seen that the time delay in control forces may affect the response of the system greatly, and may even cause phenomenological bifurcation.

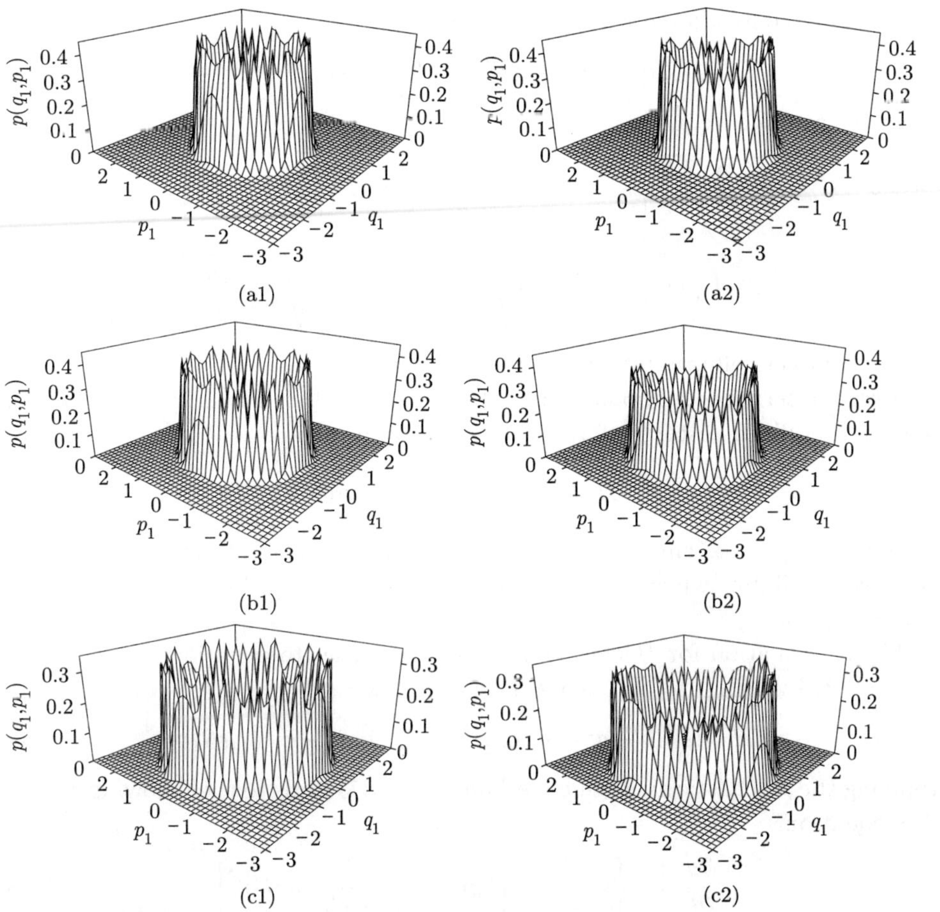

Fig. 2. Stationary probability density of system (67) with velocity feedback in resonant case. (a1) $\tau = 0$; (a2) $\tau = 0$; (b1) $\tau = 1$; (b2) $\tau = 1$; (c1) $\tau = 2$; (c2) $\tau = 2$.

Example 2[12]

As an example of strongly nonlinear stochastic system, consider a Duffing-van der Pol oscillator with delayed linear feedback control subject to Gaussian white noise excitation. The equation of motion is

$$\ddot{X} + \omega_0'^2 X + \alpha X^3 = \varepsilon(b - X^2)\dot{X} - u_\tau + \varepsilon^{1/2} W(t) \tag{82}$$

where ε is a small positive parameter; $W(t)$ is a Gaussian white noise with intensity $2D$.

First, assume that the time delayed control $u_\tau = \varepsilon a \dot{X}_\tau = \varepsilon a \dot{X}(t-\tau)$, where εa is feedback control gain. Note that there is no Wong-Zakai correction term for this example. Let $X = Q$, $\dot{X} = P$. The Hamiltonian associated with the system (82) is

$$H' = \frac{1}{2}p^2 + \frac{1}{2}\omega_0'^2 q^2 + \frac{1}{4}\alpha q^4. \tag{83}$$

The time-delayed feedback control force in system (82) can be approximately converted into a control force without time delay, i.e.,

$$\varepsilon a P_\tau \doteq \varepsilon a (P\cos\omega'\tau + Q\omega'\sin\omega'\tau) \tag{84}$$

where the average frequency ω' is

$$\begin{aligned} \omega'(H') &= \frac{\pi\sqrt{\alpha}}{2\sqrt{2}}\frac{\sqrt{A^2+B^2}}{K(r)}, \quad r = A/\sqrt{A^2+B^2}, \\ A^2 &= \frac{\omega_0'^2}{\alpha}\left(\sqrt{1+\frac{4\alpha H'}{\omega_0'^4}} - 1\right), \quad B^2 = \frac{\omega_0'^2}{\alpha}\left(\sqrt{1+\frac{4\alpha H'}{\omega_0'^4}} + 1\right). \end{aligned} \tag{85}$$

$K(r)$ is complete elliptic integral of the first kind.

After the term proportional to Q in Eq.(84) is combined with the restoring force to form a modified restoring force, the new Hamiltonian is

$$H = \frac{1}{2}p^2 + \frac{1}{2}\omega_0^2 q^2 + \frac{1}{4}\alpha q^4 \tag{86}$$

where $\omega_0^2 = \omega_0'^2 + \varepsilon\omega'\sin\omega'\tau$. Also, we can obtain modified frequency $\omega(H)$ of the nonlinear oscillator, which is of the same form as $\omega'(H')$ in Eq. (85) but with ω_0' is replaced by ω_0.

The Itô equation for H can be obtained from the Itô equation associated with system (82) by using Itô differential rule as follows:

$$dH = \varepsilon\left[(b - q^2 - a\cos\omega\tau)p^2 + D\right]dt + \varepsilon^{1/2} p\, dB(t). \tag{87}$$

Applying the stochastic averaging method to Eq.(87) leads to the following averaged FPK equation

$$\frac{\partial p}{\partial t} = \varepsilon\left\{-\frac{\partial}{\partial H}[a(H)p] + \frac{1}{2}\frac{\partial^2}{\partial H \partial H}[b(H)p]\right\} \tag{88}$$

where the drift and diffusion coefficients are

$$\begin{aligned} a(H) &= \frac{1}{T(H)}\oint_\Omega \left[(b - q^2 - a\cos\omega\tau)p^2 + D\right]/p\,dq, \\ b(H) &= \frac{1}{T(H)}\oint_\Omega \left[2Dp^2\right]/p\,dq, \\ T(H) &= \oint_\Omega (1/p)\,dq, \quad \Omega = \{q | H = \omega_0^2 q^2/2 + \alpha q^4/4\}. \end{aligned} \tag{89}$$

The stationary solutions to the averaged FPK equation (88) is

$$p(H) = C\exp[-\lambda(H)] \tag{90}$$

where

$$\lambda(H) = \lambda_0 + \int_0^H \frac{1}{b(H)}\left[\frac{db(H)}{dH} - 2a(H)\right] dH. \tag{91}$$

The approximate stationary probability density of the original system (82) is then

$$p(q,p) = \frac{1}{2\pi} p(H)\Big|_{H=\frac{1}{2}p^2+\frac{1}{2}\omega_0^2q^2+\frac{1}{4}\alpha q^4}. \tag{92}$$

Some numerical results for stationary marginal probability density $p(x)$ obtained by using the proposed stochastic averaging method (solid lines) and from digital simulation (dots) are shown in Fig.3. From these figures it is seen that the analytical results obtained by using the proposed method agree well with those from

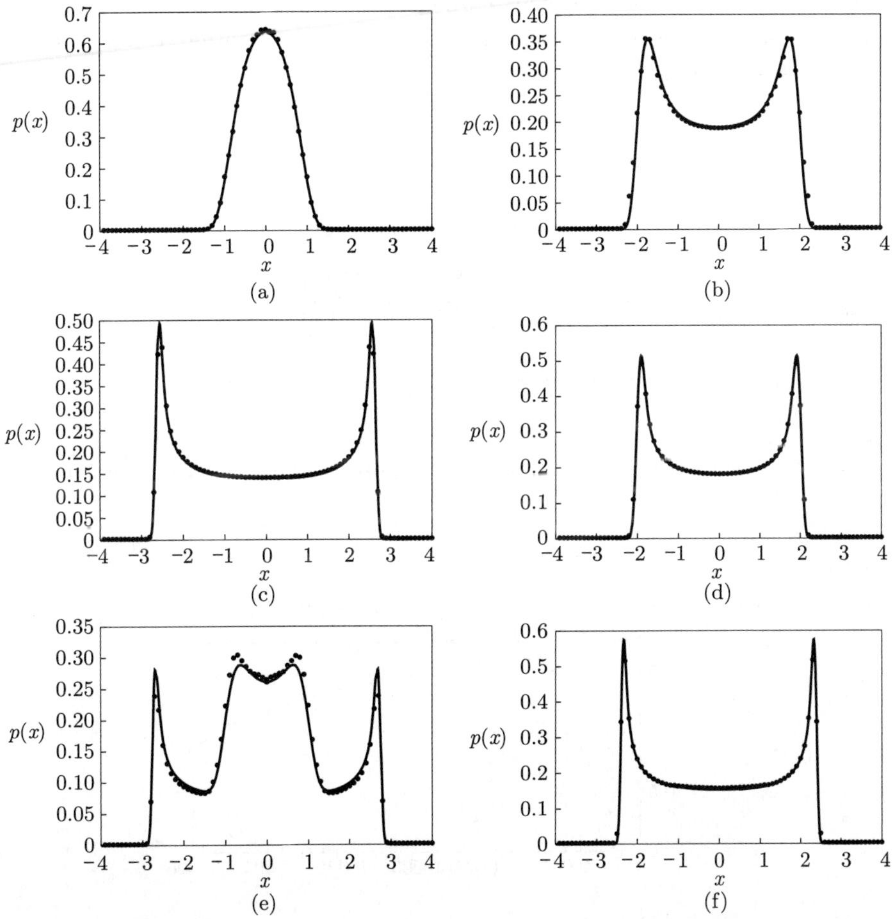

Fig. 3. Stationary marginal probability density $p(x)$ of system (82) with velocity feedback. (a) $\tau = 0$, (b) $\tau = 1$, (c) $\tau = 2$, (d) $\tau = 3$, (e) $\tau = 5$, (f) $\tau = 6$.

digital simulation even for long delay time. From the figures we can also see the phenomenological bifurcation in the response of the system caused by the delayed feedback control.

The second case considered in Ref.13 is time delayed feedback bang-bang control $u_\tau = \varepsilon c \,\mathrm{sgn}\left(\dot{X}(t-\tau)\right)$, where εc is control gain. The delayed bang-bang control force u_τ can be approximately expressed as

$$u_\tau = \varepsilon c \cos \omega\tau \,\mathrm{sgn}\,(P)\,. \tag{93}$$

Some numerical results for stationary marginal probability $p(q)$ for system (82) with different time delays obtained by using the stochastic averaging method (solid lines) and from digital simulation (dots) are shown in Fig.4 and those for the corresponding mean-square value $E[Q^2]$ are shown in Fig.5. It is seen that the time delay deteriorates the control effectiveness remarkably. When $\tau = 0$, the control force can reduce the response of the system effectively. The control effectiveness is weakened when $\tau = 1$ and even worse when $\tau = 2$. It is also seen that the results obtained by using the proposed stochastic averaging method agree well with those from digital simulations.

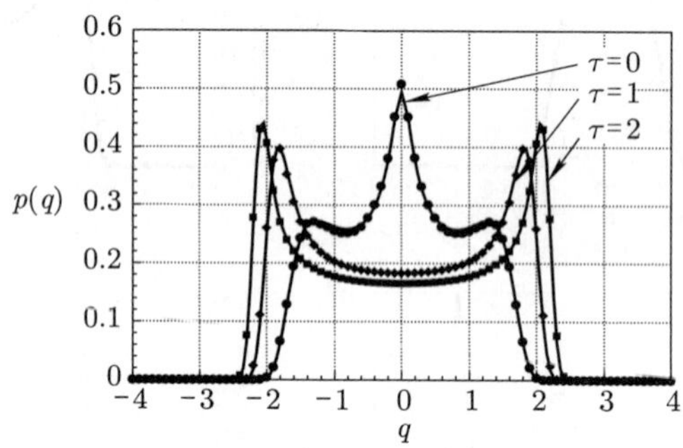

Fig. 4. Stationary marginal probability density $p(q)$ of the displacement of system (82).

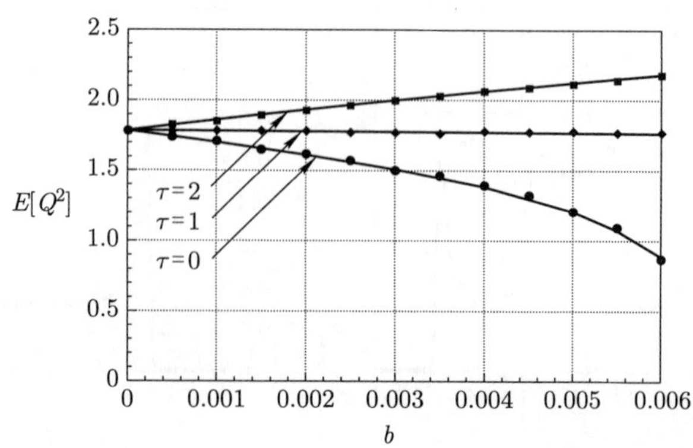

Fig. 5. Stationary mean-square value $E[Q^2]$ of the displacement of system (82).

The procedure to obtain the response of quasi integrable Hamiltonian systems with time delay feedback control under non-Gaussian white noise excitations is similar to that under Gaussian white noise excitations. In the following some examples are given to illustrate the stochastic responses for quasi integrable Hamiltonian systems with time delay feedback control under non-Gaussian white noise excitations.

Example for Wide-band Random Excitation[14]

Consider two linear oscillators coupled by linear and nonlinear dampings subject to external excitations of two wide-band random excitations and 2-time-delayed feedback controls. The equations of the system are of the form

$$\begin{aligned} &\ddot{X}_1 + \varepsilon[\alpha'_{11}\dot{X}_1 + \alpha_{12}\dot{X}_2 + \beta_1\dot{X}_1(X_1^2 + X_2^2)] + \omega_1'^2 X_1 = u_1 + \varepsilon^{1/2}\xi_1(t),\\ &\ddot{X}_2 + \varepsilon[\alpha_{21}\dot{X}_1 + \alpha'_{22}\dot{X}_2 + \beta_2\dot{X}_2(X_1^2 + X_2^2)] + \omega_2'^2 X_2 = u_2 + \varepsilon^{1/2}\xi_2(t) \end{aligned} \tag{94}$$

where α'_{ii}, α_{ij}, β_i, η_i, ω'_i $(i, j = 1, 2)$ are constants; $u_1 = -\varepsilon(\eta_{11}X_{1\tau_1} + \eta_{12}\dot{X}_{1\tau_2})$, $u_2 = -\varepsilon(\eta_{21}X_{2\tau_1} + \eta_{22}\dot{X}_{2\tau_2})$ are the time-delayed control forces; $X_{i\tau_1} = X_i(t - \tau_1)$ are time-delayed displacement, $\dot{X}_{i\tau_2} = \dot{X}_i(t - \tau_2)$are time-delayed velocity; $\xi_i(t)$ are independent wide-band random excitations. Let $X_i = Q_i$, $\dot{X}_i = P_i$.

The results for stationary probability density $p(q_1, p_1)$ are shown in Figs.6 and 7, where (a1), (b1) denote the results obtained by using the proposed stochastic averaging technique and (a2), (b2) denote those from digital simulation. In general,

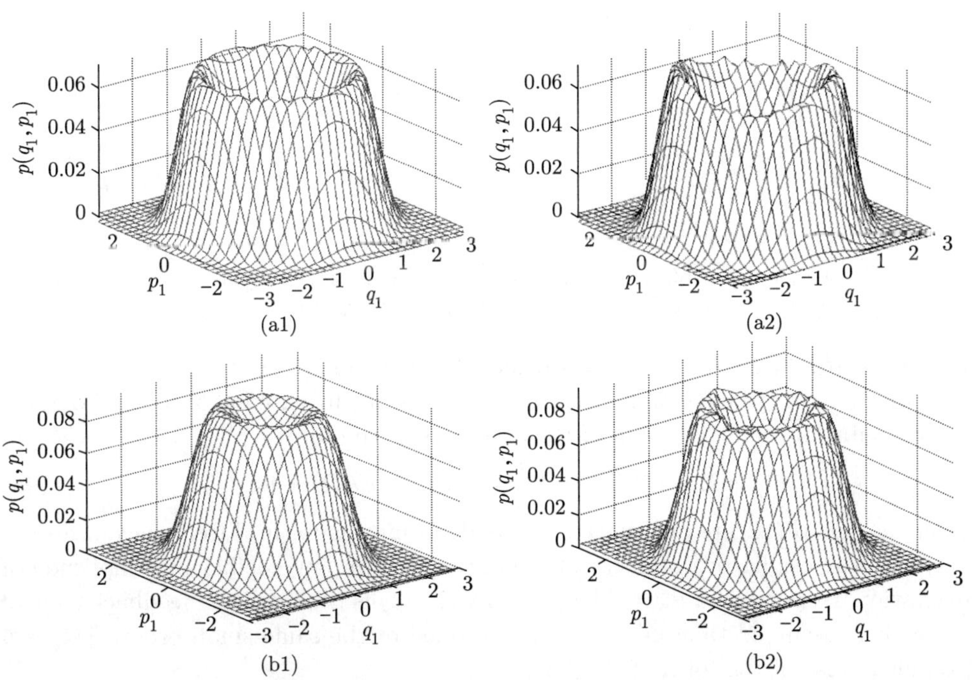

Fig. 6. Stationary probability density $p(q_1, p_1)$ of system (94) in non-resonant case. (a1) $\tau_1 = 0$, $\tau_2 = 0$; (a2) $\tau_1 = 0$, $\tau_2 = 0$; (b1) $\tau_1 = 0$, $\tau_2 = 1$; (b2) $\tau_1 = 0$, $\tau_2 = 1$.

the compatibility condition is not satisfied. In this case, the averaged FPK equation has no analytical stationary solution. However, the proposed technique is still applicable and the stationary probability densities of the system can be obtained from solving the averaged FPK equation (20) numerically. It is seen that the results obtained by using the proposed technique are in good agreement with those from simulation.

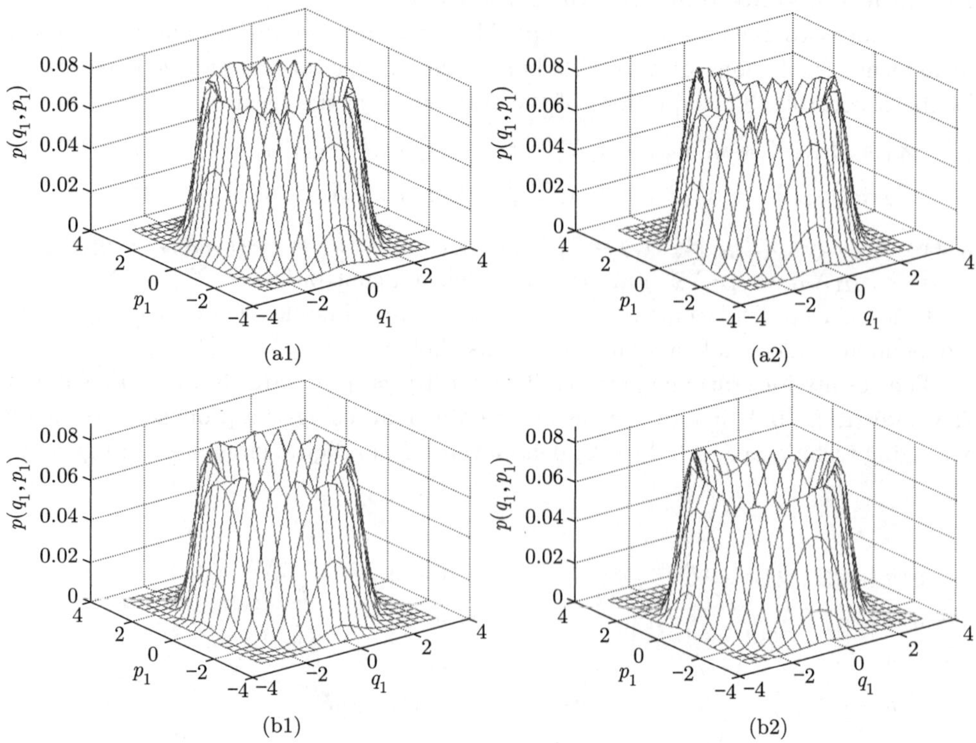

Fig. 7. Stationary probability density $p(q_1,\, p_1)$ of system (94) in resonant case. (a1) $\tau_1 = 0$, $\tau_2 = 0$; (a2) $\tau_1 = 0$, $\tau_2 = 0$; (b1) $\tau_1 = 0$, $\tau_2 = 1$; (b2) $\tau_1 = 0$, $\tau_2 = 1$.

Example for Narrow-band Bounded Noise Excitation[15]

Consider a Duffing system with time-delayed feedback control under bounded noise excitation. The equation of the system is of the form

$$\ddot{X} + \omega_0'^2 X + \alpha X^3 = -\beta' \dot{X} + E\xi(t) + u_\tau \tag{95}$$

where ω_0', α, β', E are positive constants denoting the natural frequency of degenerated linear oscillator, intensity of nonlinearity, damping coefficient, amplitude of excitation, respectively; $u_\tau = -b\dot{X}_\tau = -b\dot{X}(t-\tau)$ is time-delayed feedback control force. It is assumed that β', E, b are assumed of the same small order. $\xi(t)$ is a bounded noise of the form of Eq.(35).

Some numerical results of the stationary mean amplitude value $E[A]$ as a function of control force b obtained by using the stochastic averaging method (solid line) and from digital simulation of Eq.(95) (dots) are shown in Fig.8. It is seen that the results obtained by using the proposed method agree well with those from digital simulation. It is also seen that the time delay deteriorates the control effectiveness remarkably. When $\tau = 0$, the control force can reduce the response of the system effectively. However, the control effectiveness is weakened as time delay τ increases.

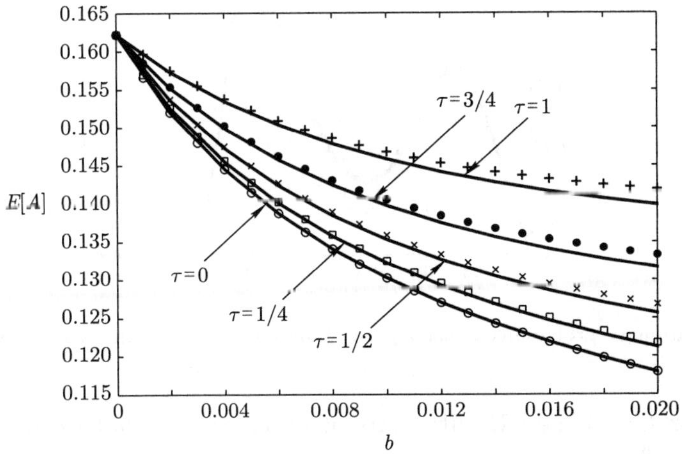

Fig. 8. Stationary mean amplitude $E[A]$ of system (95) in primary external resonance.

It is well-known that a Duffing oscillator with hardening stiffness subject to harmonic excitation may exhibit the phenomenon of sharp jumps in amplitude. The jump phenomenon may also occur when the Duffing oscillator is subjected to bounded noise excitations. Hero, thc cffcct of time delay in feedback control on the jump and its bifurcation of system (95) can also be seen in Figs.9 and 10. Figure 9 shows that the probability density $p(a)$ is unimodal and no jump may occur in system (95) with $\tau = 0$. Thc probability density $p(a)$becomes bimodal for the same

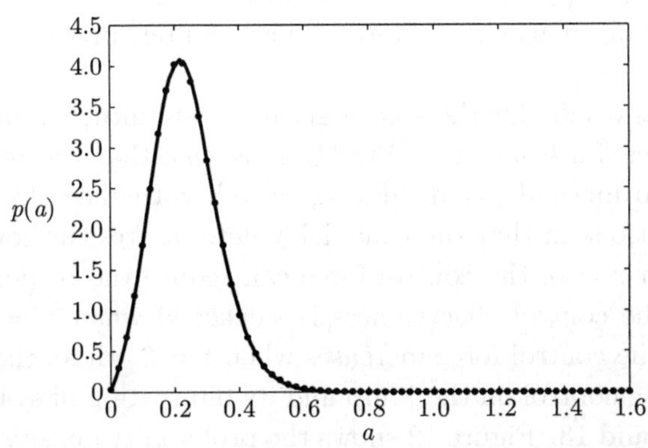

Fig. 9. Stationary probability density $p(a)$ of amplitude of system (95) in primary external resonance. $b = 0.03$, $\tau = 0$.

control force with $\tau = 1$ (see Fig.10), and the stochastic jump may occur. It is also seen that the results obtained by using the proposed method agree well with those from digital simulation.

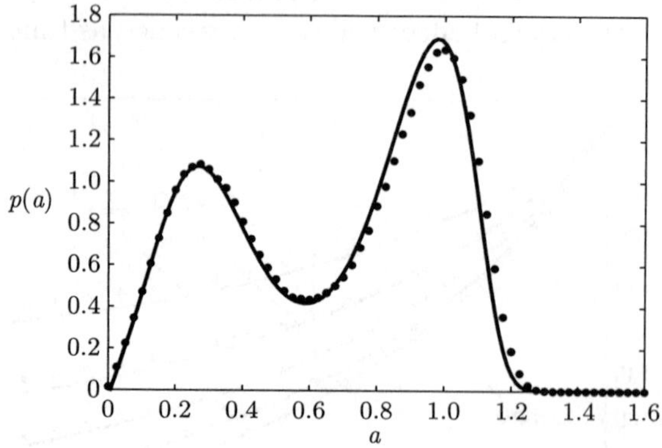

Fig. 10. Stationary probability density $p(a)$ of amplitude of system (95) in primary external resonance. $\tau = 1$.

Example for Combined Excitations of Harmonic Function and Gaussian White Noise[16]

Consider a Duffing oscillator subject to external harmonic excitation and external and parametric white noise excitations with time-delayed feedback bang-bang control. The equation of the system is of the form

$$\ddot{X} + \omega_0^2 X + \alpha X^3 = -\beta \dot{X} + E \cos \Omega t + \xi_1(t) + X\xi_2(t) + u_\tau \tag{96}$$

where ω_0, α, β, E, Ω are positive constants denoting the natural frequency of degenerated linear oscillator, intensity of nonlinearity, damping coefficient, amplitude and frequency of harmonic excitation, respectively; $\xi_k(t)\,(k = 1,\ 2)$ are independent Gaussian white noise in the sense of Stratonovich with intensities $2D_k$. It is assumed that β, E, D_k are of the same order of ε. $u_\tau = -b\,\mathrm{sgn}\,(\dot{X}(t-\tau))$is delayed feedback bang-bang control force, which can be approximately expressed as that in Eq.(12).

Some numerical results for the stationary mean amplitude value $E[A]$ as a function of control force b are shown in Fig.11. It is seen that the results obtained by using the proposed method (solid line) agree well with those from digital simulation (dots). It is also seen that the time delay deteriorates the control effectiveness remarkably. When $\tau = 0$, the control force can reduce the response of the system effectively while the control effectiveness is weakened when $\tau = 1$. The response even increases as the control force increases when $\tau = 2$. Here, the effect of delayed feedback bang-bang control on the jump and its bifurcation of system (96) can also be seen in Figs.12 and 13. Figure 12 shows the probability density $p(a)$ is unimodal and no jump may occur in system (96) for $\tau = 0$ with appropriate control force. In Fig.13, the probability density $p(a)$ becomes bimodal when $\tau = 1$ and the stochas-

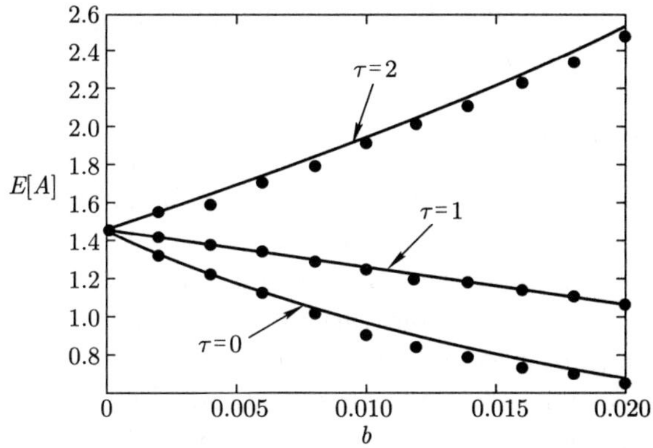

Fig. 11. Stationary mean amplitude $E[A]$ of system (96) in primary external resonance.

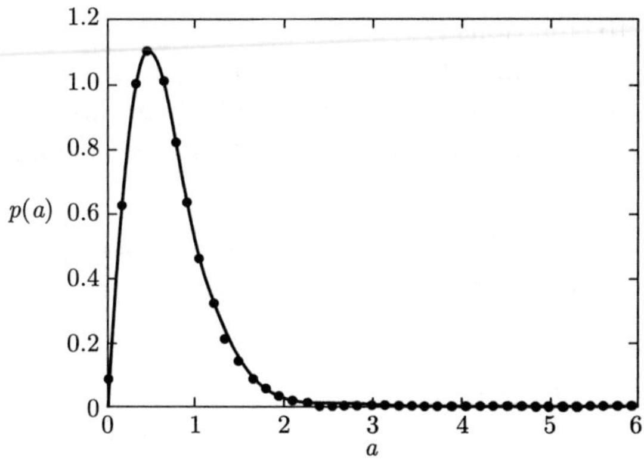

Fig. 12. Stationary probability density $p(a)$ of amplitude of controlled system (96) in primary external resonance. $b = 0.015$, $\tau = 0$.

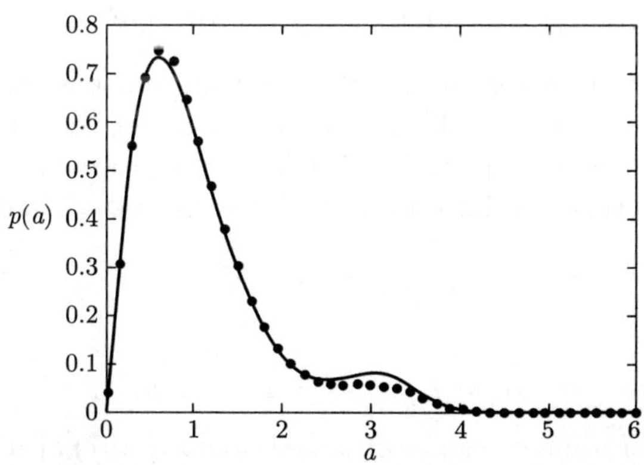

Fig. 13. Stationary probability density $p(a)$ of amplitude of controlled system (96) in primary external resonance. $b = 0.015$, $\tau = 1$.

tic jump may occur. The results obtained by using the proposed method agree well with those from digital simulation.

3.2. *Stochastic Stability*

The theory of stochastic stability deals with the stability of the trivial solution of dynamical systems under random parameter perturbation. Stability is defined in terms of boundedness and convergence of a suitable norm of the response. Since the convergence of a series of random variables can be interpreted in several ways, there are many definitions of stochastic stability. Among them, Lyapunov stability with probability one or almost sure Lyapunov stability, stability in probability, and stability in the pth moment are most often used. In the following, the asymptotical Lyapunov stability with probability one is studied by using the largest Lyapunov exponent.

In the averaged Itô equations (16) for Gaussian white noise excitation, the terms $\varepsilon U_r(\mathbf{I};\tau)$ and $\varepsilon^{1/2}V_{rk}(\mathbf{I};\tau)$ may be linear or homogeneous of degree one in I_r. If this is not the case, then $\varepsilon U_r(\mathbf{I};\tau)$ and $\varepsilon^{1/2}V_{rk}(\mathbf{I};\tau)$ are linearized with respect to I_r at $\mathbf{I}=0$. The linearized (or homogenous of degree one) equations will be of the form

$$\begin{gathered} dI_r = \varepsilon F_r(\mathbf{I};\tau)dt + \varepsilon^{1/2}G_{rk}(\mathbf{I};\tau)dB_k(t), \\ r=1,2,\ldots,n; \quad k=1,2,\ldots,m, \end{gathered} \tag{97}$$

where

$$\begin{gathered} F_r(\mathbf{0};\tau) = G_{rk}(\mathbf{0};\tau) = 0, \\ nF_r(\mathbf{I};\tau) = F_r(n\mathbf{I};\tau), \quad nG_{rk}(\mathbf{I};\tau) = G_{rk}(n\mathbf{I};\tau), \quad n>0, \\ r=1,2,\ldots,n; \quad k=1,2,\ldots,m. \end{gathered} \tag{98}$$

Assumption that the diffusion matrix of Itô equation (97) is non-singular, i.e.,

$$(\mathbf{G}(\mathbf{I};\tau)\mathbf{G}(\mathbf{I};\tau)^T\mathbf{a},\mathbf{a}) \geq c\,|\mathbf{I}|^2\,|\mathbf{a}|^2\,, \tag{99}$$

where $\mathbf{a}$ is an arbitrary vector and $c>0$ is a scalar, $|\cdot|$means the norm of a vector. Under this assumption, we can derive the expression for the largest Lyapunov exponent for the averaged Itô equation (97) following a procedure similar to that due to Khasminskii. To this end, introduce the following new variables:

$$\rho = \frac{1}{2}\log I, \quad I = \sum_{r=1}^{n} I_r, \tag{100}$$

$$\alpha_r = I_r/I, \quad r=1,2,\ldots,n. \tag{101}$$

The Itô equations for ρ and α_r can be obtained from Eq. (97) by using Itô differential rule as follows:

$$d\rho = \varepsilon Q(\mathbf{a};\tau)dt + \varepsilon^{1/2}\Sigma_k(\mathbf{a};\tau)dB_k(t), \tag{102}$$

$$d\alpha_r = \varepsilon m_r(\mathbf{a};\tau)dt + \varepsilon^{1/2}\sigma_{rk}(\mathbf{a};\tau)dB_k(t), \tag{103}$$

where $\mathbf{a} = [\alpha_1, \alpha_2, \ldots, \alpha_n]^T$ and

$$Q(\mathbf{a};\tau) = \frac{1}{2}\sum_{s=1}^{n} F_r(\mathbf{a};\tau) - \frac{1}{4}\sum_{s,s'=1}^{n}\sum_{u=1}^{m} G_{su}(\mathbf{a};\tau)G_{s'u}(\mathbf{a};\tau), \tag{104}$$

$$\begin{aligned} m_r(\mathbf{a};\tau) = & -\alpha_r\sum_{s=1}^{n} F_s(\mathbf{a};\tau) + F_r(\mathbf{a};\tau) + \frac{1}{2}\alpha_r\sum_{s,s'=1}^{n}\sum_{u=1}^{m} G_{su}(\mathbf{a};\tau)G_{s'u}(\mathbf{a};\tau) \\ & -\frac{1}{2}\sum_{s=1}^{n}\sum_{u=1}^{m} G_{ru}(\mathbf{a};\tau)G_{su}(\mathbf{a};\tau), \end{aligned} \tag{105}$$

$$\sigma_{ru}(\mathbf{a};\tau) = G_{ru}(\mathbf{a};\tau) - \alpha_r\sum_{s=1}^{n} G_{su}(\mathbf{a};\tau). \tag{106}$$

Note that $\sum_{r=1}^{n}\alpha_r = 1$, so only $n-1$ equations for α_r in Eq.(103) are independent. In the following we take the first $n-1$ equations for $\alpha_1, \alpha_2, \ldots, \alpha_{n-1}$ as independent equations. Let $\mathbf{a}' = [\alpha_1, \alpha_2, \ldots, \alpha_{n-1}]$ be an $(n-1)$-dimensional vector diffusing process with α_n replaced by $\alpha_n = 1 - \sum_{r=1}^{n-1}\alpha_r$.

Integrating Eq.(102) from 0 to t and dividing it by t, we obtain

$$\frac{1}{2t}\log I(t) = \frac{1}{2t}\log I(0) + \frac{1}{t}\int_0^t \varepsilon Q[\mathbf{a}'(T)]dT + \frac{1}{t}\int_0^t \varepsilon^{1/2}\Sigma_k[\mathbf{a}'(T)]dB_k(T). \tag{107}$$

As $t \to \infty$, the first term on the right hand side of Eq.(107) vanishes. The last term also vanishes, since $\Sigma_k[\mathbf{a}'(T)]$ is bounded while $B_k(T)$ grows as $[T\log(\log T)]^{1/2}$.

The Lyapunov exponent of averaged system (97) is defined as the asymptotic rate of exponential growth of the square-root of I. Thus,

$$\lambda = \lim_{t\to\infty}\frac{1}{2t}\log I = \lim_{t\to\infty}\frac{1}{t}\int_0^t \varepsilon Q[\mathbf{a}'(T);\tau]dT. \tag{108}$$

Assume that (n-1)-dimensional vector diffusion process $\mathbf{a}'$ is ergodic in the interval $0 \le \alpha_r \le 1$ $(r = 1, 2, \ldots, n-1)$. In this case, the time averaging in Eq.(108) can be replaced by assembly averaging and λ is the largest Lyapunov exponent, i.e.,

$$\lambda_{\max} = \varepsilon\int Q(\mathbf{a}';\tau)p(\mathbf{a}';\tau)d\mathbf{a}', \tag{109}$$

where $p(\mathbf{a}';\tau)$ is the invariant measure (stationary probability density) of $\mathbf{a}'$.

Based on Oseledec multiplicative ergodic theorem, the necessary and sufficient condition for the asymptotic Lyapunov stability with probability one of the trivial solution $I_i = 0$ $(i = 1, 2, \ldots, n)$ of averaged system (97) is $\lambda_{\max} < 0$. Considering the limit theorem due to Khasminskii, $\lambda_{\max} < 0$ is approximately the necessary and sufficient condition for the asymptotic Lyapunov stability with probability one of the trivial solution $q_i = p_i = 0$ $(i = 1, 2, \ldots, n)$ of original system (1). The boundary between stable and unstable regions of system (1) in parametric space is determined approximately by $\lambda_{\max} = 0$.

Example for Gaussian White Noise Excitation[18]

Consider two linear oscillators coupled by nonlinear dampings subjected to time-delayed feedback control forces and Gaussian white noise parametric excitations. The equations of the system are of the form

$$\begin{aligned} &\ddot{X}_1 + \alpha'_{11}\dot{X}_1 + \alpha'_{12}\dot{X}_2 + \beta_1\dot{X}_1(X_1^2 + X_2^2) + \omega_1'^2 X_1 \\ &= -\eta_1\dot{X}_{1\tau} + k_{11}X_1W_1(t) + k_{12}X_2W_2(t), \\ &\ddot{X}_2 + \alpha'_{21}\dot{X}_1 + \alpha'_{22}\dot{X}_2 + \beta_2\dot{X}_2(X_1^2 + X_2^2) + \omega_2'^2 X_2 \\ &= -\eta_2\dot{X}_{2\tau} + k_{21}X_1W_1(t) + k_{22}X_2W_2(t), \end{aligned} \tag{110}$$

where X_i are generalized coordinates; α'_{ij} and β_i are damping coefficients; k_{ij} are constants; ω'_i are the natural frequencies of the two linear oscillators; $-\eta_i\dot{X}_{i\tau}$ represent the time-delayed feedback control forces, where η_i are control gains and $\dot{X}_{i\tau} = \dot{X}_i(t-\tau)$ with time delay τ; $W_j(t)$ are independent Gaussian white noises in the senses of Stratonovich with intensities $2D_j$. Assume that α'_{ij}, β_i, D_j, and η_i are of the order ε.

Following Eq.(8), the time-delayed feedback control forces in system (110) can be expressed in terms of system state variables without time delay as follows:

$$\eta_i\dot{X}_{i\tau} = \eta_i\dot{X}\cos(\omega'_i\tau) + \eta_i\omega'_i X\sin(\omega'_i\tau). \tag{111}$$

With these approximations, equation (110) can be converted into Itô stochastic differential equation (15) with

$$\begin{aligned} &H = H_1 + H_2, \quad H_i = \left(\dot{X}_i^2 + \omega_i^2 X_i^2\right)/2, \\ &[\varepsilon m_{ij}] = \begin{bmatrix} \alpha_{11} + \beta_1(X_1^2 + X_2^2) & \alpha'_{12} \\ \alpha'_{21} & \alpha_{22} + \beta_2(X_1^2 + X_2^2) \end{bmatrix}, \\ &\left[\varepsilon^{1/2}\sigma_{ik}\right] = \begin{bmatrix} k_{11}X_1 & k_{12}X_2 \\ k_{21}X_1 & k_{22}X_2 \end{bmatrix}, \\ &\omega_i^2 = \omega_i'^2 + \eta_i\omega'_i\sin\omega'_i\tau, \quad \alpha_{ii} = \alpha'_{ii} + \eta_i\cos\omega'_i\tau, \quad i,j,k = 1,2. \end{aligned} \tag{112}$$

In the case where $k_1\omega_1 + k_2\omega_2 \neq o(\varepsilon)$, where k_1, k_2 are integers, this quasi integrable Hamiltonian system is non-resonant. The averaged Itô equation of the form of Eq.(16) can be derived and it is nonlinear. The linearized averaged Itô equations are of the form

$$\begin{aligned} dI_1 &= F_1(\mathbf{I};\tau)dt + G_{11}(\mathbf{I};\tau)dB_1(t) + G_{12}(\mathbf{I};\tau)dB_2(t), \\ dI_2 &= F_2(\mathbf{I};\tau)dt + G_{21}(\mathbf{I};\tau)dB_1(t) + G_{22}(\mathbf{I};\tau)dB_2(t), \end{aligned} \tag{113}$$

where

$$F_1(\mathbf{I};\tau) = F_{11}(\tau)I_1 + F_{12}(\tau)I_2, \quad F_2(\mathbf{I};\tau) = F_{21}(\tau)I_1 + F_{22}(\tau)I_2,$$

$$[GG^T]_{11} = b_{11}(\mathbf{I};\tau) = b_{11}^{(1)}(\tau)I_1^2 + b_{11}^{(2)}(\tau)I_1I_2, \quad [GG^T]_{12} = b_{12}(\mathbf{I};\tau) = 0,$$

$$[GG^T]_{22} = b_{22}(\mathbf{I};\tau) = b_{22}^{(1)}(\tau)I_1I_2 + b_{22}^{(2)}(\tau)I_2^2, \quad [GG^T]_{21} = b_{21}(\mathbf{I};\tau) = 0,$$

$$F_{11}(\tau) = -\alpha_{11} + \frac{D_{11}k_{11}^2}{\omega_1^2}, \quad F_{12}(\tau) = \frac{D_{22}k_{12}^2}{\omega_1\omega_2}, \quad F_{21}(\tau) = \frac{D_{11}k_{21}^2}{\omega_1\omega_2}, \tag{114}$$

$$F_{22}(\tau) = -\alpha_{22} + \frac{D_{22}k_{22}^2}{\omega_2^2}, \quad b_{11}^{(1)}(\tau) = \frac{D_{11}k_{11}^2}{\omega_1^2}, \quad b_{11}^{(2)}(\tau) = \frac{2D_{22}k_{12}^2}{\omega_1\omega_2},$$

$$b_{22}^{(1)}(\tau) = \frac{2D_{11}k_{21}^2}{\omega_1\omega_2}, \quad b_{22}^{(2)}(\tau) = \frac{D_{22}k_{22}^2}{\omega_2^2}.$$

Defining ρ and α_1 as those in Eqs.(102) and (103), the Itô equations for ρ and α_1 are

$$d\rho = Q(\alpha_1;\tau)dt + \Sigma(\alpha_1;\tau)dB(t), \tag{115}$$

$$d\alpha_1 = m(\alpha_1;\tau)dt + \sigma(\alpha_1;\tau)dB(t), \tag{116}$$

where

$$\begin{aligned}
&Q(\alpha_1;\tau) = \frac{1}{4}\varphi(\alpha_1) + \frac{1}{2}\alpha_1(\lambda_1 - \lambda_2) + \frac{1}{2}\lambda_2,\\
&\sigma^2(\alpha_1;\tau) = \alpha_1(1-\alpha_1)\varphi(\alpha_1;\tau),\\
&m(\alpha_1;\tau) = \left(\frac{1}{2} - \alpha_1\right)\varphi(\alpha_1) + \alpha_1(1-\alpha_1)(\lambda_1 - \lambda_2),\\
&\varphi(\alpha_1;\tau) = a\alpha_1^2 + b\alpha_1 + c,\\
&\qquad a = b_{11}^{(2)} + b_{22}^{(1)} - b_{11}^{(1)} - b_{22}^{(2)}, \quad b = b_{11}^{(1)} + b_{22}^{(2)} - 2b_{11}^{(2)}, \quad c = b_{11}^{(2)},\\
&\qquad \lambda_1 = F_{11} - \frac{1}{2}b_{11}^{(1)}, \quad \lambda_2 = F_{22} - \frac{1}{2}b_{22}^{(2)}.
\end{aligned} \tag{117}$$

Generally, $\sigma(\alpha_1;\tau)$ does not vanish in interval $0 < \alpha_1 < 1$ and vanishes at boundaries $\alpha_1 = 0, 1$. It can be shown that both the two boundaries are entrance. So, diffusion process α_1 is ergodic throughout the interval $0 \le \alpha_1 \le 1$. The invariant measure is obtained from solving the reduced Fokker-Planck equation associated with Itô equation (116) as follows:

$$p(\alpha_1;\tau) = \frac{C}{\varphi(\alpha_1)}F(\alpha_1;\tau), \tag{118}$$

where C is a normalization constant with its value

$$C = \frac{2(\lambda_1 - \lambda_2)}{F(1) - F(0)} \tag{119}$$

and

$$
\begin{aligned}
&F(\alpha_1;\tau) = \exp\left[\frac{2(\lambda_1-\lambda_2)}{\sqrt{\Delta}}\ln\left(\frac{2a\alpha_1+b-\sqrt{\Delta}}{2a\alpha_1+b+\sqrt{\Delta}}\right)\right], \quad \Delta>0,\\
&F(\alpha_1;\tau) = \exp\left[\frac{4(\lambda_1-\lambda_2)}{\sqrt{-\Delta}}\arctan\left(\frac{2a\alpha_1+b}{\sqrt{-\Delta}}\right)\right], \quad \Delta<0,\\
&F(\alpha_1;\tau) = \exp\left[\frac{-4(\lambda_1-\lambda_2)}{2a\alpha_1+b}\right], \quad \Delta=0,\\
&\qquad \Delta = b^2-4ac.
\end{aligned}
\tag{120}
$$

The largest Lyapunov exponent is then obtained using formula (109) as follows:

$$
\lambda_{\max} = \int_0^1 Q(\alpha_1;\tau)p(\alpha_1;\tau)d\alpha_1 = \frac{1}{2}\frac{F(1)\lambda_1 - F(0)\lambda_2}{F(1)-F(0)}. \tag{121}
$$

Numerical results have been obtained for system (110) with parameter values: $\alpha'_{11} = 0.02, \alpha'_{12} = 0.01, \alpha'_{21} = 0.01, \alpha'_{22} = 0.01, \beta_1 = 0.01, \beta_2 = 0.01, k_{11} = 1, k_{12} = 1, k_{21} = -1, k_{22} = 1, \omega'_1 = 1, \omega'_2 = 1.414, \eta_1 = 0.03, \eta_2 = 0.03, D_{11} = 0.02, D_{22} = 0.02$ unless otherwise mentioned, and given in Figs.14–16. The largest Lyapunov exponent $\lambda_{\max}$ versus stochastic excitation intensity $D_{11} = D_{22} = D$ for different time delay τ of feedback control forces is shown in Fig.14. It is seen that the largest Lyapunov exponent $\lambda_{\max}$ increases as the excitation intensity D or time delay τ increases. Note that $\lambda_{\max}$ is a measure of the asymptotic Lyapunov stability with probability one of system (110). It is stable if $\lambda_{\max} < 0$ and unstable if $\lambda_{\max} > 0$. Obviously, for larger time delay, such as $\tau = 1.5$, the system is always unstable no mater what excitation intensity D is.

The largest Lyapunov exponent $\lambda_{\max}$ versus control gain $\eta_1 = \eta_2 = \eta$ for different time delay τ of feedback control forces is shown in Fig.15. It is also seen that time delay τ remarkably deteriorates system stability. For smaller time delay τ, such as $\tau \le 1$, $\lambda_{\max}$ decreases as control gain η increases but for larger time delay, such as $\tau = 1.5$, $\lambda_{\max}$ even increases as η increases.

The stable region of system (110) in damping parameter plane (α'_{11}, α'_{22}) for different time delay τ is shown in Fig.16. Note that the boundary of stable region is determined by $\lambda_{\max} = 0$. It is seen that the stable region reduces as τ increases.

To verify the largest Lyapunov exponent calculated by using Eq.(109) and shown in Figs.14–16 by solid lines, it is also obtained by using digital simulation of Eq.(110) and shown in Figs.14–16 by using symbols ● ◆ ▲ ▼. It is seen that the two results well coincide.

Example for Wide-band Random Excitation[19]

Consider two linear oscillators coupled by linear and nonlinear dampings subject to parametric excitations of two wide-band random excitations and two time-delayed

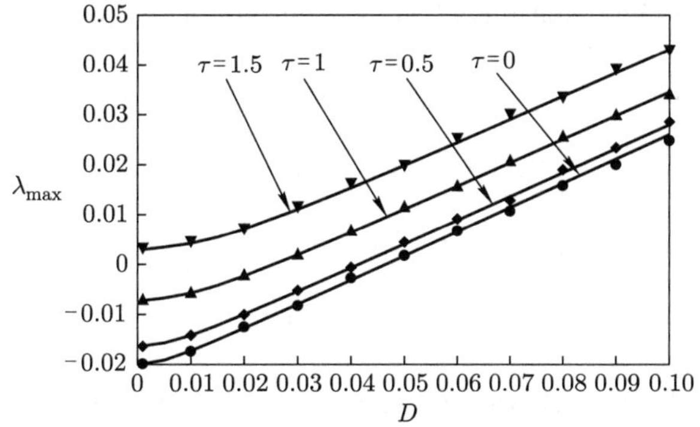

Fig. 14. The largest Lyapunov exponent λ_{max} of system (110) versus noise intensity $D_{11} = D_{22} = D$ for different values of time delay τ. — Analytical result. ● ◆ ▲ ▼ Digital simulation.

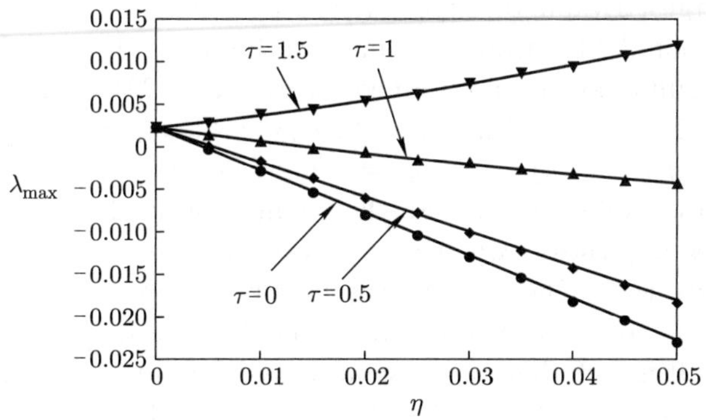

Fig. 15. The largest Lyapunov exponent λ_{max} of system (110) versus control gain $\eta_1 = \eta_2 = \eta$ for different values of time delay τ. — Analytical result. ● ◆ ▲ ▼ Digital simulation.

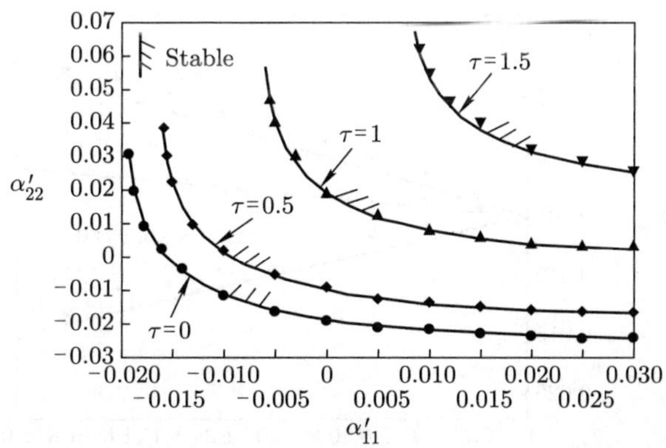

Fig. 16. Stability boundaries in the plane of $\alpha'_{11} - \alpha'_{22}$ for different values of time delay τ. — Analytical result. ● ◆ ▲ ▼ Digital simulation.

feedback controls. The equations of the system are of the form

$$\begin{aligned}
&\ddot{X}_1 + \varepsilon[\alpha'_{11}\dot{X}_1 + \alpha_{12}\dot{X}_2 + \beta_1\dot{X}_1(X_1^2 + X_2^2)] + \omega_1'^2 X_1 \\
&= u_1 + \varepsilon^{1/2}[X_1\xi_1(t) + X_2\xi_2(t)], \\
&\ddot{X}_2 + \varepsilon[\alpha_{21}\dot{X}_1 + \alpha'_{22}\dot{X}_2 + \beta_2\dot{X}_2(X_1^2 + X_2^2)] + \omega_2'^2 X_2 \\
&= u_2 + \varepsilon^{1/2}[X_2\xi_1(t) + X_1\xi_2(t)],
\end{aligned} \tag{122}$$

where α'_{ii}, α_{ij}, β_i, η_i, ω'_i are constants; $u_1 = -\varepsilon(\eta_{11}X_{1\tau_1} + \eta_{12}\dot{X}_{1\tau_2})$, $u_2 = -\varepsilon(\eta_{21}X_{2\tau_1} + \eta_{22}\dot{X}_{2\tau_2})$ are the time-delayed feedback control forces; $X_{i\tau_1} = X_i(t-\tau_1)$ are time-delayed displacement, $\dot{X}_{i\tau_2} = \dot{X}_i(t-\tau_2)$ are time-delayed velocity; $\xi_i(t)$ are independent wide-band random processes with spectral densities of the form in Eq.(27). The largest Lyapunov exponent can be calculated from the averaged Itô equations in Section 2.2 by using the similar technique as that in the last example.

Numerical results have been obtained for system (122) with parameter values: $\varepsilon = 0.01$, $\omega'_1 = 1$, $\omega'_2 = 1.414$, $\alpha'_{11} = 2$, $\alpha_{12} = 2$, $\alpha_{21} = 2$, $\alpha'_{22} = 2$, $\beta_1 = 2$, $\beta_2 = 2$, $\eta_{11} = 2$, $\eta_{12} = 2$, $\eta_{21} = 2$, $\eta_{22} = 2$ unless otherwise mentioned, and given in Figs.17 and 18. Solid lines denote the results obtained by using the proposed approach while • ∘ ∗ those from Monte Carlo simulation. It is seen that the two results agree quite well. The largest Lyapunov exponent $\lambda_{\max}$ versus control gain $\eta_{11} = \eta_{12} = \eta_{21} = \eta_{22} = \eta$ for different delay times in feedback control forces is shown in Fig.17. It is seen that $\lambda_{\max}$ increases as delay time increases and as control gain η increases for larger time delay $\tau_1 = \tau_2 = 1$ and $\tau_1 = \tau_2 = 2$. The stable region of system (122) in damping parameter plane (α'_{11}, α'_{22}) is shown in Fig.18 for $\tau_1 = \tau_2 = 1$.

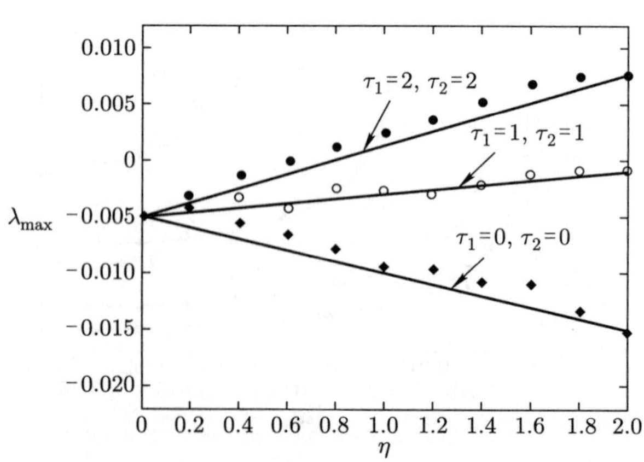

Fig. 17. The largest Lyapunov exponent $\lambda_{\max}$ of system (122) versus control gain $\eta_{11} = \eta_{12} = \eta_{21} = \eta_{22} = \eta$ for different values of time delay τ_1, τ_2.

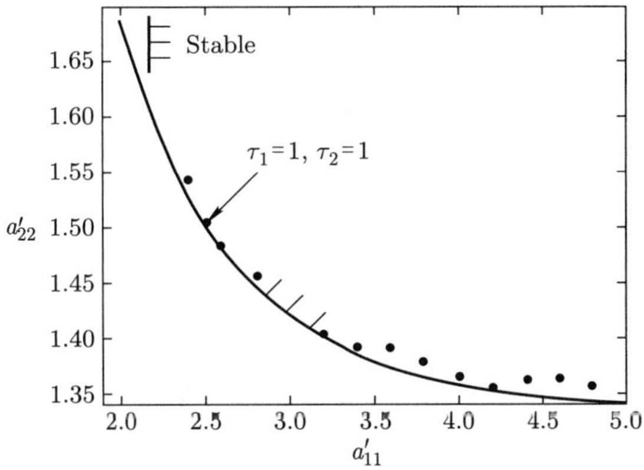

Fig. 18. Stability boundaries of the system (122) in the plane $(\alpha'_{11}, \alpha'_{22})$ for time delay $\tau_1 = 1$, $\tau_2 = 1$.

Example for Narrow-band Bounded Noise Excitation[20]

Consider a Duffing oscillator with time-delayed feedback control under bounded noise parametric excitation. The equation of the system is of the form

$$\ddot{X} + \omega_0'^2 X + \alpha X^3 = -\beta' \dot{X} + EX\xi(t) + u_\tau, \tag{123}$$

where ω'_0, α, β', E are positive constants denoting the natural frequency of degenerated linear oscillator, intensity of nonlinearity, damping coefficient, amplitude of excitation, respectively; $u_\tau = -\eta\dot{X}_\tau = -\eta\dot{X}(t-\tau)$ is time-delayed feedback control force. It is assumed that β', E, η are of the same small order. $\xi(t)$ is a narrow-band bounded noise excitation as in Eq.(35). The largest Lyapunov exponent can be calculated from the averaged Itô equation in Section 2.3 by using similar technique.

Some results have been obtained for system (123) with parameter values $\omega'_0 = 1$, $\alpha = 0.3$, $\beta' = 0.01$, $E = 0.08$, $\Omega = 2$, $\sigma^2 = 0.09$, $\eta = 0.05$ unless otherwise mentioned. The largest Lyapunov exponent λ versus control gain η for different time delay τ is shown in Fig.19. It is seen that time delay τ remarkably deteriorates the system stability. For smaller time delay τ, such as $\tau = 0.5$, λ decreases as control gain η increases but for larger time delay, such as $\tau = 2$, λ even increases as η increases.

The stable region of system (123) in plane $(\Omega/2\omega'_0, E)$ for different time delay τ is shown in Fig.20. It is seen that the stable region reduces as τ increases, and for larger time delay, such as $\tau = 2$, there is even no stable region.

The results from digital simulation are also obtained and shown in Figs.19 and 20 to verify the theoretical results. It is seen that the two results well coincide.

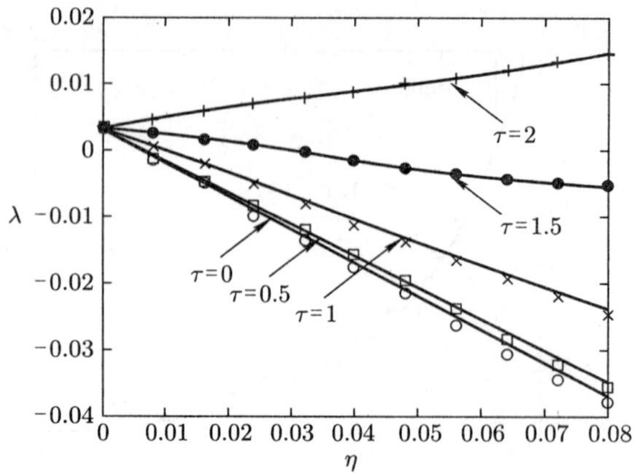

Fig. 19. The largest Lyapunov exponent λ of system (123) versus control gain η for different values of time delay τ. — results obtained by using the proposed method; ○ □ × • + results from digital simulation.

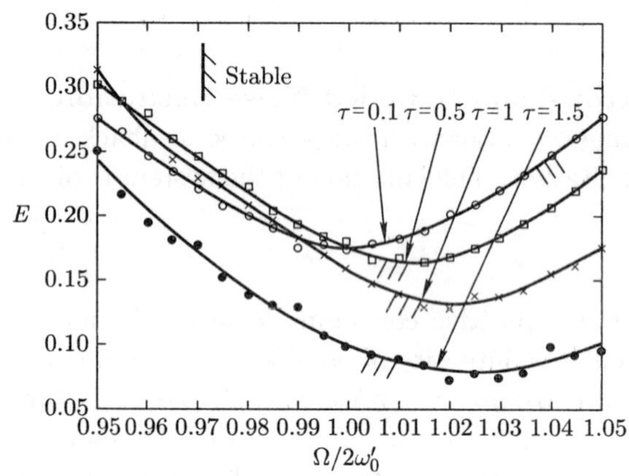

Fig. 20. Stability boundaries in the plane $(\Omega/2\omega_0', E)$ of system (123) for different values of time delay τ. — results obtained by using the proposed method; ○ □ × • results from digital simulation.

Example for Combined Excitations of Harmonic Function and Gaussian White Noise[21]

Consider a Duffing-Mathieu system with time-delayed feedback control under white noise parametric excitation. The equation of the system is of the form

$$\ddot{X} + \omega_0'^2 X + \alpha X^3 = -\beta' \dot{X} + EX\cos\Omega t + X\xi(t) + u_\tau, \tag{124}$$

where ω_0', α, β', E, Ω are positive constants denoting the natural frequency of degenerated linear oscillator, intensity of nonlinearity, damping coefficient, amplitude and frequency of harmonic excitation, respectively; $\xi(t)$ is Gaussian white noise in the sense of Stratonovich with intensity $2D$. $u_\tau = -\eta\dot{X}_\tau = -\eta\dot{X}(t-\tau)$ is time-delayed feedback control force. It is assumed that β', E, D, η are of the same small order. The largest Lyapunov exponent can be calculated from the averaged Itô equation in Section 2.4 by using similar technique.

Some results have been obtained for system (124) with parameter values: $\omega_0' = 1$, $\alpha = 0.3$, $\beta' = 0.01$, $E = 0.05$, $\Omega = 2$, $D = 0.05$, $\eta = 0.03$, $X(0) = 0.001$, $\dot{X}(0) = 0.001$ unless otherwise mentioned, and given in Figs.21–23. The largest Lyapunov exponent λ versus amplitude of harmonic excitation E for different time delay τ is shown in Fig.21. It is seen that the largest Lyapunov exponent λ increases as amplitude of harmonic excitation E or time delay τ increases. For larger time delay, such as $\tau = 1.5$, the system is always unstable no matter what amplitude of harmonic excitation E is.

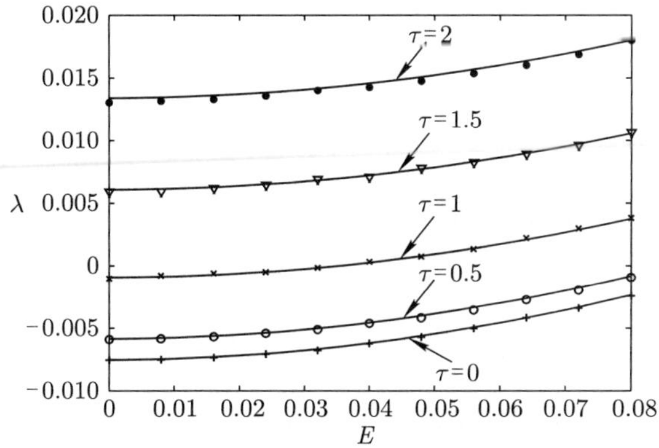

Fig. 21. The largest Lyapunov exponent λ of system (124) versus amplitude E of harmonic excitation for different values of time delay τ. — results obtained by using the proposed method; $+ \circ \times \nabla \bullet$ results from digital simulation of system (124).

The stable region of system (124) in plane $(\Omega/2\omega_0', E)$ for different time delay τ is shown in Fig.22. It is seen that the stable region reduces as τ increases, and for larger time delay, such as $\tau = 1.5$, there is even no stable region.

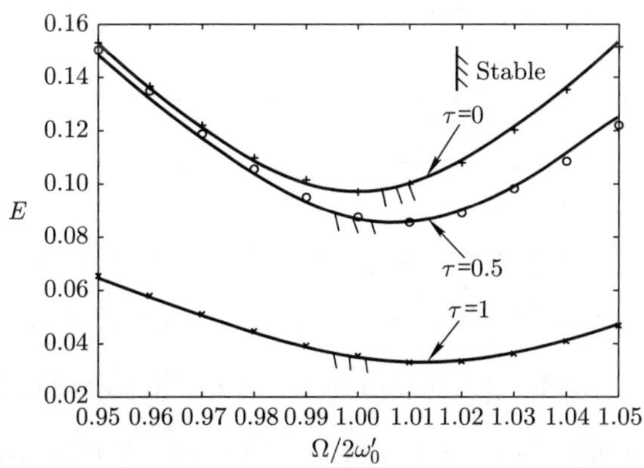

Fig. 22. Stability boundaries in the plane $(\Omega/2\omega_0', E)$ of system (124) for different values of time delay τ. — results obtained by using the proposed method; $+ \circ \times$ results from digital simulation of system (124).

With $\tau = 2$, $\lambda = 0.015225 > 0$ and the trivial solution of the system is unstable. It is verified in Fig.23 by using a typical sample of original system (124).

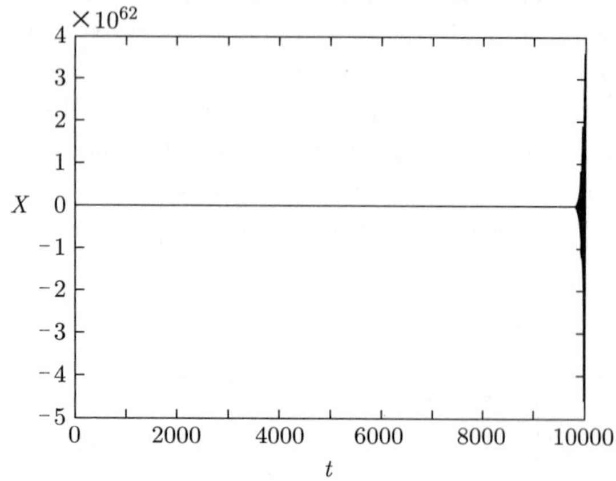

Fig. 23. A sample of displacement $X(t)$ from digital simulation of the original system (124) for $\tau = 2$.

3.3. *Stochastic Bifurcation*[23]

The theory of stochastic bifurcation studies the qualitative changes in the asymptotic behavior of random dynamical systems when system parameters change. The stochastic bifurcation can be classified into dynamical bifurcation (D-bifurcation) and phenomenological bifurcation (P-bifurcation). D-bifurcation is related to sign change of the Lyapunov exponent and it reduces to the deterministic bifurcation in the absence of random noise. P-bifurcation is associated with qualitative change of the stationary probability density of system response, e.g., from unimodal to bimodal or multi-modal or vice versa. The stochastic bifurcation has been studied by using the proposed stochastic averaging method for quasi integrable Hamiltonian system with time delay.

For studying stochastic Hopf bifurcation, introduce the following new variable

$$\alpha_r = \frac{H_r}{H}, \quad r = 1, 2, \ldots, n. \tag{125}$$

Note that $\sum_{r=1}^{n} \alpha_r = 1$, so only $n-1$ variables for α_r in Eq.(125) are independent. In the following, we take the first $n-1$ variables for $\mathbf{a}' = [\alpha_1, \alpha_2, \ldots, \alpha_{n-1}]$ as independent variables with α_n replaced by $\alpha_n = 1 - \sum_{r=1}^{n-1} \alpha_r$. The Itô equastions for H and α_r in non-resonant case can be obtained from Eq.(19) by using Itô differential rule as follows:

$$dH = Q(\mathbf{a}', H; \tau)dt + \Sigma_k(\mathbf{a}', H; \tau)d\bar{B}_k(t), \quad r = 1, 2, \ldots, n-1, \tag{126}$$

$$\begin{aligned} d\alpha_r &= \tilde{m}_r(\mathbf{a}', H; \tau)dt + \tilde{\sigma}_{rk}(\mathbf{a}', H; \tau)d\bar{B}_k(t), \\ r &= 1, 2, \ldots, n-1; \quad k = 1, 2, \ldots, m, \end{aligned} \tag{127}$$

where

$$\begin{aligned} Q(\mathbf{a}', H; \tau) &= \varepsilon \sum_{r=1}^{n} \bar{m}_r(\mathbf{a}', H; \tau), \\ \Sigma_k(\mathbf{a}', H; \tau) &= \varepsilon^{1/2} \sum_{r=1}^{n} \bar{\sigma}_{rk}(\mathbf{a}', H; \tau), \\ \tilde{m}_r(\mathbf{a}', H; \tau) &= \varepsilon \Bigg[-\alpha_r \sum_{s=1}^{n} \bar{m}_s(\mathbf{a}', H; \tau)/H + \bar{m}_r(\mathbf{a}', H; \tau)/H \\ &\quad - \frac{1}{2} \sum_{s=1}^{n} \sum_{k=1}^{m} \bar{\sigma}_{rk}(\mathbf{a}', H; \tau) \bar{\sigma}_{sk}(\mathbf{a}', H; \tau)/H^2 \\ &\quad + \frac{1}{2} \alpha_r \sum_{s,s'=1}^{n} \sum_{k=1}^{m} \sigma_{sk}(\mathbf{a}', H; \tau) \bar{\sigma}_{s'k}(\mathbf{a}', H; \tau)/H^2 \Bigg], \\ \tilde{\sigma}_{rk}(\mathbf{a}', H; \tau) &= \varepsilon^{1/2} \left[\bar{\sigma}_{rk}(\mathbf{a}', H; \tau)/H - \alpha_r \sum_{s=1}^{n} \bar{\sigma}_{sk}(\mathbf{a}', H; \tau)/H \right]. \end{aligned} \tag{128}$$

Stochastic Hopf bifurcation consists of a D-bifurcation and a P-bifurcation. Before D-bifurcation the trivial solution is stable and the stationary probability density is a delta function. After D-bifurcation and before P-bifurcation the trivial solution is unstable and the stationary probability density has a peak at the origin. After P-bifurcation the trivial solution is still unstable and the stationary probability density has a peak away from the origin. For one-dimensional diffusion process $H(t)$ governed by Eq.(126), the boundary $H = \infty$ must be either an entrance or a repulsively natural in order that the trivial solution $H = 0$ is stable in probability or $H(t)$ has a stationary probability density, i.e., the boundary $H = \infty$ must be either an entrance or repulsively natural during the first and second bifurcations. In the following we will focus our attention on the qualitative change in sample behavior of $H(t)$ near the boundary $H = 0$ during the first and second bifurcation.

For the one-dimensional diffusion process reduced from higher-dimensional systems under parametric excitations by using the stochastic averaging, the boundary $H = 0, \infty$ are often singular and the sample behaviors of the process near the boundaries are characterized by the diffusion exponent, the drift exponent and the character value. For singular left boundary of the first kind, i.e., $\Sigma_k(\mathbf{a}, 0; \tau) = 0$, the diffusion exponent α_l, the drift exponent β_l and the character value c_l are defined as follows:

$$b'(\mathbf{a}', H; \tau) = (\Sigma_k(\mathbf{a}', H; \tau))^2 = O(H^{\alpha_l}), \quad \alpha_l > 0 \text{ as } H \to 0, \tag{129}$$

$$Q(\mathbf{a}', H; \tau) = O(H^{\beta_l}), \quad \beta_l > 0 \text{ as } H \to 0, \tag{130}$$

$$c_l(\mathbf{a}'; \tau) = \lim_{H \to 0^+} \frac{2Q(\mathbf{a}', H; \tau) H^{\alpha_l - \beta_l}}{b'(\mathbf{a}', H; \tau)}, \tag{131}$$

where $O(\cdot)$denotes the order of magnitude of $(\cdot)$. For singular right boundary of the second kind, i.e., $\tilde{m}(\infty) = \infty$, the diffusion exponent α_r, the drift exponent β_r and the character value c_r are defined as follows:

$$b'(\mathbf{a}', H; \tau) = (\Sigma_k(\mathbf{a}', H; \tau))^2 = O(H^{\alpha_r}), \quad \alpha_r > 0 \text{ as } H \to \infty, \tag{132}$$

$$Q(\mathbf{a}', H; \tau) = O(H^{\beta_r}), \quad \beta_r > 0 \text{ as } H \to \infty, \tag{133}$$

$$c_r(\mathbf{a}'; \tau) = \lim_{H\to\infty} \frac{2Q(\mathbf{a}', H; \tau)H^{\alpha_r - \beta_r}}{b'(\mathbf{a}', H; \tau)}. \tag{134}$$

The following asymptotic expression for the stationary probability density of $H(t)$can be obtained:

$$p(H; \tau) = O\left\{H^{-\alpha_l} \exp\left[c_l \int_0^H x^{(\beta_l - \alpha_l)} dx\right]\right\} \text{ as } H \to 0. \tag{135}$$

Two cases can be identified.

Case 1. $\beta_l - \alpha_l = -1$. In this case

$$p(H; \tau) = O\left(H^v\right), \text{ as } H \to 0 \tag{136}$$

with

$$v(\mathbf{a}'; \tau) = c_l(\mathbf{a}'; \tau) - \alpha_l. \tag{137}$$

Particularly, when $\beta_l = 1$ and $\alpha_l = 2$, the diffusion and drift coefficients in Eq.(127) are linear. Introduce an average bifurcation parameter $\bar{v}(\tau)$ defined by

$$\bar{v}(\tau) = \int_\Omega v(\mathbf{a}'; \tau)p(\mathbf{a}'; \tau)d\mathbf{a}', \quad \Omega = \left\{\mathbf{a}' \middle| \sum_{r=1}^n \alpha_i = 1, 0 \le \alpha_i \le 1\right\}, \tag{138}$$

where $p(\mathbf{a}'; \tau)$ is the stationary solution of the Fokker-Plank-Kolmogorov (FPK) equation associated with the Itô differential equations in Eq.(127).

Equation (136) is non-integrable and the probability density $p(H; \tau)$ is a delta function if $\bar{v}(\tau) < -1$. When $-1 < \bar{v}(\tau) < 0$, equation (136) is integrable and a stationary probability density $p(H; \tau)$ exists with a peak at $H = 0$. If $\bar{v}(\tau) > 0$, equation (136) is integrable and $p(H; \tau)$ exists with a peak away from $H = 0$. Thus, the first bifurcation (D-bifurcation) occurs at $\bar{v}(\tau) = -1$ and the second bifurcation (P-bifurcation) at $\bar{v}(\tau) = 0$ provided that the right boundary $H = \infty$ is an entrance or repulsively natural. It is interesting to note that the condition for the first bifurcation here is consistent with the necessary and sufficient condition for the asymptotic Lyapunov stability with probability one of the trivial solution.

Case 2. $\beta_l - \alpha_l \ne -1$. In this case

$$p(H) = O\left\{H^{-\alpha_l} \exp\left[\frac{c_l}{1 + \beta_l - \alpha_l} H^{(\beta_l - \alpha_l + 1)}\right]\right\} \text{ as } H \to 0, \tag{139}$$

which cannot be expressed in the form of Eq.(136). It can be shown that it is impossible for $p(H; \tau)$ to have a peak at or near $H = 0$ when it exists. In other words, in this case, although the first bifurcation may occur it is impossible for the second bifurcation to occur.

Thus, the stochastic Hopf bifurcation of a quasi integrable Hamiltonian system with time-delayed feedback control can be determined by examing the sample behavior of one-dimensional averaged diffusion process at its boundaries $H = 0$ and ∞.

Example[23]

To illustrate the above criterion for stochastic Hopf bifurcation, consider two coupled Rayleigh oscillators with time-delayed feedback control subject to parametric excitations of Gaussian white noises. The equations of the system are of the form

$$\begin{aligned} &\ddot{X}_1 + \left(-\beta'_{10} + \beta_{11}\dot{X}_1^2 + \beta_{12}\dot{X}_2^2\right)\dot{X}_1 + \omega_1'^2 X_1 \\ &= -\eta_1 \dot{X}_{1\tau} + f_{11}\dot{X}_1 W_1(t) + f_{12}\dot{X}_2 W_2(t), \\ &\ddot{X}_2 + \left(-\beta'_{20} + \beta_{21}\dot{X}_1^2 + \beta_{22}\dot{X}_2^2\right)\dot{X}_2 + \omega_2'^2 X_2 \\ &= -\eta_2 \dot{X}_{2\tau} + f_{21}\dot{X}_1 W_1(t) + f_{22}\dot{X}_2 W_2(t), \end{aligned} \tag{140}$$

where X_i are generalized coordinates; β'_{i0} and β_{ij} $(i, j = 1, 2)$ are damping coefficients; ω'_i are the natural frequencies of the two linear oscillators; $W_k(t)$ $(k = 1, 2)$ are independent Gaussian white noises with intensities $2D_{kk}$; $-\eta_i \dot{X}_{i\tau}$ represent the time-delayed feedback control forces. Here we study the effects of time delay in feedback control forces on the stochastic Hopf bifurcation of system (140).

Following Eq.(8), the time-delayed feedback control forces can be expressed in terms of system state variables without time delay as follows:

$$-\eta_i \dot{X}_{i\tau} = -\eta_i \dot{X}_i \cos\omega'_i\tau - \eta_i\omega'_i X_i \sin\omega'_i\tau, \quad i = 1, 2. \tag{141}$$

Let $X_1 = Q_1, X_2 = Q_2, \dot{X}_1 = P_1, \dot{X}_2 = P_2$. By applying the stochastic averaging method for quasi integrable Hamiltonian systems in non-resonant case, the following averaged Itô equations can be obtained:

$$\begin{aligned} dH_1 &= \left[(\beta_{10} + 2f_{11}^2 D_{11})H_1 - \frac{3}{2}\beta_{11}H_1^2 - \beta_{12}H_1H_2 + f_{12}^2 D_{22}H_2\right]dt \\ &\quad + \sigma_{11}dB_1(t) + \sigma_{12}dB_2(t), \\ dH_2 &= \left[(\beta_{20} + 2f_{22}^2 D_{22})H_2 - \frac{3}{2}\beta_{22}H_2^2 - \beta_{21}H_1H_2 + f_{21}^2 D_{11}H_1\right]dt \\ &\quad + \sigma_{21}dB_1(t) + \sigma_{22}dB_2(t), \end{aligned} \tag{142}$$

where

$$\begin{aligned} &H_i = (P_i^2 + \omega_i^2 Q_i^2)/2, \quad \omega_i^2 = \omega_i'^2 + \eta_i\omega'_i \sin\omega'_i\tau, \quad \beta_{i0} = \beta'_{i0} + \eta_i\cos\omega'_i\tau, \\ &b_{11} = \sigma_{1j}\sigma_{1j} = 3D_{11}f_{11}^2 H_1^2 + 2f_{12}^2 D_{22}H_1H_2, \quad b_{12} = b_{21} = \sigma_{1j}\sigma_{2j} = 0, \\ &b_{22} = \sigma_{2j}\sigma_{2j} = 3D_{22}f_{22}^2 H_2^2 + 2f_{21}^2 D_{11}H_1H_2. \end{aligned} \tag{143}$$

The Itô differential equations associated with $H = H_1 + H_2$ and $\alpha_1 = H_1/H$ can be obtained from Eq.(142) by using Itô differential rule as follows:

$$\begin{aligned} dH &= (Q_1 H + Q_2 H^2)dt + \Sigma_1 H dB_1(t), \\ d\alpha_1 &= \tilde{m}_1 dt + \tilde{\sigma}_1 dB_1(t), \end{aligned} \tag{144}$$

where

$$\begin{aligned} Q_1 &= (\beta_{10} + 2f_{11}^2 D_{11} + f_{21}^2 D_{11})\alpha_1 + (\beta_{20} + 2f_{22}^2 D_{22} + f_{12}^2 D_{22})(1-\alpha_1), \\ Q_2 &= -\frac{3}{2}\beta_{11}\alpha_1^2 - \frac{3}{2}\beta_{22}(1-\alpha_1)^2 - (\beta_{12}+\beta_{21})\alpha_1(1-\alpha_1), \\ \Sigma_1^2 &= 3f_{11}^2 D_{11}\alpha_1^2 + 3f_{22}^2 D_{22}(1-\alpha_1)^2 + 2(f_{21}^2 D_{11} + f_{12}^2 D_{22})\alpha_1(1-\alpha_1), \\ \tilde{m}_1 &= (1/2 - \alpha_1)\varphi(\alpha_1) + 2\alpha_1(1-\alpha_1)(\lambda_1 - \lambda_2), \quad \tilde{\sigma}_1^2 = \alpha_1(1-\alpha_1)\varphi(\alpha_1), \\ \varphi(\alpha_1) &= a\alpha_1^2 + b\alpha_1 + c, \quad a = G_{12} + G_{21} - G_{11} - G_{22}, \quad b = G_{11} + G_{22} - 2G_{12}, \\ c &= G_{12}, \quad \lambda_1 = \beta_{10}/2 + f_{11}^2 D_{11}/4, \quad \lambda_2 = \beta_{20}/2 + f_{22}^2 D_{22}/4, \\ G_{11} &= 3f_{11}^2 D_{11}, \quad G_{22} = 3f_{22}^2 D_{22}, \quad G_{12} = 2f_{12}^2 D_{22}, \quad G_{21} = 2f_{21}^2 D_{11}. \end{aligned} \tag{145}$$

For $H(t)$ governed by Eq.(144) at $H \to \infty$, the diffusion exponent $\alpha_r = 2$, the drift exponent $\beta_r = 2$. If $\beta_{ij} > 0\,(i,j = 1,2)$, then $Q_2 < 0$, the boundary $H = \infty$ is an entrance. At boundary $H = 0$, the diffusion exponent, draft exponent, and character value are

$$\alpha_l = 2, \quad \beta_l = 1, \quad c_l = \frac{2Q_1}{\Sigma_1^2} = c_l(\alpha_1;\tau), \quad v(\alpha_1;\tau) = c_l(\alpha_1;\tau) - 2. \tag{146}$$

The stationary solution $p(\alpha_1;\tau)$ of the FPK equation associated with the second equation of Itô equation (144) is

$$p(\alpha_1;\tau) = \frac{C}{\varphi(\alpha_1)}F(\alpha_1), \tag{147}$$

where

$$F(\alpha_1) = \begin{cases} \exp\left[\dfrac{4(\lambda_1-\lambda_2)}{\sqrt{\Delta}} \ln\left|\dfrac{2a\alpha_1 + b - \sqrt{\Delta}}{2a\alpha_1 + b + \sqrt{\Delta}}\right|\right], & \Delta > 0, \\ \exp\left[\dfrac{8(\lambda_1-\lambda_2)}{\sqrt{-\Delta}} \arctan\left|\dfrac{2a\alpha_1 + b}{\sqrt{-\Delta}}\right|\right], & \Delta < 0, \\ \exp\left[\dfrac{8(\lambda_1-\lambda_2)}{2a\alpha_1 + b}\right], & \Delta = 0, \end{cases} \tag{148}$$

$$C = \frac{4(\lambda_1-\lambda_2)}{F(1)-F(0)}, \quad \Delta = b^2 - 4ac.$$

The average bifurcation parameter $\bar{v}(\tau)$ can be obtained as follows:

$$\bar{v}(\tau) = \int_0^1 v(\alpha_1;\tau)p(\alpha_1;\tau)d\alpha_1. \tag{149}$$

The stochastic Hopf bifurcation of system (140) can be determined by using the average bifurcation parameter $\bar{v}(\tau)$. If $\bar{v}(\tau) < -1$, the probability density $p(H;\tau)$ is a delta function and the system is stable; if $-1 < \bar{v}(\tau) < 0$, the probability density $p(H;\tau)$ exists with a peak at $H = 0$; if $\bar{v}(\tau) > 0$, the probability density $p(H;\tau)$ exists with a peak away from $H = 0$. Thus, the first bifurcation (D-bifurcation) occurs at $\bar{v}(\tau) = -1$ and the second bifurcation (P-bifurcation) occurs at $\bar{v}(\tau) = 0$.

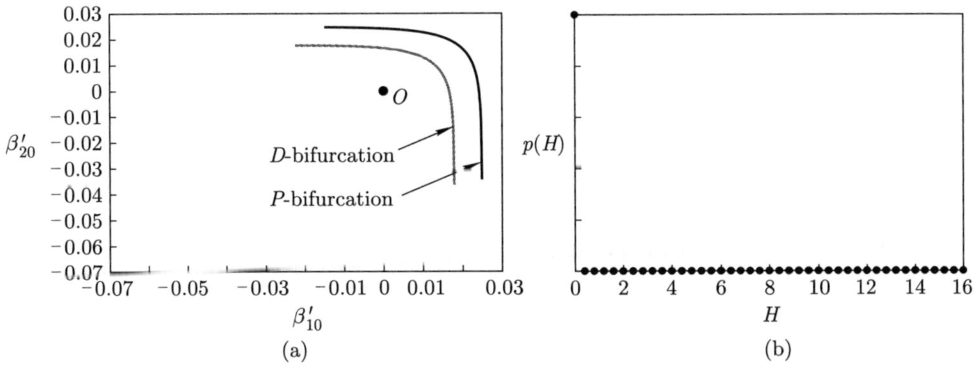

Fig. 24. Results for $\tau = 0$. (a) The D-bifurcation and P-bifurcation curves in plane $(\beta'_{10}, \beta'_{20})$ and point $O(0,0)$. (b) The stationary probability density $p(H)$ at point $O(0,0)$.

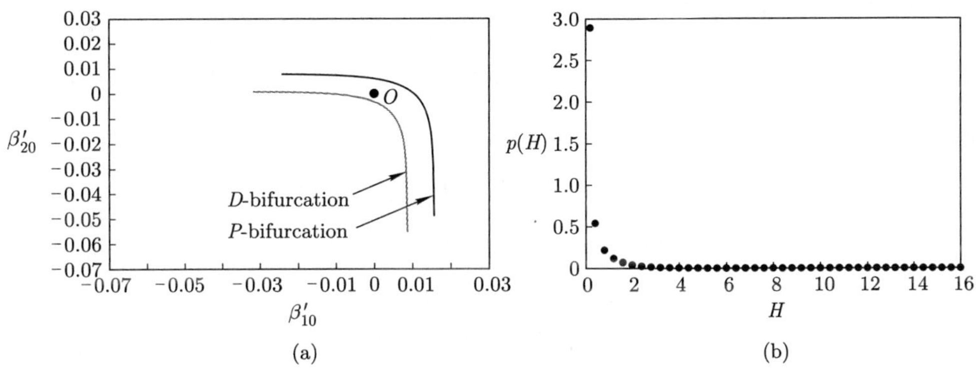

Fig. 25. Results for $\tau = 1$. (a) The D-bifurcation and P-bifurcation curves in plane $(\beta'_{10}, \beta'_{20})$ and point $O(0,0)$. (b) The stationary probability density $p(H)$ at point $O(0,0)$.

Some numerical results for the stochastic Hopf bifurcation of system (140) caused by time-delayed feedback control are shown in Figs.24–26. The result for $\tau = 0$ is shown in Fig.24. It is seen that without time delay system (140) is stable and stationary probability densities $p(H)$ is Dirac delta function. The result for $\tau = 1$ is shown in Fig.25. It is seen that in this case system (140) is unstable and the stationary probability density $p(H)$ is a normalizable function with peak at the origin. It implies that D-bifurcation occurs in system (140) with τ_D value between 0 and 1. This inference is verified by the value $\tau_D = 0.9107$ determined by $\bar{v}(\tau_D) =$

−1. The result for $\tau = 1.5$ is shown in Fig.26. It is seen that system (140) is unstable and the stationary probability density $p(H)$ is a nomalizable function with peak away from the origin. The stationary probability density of the displacement and velocity of the second oscillator is crater-like. It implies that P-bifurcation occurs in the second oscillator of system (140) at τ_P value between 1 and 1.5. This inference is verified by $\tau_P = 1.1803$ determined by $\bar{v}(\tau_P) = 0$.

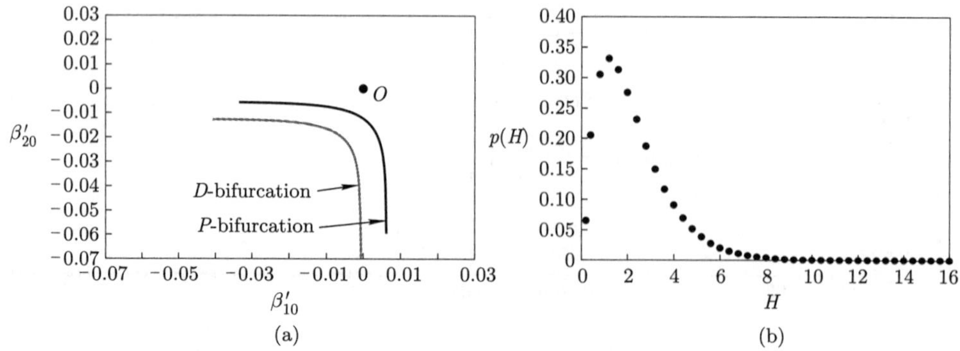

Fig. 26. Results for$\tau = 1.5$. (a) The D-bifurcation and P-bifurcation curves in plane $(\beta'_{10}, \beta'_{20})$ and point $O(0, 0)$. (b) The stationary probability density $p(H)$ at point $O(0, 0)$.

Example for Wide-band Random Excitation[24]

To illustrate the proposed criterion for stochastic Hopf bifurcation, consider two coupled Rayleigh oscillators with time-delayed feedback control subject to parametric excitations of wide-band random processes. The equations of the system are of the form

$$\begin{aligned}
&\ddot{X}_1 + \left(\beta'_{10} + \beta_{11}\dot{X}_1^2 + \beta_{12}\dot{X}_2^2\right)\dot{X}_1 + \omega_1'^2 X_1 \\
&= -\eta_1 \dot{X}_{1\tau_1} + f_{11}\dot{X}_1\xi_1(t) + f_{12}\dot{X}_2\xi_2(t), \\
&\ddot{X}_2 + \left(\beta'_{20} + \beta_{21}\dot{X}_1^2 + \beta_{22}\dot{X}_2^2\right)\dot{X}_2 + \omega_2'^2 X_2 \\
&= -\eta_2 \dot{X}_{2\tau_2} + f_{21}\dot{X}_1\xi_1(t) + f_{22}\dot{X}_2\xi_2(t),
\end{aligned} \tag{150}$$

where X_i are generalized coordinates; β'_{i0} and β_{ij} $(i, j = 1, 2)$ are damping coefficients; ω'_i are the natural frequencies of the two linear oscillators; $-\eta_i \dot{X}_{i\tau}$ represent the time-delayed feedback control forces with time delays τ_1 and τ_2; $\xi_k(t)$ $(k = 1, 2)$ are independent wide-band processes. Here we study the effect of time delay in feedback control forces on the stochastic Hopf bifurcation of system (150).

Some numerical results for the stochastic Hopf bifurcation of system (150) induced by time-delayed feedback control are shown in Figs.27–30. The system parameters are $\beta'_{10} = \beta'_{20} = 0$, $\beta_{11} = \beta_{12} = \beta_{21} = \beta_{22} = 0.005$, $\omega'_1 = 1$, $\omega'_2 = 1.414$, $f_{11} = f_{12} = f_{21} = f_{22} = 1$, $\eta_1 = \eta_2 = 0.02$, $\bar{\omega}_1 = 2$, $\bar{\omega}_2 = 1$, $D_1 = D_2 = D$. The D-bifurcation and P-bifurcation curves in parameter plane (τ_1, τ_2) for $D = 0.01$are shown in Fig.27. It is seen that system (150) is stable when time delays (τ_1, τ_2)

are inside the domain with D-bifurcation curve as boundary. This is verified by the simulation results for $(\tau_1, \tau_2) = (0, 0)$ in Fig.28, where stationary probability densities $p(q_1, p_1)$ and $p(q_2, p_2)$ are all Dirac delta functions. As τ_1 and/or τ_2 become larger and (τ_1, τ_2) locate inside bifurcation interval, the D-bifurcation occurs. This is verified by the simulations results for $(\tau_1, \tau_2) = (1.5, 1)$ in Fig.29, where all the stationary probability densities are normalizable functions with peak at (q_i, p_i) = (0, 0). As τ_1 and τ_2 become even larger and (τ_1, τ_2) are outside the bifurcation interval and the P-bifurcation occurs. This is verified by the simulations results for $(\tau_1, \tau_2) = (2, 2)$ in Fig.30, where the stationary probability densities $p(H)$ is nomalizable with peak away from the origin and stationary probability densities $p(q_1, p_1)$ and $p(q_2, p_2)$ are crater-like.

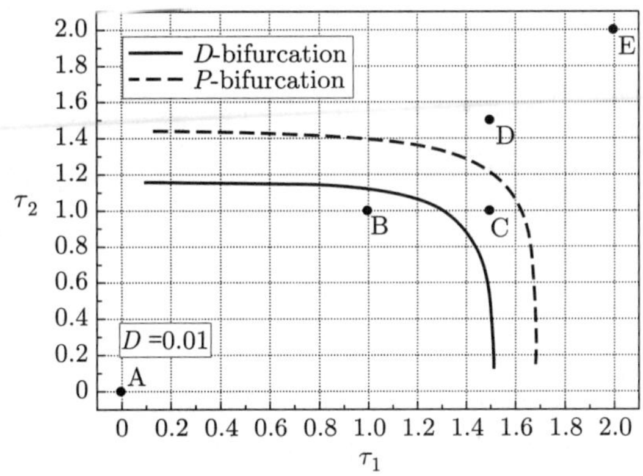

Fig. 27. D-bifurcation and P-bifurcation curves in plane (τ_1, τ_2) for $D = D_1 = D_2 = 0.01$.

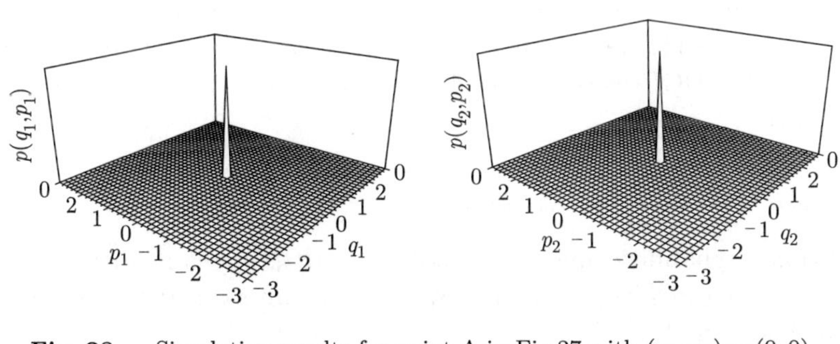

Fig. 28. Simulation results for point A in Fig.27 with $(\tau_1, \tau_2) = (0, 0)$.

3.4. *First Passage Failure*

Stochastic stability is the probabilistic or statistical measure of a stochastic system staying in the vicinity of an equilibrium /stationary motion or returning to the equilibrium /stationary motion in a semi-infinite time interval after an initial

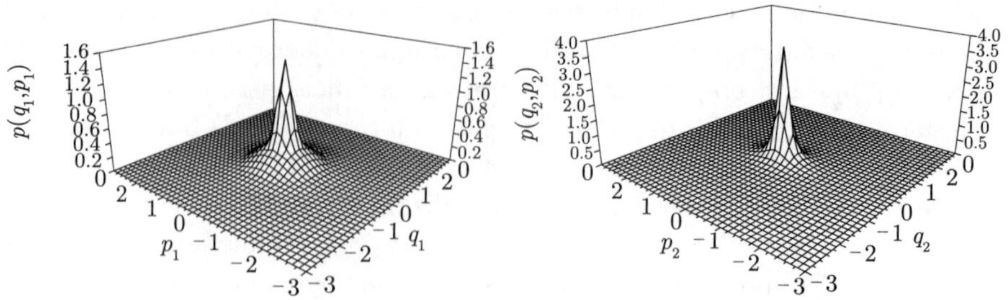

Fig. 29. Simulation results for point C in Fig.27 with $(\tau_1, \tau_2) = (1.5, 1)$.

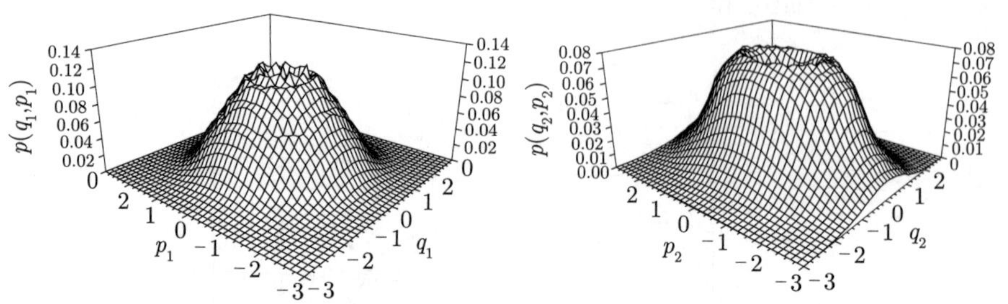

Fig. 30. Simulation results for point E in Fig.27 with $(\tau_1, \tau_2) = (2, 2)$.

perturbation. When the system is unstable, it will move away from the equilibrium randomly in the state space. The first-passage time is the random time when the system leaves certain domain of the state space for the first time. The first-passage time is related to the phase transition of multi-steady-states physical systems and to the reliability of mechanical/ structural systems under random excitation. The combination of the stochastic averaging method for quasi Hamiltonian systems and the diffusion process method for first-passage time can be applied to study the first-passage time of MDOF strongly nonlinear stochastic systems with time-delayed feedback control.

3.4.1. *Gaussian White Noise Excitation*[25]

The conditional reliability function, denoted by $R(t|\mathbf{H}_0)$, is defined as the probability of $\mathbf{H}(t)$ being in safety domain Ω within time interval $(0, t]$ given initial state $\mathbf{H}_0 = \mathbf{H}(0)$ being in Ω, i.e.,

$$R(t|\mathbf{H}_0) = P\{\mathbf{H}(s) \in \Omega, s \in (0, t]|\mathbf{H}_0 \in \Omega\}. \tag{151}$$

It is the integral of the conditional transition probability density in Ω, which is the transition probability density of the sample functions that remain in Ω in time interval $[0, t]$. For an averaged system, the conditional transition probability density satisfies the backward Kolmogorov equation with drift and diffusion coefficients defined by Eq.(21). Thus, the following backward Kolmogorov equation can be derived for the conditional reliability function:

$$\frac{\partial R}{\partial t} = a_r(H_0)\frac{\partial R}{\partial H_{r0}} + \frac{1}{2}b_{rs}(H_0)\frac{\partial^2 R}{\partial H_{r0}\partial H_{s0}}, \quad r, s = 1, 2, \ldots, n, \tag{152}$$

where $a_r(\mathbf{H}_0)$ and $b_{rs}(\mathbf{H}_0)$ are defined by Eq.(21) with $\mathbf{H}$ replaced by $\mathbf{H}_0$. The initial condition is

$$R(0|\mathbf{H}_0) = 1, \quad \mathbf{H}_0 \in \Omega, \tag{153}$$

which implies that the system is initially in the safety domain. The boundary conditions are

$$R(t|\Gamma_0) = finite, \tag{154}$$

$$R(t|\Gamma_c) = 0. \tag{155}$$

Here Γ_0 is reflecting boundary while Γ_c is absorbing boundary.

The first-passage time T is defined as the time when the system reaches critical boundary Γ_c for the first time given $\mathbf{H}_0$ being in Ω. Noting that the conditional probability of the first-passage failure $F(t|\mathbf{H}_0) = 1 - R(t|\mathbf{H}_0)$, the conditional probability density of the first-passage time can be obtained from the conditional reliability function as follows:

$$p(T|\mathbf{H}_0) = \left.\frac{-\partial R(t|\mathbf{H}_0)}{\partial t}\right|_{t=T}. \tag{156}$$

The conditional moments of first-passage time are defined as

$$\mu_l(H_0) = \int_0^\infty T^l p(T|\mathbf{H}_0)dt, \quad l = 1, 2, \ldots. \tag{157}$$

The equations governing the conditional moments of first-passage time can be obtained as follows:

$$\frac{1}{2}b_{rs}(\mathbf{H}_0)\frac{\partial^2 \mu_{l+1}}{\partial H_{r0}\partial H_{s0}} + a_r(\mathbf{H}_0)\frac{\partial \mu_{l+1}}{\partial H_{r0}} = -(l+1)\mu_l, \\ r, s = 1, 2, \ldots, n; \quad l = 0, 1, 2, \ldots. \tag{158}$$

It is easily seen from Eq.(157) that $\mu_0 = 1$. The boundary conditions associated with Eq.(158) are obtained as

$$\mu_l(\Gamma_0) = finite, \tag{159}$$

$$\mu_l(\Gamma_c) = 0. \tag{160}$$

The conditional reliability function is obtained from solving backward Kolmogorov equation (152) together with initial condition (153) and boundary conditions (154) and (155). The conditional probability density of first-passage time is obtained from the conditional reliability function by using Eq.(156). The conditional moments of first-passage time are obtained either from the conditional probability density of first-passage time by using definition (157) or directly from solving generalized Pontryagin equation (158) together with boundary conditions (159) and (160).

The reliability function, the probability density and mean of first-passage time of quasi integrable Hamilton systems subject to time-delayed feedback control under wide-band random excitation, narrow-band bounded noise excitation, combined excitations of harmonic function and Gaussian white noise can be obtained similarly. The only difference is that the averaged drift and diffusion coefficients are from Sections 2.2, 2.3, and 2.4, respectively, rather than Section 2.1. See the following examples.

Example for Gaussian White Noise Excitation[25]

Consider linearly and nonlinearly coupled two linear oscillators with time-delayed feedback control subject to external and parametric excitations of Gaussian white noises. The equations of the system are of the form

$$\begin{aligned} &\ddot{X}_1 + \alpha'_{11}\dot{X}_1 + \alpha'_{12}\dot{X}_2 + \beta_1(X_1^2 + X_2^2)\dot{X}_1 + \omega_1'^2 X_1 \\ &= F_{1\tau} + W_1(t) + X_1 W_2(t), \\ &\ddot{X}_2 + \alpha'_{21}\dot{X}_1 + \alpha'_{22}\dot{X}_2 + \beta_2(X_1^2 + X_2^2)\dot{X}_2 + \omega_2'^2 X_2 \\ &= F_{2\tau} + W_3(t) + X_2 W_4(t), \end{aligned} \tag{161}$$

where X_i are generalized coordinates; α'_{ij} and β_i are damping coefficients; ω'_i are the natural frequencies of the two linear oscillators; $W_k(t)$ $(k = 1, 2, 3, 4)$ are independent Gaussian white noises with intensities $2D_k$; $F_{i\tau}$ represent the time-delayed feedback control forces; $\alpha'_{i,j}$,β_i, D_k, and $F_{i\tau}$ are assumed of the same order of ε. Here we study the effect of time delay in feedback control forces on the statistics of the first-passage failure of system (161). Two different cases of $F_{i\tau}$ are considered.

Case 1. $F_{i\tau}$ are time-delayed linear feedback controls, i.e., $F_{i\tau} = -\eta_i \dot{X}_{i\tau}$. Following Eq.(8), the time-delayed feedback control forces $F_{i\tau}$ can be expressed in terms of system state variables without time delay as follows:

$$F_{i\tau} = -\eta_i \dot{X}_{i\tau} = -\eta_i \cos\omega'_i\tau \dot{X}_i - \eta_i\omega'_i \sin\omega'_i\tau X_i. \tag{162}$$

Let $X_1 = Q_1, X_2 = Q_2, \dot{X}_1 = P_1, \dot{X}_2 = P_2$, equation (161) can be converted into Itô stochastic differential equation (15) with

$$\begin{aligned} H &= H_1 + H_2, \quad H_i = \left(P_i^2 + \omega_i^2 Q_i^2\right)/2, \\ \varepsilon m_{11} &= \alpha_{11} + \beta_1(Q_1^2 + Q_2^2), \quad \varepsilon m_{12} = \alpha'_{12}, \\ \varepsilon m_{21} &= \alpha'_{21}, \quad \varepsilon m_{22} = \alpha_{22} + \beta_2(Q_1^2 + Q_2^2), \\ \varepsilon^{1/2}\sigma_{11} &= 1, \quad \varepsilon^{1/2}\sigma_{12} = Q_1, \quad \varepsilon^{1/2}\sigma_{23} = 1, \\ \varepsilon^{1/2}\sigma_{24} &= Q_2, \quad \omega_i^2 = \omega_i'^2 + \eta_i\omega'_i \sin\omega'_i\tau, \\ \alpha_{ii} &= \alpha'_{ii} + \eta_i \cos\omega'_i\tau, \quad i, j = 1, 2. \end{aligned} \tag{163}$$

The Hamiltonian system with Hamiltonian H is integrable. Thus, system (161) is a quasi integrable Hamiltonian system. By using the stochastic averaging method for quasi integrable Hamiltonian systems in nonresonant case, the following averaged Itô equations can be obtained:

$$dH_r = \bar{m}_r(H_1, H_2)dt + \bar{\sigma}_{rk}(H_1, H_2)dB_k(t), \quad r = 1, 2; \quad k = 1, 2, 3, 4. \tag{164}$$

where

$$\begin{aligned}
a_1 &= \bar{m}_1 = -\alpha_{11}H_1 - \frac{\beta_1}{2\omega_1^2}H_1^2 - \frac{\beta_1}{\omega_2^2}H_1H_2 + D_1 + \frac{D_2}{\omega_1^2}H_1, \\
a_2 &= \bar{m}_2 = -\alpha_{22}H_2 - \frac{\beta_2}{2\omega_2^2}H_2^2 - \frac{\beta_2}{\omega_1^2}H_1H_2 + D_3 + \frac{D_4}{\omega_2^2}H_2, \\
b_{11} &= \bar{\sigma}_{1k}\bar{\sigma}_{1k} = 2D_1H_1 + D_2\frac{H_1^2}{\omega_1^2}, \\
b_{22} &= \bar{\sigma}_{2k}\bar{\sigma}_{2k} = 2D_3H_2 + D_4\frac{H_2^2}{\omega_2^2}.
\end{aligned} \tag{165}$$

It is seen from Eq.(163) that H_i varies from 0 to ∞. So, the state of averaged system (164) varies randomly in the first quadrant of plane (H_1, H_2). Suppose that the limit state of the system is $H = H_1 + H_2 = H_c$, i.e.,

$$\Gamma_c : H_1 + H_2 = H_c, \quad H_1,\ H_2 \geq 0. \tag{166}$$

The safety domain of the system is the inside of a right triangle with boundaries Γ_c defined by Eq.(166) and Γ_0 defined by

$$\begin{aligned}
&\Gamma_0 = \Gamma_{01} + \Gamma_{02}, \quad \Gamma_{01} : H_1 = 0, \quad 0 \leq H_2 < H_c, \\
&\Gamma_{02} : H_2 = 0, \quad 0 \leq H_1 < H_c.
\end{aligned} \tag{167}$$

See Fig.31.

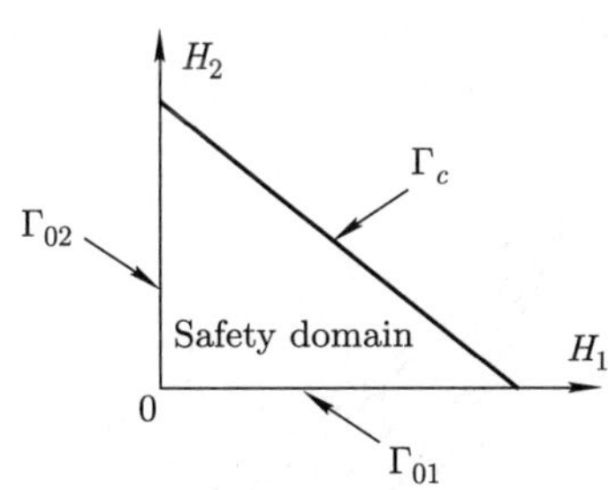

Fig. 31. Safety domain Ω and its boundary on plane H_1 and H_2 for system (161).

Following Eq.(152), the conditional reliability function $R(t|H_{10}, H_{20})$ of system (161) is governed by the following backward Kolmogorov equation:

$$\frac{\partial R}{\partial t} = a_1\frac{\partial R}{\partial H_{10}} + a_2\frac{\partial R}{\partial H_{20}} + \frac{1}{2}b_{11}\frac{\partial^2 R}{\partial H_{10}^2} + \frac{1}{2}b_{22}\frac{\partial^2 R}{\partial H_{20}^2}, \tag{168}$$

where a_1, a_2, b_{11}, and b_{22} are defined by Eq.(165) with H_1, H_2 replaced by H_{10} and H_{20}, respectively. The initial condition is Eq.(153) with $\mathbf{H}_0 = [H_{10}, H_{20}]^T$. One boundary condition is Eq.(155) with Γ_c defined by Eq.(166). The other qualitative boundary condition is Eq.(154) with Γ_0 defined by Eq.(167), which can be converted into a quantitative one using Eq.(168).

Equation (168) with initial and boundary conditions can be solved numerically by using the finite difference method to yield the conditional reliability function of system (161). The conditional probability density of the first-passage time of the system is then obtained from the conditional reliability function by using Eq.(156).

Similarly, the generalized Pontryagin equations for the conditional moments of the first passage time of system (161) can be derived from the averaged Itô equations

$$\frac{1}{2}b_{11}\frac{\partial^2 \mu_{l+1}}{\partial H_{10}^2} + \frac{1}{2}b_{22}\frac{\partial^2 \mu_{l+1}}{\partial H_{20}^2} + a_1\frac{\partial \mu_{l+1}}{\partial H_{10}} + a_2\frac{\partial \mu_{l+1}}{\partial H_{20}} = -(l+1)\mu_l, \tag{169}$$

where a_1, a_2, b_{11}, and b_{22} are defined by Eq.(165) with H_1 and H_2 replaced by H_{10} and H_{20}, respectively. Boundary conditions are Eqs.(159) and (160) with Γ_0 and Γ_c defined by Eqs.(167) and (166), respectively.

Equation (169) with boundary conditions can be solved numerically by using the finite difference method to yield the conditional moments of first-passage time of system (161).

Some numerical results for the conditional reliability function, the conditional probability density and the conditional mean of the first passage time of system (161) obtained by using the above procedure (solid line) are shown in Figs.32–34. The results from digital simulations (dots) for system (161) are also shown for comparison. The parameters are: $\alpha'_{11} = 0.02$, $\alpha'_{12} = 0.01$, $\beta_1 = 0.02$, $\omega'_1 = 1$, $2D_1 = 0.02$, $2D_2 = 0.02$, $\eta_1 = 0.03$, $\alpha'_{21} = 0.01$, $\alpha'_{22} = 0.02$, $\beta_2 = 0.02$, $\omega'_2 = 1.414$, $2D_3 = 0.04$, $2D_4 = 0.04$, $\eta_2 = 0.03$, $H_{10} = H_{20} = 0$. It is seen that the two results are in excellent agreement. Note that the conditional reliability function is a monotonously decreasing function of time. As the delay time τ increases, the conditional reliability function and the mean of first-passage time of system (161)

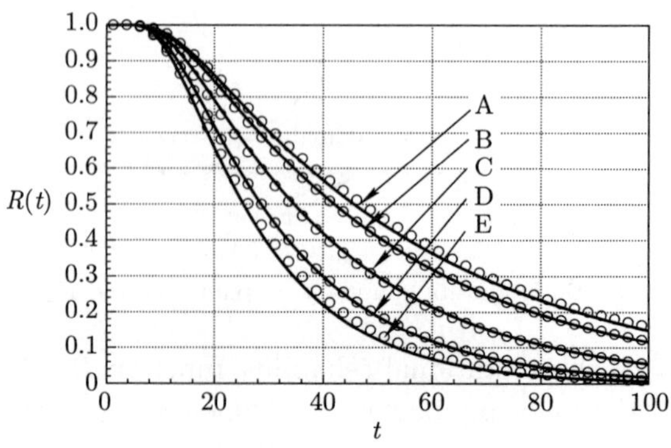

Fig. 32. Reliability function of system (161) with different delay time τ. A: $\tau = 0$; B: $\tau = 0.5$; C: $\tau = 1$; D: $\tau = 1.5$; E: $\tau = 2$.

decrease. It means that time delay in feedback control forces can remarkably reduce the reliability of controlled systems.

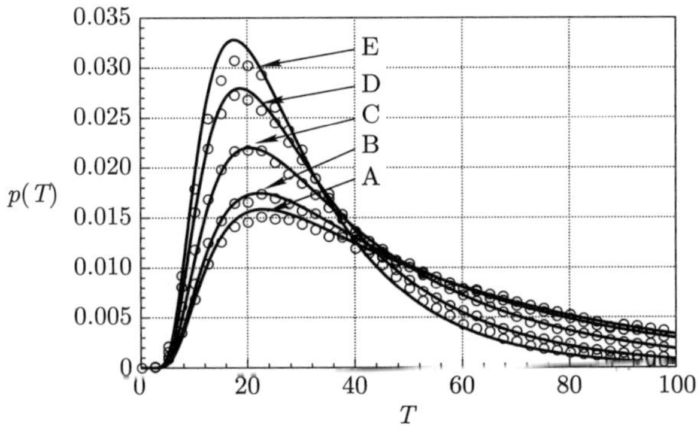

Fig. 33. Probability density of first-passage time of system (161) with different time delay τ. A: $\tau = 0$; B: $\tau = 0.5$; C: $\tau = 1$; D: $\tau = 1.5$; E: $\tau = 2$.

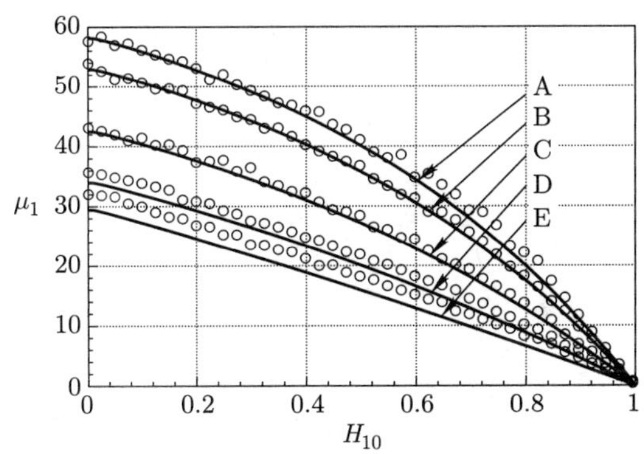

Fig. 34. Mean first-passage time of system (161) as function of H_{10} for different delay time τ. A: $\tau = 0$, B: $\tau = 0.5$; C: $\tau = 1$; D: $\tau = 1.5$; E: $\tau = 2$.

Case 2. $F_{i\tau}$ are time-delayed feedback bang-bang control forces, i.e., $F_{i\tau} = -\eta_i \operatorname{sgn}(\dot{X}_1)$. Following Eq.(12), the time-delayed feedback control forces $F_{i\tau}$ can be expressed in terms of system state variables without time delay as follows:

$$F_{i\tau} = -\eta_i \operatorname{sgn}\left(\dot{X}_{i\tau}\right) = -\eta_i \cos\omega_i'\tau \operatorname{sgn}\left(\dot{X}_i\right). \tag{170}$$

Then the reliability function, the probability density and mean first-passage time can be obtained similarly as in case one. Some numerical results for the reliability function, probability density and mean of the first passage time of system (161) obtained are shown in Figs.35–37. The results from digital simulation for system (161) are also obtained for comparison. It is seen that the two results are in excellent agreement. It is also seen that, as the delay time τ increases, the conditional reliability function and the mean of first-passage time of system (161) decrease. Comparing with case 1, the bang-bang control is more effective in increasing the reliability and mean first-passage than the linear feedback control for the case of without time delay. However, the performance of bang-bang control is easier affected by time delay than the linear feedback control.

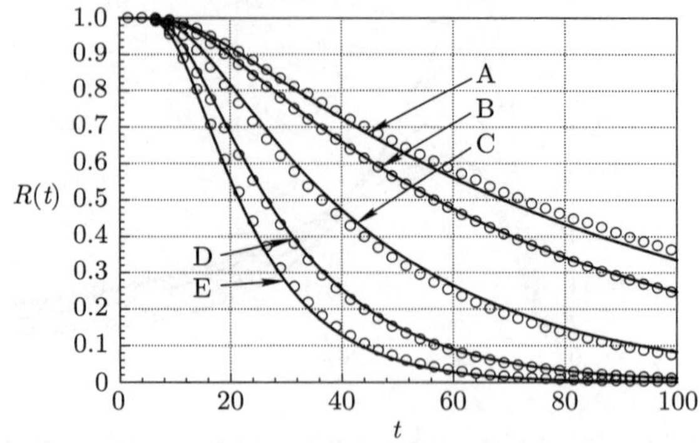

Fig. 35. Reliability function of system (161) with different delay time τ. A: $\tau = 0$; B: $\tau = 0.5$; C: $\tau = 1$; D: $\tau = 1.5$; E: $\tau = 2$.

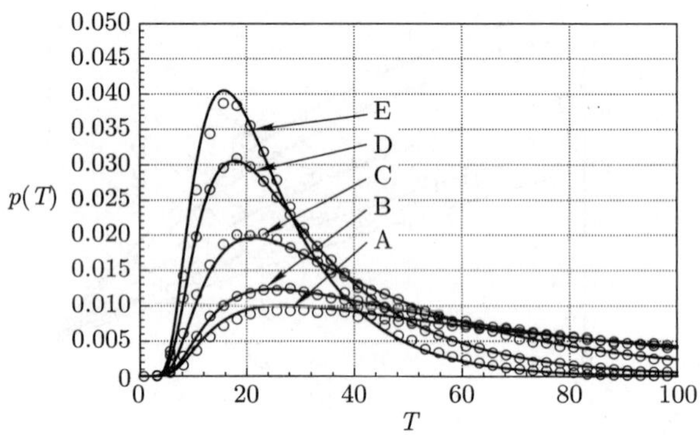

Fig. 36. Probability density of first-passage time of system (161) with different delay time τ. A: $\tau = 0$; B: $\tau = 0.5$; C: $\tau = 1$; D: $\tau = 1.5$; E: $\tau = 2$.

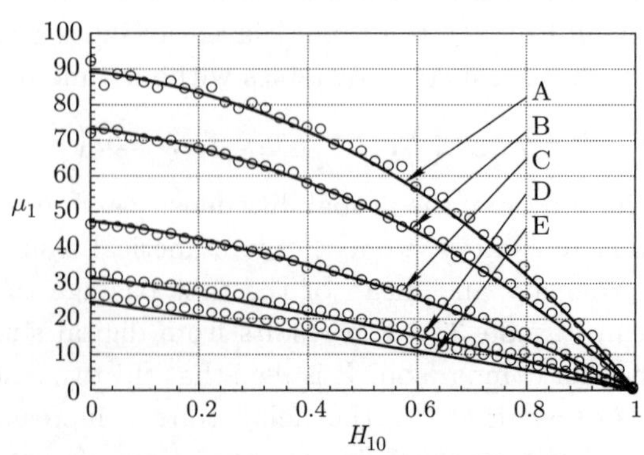

Fig. 37. Mean first-passage time of system (161) as function of H_{10} for different delay time τ. A: $\tau = 0$; B: $\tau = 0.5$; C: $\tau = 1$; D: $\tau = 1.5$; E: $\tau = 2$.

Example for Wide-band Random Excitation[26]

Consider two linear oscillators coupled by linear and nonlinear dampings subject to external excitations of two wide-band random excitations and two time-delayed feedback controls. The equations of the system are of the form

$$\begin{aligned} &\ddot{X}_1 + \varepsilon[\alpha'_{11}\dot{X}_1 + \alpha_{12}\dot{X}_2 + \beta_1\dot{X}_1(X_1^2 + X_2^2)] + \omega_1'^2 X_1 \\ &= u_1 + \varepsilon^{1/2}X_1\xi_1(t) + \varepsilon^{1/2}\xi_2(t), \\ &\ddot{X}_2 + \varepsilon[\alpha_{21}\dot{X}_1 + \alpha'_{22}\dot{X}_2 + \beta_2\dot{X}_2(X_1^2 + X_2^2)] + \omega_2'^2 X_2 \\ &= u_2 + \varepsilon^{1/2}X_2\xi_3(t) + \varepsilon^{1/2}\xi_4(t), \end{aligned} \tag{171}$$

where $\alpha'_{ii}, \alpha_{ij}, \beta_i, \eta_i, \omega'_i$ are constants;

$$u_1 = -\varepsilon(\eta_{11}X_{1\tau_1} + \eta_{12}\dot{X}_{1\tau_2}), \quad u_2 = -\varepsilon(\eta_{21}X_{2\tau_3} + \eta_{22}\dot{X}_{2\tau_4})$$

are the time-delayed feedback control forces; $X_{i\tau_j} = X_i(t - \tau_j)$ are time-delayed displacement, $\dot{X}_{i\tau_k} = \dot{X}_i(t - \tau_k)$ are time-delayed velocity; $\xi_i(t)$ are independent wide-band random excitations with spectral densities of the form in Eq.(27). Let $X_i = Q_i$, $\dot{X}_i = P_i$. The safety domain of the system is the inside of a right triangle with boundaries Γ_0 and Γ_c as shown in Fig.31. The reliability function, the probability density and mean first-passage time can be obtained by first applying the stochastic averaging method in Section 2.2 and then following Eqs.(151)–(160).

Some numerical results for the reliability function, probability density and mean of the first passage time of system (171) with multi-time-delayed feedback control obtained by using the above procedure are shown in Figs.38 and 39. The results from digital simulations for system (171) are also shown in the figures for comparison. It is seen that the two results are in excellent agreement. The system parameters in the calculation are $\varepsilon = 0.01$, $\omega'_1 = 1$, $\omega'_2 = 1.414$, $\alpha'_{11} = 2$, $\alpha_{12} = 2$, $\alpha_{21} = 2$,

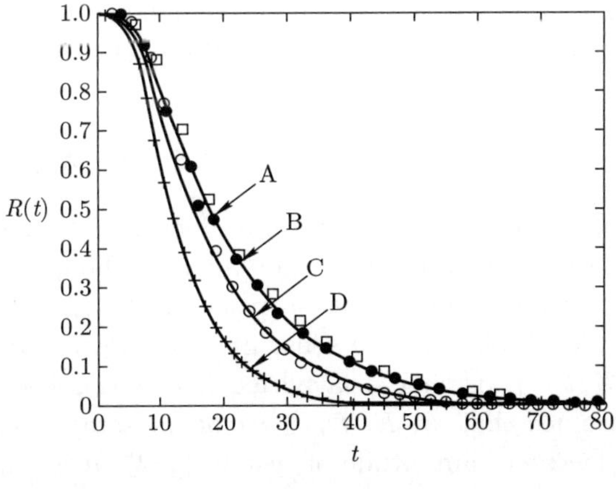

Fig. 38. Reliability function of system (171) for different delay times τ_i in feedback control. A: $\tau_1 = 0$, $\tau_2 = 0$, $\tau_3 = 0$, $\tau_4 = 0$; B: $\tau_1 = 3\pi/2\omega_1; \tau_2 = 2\pi/\omega_1; \tau_3 = 3\pi/2\omega_2; \tau_4 = 2\pi/\omega_2$; C: $\tau_1 = 3$, $\tau_2 = 4$, $\tau_3 = 3$, $\tau_4 = 4$; D: $\tau_1 = \pi/2\omega_1; \tau_2 = \pi/\omega_1; \tau_3 = \pi/\omega_2; \tau_4 = \pi/\omega_2$.

$\alpha'_{22} = 2$, $\beta_1 = 2$, $\beta_2 = 2$, $\eta_{11} = 4$, $\eta_{12} = 4$, $\eta_{21} = 4$, $\eta_{22} = 4$, $H_{10} = H_{20} = 0$, $H_c = 1$. The symbol in these figures, —— denotes the analytical results obtained by using the proposed method and $\bullet \circ * \square$ denote those from digital simulation.

Figures 38 and 39 show that the delay time in the feedback control deteriorates the conditional reliability function, the probability density of the first passage time. The best result is almost the same as that without time delay when $\tau_1 = 2\pi/\omega'_1$, $\tau_2 = 2\pi/\omega'_1\tau_3 = 2\pi/\omega'_2$. It means that the deterioration effect of the delay time can be almost eliminated if the delay time can be set correctly.

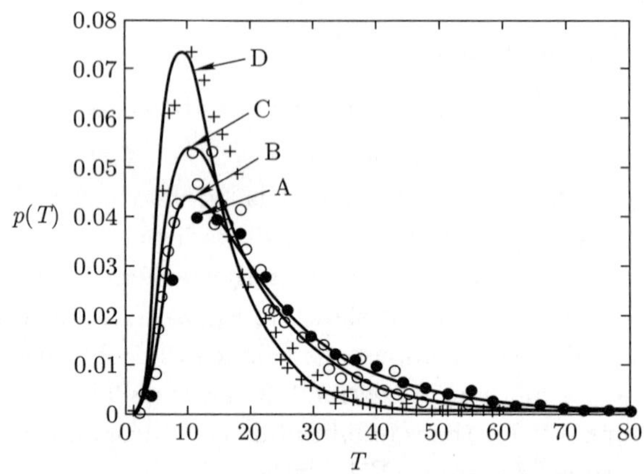

Fig. 39. Probability density of first-passage time of system (171) for different delay times τ_i in feedback control. The parameters are the same as those in Fig.38.

Example for Combinational Excitations of Harmonic Function and Gaussian White Noise[27]

Consider a Duffing oscillator subject to external harmonic excitation and external and parametric white noise excitations with time-delayed feedback control. The motion equation of the system is of the form

$$\ddot{X} + \tilde{\omega}_0^2 X + \alpha X^3 = -\tilde{\beta}\dot{X} + E\cos\Omega t + \xi_1(t) + X\xi_2(t) + u_\tau, \tag{172}$$

where $\tilde{\omega}_0$, α, $\tilde{\beta}$, E, Ω are positive constants denoting the natural frequency of degenerated linear oscillator, intensity of nonlinearity, damping coefficient, amplitude and frequency of harmonic excitation, respectively; $\xi_k(t)$ $(k = 1, 2)$ are independent Gaussian white noises in the sense of Stratonovich with intensities $2D_k$. $u_\tau = -b\dot{X}_\tau = -b\dot{X}(t-\tau)$ is time-delayed feedback control force, where τ is the delay time. It is assumed that $\tilde{\beta}$, E, D_k, b are of the same small order $\varepsilon \ll 1$.

$A(t)$ is the displacement amplitude of system (172). It is reasonable to assume that the first-passage failure occurs once $A(t)$ exceed certain critical value a_c for the first time. In phase plane (a, γ), the safe domain Ω_s is inside of the two parallel lines $a = 0$ and $a = a_c$ (Fig.40).

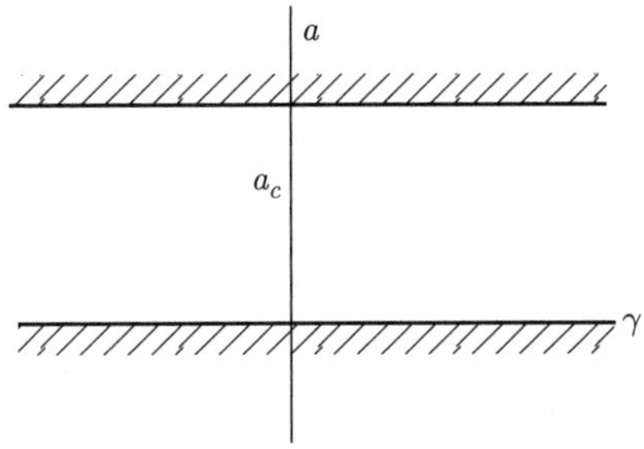

Fig. 40. Safety domain Ω_s and its boundaries for system (172).

In the case of external resonance, the averaged Itô equation for $A(t)$ can be derived by using the technique in Section 2.4.2 with **I** replaced by **A**. Then the reliability function, the probability density and mean of first-passage time can be obtained by following Eqs.(151)–(160).

Some numerical results for reliability functions, the probability density and mean first-passage time of system (172) by using the proposed procedure are shown in Figs.41–43. The results from Monte Carlo simulation for original system (172) are also obtained for comparison. The parameters are: $\omega_0 = 1$, $\alpha = 0.3$, $\beta = 0.01$, $E = 0.01$, $\Omega = 1$, $D_1 = 0.01$, $D_2 = 0.03$, $a_c = 1$, $b = 0.03$, $a_0 = \gamma_0 = 0$. — denotes result obtained by using the proposed method and $\circ\,\square\,\times$ denote results from digital simulation of original system (172). It is seen that the analytical results are in rather good agreement with those from Monte Carlo simulation.

It is also seen that, as the delay time τ increases, the reliability function and mean first-passage time of system (172) decrease, which means time delay in feedback control force can remarkably reduce the reliability of the controlled system.

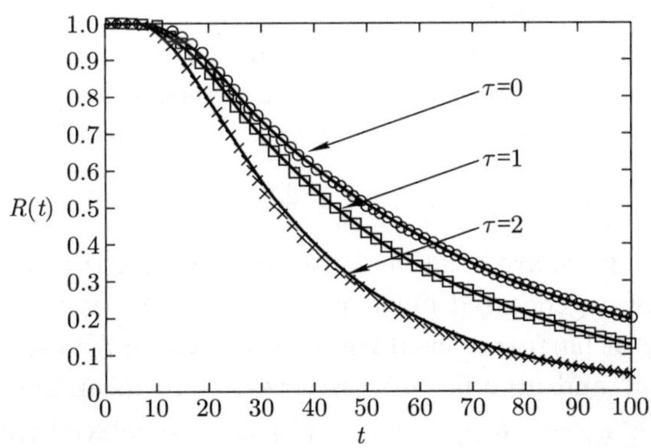

Fig. 41. Reliability function of system (172) for different delay time τ.

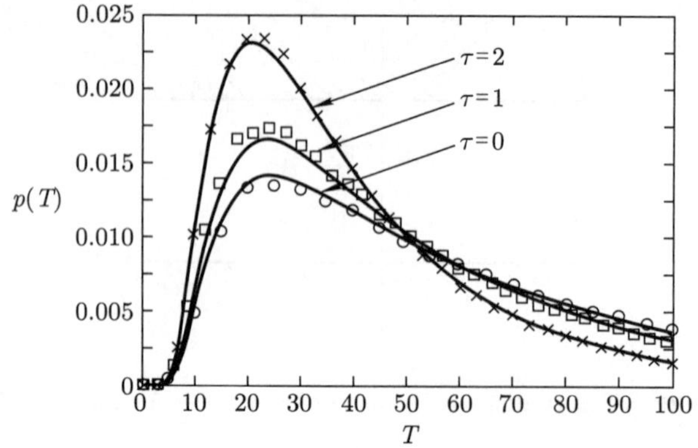

Fig. 42. Probability density of first-passage time of system (172) for different delay time τ.

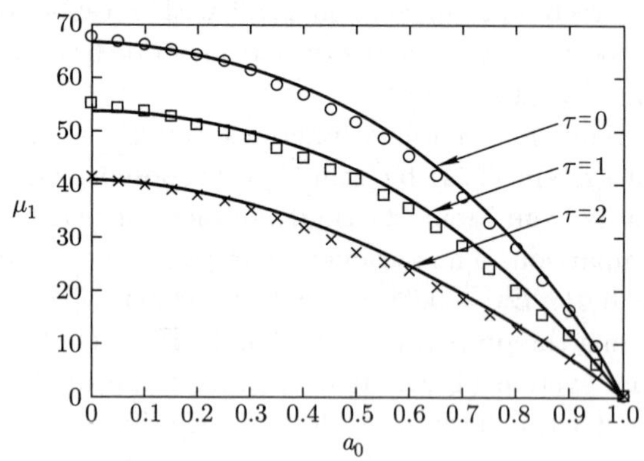

Fig. 43. Mean first-passage time of system (172) for different delay time τ.

Example for Combinational Excitations of Harmonic Function and Wide-band Random Processes[28]

Consider a Duffing-van der Pol oscillator with time-delayed feedback bang-bang control under external harmonic excitation and external and parametric wide-band noise excitations. The equation of the system is of the form

$$\ddot{X} + \left(\beta_1 + \beta_2 X^2\right)\dot{X} + \omega_0^2 X + \alpha X^3 = E\cos\Omega t + \xi_1(t) + X\xi_2(t) + u_\tau, \qquad (173)$$

where β_1, β_2, ω_0, α, E, Ω are positive constants denoting the damping coefficients, natural frequency of degenerated linear oscillator, intensity of nonlinearity, amplitude and frequency of harmonic excitation, respectively; $\xi_k(t)\,(k = 1,\ 2)$ are independently stationary and ergodic processes with zero mean and rational spectral densities in Eq.(27). $u_\tau = -b\ \mathrm{sgn}\left(\dot{X}(t-\tau)\right)$ is time-delayed feedback bang-bang control force. It is assumed that β, E, b are of the same order of ε.

The averaged Itô equation for $A(t)$ can be derived by using the techniques in Sections 2.2 and 2.4. Then the reliability function, the probability density and mean of first-passage time can be obtained by following Eqs.(151)–(160).

Some numerical results for the reliability functions, the probability density and the mean of first-passage time of system (173) obtained by using the proposed procedure (solid lines) are shown in Figs.44–46. The results from digital simulation (dots) for system (173) are also obtained for comparison. It is seen that the two results are in excellent agreement. It is also seen that, as the delay time τ increases, the reliability function and mean first-passage time of system (173) decrease, which means that time delay in feedback control forces can immensely reduce the reliability of the controlled system.

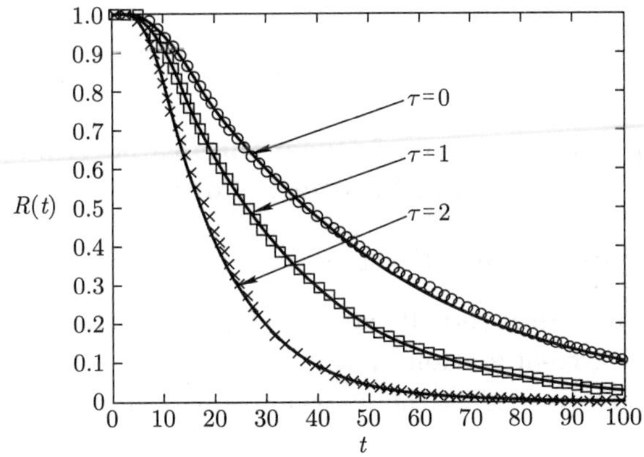

Fig. 44. Reliability function of system (173) for different delay time τ.

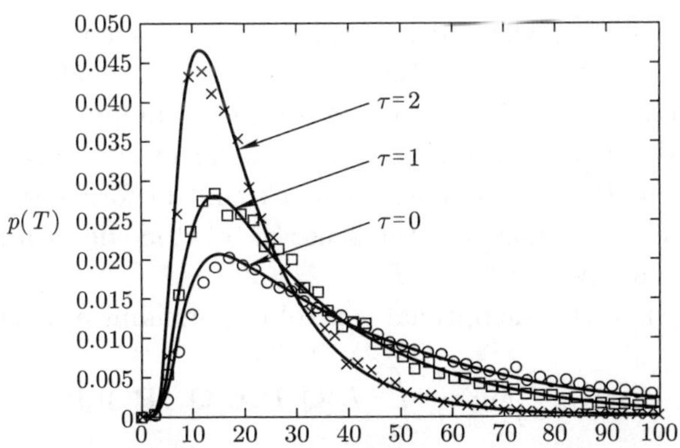

Fig. 45. Probability density of first-passage time of system (173) for different delay time τ.

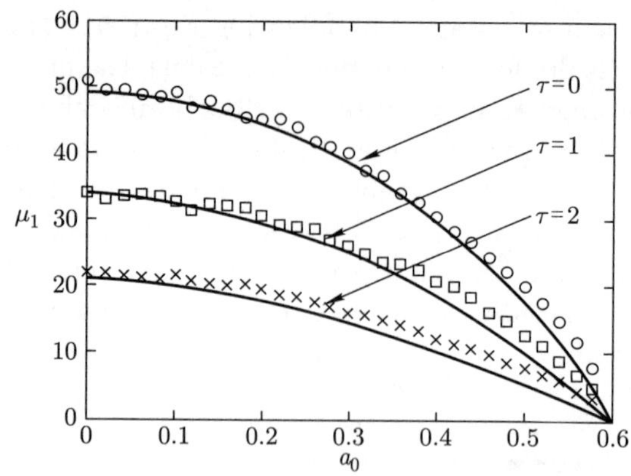

Fig. 46. Mean first-passage time of system (173) for different delay time τ.

4. Stochastic Optimal Control of Quasi Integrable Hamiltonian Systems with Time-delayed Feedback Control

4.1. *Response Minimization Control*[29]

Formulation of Time-delayed Stochastic Optimal Control Problem

Consider an n-degree-of-freedom quasi Hamiltonian system with time-delayed feedback control and governed by the following equations:

$$\dot{Q}_i = \frac{\partial H'}{\partial P_i},$$
$$\dot{P}_i = -\frac{\partial H'}{\partial Q_i} - \varepsilon c_{ij}\frac{\partial H'}{\partial P_j} + \varepsilon u_i(\mathbf{Q}_\tau, \mathbf{P}_\tau) + \varepsilon^{1/2} f_{ik} W_k(t), \tag{174}$$
$$i, j = 1, 2, \ldots, n; \quad k = 1, 2, \ldots, m,$$

where Q_i and P_i are generalized displacements and momenta, respectively; $H' = H'(\mathbf{Q}, \mathbf{P})$ is a twice differentiable Hamiltonian; ε is a small parameter; $\varepsilon c_{ij} = \varepsilon c_{ij}(\mathbf{Q}, \mathbf{P})$ represent the coefficients of quasi linear dampings; $\varepsilon^{1/2} f_{ik} = \varepsilon^{1/2} f_{ik}(\mathbf{Q}, \mathbf{P})$ represent the magnitudes of stochastic excitations; $\varepsilon u_i(\mathbf{Q}_\tau, \mathbf{P}_\tau)$ with $\mathbf{Q}_\tau = \mathbf{Q}(t-\tau)$ and $\mathbf{P}_\tau = \mathbf{P}(t-\tau)$ denote feedback control forces with delay time τ; $W_k(t)$ are Gaussian white noises in the sense of Stratonovich with zero mean and correlation functions $E[W_k(t)W_l(t+T)] = 2D_{kl}\delta(T)$.

The objective of stochastic optimal control is to minimize a performance index

$$J(\mathbf{u}) = \lim_{T\to\infty} \frac{1}{T}\int_0^T L\left[\mathbf{Q}, \mathbf{P}, \mathbf{u}(\mathbf{Q}_\tau, \mathbf{P}_\tau)\right] dt \tag{175}$$

for semi-infinite time-interval ergodic control. In Eq.(175), $L\left[\mathbf{Q}, \mathbf{P}, \mathbf{u}(\mathbf{Q}_\tau, \mathbf{P}_\tau)\right]$ is cost function, which is a continuous, differentiable and convex function. Equations (174) and (175) constitute a time-delayed stochastic optimal control problem for quasi integrable Hamiltonian systems.

Converted Stochastic Optimal Control Problem Without Time Delay

Following Eqs.(8) or (12) the time-delayed feedback control forces can be expressed approximately in terms of the state variables without time delay, i.e., $u_i(\mathbf{Q}_\tau, \mathbf{P}_\tau) = u_i(\mathbf{Q}, \mathbf{P}; \tau)$. By using Eqs.(8) or (12) and adding possible Wong-Zakai correction terms, equation (174) can be rewritten as

$$\begin{aligned} dQ_i &= \frac{\partial H}{\partial P_i} dt, \\ dP_i &= \left[-\frac{\partial H}{\partial Q_i} - \varepsilon m_{ij} \frac{\partial H}{\partial P_j} + \varepsilon u_i(\mathbf{Q}, \mathbf{P}; \tau) \right] dt + \varepsilon^{1/2} \sigma_{ik}(\mathbf{Q}, \mathbf{P}) dB_k(t), \\ & i, j = 1, 2, \ldots, n; \quad k = 1, 2, \ldots, m, \end{aligned} \tag{176}$$

where $H = H(\mathbf{Q}, \mathbf{P}; \tau)$ and $m_{ij} = m_{ij}(\mathbf{Q}, \mathbf{P}; \tau)$ are the modified Hamiltonian and coefficients of quasi linear dampings, respectively. $B_k(t)$ are standard Wiener processes. $\sigma_{ik}\sigma_{jk} = 2D_{kl} f_{ik} f_{jl}$. Accordingly, the performance index in Eq.(175) is modified to

$$J(\mathbf{u}) = \lim_{T \to \infty} \frac{1}{T} \int_0^T L\left[\mathbf{Q}, \mathbf{P}, \mathbf{u}(\mathbf{Q}, \mathbf{P}; \tau)\right] dt. \tag{177}$$

Equations. (176) and (177) constitute a converted stochastic optimal control problem without time delay.

Stochastic Optimal Control

Assume that H is still separable, i.e., $H = H_1 + H_2 + \cdots + H_n$. In the non-resonant case, applying the stochastic averaging method for quasi integrable Hamiltonian systems to system (176) yields the following partially averaged Itô stochastic differential equations:

$$dH_r = \left[\bar{m}_r(\mathbf{H}) + \left\langle u_i \frac{\partial H_r}{\partial P_i} \right\rangle \right] dt + \bar{\sigma}_{rk}(\mathbf{H}) dB_k(t), \quad r = 1, 2, \ldots, n, \tag{178}$$

where $\mathbf{H} = [H_1, H_2, \ldots, H_n]^T$. $\langle \cdot \rangle$ denotes averaging operation. $\bar{m}_r(\mathbf{H})$ and $\bar{\sigma}_{rk}(\mathbf{H})$ are, respectively, the averaged drift coefficients and diffusion coefficients. Correspondingly, the partial averaged performance index is of the form

$$J(\mathbf{u}) = \lim_{T \to \infty} \frac{1}{T} \int_0^T L(\mathbf{H}, \langle \mathbf{u} \rangle) dt. \tag{179}$$

Equation (178) implies that $\mathbf{H}$ is a controlled diffusion process. By applying the stochastic dynamical programming principle to Eqs. (178) and (179), the following dynamical programming equation can be established:

$$\inf_{\mathbf{u}} \left\{ \frac{1}{2} \bar{\sigma}_{rk} \bar{\sigma}_{sk} \frac{\partial^2 V}{\partial H_r \partial H_s} + \left[\bar{m}_r(\mathbf{H}) + \left\langle u_i \frac{\partial H_r}{\partial P_i} \right\rangle \right] \frac{\partial V}{\partial H_r} + L(\mathbf{H}, \langle \mathbf{u} \rangle) \right\} = \gamma, \tag{180}$$

where $V(\mathbf{H})$ is called value function,

$$\gamma = \lim_{T \to \infty} \frac{1}{T} \int_0^T L(\mathbf{H}(t), \langle \mathbf{u}^*(t) \rangle) dt \tag{181}$$

is the optimal average cost and $\mathbf{u}^*(t)$ is the optimal control.

Let L be quadratic with respect to $\mathbf{u}$, i.e.,

$$L(\mathbf{H}, \langle \mathbf{u} \rangle) = f_1(\mathbf{H}) + \langle \mathbf{u}^T \mathbf{R} \mathbf{u} \rangle, \tag{182}$$

where $f_1(\mathbf{H}) \geq 0$ and $\mathbf{R}$ is a positive definite symmetric matrix. Minimizing the left-hand side of Eq.(180) with respect to $\mathbf{u}$ yields

$$u_i^* = -\frac{1}{2}(R^{-1})_{ij} \frac{\partial V}{\partial H_r} \frac{\partial H_r}{\partial P_j}. \tag{183}$$

Substituting Eq.(183) into Eq.(180) and averaging the terms involving u_i^* lead to the final dynamical programming equation

$$\frac{1}{2}\bar{\sigma}_{rk}\bar{\sigma}_{sk}\frac{\partial^2 V}{\partial H_r \partial H_s} + \left[\bar{m}_r(\mathbf{H}) + \left\langle u_i^* \frac{\partial H_r}{\partial P_i} \right\rangle \right] \frac{\partial V}{\partial H_r} + L(\mathbf{H}, \langle \mathbf{u}^* \rangle) = \gamma. \tag{184}$$

Solving this equation and substituting the resultant $V(\mathbf{H})$ into Eq. (183) yields the optimal control. Note that $\mathbf{u}^*$ is generally nonlinear in Q_i and P_i. The reverse of Eq.(8) is

$$\begin{aligned} Q_i(t) &\doteq Q_i(t-\tau)\cos(\omega_i\tau) + P_i(t-\tau)\sin(\omega_i\tau)/\omega_i, \\ P_i(t) &\doteq P_i(t-\tau)\cos(\omega_i\tau) - Q_i(t-\tau)\omega_i\sin(\omega_i\tau). \end{aligned} \tag{185}$$

Substituting Eq.(185) into Eq.(183), the following time-delayed optimal control can be obtained:

$$u_i^*(\mathbf{Q}_\tau, \mathbf{P}_\tau) = -\frac{1}{2}(R^{-1})_{ij} \left. \frac{\partial V}{\partial H_r} \frac{\partial H_r}{\partial P_j} \right|_{\mathbf{Q}=\mathbf{Q}(\mathbf{Q}_\tau,\mathbf{P}_\tau),\ \mathbf{P}=\mathbf{P}(\mathbf{Q}_\tau,\mathbf{P}_\tau)}. \tag{186}$$

Response of Controlled System

Substituting $u_i^*(\mathbf{Q}_\tau, \mathbf{P}_\tau)$ in Eq. (186) into Eq.(178) to replace u_i and averaging the terms involving u_i^* yield

$$dH_r = [\bar{\bar{m}}_r(\mathbf{H})]dt + \bar{\sigma}_{rk}(\mathbf{H})dB_k(t), \;\; \bar{\bar{m}}_r(\mathbf{H}) = \bar{m}_r(\mathbf{H}) + \left\langle u_i^* \frac{\partial H_r}{\partial P_i} \right\rangle. \tag{187}$$

The response of the controlled quasi integrable Hamiltonian system can be predicted by solving the Fokker-Planck-Kolmogorov (FPK) equation associated with fully averaged Itô equation (187). The standard deviations, σ_h^u and σ_h^c, of displacements of uncontrolled and controlled systems and the standard deviations σ_{u^*} of optimal control forces can be calculated.

To characterize the performance of a controller, two criteria are introduced. One is the control effectiveness

$$K = \frac{\sigma_h^u - \sigma_h^c}{\sigma_h^u} \tag{188}$$

which is relative reduction of the standard deviation of displacement due to the control. The other is control efficiency, which is defined as the ratio of control effectiveness to the standard deviation of control force, i.e.,

$$\mu = K/\sigma_{u^*} \tag{189}$$

Obviously, the higher K and μ are, the better the controller is.

Example[29]

Consider two linear oscillators coupled by linear and polynomial nonlinear dampings subject to external excitations of two uncorrelated Gaussian white noises and time-delayed feedback controls. The equations of the system are of the form:

$$\begin{aligned} &\ddot{X}_1 + \alpha_{11}\dot{X}_1 + \alpha_{12}\dot{X}_2 + \beta_1\dot{X}_1(X_1^2 + X_2^2) + \omega_1^2 X_1 \\ &= u_{1\tau} + k_{11}(\mathbf{X})W_1(t) + k_{12}(X)W_2(t), \\ &\ddot{X}_2 + \alpha_{21}\dot{X}_1 + \alpha_{22}\dot{X}_2 + \beta_2\dot{X}_2(X_1^2 + X_2^2) + \omega_2^2 X_2 \\ &= u_{2\tau} + k_{21}(\mathbf{X})W_1(t) + k_{22}(X)W_2(t), \end{aligned} \tag{190}$$

where $W_i(t)$ are uncorrelated Gaussian white noises with small intensities $2D_i$; $u_{i\tau}$ are time-delayed feedback control forces. The Hamiltonian system associated with Eq.(190) is linear and integrable.

Consider the case of purely external stochastic excitations, i.e., $k_{ij}(\mathbf{X}) = k_{ij}$. Assume that $\omega_1/\omega_2 \neq r/s + O(\varepsilon)$, where r, s are prime integers. Then system (190) is not internally resonant. The partially averaged Itô differential equations are

$$dH_i = \left[m_i(\mathbf{H}) + \left\langle u_i \frac{\partial H_i}{\partial P_i} \right\rangle\right] dt + \sigma_i(\mathbf{H})dB_i(t), \quad i = 1, 2, \tag{191}$$

where $\mathbf{H} = [H_1, H_2]^T$ with $H_i = (\dot{X}_i^2 + \omega_i^2 X_i^2)/2$ representing energies of subsystems,

$$\begin{aligned} m_1(\mathbf{H}) &= k_{11}^2 D_1 + k_{12}^2 D_2 - \alpha_{11}H_1 - \beta_1 H_1\left(\frac{H_1}{2\omega_1^2} + \frac{H_2}{2\omega_2^2}\right), \\ m_2(\mathbf{H}) &= k_{21}^2 D_1 + k_{22}^2 D_2 - \alpha_{22}H_2 - \beta_2 H_2\left(\frac{H_1}{2\omega_1^2} + \frac{H_2}{2\omega_2^2}\right), \\ \sigma_1^2(\mathbf{H}) &= 2(k_{11}^2 D_1 + k_{12}^2 D_2)H_1, \quad \sigma_2^2(\mathbf{H}) = 2(k_{21}^2 D_1 + k_{22}^2 D_2)H_2. \end{aligned} \tag{192}$$

The partially averaged performance index is

$$J(\mathbf{u}) = \lim_{T\to\infty} \frac{1}{T}\int_0^T L(H_1, H_2, \langle\mathbf{u}\rangle)dt. \tag{193}$$

The dynamical programming equation is of the form

$$\inf_{\mathbf{u}}\left\{\frac{1}{2}\sigma_1\sigma_2\frac{\partial^2 V}{\partial H_1 \partial H_2} + \left[m_r(\mathbf{H}) + \left\langle u_i\frac{\partial H_r}{\partial P_i}\right\rangle\right]\frac{\partial V}{\partial H_r} + L(\mathbf{H}, \langle\mathbf{u}\rangle)\right\} = \gamma, \tag{194}$$

where $V(\mathbf{H})$ is the value function, and

$$\gamma = \lim_{T\to\infty}\frac{1}{T}\int_0^T L(\mathbf{H}(t), \langle\mathbf{u}^*(t)\rangle)dt \tag{195}$$

is the optimal average cost. The cost function is of the form $L(\mathbf{H}, \langle\mathbf{u}\rangle) = f_1(\mathbf{H}) + \langle\mathbf{u}^T\mathbf{R}\mathbf{u}\rangle$ with $\mathbf{u} = [u_1, u_2]^T$ and

$$\begin{aligned} f_1(\mathbf{H}) = {}& s_{00} + s_{10}H_1 + s_{01}H_2 + s_{20}H_1^2 + s_{11}H_1H_2 + s_{02}H_2^2 + s_{30}H_1^3 \\ &+ s_{21}H_1^2H_2 + s_{12}H_1^2H_2 + s_{03}H_2^3. \end{aligned} \tag{196}$$

The solution of the dynamic programming equation (194) is assumed to be of the form

$$V(\mathbf{H}) = c_1 H_1 + c_2 H_2 + c_3 H_1^2 + c_4 H_1 H_2 + c_5 H_2^2. \tag{197}$$

where

$$\begin{aligned}
c_1 &= 2[(\alpha_{11}^2 + (R^{-1})_{11}(4D_1 c_3 k_1^2 + s_{11}))^{1/2} - \alpha_{11}]/(R^{-1})_{11},\\
c_2 &= 2[(\alpha_{22}^2 + (R^{-1})_{22}(4D_2 c_5 k_2^2 + s_{12}))^{1/2} - \alpha_{22}]/(R^{-1})_{22},\\
c_3 &= [(\beta_1^2 + 4s_{31}k_1^2(R^{-1})_{11})^{1/2} - \beta_1]/(2k_1(R^{-1})_{11}),\\
c_4 &= 0,\ c_5 = [(\beta_2^2 + 4s_{34}k_2^2(R^{-1})_{22})^{1/2} - \beta_2]/(2k_2(R^{-1})_{22}),\\
R_{11}^{-1} &= R_{22}/(R_{11}R_{22} - R_{12}^2),\\
R_{22}^{-1} &= R_{11}/(R_{11}R_{22} - R_{12}^2),\\
s_{00} &= \gamma - (c_1 D_1 k_{11}^2 + c_2 D_2 k_{22}^2),\\
s_{21} &= 2\alpha_{11}c_3 + c_1(\beta_1^2 + 4s_{31}k_1^2(R^{-1})_{11})^{1/2}/(2k_1),\\
s_{22} &= \beta_1(c_1 + c_2)/k_2,\\
s_{23} &= 2\alpha_{22}c_5 + c_2(\beta_2^2 + 4s_{34}k_2^2(R^{-1})_{22})^{1/2}/(2k_2),\\
s_{32} &= 2\beta_1 c_3/k_2,\\
s_{33} &= 2\beta_2 c_5/k_1,\\
k_1 &= k_{11}, \quad k_2 = k_{22}, \quad k_{12} = k_{21} = 0.
\end{aligned} \tag{198}$$

The optimal control forces u_i^* can be obtained by minimizing the left-hand side of Eq.(194) with respect to $\mathbf{u}$ as follows

$$u_i^* = -\frac{1}{2}(R^{-1})_{ij}\frac{\partial V}{\partial H_r}\frac{\partial H_r}{\partial P_j}. \tag{199}$$

Then the time-delayed optimal control forces can be obtained by using Eqs.(199) and (185) as

$$u_{i\tau}^* = -\frac{1}{2}(R^{-1})_{ij}\frac{\partial V}{\partial H_j}\left(\cos(\omega_j\tau)P_{j\tau} - \omega_j\sin(\omega_j\tau)Q_{j\tau}\right). \tag{200}$$

The control effectiveness K_i and control efficiency μ_i can be evaluated by substituting $u_i^*(\mathbf{Q}_\tau, \mathbf{P}_\tau)$ into the original system (190), solving this equation and using Eqs. (188) and (189).

Numerical calculations are carried out for system (190) with the following parameter values: $\alpha_{11} = \alpha_{22} = 0.01$, $\alpha_{12} = \alpha_{21} = 0$, $\beta_1 = \beta_2 = 0.01$, $\omega_1^2 = 1$, $\omega_2^2 = 2$, $k_{11} = k_{22} = 1$, $k_{12} = k_{21} = 0$, $D_1 = D_2 = 0.05$, $R_{11} = R_{22} = 10$, $R_{12} = R_{21} = 0$, $s_{30} = s_{03} = 1$, and $s_{10} = s_{01} = 0$. The results are shown in Tables 1 and 2, where $\sigma(x_i)$, $\sigma(u_i)$, K_i, μ_i denote standard deviation of displacement x_i $(i = 1, 2)$, standard deviation of control force $u_{i\tau}^*$, control effectiveness, and control efficiency, respectively, obtained by using the proposed time-delayed optimal control, while $\bar{\sigma}(x_i)$, $\bar{\sigma}(u_i)$, $\bar{K}_i$, $\bar{\mu}_i$ are those obtained by using a control strategy without considering time delay effect, i.e., letting $\cos(\omega_j\tau) = 1$ and $\sin(\omega_j\tau) = 0$ in Eq.(200). It is seen from Tables 1 and 2 that the proposed controller is better in the sense of having higher control effectiveness and efficiency. Furthermore, the system with the proposed controller is stable even for larger time delay, e.g., 0.8, 1, while the system is instable if the control strategy without considering time delay effect is used.

Table 1. Results for the first degree of freedom of the system (190).

τ	0	0.2	0.4	0.6	0.8	1.0
$\sigma(x_1)$	0.385	0.390	0.396	0.403	0.408	0.411
$\sigma(u_1)$	0.125	0.133	0.141	0.151	0.163	0.176
K_1	0.635	0.631	0.625	0.619	0.614	0.611
μ_1	5.068	4.758	4.425	4.094	3.765	3.470
$\bar{\sigma}(x_1)$	0.385	0.391	0.401	0.412	Unstable	
$\bar{\sigma}(u_1)$	0.125	0.133	0.145	0.164		
$\bar{K}_1$	0.635	0.629	0.622	0.610		
$\bar{\mu}_1$	5.068	4.728	4.302	3.719		

Table 2. Results for the second degree of freedom of the system (190).

τ	0	0.2	0.4	0.6	0.8	1.0
$\sigma(x_2)$	0.271	0.273	0.280	0.281	0.281	0.277
$\sigma(u_2)$	0.126	0.134	0.143	0.151	0.160	0.168
K_2	0.713	0.711	0.704	0.703	0.703	0.707
μ_2	5.626	5.281	4.924	4.652	4.391	4.208
$\bar{\sigma}(x_2)$	0.271	0.274	0.287	0.322	Unstable	
$\bar{\sigma}(u_2)$	0.126	0.135	0.154	0.200		
$\bar{K}_2$	0.713	0.710	0.695	0.659		
$\bar{\mu}_2$	5.626	5.262	4.500	3.287		

4.2. *Stabilization*[29]

Now consider the case in which the stochastic excitations in Eq.(174) are pure parametric. In this case the trivial solution may be unstable without control and the feedback control may stabilize the system. The time-delayed feedback stabilization is formulated as an ergodic control problem with undetermined cost function, i.e., Equations (174) and (175) with undetermined cost function L. Following the same derivation as that in Section 4, we obtain Eq.(178) for the controlled system. Since the stochastic excitations are pure parametric, the drift and diffusion coefficients of Eq.(178) would satisfy the following conditions:

$$\bar{\bar{m}}_r(0) = 0, \quad \bar{\sigma}_{rk}(0) = 0, \tag{201}$$

which indicates that $\mathbf{H} = \mathbf{0}$ is the trivial solution of the controlled system. To study the stability of the trivial solution, equation (178) is linearized at the trivial solution. So $\bar{\bar{m}}_r$ and $\bar{\sigma}_{rk}\bar{\sigma}_{sk}$ will be homogeneous in H_s of degree one. We assume that diffusion process $\mathbf{H}(t)$ is nonsingular. Introduce the following new variables

$$\rho = (\ln H)/2, \quad \alpha_r = H_r/H. \tag{202}$$

The Itô equations for ρ and α_r are obtained from linearized version of Eq.(178) by using Itô differential rule as follows:

$$d\rho = Q^c(\boldsymbol{\alpha})dt + \Sigma_k(\boldsymbol{\alpha})dB_k(t), \tag{203}$$

$$d\alpha_r = m_r^c(\boldsymbol{\alpha})dt + \sigma_{rk}(\alpha)dB_k(t), \tag{204}$$

where $\boldsymbol{\alpha} = [\alpha_1, \ldots, \alpha_{n-1}, \alpha_n]^T$, $r = 1, 2, \ldots, n$; $k = 1, 2, \ldots, m$. Note that only $n-1$ equations for α_r in Eq.(204) are independent. Thus, we can use $\boldsymbol{\alpha}' = [\alpha_1, \ldots, \alpha_{n-1}]^T$ to replace α in (203) and (204). It is noted that Eq.(204) is independent of Eq.(203).

Define the Lyapunov exponent of controlled system (178) as the asymptotic rate of the exponential growth of the square root of H, i.e.,

$$\lambda = \lim_{t\to\infty} \frac{1}{t} \ln H^{1/2}. \tag{205}$$

The largest Lyapunov exponent of controlled system (179) can be calculated by using formula

$$\lambda_1^c = \int Q^c(\boldsymbol{\alpha}')p^c(\boldsymbol{\alpha}')d\boldsymbol{\alpha}' \tag{206}$$

where $Q^c(\boldsymbol{\alpha}')$ is obtained from $Q^c(\boldsymbol{\alpha})$ in Eq.(203) with α_n replaced by $\boldsymbol{\alpha}' = [\alpha_1, \alpha_2, \ldots, \alpha_{n-1}]$ and $p^c(\boldsymbol{\alpha}')$ is the stationary probability density of $\boldsymbol{\alpha}'$ obtained from solving the reduced FPK equations associated with Itô equation (204). The necessary and sufficient condition for asymptotic Lyapunov stability with probability one of the trivial solutions of system (178) is $\lambda_1^c < 0$. It can also be considered as the approximate condition for asymptotic Lyapunov stability with probability one of the trivial solutions of original system (174). Let $u_i^* = 0$, the largest Lyapunov exponent λ_1^u of the uncontrolled system can also be calculated by using Eq.(206). The difference $\lambda^c - \lambda^u$ of the Lyapunov exponent between controlled and uncontrolled systems depends on the optimal control forces. Thus, the feedback stabilization can be achieved by proper setting the cost function L in Eq.(175) so that the optimal control forces make λ_1^c negative and minimized. For example, if the cost function is of the form of Eq.(182), then the feedback stabilization is actually to select $f_1(\mathbf{H})$ and $\mathbf{R}$ to make λ_1^c negative and minimized. In the following, an example is given to illustrate the designing procedure in detail.

Example[29]

Consider system (189) with pure parametric excitations, i.e., $k_{ij}(\mathbf{X}) = k_{ij}X_j$. The partially averaged performance index is of the form of Eq.(178) with cost function of the form of Eq.(181), where $f_1(\mathbf{H})$ and $\mathbf{R}$ are to be determined. For stabilization, only the asymptotic behavior of averaged equations near $H = 0$ is of interest. The linearized version of averaged Itô equations is of the form

$$\begin{aligned} dH_1 &= \left[F_1(\mathbf{H}) + \left\langle u_1 \frac{\partial H_1}{\partial P_1} \right\rangle\right] dt + G_{11}(\mathbf{H})dB_1(t) + G_{12}(\mathbf{H})dB_2(t), \\ dH_2 &= \left[F_2(\mathbf{H}) + \left\langle u_2 \frac{\partial H_2}{\partial P_2} \right\rangle\right] dt + G_{21}(\mathbf{H})dB_1(t) + G_{22}(\mathbf{H})dB_2(t), \end{aligned} \tag{207}$$

where

$$\begin{aligned} F_1(\mathbf{H}) &= F_{11}H_1 + F_{12}H_2, \quad F_2(\mathbf{H}) = F_{21}H_1 + F_{22}H_2, \\ F_{11} &= -\alpha_{11} + D_{11}k_{11}^2, \quad F_{12} = D_{22}k_{12}^2, \\ F_{21} &= D_{11}k_{21}^2, \quad F_{22} = -\alpha_{22} + D_{22}k_{22}^2, \\ [GG^T]_{11} &= b_{11}(\mathbf{H}) = b_{11}^{(1)}H_1^2 + b_{11}^{(2)}H_1H_2, \quad [GG^T]_{12} = b_{12}(\mathbf{H}) = 0, \\ [GG^T]_{21} &= b_{21}(\mathbf{H}) = 0, \quad [GG^T]_{22} = b_{22}(\mathbf{H}) = b_{22}^{(1)}H_1H_2 + b_{22}^{(2)}H_2^2, \\ b_{11}^{(1)} &= D_{11}k_{11}^2, \quad b_{11}^{(2)} = 2D_{22}k_{12}^2, \\ b_{22}^{(1)} &= 2D_{11}k_{21}^2, \quad b_{22}^{(2)} = D_{22}k_{22}^2. \end{aligned} \tag{208}$$

The optimal control strategy is of the form of Eq.(200). The value function $V(\mathbf{H})$ is linear function of H_1 and H_2. It is seen from dynamical programming equation (193) that $f_1(H) - \gamma$ should also be linear function of H_1 and H_2. Let

$$V(H) = C_1H_1 + C_2H_2, \quad f_1(H) - \gamma = k_1H_1 + k_2H_2. \tag{209}$$

Substituting Eqs.(208) and (209) into final dynamical programming equation (194) leads to the following equations:

$$\begin{aligned} k_1 + F_{11}C_1 + F_{21}C_2 - C_1^2/4R_1 = 0, \\ k_2 + F_{12}C_1 + F_{22}C_2 - C_2^2/4R_2 = 0, \end{aligned} \tag{210}$$

where $\mathbf{R} = \text{diag}[R_1, R_2]$. C_1 and C_2 can be solved for given k_i and R_i. The optimal control u_i^* are then obtained from Eq.(194) as follows:

$$u_i^* = -\frac{C_i}{2R_i}P_i. \tag{211}$$

Following Eq.(206), the expression for the largest Lyapunov exponent of the controlled system (190) is obtained as follows

$$\lambda_1^c = \int_0^1 Q^c(\alpha_1)p^c(\alpha_1)d\alpha_1. \tag{212}$$

When $u_i = 0$, the largest Lyapunov exponent λ_1^u of uncontrolled system (190) is obtained.

The objective of the feedback stabilization of system (190) is to determine k_i, R_i such that the largest Lyapunov exponent determined by Eq.(212) is negative and minimized. After k_i and R_i are determined, the optimal time-delayed control is then

$$u^*_{i\tau} = -\frac{C_i}{2R_i}\left(\cos(\omega_i\tau)P_{i\tau} - \omega_i\sin(\omega_i\tau)Q_{i\tau}\right). \tag{213}$$

Numerical results have been obtained for system (190) with the following parameters: $\alpha_{11} = -0.02, \alpha_{22} = -0.01$, $\alpha_{12} = \alpha_{21} = 0.01$, $\beta_1 = 0.01$, $\beta_2 = 0.01$, $\omega_1^2 = 1$, $\omega_2^2 = 2$, $k_{11} = k_{12} = k_{21} = k_{22} = 1$, $D_1 = D_2 = D$, $k_1 = k_2 = 0.0005$, $R_1 = R_2 = 1$. The initial condition for simulations are $X_1 = X_2 = 0.01$ and $\dot{X}_1 = \dot{X}_2 = 0$. The largest Lyapunov exponents are shown in Fig.47, where the symbol US denotes the uncontrolled system ($u_i = 0$), WTD the controlled system without time delay ($u_i^* = -(C_i/2R_i)P_i$), NCT the controlled system with control strategy designed without considering time-delayed effects ($u_{i\tau} = -(C_i/2R_i)P_i(t-\tau)$), and PCS the controlled system using the proposed control strategy. It is seen from Fig.1 that the largest Lyapunov exponent of uncontrolled system (λ_1^u) is positive and the trivial solution of uncontrolled system is unstable. For time-delayed controlled system, the largest Lyapunov exponent calculated by using PCS is generally less than that by using NCT control strategy. This means that the controlled system designed by using PCS is more stable than that by using NCT. The stabilization effect of the proposed control strategy is remarkable, especially for weak noise excitations and small time delay. However, the results obtained by using PCS becomes worse when excitation intensity D becomes larger or time delay τ is longer because the averaging method is based on the assumption of weak excitation and small time delay.

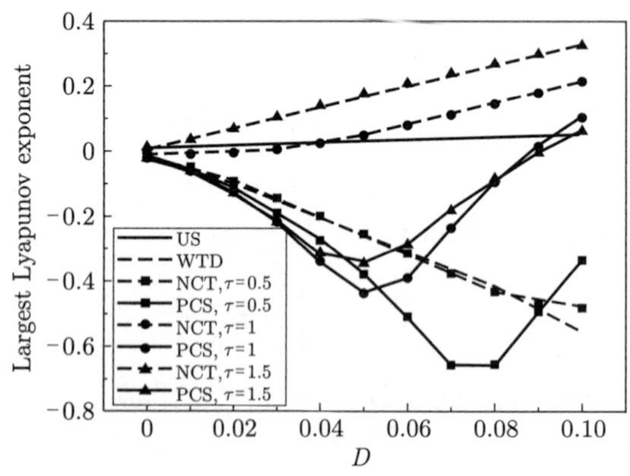

Fig. 47. The largest Lyapunov exponents λ_1 versus stochastic excitations intensity D.

4.3. *Minimax Optimal Bounded Control*[30]

The capacity of control actuators in practice is always bounded. In view of the stochastic nature of the excitations, it is conceivable that the required control force may well exceed the capacity which will result in actuator saturation. The actuator saturation may cause deterioration of the control performance. Consequently, special attention should be paid to the actuator saturation problem. Wang, Zhu, and Ying[32] have proposed a combination control strategy for quasi Hamiltonian systems with actuator saturation. It has been proved that this control strategy has higher control effectiveness and control efficiency, and the chattering is reduced significantly compared with bang-bang control strategy.

For practical engineering systems, parameters and external disturbances are usually uncertain, which may also degenerate the performance of the controller. In the past several decades, the robust control of the deterministic linear and nonlinear systems with uncertain disturbances has been studied extensively. The robust control of linear and nonlinear stochastic systems with uncertain parameters and external disturbances has been less investigated.

In the following, the time-delayed stochastic optimal control of quasi integrable Hamiltionian systems with considering both actuator saturation and the uncertainties of system parameters and external disturbances is studied.

Formulation of Problem

Consider an n-degree-of-freedom (nDOF) uncertain quasi Hamiltonian system with time-delayed bounded feedback control. The equations of the system are

$$\begin{aligned}
\dot{Q}_i &= \frac{\partial H'(\mathbf{Q},\mathbf{P})}{\partial P_i}, \\
\dot{P}_i &= -\frac{\partial H'(\mathbf{Q},\mathbf{P})}{\partial Q_i} - [\bar{c}_{ij}(\mathbf{Q},\mathbf{P}) + \tilde{c}_{ij}(t)]\frac{\partial H'(\mathbf{Q},\mathbf{P})}{\partial P_j} + u_i(\mathbf{Q}_\tau,\mathbf{P}_\tau) \\
&\quad + f_{ik}(\mathbf{Q},\mathbf{P})\xi_k(t) - \tilde{s}_l(t)g_{il}(\mathbf{Q}) + \tilde{w}_i(t), \\
|u_i| &\le U_i^0, \quad i,j = 1,2,\ldots,n; \quad k = 1,2,\ldots,m; \quad l = 1,2,\cdots,p,
\end{aligned} \tag{214}$$

where Q_i and P_i are generalized displacements and momenta, respectively, $\mathbf{Q} = [Q_1, Q_2, \ldots, Q_n]^T$, $\mathbf{P} = [P_1, P_2, \ldots, P_n]^T$; $H'(\mathbf{Q},\mathbf{P},\bar{\mathbf{s}})$ is the Hamiltonian generally representing total energy of system; $\bar{c}_{ij}$ are the nominal values of damping coefficients; f_{ik} are the amplitudes of excitations; $\xi_k(t)$ are Gaussian white noises with zero mean and correlation function $2D_{kl}\delta(s)$; $u_i(\mathbf{Q}_\tau,\mathbf{P}_\tau)$ with $\mathbf{Q}_\tau = \mathbf{Q}(t-\tau)$ and $\mathbf{P}_\tau = \mathbf{P}(t-\tau)$ represent time-delayed feedback control, which is usually bounded due to actuator saturation; $\tilde{s}_l(t)$, $\tilde{c}_{ij}(t)$, and $\tilde{w}_i(t)$ represent the uncertain disturbances. It is assumed that $\tilde{c}_{ij}(t)$, $\tilde{s}_l(t)$, and $\tilde{w}_i(t)$ are all bounded, i.e., $\tilde{c}_{ij}(t) \in [-c_{ij}^0, c_{ij}^0]$, $\tilde{s}_l(t) \in [-s_l^0, s_l^0]$, and $\tilde{w}_i(t) \in [-w_i^0, w_i^0]$.

The objective of control is to minimize the response of system (214) which can be expressed in terms of minimizing a performance index depending on control time interval. For finite time-interval control, the performance index is of the form

$$J_1(\tilde{s}_l, \tilde{c}_{ij}, \tilde{w}_i, u_i) = E\left[\int_0^{t_f} f_1(\mathbf{Q}(t),\mathbf{P}(t),\mathbf{u}(\mathbf{Q}_\tau,\mathbf{P}_\tau))dt + g(\mathbf{Q}(t_f),\mathbf{P}(t_f))\right] \tag{215}$$

and for infinite time-interval ergodic control, the index is of the form

$$J_2(\tilde{s}_l, \tilde{c}_{ij}, \tilde{w}_i, u_i) = \lim_{T\to\infty} \frac{1}{T} \int_0^T f_2(\mathbf{Q}(t), \mathbf{P}(t), \mathbf{u}(\mathbf{Q}_\tau, \mathbf{P}_\tau))dt, \tag{216}$$

where $E[\cdot]$ denotes an expectation operation; t_f is the terminal time of control. Equations (214), (215) or (216) constitute the mathematical formulation of the optimal control problem of an nDOF uncertain quasi Hamiltonian system with time-delayed bounded feedback control.

The time-delayed control system can be converted partially averaged system without time delay. Then, by adding Wong-Zakai correction terms, the Itô differential equations of system (214) without time delay can be obtained:

$$\begin{aligned}
dQ_i &= \frac{\partial H(\mathbf{Q},\mathbf{P})}{\partial P_i} dt, \\
dP_i &= -\left[\frac{\partial H(\mathbf{Q},\mathbf{P})}{\partial Q_i} + (m_{ij}(\mathbf{Q},\mathbf{P}) + \tilde{c}_{ij}(t))\frac{\partial H(\mathbf{Q},\mathbf{P})}{\partial P_j} - \bar{u}_i(\mathbf{Q},\mathbf{P},\tau)\right. \\
&\quad \left. + \tilde{s}_l(t) g_{il}(\mathbf{Q}) - \tilde{w}_i(t)\right] dt + \sigma_{ik}(\mathbf{Q},\mathbf{P}) dB_k(t), \\
|\bar{u}_i| &\le U_i^0,
\end{aligned} \tag{217}$$

where H and m_{ij} represent the new Hamiltonian and damping coefficients; σ_{ik} are the elements of matrix $\boldsymbol{\sigma}$ with $\boldsymbol{\sigma}\boldsymbol{\sigma}^T = 2\mathbf{f}\mathbf{D}\mathbf{f}^T$; $\mathrm{B}_k(t)$ are the standard Wiener processes; $\bar{u}_i$ are bounded feedback control without time delay.

For integrable and nonresonant case, the partially averaged Itô differential equations of system (217) can be obtained by using the stochastic averaging method for quasi integrable Hamiltonian systems[31] as follows

$$\begin{aligned}
dH_r &= \left[\bar{m}_r(\mathbf{H}) + \left\langle \frac{\partial H_r}{\partial P_i}(\bar{u}_i(\mathbf{Q},\mathbf{P},\tau) - \tilde{s}_l(t) g_{il} - \tilde{c}_{ij}(t)\frac{\partial H}{\partial P_j} + \tilde{w}_i(t))\right\rangle\right] dt \\
&\quad + \bar{\sigma}_{rk}(\mathbf{H}) dB_k(t), \quad |\bar{u}_i| \le U_i^0, \quad r = 1, 2, \ldots, n,
\end{aligned} \tag{218}$$

where H_r are the independent first integrals; $\bar{m}_r(\mathbf{H})$ and $\bar{\sigma}_{rk}(\mathbf{H})$ are the drift and diffusion coefficients determined by

$$\begin{aligned}
&\bar{m}_r(\mathbf{H}) \\
&= \frac{1}{T(\mathbf{H})} \oint \left[\left(-m_{ij}\frac{\partial H}{\partial p_j}\frac{\partial H_r}{\partial p_i} + D_{kl} f_{ik} f_{jl} \frac{\partial^2 H_r}{\partial p_i \partial p_j}\right) \Big/ \right. \\
&\quad \left. \prod_{u=1}^{n} \left(\frac{\partial H_u}{\partial p_u}\right)\right] dq_1 dq_2 \cdots dq_n,
\end{aligned}$$

$$\begin{aligned}
&\bar{\sigma}_{ru}(\mathbf{H})\bar{\sigma}_{su}(\mathbf{H}) \\
&\quad = \frac{1}{T(\mathbf{H})}\oint\left[\left(2D_{kl}f_{ik}f_{jl}\frac{\partial H_r}{\partial p_i}\frac{\partial H_s}{\partial p_j}\right)\Big/\right. \\
&\quad\quad \left.\prod_{u=1}^{n}\left(\frac{\partial H_u}{\partial p_u}\right)\right]dq_1dq_2\cdots dq_n, \\
&T(\mathbf{H}) = \oint\left[1\Big/\prod_{u=1}^{n}\left(\frac{\partial H_u}{\partial p_u}\right)\right]dq_1dq_2\cdots dq_n, \\
&\langle\cdot\rangle = \frac{1}{T(\mathbf{H})}\oint\left[\cdot\Big/\prod_{u=1}^{n}\left(\frac{\partial H_u}{\partial p_u}\right)\right]dq_1dq_2\cdots dq_n.
\end{aligned} \tag{219}$$

Note that the second terms on the right hand side of Eq.(218) has not been averaged since $\bar{u}_i$, $\tilde{s}_l$, $\tilde{c}_{ij}$, and $\tilde{w}_i$ are unknown at this stage. It is seen from Eq.(218) that the dimension of system (217) is reduced from $2n$ to n after using the stochastic averaging method.

To be consistent with partially averaged equation (218), performance index (215) and (216) are also partially averaged, i.e., Eqs.(215) and (216) are replaced with

$$J_3(\tilde{s}_l, \tilde{c}_{ij}, \tilde{w}_i, u_i) = E\left[\int_0^{t_f} f_3(H(s), \langle\bar{\mathbf{u}}(\mathbf{Q}, \mathbf{P}, \tau)\rangle)ds + g(H(t_f))\right] \tag{220}$$

and

$$J_4(\tilde{s}_l, \tilde{c}_{ij}, \tilde{w}_i, u_i) = \lim_{T\to\infty}\frac{1}{T}\int_0^T f_4(H(s), \langle\bar{\mathbf{u}}(\mathbf{Q}, \mathbf{P}, \tau)\rangle)ds \tag{221}$$

respectively.

Equations (218), (220) or (221) constitute the mathematical formulation of the optimal bounded control problem for an uncertain partially averaged quasi Hamiltonian system without time delay.

Minimax Optimal Bounded Control

The proposed control strategy is based on searching for a worst-case optimal control by solving the following stochastic differential game problem:

$$\inf_{\bar{u}_i\in U^0}\ \sup_{\tilde{s}_l, \tilde{c}_{ij}, \tilde{w}_i} J(\tilde{s}_l, \tilde{c}_{ij}, \tilde{w}_i, \bar{u}_i), \tag{222}$$

i.e., selecting $\tilde{s}_l^*$, $\tilde{c}_{ij}^*$, $\tilde{w}_i^*$, and $\bar{u}_i^*$ so that

$$J(\tilde{s}_l, \tilde{c}_{ij}, \tilde{w}_i, \bar{u}_i^*) \le J(\tilde{s}_l^*, \tilde{c}_{ij}^*, \tilde{w}_i^*, \bar{u}_i^*) \le J(\tilde{s}_l^*, \tilde{c}_{ij}^*, \tilde{w}_i^*, \bar{u}_i). \tag{223}$$

By applying the principle of optimality to system (218) with performance index (220), the following Hamilton-Jacobi-Issacs (HJI) equation is established

$$\begin{aligned}
\frac{\partial V_1}{\partial t} = \inf_{\bar{\mathbf{u}}\in U^0}\ \sup_{\tilde{s}_l, \tilde{c}_{ij}, \tilde{w}_i}&\left\{f_3(\mathbf{H}, \langle\bar{\mathbf{u}}\rangle) + \left[\bar{m}_r(\mathbf{H}) + \left\langle\frac{\partial H_r}{\partial P_i}\left(\bar{u}_i - \tilde{s}_l(t)g_{il}(\mathbf{Q})\right.\right.\right.\right. \\
&\left.\left.\left.\left. -\tilde{c}_{ij}(t)\frac{\partial H}{\partial P_j} + \tilde{w}_i(t)\right)\right\rangle\right]\frac{\partial V_1}{\partial H_r} + \frac{1}{2}\bar{\sigma}_{rk}(\mathbf{H})\bar{\sigma}_{sk}(\mathbf{H})\frac{\partial^2 V_1}{\partial H_r\partial H_s}\right\}.
\end{aligned} \tag{224}$$

For system (218) with performance index (221), the associated HJI equation is

$$\gamma = \inf_{\bar{\mathbf{u}}\in U^0} \sup_{\tilde{s}_l,\tilde{c}_{ij},\tilde{w}_i} \left\{ f_4(\mathbf{H},\langle\bar{\mathbf{u}}\rangle) + \left[\bar{m}_r(\mathbf{H}) + \left\langle \frac{\partial H_r}{\partial P_i}\left(\bar{u}_i - \tilde{s}_l(t) g_{il}(\mathbf{Q}) \right.\right.\right.\right. \tag{225}$$

$$\left.\left.\left.\left. -\tilde{c}_{ij}(t)\frac{\partial H}{\partial P_j} + \tilde{w}_i(t)\right)\right\rangle\right] \frac{\partial V_2}{\partial H_r} + \frac{1}{2}\bar{\sigma}_{rk}(\mathbf{H})\bar{\sigma}_{sk}(\mathbf{H})\frac{\partial^2 V_2}{\partial H_r \partial H_s} \right\},$$

where $V_1 = V(\mathbf{H},t)$ and $V_2 = V_2(\mathbf{H},t)$ are the value functions, γ is a constant and U^0 denotes the following domain of bounded control force $\bar{\mathbf{u}}$:

$$U: \quad |\bar{u}_i| \le U_i^0, \quad U_i^0 > 0, \quad i = 1,2,\ldots,n. \tag{226}$$

Let function f_4 (or f_3) be of the form

$$f_4(\mathbf{H},\langle\bar{\mathbf{u}}\rangle) = f_c(\mathbf{H}) + \langle \bar{\mathbf{u}}^T\mathbf{R}\bar{\mathbf{u}}\rangle, \tag{227}$$

where $\mathbf{R}$ is a positive-definite diagonal matrix, $\mathbf{R}$=diag(R_i) and $f_c(\mathbf{H})$ >0 is a convex function. Then, the worst-case disturbances can be determined by maximizing the right-hand side of Eqs.(224) or (225) with respect to $\tilde{c}_{ij}(t)$, $\tilde{s}_l(t)$, and $\tilde{w}_i(t)$. Due to the boundness of disturbances, the worst-case disturbances are determined by the following expressions:

$$\begin{aligned} \tilde{s}_l^*(t) &= -s_l^0 \mathrm{sgn}\left[g_{il}(\mathbf{Q})\frac{\partial H_r}{\partial p_i}\frac{\partial V}{\partial H_r}\right], \\ \tilde{c}_{ij}^*(t) &= -c_{ij}^0 \mathrm{sgn}\left[\frac{\partial H}{\partial p_j}\frac{\partial H_r}{\partial p_i}\frac{\partial V}{\partial H_r}\right], \\ \tilde{w}_i^*(t) &= w_i^0 \mathrm{sgn}\left[\frac{\partial H_r}{\partial p_i}\frac{\partial V}{\partial H_r}\right], \end{aligned} \tag{228}$$

where sgn$(\cdot)$ is the sign function. Minimizing the right-hand side of Eqs.(224) or (225) with respect to $\bar{u}_i$ yields the optimal control forces. Due to the control constraints in Eq.(226), the optimal control forces for system (218) are of the form

$$\bar{u}_i^* = \begin{cases} -\dfrac{1}{2R_i}\dfrac{\partial H_r}{\partial P_i}\dfrac{\partial V}{\partial H_r}, & \left|\dfrac{1}{2R_i}\dfrac{\partial H_r}{\partial P_i}\dfrac{\partial V}{\partial H_r}\right| < U_i^0, \\ -U_i^0 \mathrm{sgn}\left(\dfrac{\partial V}{\partial H_r}\dfrac{\partial H_r}{\partial P_i}\right), & \left|\dfrac{1}{2R_i}\dfrac{\partial H_r}{\partial P_i}\dfrac{\partial V}{\partial H_r}\right| \ge U_i^0, \end{cases} \quad i = 1,2,\ldots,n. \tag{229}$$

Substituting the worst-case disturbances $\tilde{s}_l^*, \tilde{c}_{ij}^*, \tilde{w}_i^*$ in Eq.(228) and the optimal bounded control forces $\bar{u}_i^*$ in Eq.(229) into HJI equation (224) or (225) and completing the averaging yield the final HJI equations. $\partial V/\partial H_r$ in Eqs.(228) and (229) can be obtained from solving these final HJI equations. Then, the worst-case disturbances $\tilde{s}_l^*, \tilde{c}_{ij}^*, \tilde{w}_i^*$ and the optimal controls $\bar{u}_i^*$ for system (218) are determined by substituting $\partial V/\partial H_r$ into Eqs.(228) and (229).

By using Eq.(8), the system states without time delay in optimal control forces $\bar{u}_i^*$ can be expressed approximately in terms of system state variables with time delay as follows:

$$Q_i(t) \approx Q_{\tau i}\cos\omega_i\tau_i + \frac{P_{\tau i}}{\omega_i}\sin\omega_i\tau_i,$$
$$P_i(t) \approx -Q_{\tau i}\omega_i\sin\omega_i\tau_i + P_{\tau i}\cos\omega_i\tau_i. \tag{230}$$

By substituting Eq.(230) into Eq.(229), the following optimal time-delayed controls u_i^* for system (214) can be obtained.

$$u_i^*(\mathbf{Q}_\tau, \mathbf{P}_\tau) = \begin{cases} -F_i(\mathbf{Q}_\tau, \mathbf{P}_\tau), & |F_i(\mathbf{Q}_\tau, \mathbf{P}_\tau)| < U_i^0, \\ -U_i^0\mathrm{sgn}(F_i(\mathbf{Q}_\tau, \mathbf{P}_\tau)), & |F_i(\mathbf{Q}_\tau, \mathbf{P}_\tau)| \geq U_i^0, \end{cases}$$

$$\begin{aligned} F_i(\mathbf{Q}_\tau, \mathbf{P}_\tau) &= \frac{1}{2R_i}\frac{\partial V_\tau}{\partial H_{\tau r}}\frac{\partial H_{\tau r}}{\partial P_{\tau i}} \\ &= \frac{1}{2R_i}\left(\frac{\partial V}{\partial H_r}\frac{\partial H_r}{\partial P_i}\right)\bigg|_{Q_{\tau i}\cos\omega_i\tau_i + \frac{P_{\tau i}}{\omega_i}\sin\omega_i\tau_i,\, -Q_{\tau i}\omega_i\sin\omega_i\tau_i + P_{\tau i}\cos\omega_i\tau_i}. \end{aligned} \tag{231}$$

$\partial V/\partial H_r$ in Eq.(228) are functions of Hamiltonian $\mathbf{H}$. For quasi integrable Hamiltonian system, $\mathbf{H}$ generally represents slowly varying quantitatives. Since the time delay is small so that $\partial V_\tau/\partial H_{\tau r}$ can be approximately replaced by $\partial V/\partial H_r$. Then, the second equation of Eq.(231) can be rewritten as

$$F_i(\mathbf{Q}_\tau, \mathbf{P}_\tau) = \frac{1}{2R_i}\frac{\partial V}{\partial H_r}\left(\frac{\partial H_r}{\partial P_i}\right)\bigg|_{Q_{\tau i}\cos\omega_i\tau_i + \frac{P_{\tau i}}{\omega_i}\sin\omega_i\tau_i,\, -Q_{\tau i}\omega_i\sin\omega_i\tau_i + P_{\tau i}\cos\omega_i\tau_i}. \tag{232}$$

For comparison, the optimal control strategy without considering time delay is also presented. Using this control strategy the control forces are of the form

$$u_{\mathrm{wod}}^*(\mathbf{Q}_\tau, \mathbf{P}_\tau) = \begin{cases} -\dfrac{1}{2R_i}P_\tau\dfrac{\partial V}{\partial H_r}, & \left|\dfrac{1}{2R_i}P_\tau\dfrac{\partial V}{\partial H_r}\right| < U_i^0, \\ -U_i^0\mathrm{sgn}\left(P_\tau\dfrac{\partial V}{\partial H_r}\right), & \left|\dfrac{1}{2R_i}P_\tau\dfrac{\partial V}{\partial H_r}\right| \geq U_i^0. \end{cases} \tag{233}$$

Substituting the worst-case disturbances $\tilde{s}_l^*, \tilde{c}_{ij}^*, \tilde{w}_i^*$ and the optimal control forces u_i^* in Eqs.(231) and (232) (or u_{wod}^* in Eq.(233)) into the partially averaged Itô differential equation (218) and completing the averaging yield the fully averaged Itô equations. By solving the associated Fokker-Planck-Kolmogorov equation (FPK), the stationary probability densities ($p_c(\mathbf{H})$, $p_u(\mathbf{H})$) of controlled and uncontrolled systems can be calculated. Then, the mean square values of the controlled and uncontrolled displacements can be calculated as follows:

$$E[Q_{ci}^2] = \int_0^\infty \langle Q_i^2\rangle p_c(\mathbf{H})d\mathbf{H}, \quad E[Q_{ui}^2] = \int_0^\infty \langle Q_i^2\rangle p_u(\mathbf{H})d\mathbf{H}. \tag{234}$$

The control effectiveness k_i and control efficiency μ_i are

$$k_i = \frac{E[Q_{ui}^2] - E[Q_{ci}^2]}{E[Q_{ui}^2]}, \quad \mu_i = \frac{k_i}{E[u_i^{*2}]}, \quad i = 1, 2, \cdots, n, \tag{235}$$

where $E[u_i^{*2}]$ is mean-square optimal controls for the worst case. k_i represents the percentage reduction in the mean square displacement of the controlled systems while μ_i denotes the relative reduction per unit of the mean-square control. $E[Q_{ui}^2]$, $E[Q_{ci}^2]$, and $E[u_i^{*2}]$ can be calculated by using Eq.(234). Obviously, higher k and μ indicate a better control strategy.

Example[30]

Consider the controlled system of non-linearly coupled two oscillators

$$\begin{aligned}
\dot{Q}_i &= P_i, \\
\dot{P}_i &= -[\bar{a}_i + \tilde{a}_i]Q_i - [\bar{c}_{ij} + \tilde{c}_{ij}]P_j - [\bar{d}_i + \tilde{d}_i]Q_j Q_j P_i \\
&\quad + u_i(\mathbf{Q}_\tau, \mathbf{P}_\tau) + e_i \xi_i(t) + \tilde{w}_i(t), \\
|u_i| &\le b_i, \quad i, j = 1, 2,
\end{aligned} \tag{236}$$

where $\bar{a}_i$, $\bar{c}_{ij}$, and $\bar{d}_i$ are the nominal values of linear stiffness, linear and nonlinear damping coefficients; e_i are amplitude of excitations; ξ_i are independent Gaussian white noises with intensities $2D_i$; $u_i(\mathbf{Q}_\tau, \mathbf{P}_\tau)$ are time-delayed bounded feedback controls due to actuator saturation. $\tilde{a}_i$, $\tilde{c}_{ij}$, $\tilde{d}_i$, and $\tilde{w}_i(t)$ are parameter and external disturbances, which are bounded, i.e., $\tilde{a}_i(t) \in [-a_i^0, a_i^0]$, $\tilde{c}_{ij}(t) \in [-c_{ij}^0, c_{ij}^0]$, $\tilde{d}_i(t) \in [-d_i^0, d_i^0]$, and $\tilde{w}_i(t) \in [-w_i^0, w_i^0]$.

Following Eq.(8), the time-delayed feedback control forces in system (236) can be expressed in terms of the state variables without time delay as follows

$$\begin{aligned}
u_i(\mathbf{Q}_\tau, \mathbf{P}_\tau) &\doteq u_i\left(Q_i \cos\omega_i\tau_i - \frac{P_i}{\omega_i}\sin\omega_i\tau_i, P_i\cos\omega_i\tau_i + Q_i\omega_i\sin\omega_i\tau_i\right) \\
&= \bar{u}_i(\mathbf{Q}, \mathbf{P}, \tau).
\end{aligned} \tag{237}$$

Substituting Eq.(237) into system (236), the optimal control problem of system (236) with time-delayed bounded feedback control is converted into the one without time delay. Then, by using the stochastic averaging method for quasi integrable Hamiltonian systems, the partially averaged Itô equation (218) can be obtained with

$$\begin{aligned}
\bar{m}_1(\mathbf{H}) &= -\bar{c}_{11}H_1 - \frac{\bar{d}_1}{2\bar{a}_1}H_1^2 - \frac{\bar{d}_1}{\bar{a}_2}H_1H_2 + D_1, \\
\bar{m}_2(\mathbf{H}) &= -\bar{c}_{22}H_2 - \frac{\bar{d}_2}{2\bar{a}_2}H_2^2 - \frac{\bar{d}_2}{\bar{a}_1}H_1H_2 + D_2, \\
\bar{\sigma}_1^2 &= 2D_1H_1, \quad \bar{\sigma}_2^2 = 2D_2H_2,
\end{aligned} \tag{238}$$

$$H = \sum_{r=1}^{2} H_r, \quad H_r = (p_r^2 + \bar{a}_r q_r^2)/2, \quad r = 1, 2.$$

For the proposed control strategy, the partially averaged cost function f_4 is of the form of Eq.(227) with $\bar{\mathbf{u}} = [\bar{u}_1, \bar{u}_2]^T$, $\mathbf{R}$=diag$(R_1,\ R_2)$ and

$$f_c(\mathbf{H}) = s_0 + s_{11}H_1 + s_{12}H_2 + s_{21}H_1^2 + s_{22}H_1H_2 + s_{23}H_2^2 \\ +s_{31}H_1^3 + s_{32}H_1^2H_2 + s_{33}H_2^2H_1 + s_{34}H_2^3. \tag{239}$$

Following Eqs.(228) and (229), the following worst-case disturbances for system (236) can be obtained

$$\begin{aligned} \tilde{a}_i^* &= -a_i^0 \mathrm{sgn}\left(\frac{dV}{dH_i} Q_i P_i\right), \quad \tilde{c}_{ij}^* = -c_{ij}^0 \mathrm{sgn}\left(\frac{dV}{dH_i} P_i P_j\right), \\ \tilde{d}_i^* &= -d_i^0 \mathrm{sgn}\left(\frac{dV}{dH_i} Q_j^2 P_i^2\right), \quad \tilde{w}_i^* = w_i^0 \mathrm{sgn}\left(\frac{dV}{dH_i} P_i\right), \end{aligned} \tag{240}$$

and the optimal control forces without time delay are of the form

$$\bar{u}_i^* = \begin{cases} -\dfrac{1}{2R_i}\dfrac{\partial V}{\partial H_i}P_i, & \left|\dfrac{1}{2R_i}\dfrac{\partial V}{\partial H_i}P_i\right| < b_i, \\ -U_i^0 \mathrm{sgn}(P_i), & \left|\dfrac{1}{2R_i}\dfrac{\partial V}{\partial H_i}P_i\right| \geq b_i, \end{cases} \tag{241}$$

where b_i are the control bounds due to the saturation. Substituting the worst-case disturbances $\tilde{a}_i^*, \tilde{c}_{ij}^*, \tilde{d}_i^*, \tilde{w}_i^*$ and the optimal controls $\bar{u}^*$ into HJI equation (225) and completing the averaging yield the final HJI equation

$$\frac{1}{2}\sigma_{11}^2(\mathbf{H})\frac{d^2V}{dH_1^2} + \frac{1}{2}\sigma_{22}^2(\mathbf{H})\frac{d^2V}{dH_2^2} + \bar{\bar{m}}_1(\mathbf{H})\frac{dV}{dH_1} + \bar{\bar{m}}_2(\mathbf{H})\frac{dV}{dH_2} + m'(\mathbf{H}) \\ = \lambda - f_c(\mathbf{H}), \tag{242}$$

where

$$\begin{aligned} \bar{\bar{m}}_1(\mathbf{H}) &= \bar{m}_1(\mathbf{H}) + \frac{2a_1^0 H_1}{\pi\sqrt{\bar{a}_1}} + c_{11}^0 H_1 + \frac{8c_{12}^0}{\pi^2}\sqrt{H_1H_2} \\ &\quad + d_1^0\left(\frac{H_1^2}{2\bar{a}_1} + \frac{H_1H_2}{\bar{a}_2}\right) + \frac{2w_1^0}{\pi}\sqrt{2H_1}, \\ \bar{\bar{m}}_2(\mathbf{H}) &= \bar{m}_2(\mathbf{H}) + \frac{2a_2^0 H_2}{\pi\sqrt{\bar{a}_2}} + c_{22}^0 H_2 + \frac{8c_{21}^0}{\pi^2}\sqrt{H_1H_2} \\ &\quad + d_2^0\left(\frac{H_2^2}{2\bar{a}_2} + \frac{H_1H_2}{\bar{a}_1}\right) + \frac{2w_2^0}{\pi}\sqrt{2H_2}, \\ m'(\mathbf{H}) &= \frac{2b_1^2R_1}{T_1}\int_{-Rd_1}^{Rd_1}\frac{1}{p_1}dq_1 - \frac{4b_1Rd_1}{T_1}\frac{\partial V}{\partial H_1} - \frac{1}{4R_1}\left(\frac{\partial V}{\partial H_1}\right)^2 H_1 \\ &\quad + \frac{(\partial V/\partial H_1)^2}{2R_1T_1}\int_{-Rd_1}^{Rd_1} p_1 dq_1 + \frac{2b_2^2R_2}{T_2}\int_{-Rd_2}^{Rd_2}\frac{1}{p_2}dq_2 - \frac{4b_2Rd_2}{T_2}\frac{\partial V}{\partial H_2} \\ &\quad - \frac{1}{4R_2}\left(\frac{\partial V}{\partial H_2}\right)^2 H_2 + \frac{(\partial V/\partial H_2)^2}{2R_2T_2}\int_{-Rd_2}^{Rd_2} p_2 dq_2, \\ Rd_i &= \frac{\sqrt{2H_i - (2R_ib_i/(\partial V/\partial H_i))^2}}{\sqrt{\bar{a}_i}}, \quad T_1 = \frac{2\pi}{\sqrt{\bar{a}_1}}, \quad T_2 = \frac{2\pi}{\sqrt{\bar{a}_2}}. \end{aligned} \tag{243}$$

$\partial V/\partial H_i$ can be obtained from solving this equation. Then the worst-case disturbances $\tilde{a}_i^*, \tilde{c}_{ij}^*, \tilde{d}_i^*, \tilde{w}_i^*$ and the optimal controls $\bar{u}_i^*$ without time delay are determined by substituting $\partial V/\partial H_i$ into Eqs.(240) and (241).

By using the transformation expressed in Eq.(230), the following optimal time-delayed controls u_i^* for system (236) can be obtained

$$u_i^* = \begin{cases} -F_i(Q_{i\tau}, P_{i\tau}), & |F_i(Q_{i\tau}, P_{i\tau})| < b_i, \\ -U_i^0 \mathrm{sgn}(F_i(Q_{i\tau}, P_{i\tau})), & |F_i(Q_{i\tau}, P_{i\tau})| \geq b_i, \end{cases}$$

$$F_i(Q_{i\tau}, P_{i\tau}) = \frac{1}{2R_i} \frac{\partial V}{\partial H_i}\bigg|_{-Q_{i\tau}\omega_i \sin\omega_i\tau_i + P_{i\tau}\cos\omega_i\tau_i}. \tag{244}$$

The response of uncontrolled and controlled, control effectiveness and control efficiency can then be obtained by substitue Eq.(244) into Eq.(218), completing the averaging and then solving the averaged FPK equation associated with completely averaged Itô equation (218).

Numerical results are shown in Figs.48–50 for system (236) with the following parameter values: $a_1 = 0.125$, $a_2 = 0.5$, $c_{11} = c_{12} = c_{21} = c_{22} = 0.02$, $d_1 = 0.05, d_2 = 0.02$, $D_1 = D_2 = 1$, $R_1 = R_2 = 0.01$, $s_{31} = s_{34} = 2$, $b_i = 0.08$, $e_1 = e_2 = 0.1$, $a_1^0 = 0.01$, $a_2^0 = 0.02$, $c_{ij}^0 = 0.001$, $d_i^0 = 0.001$, $w_i^0 = 0.001$, $\tau_i = 0.1$ unless otherwise mentioned. It is seen from these figures that the proposed control strategy has high control effectiveness and control efficiency even for long delay time or strong intensities of excitations.

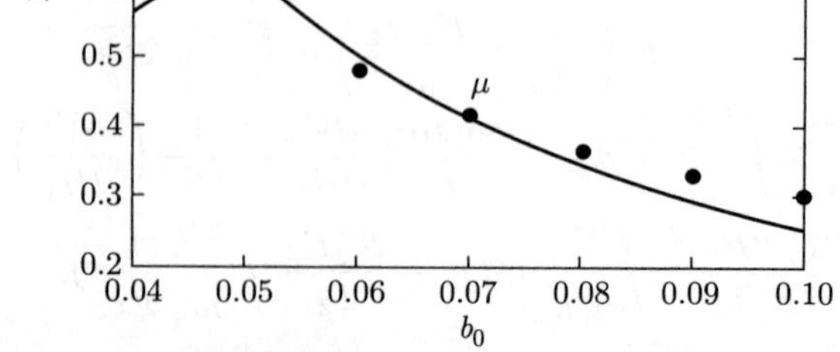

Fig. 48. Control effectiveness k and control efficiency μ for displacement of the first DOF of system (236) as functions of control bounds ($b_1 = b_2 = b_0$). — analytical result. ▼ from digital simulation.

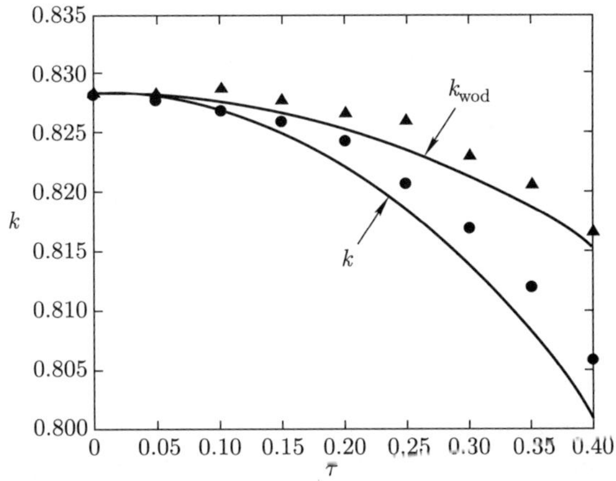

Fig. 49. Control effectiveness for displacement of the first DOF of system (236) as functions of delay time τ. k for the proposed optimal control and k_{wod} for the control without considering time delay. — analytical result. ▼ from digital simulation.

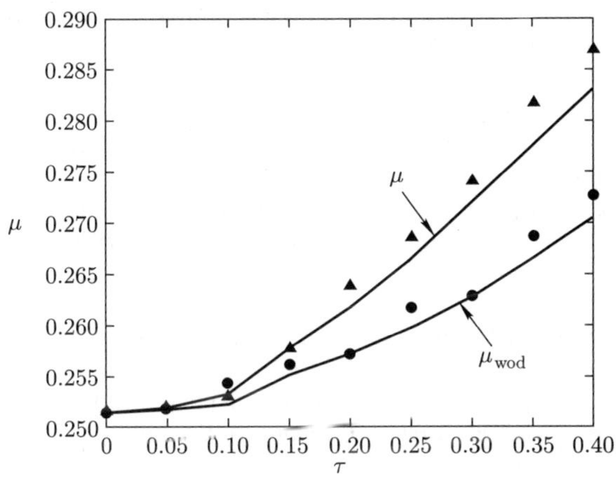

Fig. 50. Control efficiency for displacement of the first DOF of system (236) as functions of delay time τ. μ for the proposed optimal control and μ_{wod} for the control without considering time delay. — analytical result. ▼ digital simulation.

5. Concluding Remark

In this chapter, some results on the dynamics and optimal control of quasi integrable Hamiltonian systems with time-delayed feedback control under various stochastic excitations obtained in recent years by the authors and their coworkers have been reviewed. Generally, the stochastic optimal control problem for this kind of systems is of infinite dimension and it is very difficult to study. Here, we utilized the randomly periodic behavior of quasi integrable Hamiltonian systems to convert the time-delayed control forces into the control forces without time delay. The converted systems are of finite dimension and can be further simplified by applying the stochastic averaging method for quasi integrable Hamiltonian systems. Then, the methods for studying the dynamics and optimal control of quasi integrable Hamiltonian systems proposed before by the first author and his coworkers can be easily extended to the systems investigated in this chapter. All results obtained for many examples showed that our proposed procedure is feasible, effective and rather accurate, and that the time delay in the feedback control has significant effect on the dynamics and control of such systems. Finally, it is noted that more work needs to be done for the cases not covered in this review.

Acknowledgments. This work was supported by the National Natural Science Foundation of China under Grant Nos. 10772159, 10902094, 10932009, and 11072212. We would like to thank X. P. Li, R. H. Huan, and C. S. Feng for providing some meterial used in this article.

References

1. M. Malek-Zavarei and M. Jamshidi, *Time-Delay Systems: Analysis, Optimization and Applications* (North-Holland, Now York, 1987).
2. G. Stepan, *Retarded Dynamical Systems: Stability and Characteristic Functions* (Longman Scientific and Technical, Essex, 1989).
3. H. Y. Hu and Z. H. Wang, *Dynamics of Controlled Mechanical Systems with Delayed Feedback* (Springer-Verlag, Berlin, 2002).
4. J. Marshall, *Control of Time-delay System* (Peter Peregrinus, New York, 1979).
5. V. Kolmanovskiĭ and L. Shaĭkhet, Control of systems with aftereffect, *Transaction of Mathematical monographs*, **157** (Providence, RI, American Mathematical Society, 1996).
6. J. P. Richard, Time-delay systems: an overview of some recent advances and open problems, *Automatica*, **39**, 1667–1694 (2003).
7. K. Q. Gu and S. L. Niculescu, Survey on recent results in the stability and control of time-delay systems, *ASME Journal of Dynamic Systems, Measurement, and Control*, **125**, 158–165 (2003).
8. M. M. Klosek and R. Kuske, Multi-scale analysis of stochastic delay differential equations, *Multiscale Modeling and Simulation*, **3**, 706–729 (2005).
9. M. S. Fofana, Asymptotic stability of a stochastic delay equation, *Probabilistic Engineering Mechanics*, **17**, 385–392 (2002).
10. M. Grigoriu, Control of time delay linear systems with Gaussian white noise, *Probabilistic Engineering Mechanics*, **12**, 89–96 (1997).
11. M. Di Paola and A. Pirrotta, Time delay induced effects on control of linear systems under random excitation, *Probabilistic Engineering Mechanics*, **16**, 43–51(2001).
12. Z. H. Liu and W. Q. Zhu, Stochastic averaging of quasi-integrable Hamiltonian systems with delayed feedback control, *Journal of Sound and Vibration*, **299**, 178–195 (2007).
13. W. Q. Zhu and Z. H. Liu, Response of quasi-integrable Hamiltonian systems with delayed feedback bang-bang control, *Nonlinear Dynam*, **49**, 31–47 (2007).
14. X. P. Li, W. Q. Zhu, and Z. H. Liu, Stochastic averaging of quasi linear systems subject to multi-time-delayed feedback control and wide-band random Excitation, *Journal of Vibration and Control*, **15**, 1187–1205 (2009).

15. C. S. Feng and R. Liu, Response of Duffing system with delayed feedback control under bounded noise excitation, *Archive of Applied mechanics* (2012).
16. C. S. Feng and W. Q. Zhu, Response of harmonically and stochastically excited strongly nonlinear oscillators with delayed feedback bang-bang control, *Journal of Zhejiang University SCIENCE A*, **10**, 54–61 (2009).
17. C. S. Feng, Y. J. Wu, and W. Q. Zhu, Response of Duffing system with delayed feedback control under combined harmonic and real noise excitations, *Commun Nonlinear Sci Numer Simulation*, **14**, 2542–2550 (2009).
18. Z. H. Liu and W. Q. Zhu, Asymptotic Lyapunov stability with probability one of quasi-integrable Hamiltonian systems with delayed feedback control, *Automatica*, **44**, 1923–1928 (2008).
19. X.P. Li, Z. H. Liu, R. H. Huan, and W. Q. Zhu, Asymptotic Lyapunov stability with probability one of quasi linear systems subject to multi-time-delayed feedback control and wide-band parametric random excitation, *Archive of Applied Mechanics*, **79**, 1051–1061 (2009).
20. C. S. Feng and W. Q. Zhu, Asymptotic Lyapunov stability with probability one of Duffing oscillator subject to time-delayed feedback control and bounded noise excitation, *Acta mech*, **208**, 55–62 (2009).
21. C. S. Feng and S. L. Chen. Stochastic stability of Duffing-Mathieu system with delayed feedback control under white noise excitation, *Commun Nonlinear Sci Numer Simulation*, **17**, 3763–3771 (2012).
22. C. S. Feng and S. L. Chen, Stochastic stability of Mathieu-Van der Pol system with delayed feedback control, *Applied Mechanics and Materials*, **105–107**, 132–138 (2012).
23. Z. H. Liu and W. Q. Zhu, Stochastic Hopf bifurcation of quasi-integrable Hamiltonian systems with delayed feedback control, *Journal of Theoretical and Applied mechanics*, **46**, 531–550 (2008).
24. Z. H. Liu and W. Q. Zhu, Stochastic Hopf bifurcation of quasi-integrable Hamiltonian systems with multi-time-delayed feedback control and wide-band noise excitations, *Nonlinear Dynamics*, **69**, 935–947 (2012).
25. Z. H. Liu and W. Q. Zhu, First-passage failure of quasi-integrable Hamiltonian systems under time-delayed feedback control, *Journal of Sound and Vibration*, **315**, 301–317 (2008).
26. X. P. Li, Z. H. Liu, and W. Q. Zhu. First-passage failure of quasi linear systems subject to multi-time-delayed feedback control and wide-band random excitation, *Probabilistic Engineering Mechanics*, **24**, 144–150 (2009).
27. C. S. Feng and W. Q. Zhu, First-passage failure of harmonically and stochastically excited Duffing oscillator with delayed feedback control, *Science China Technological Sciences*, **5**, 1072–1077 (2011).
28. C. S. Feng, Y. J. Wu, and W. Q. Zhu, First-passage failure of strongly nonlinear oscillators with time-delayed feedback control under combined harmonic and wide-band noise excitations, *International Journal of Non-Linear Mechanics*, **44**, 269–275 (2009).
29. Z. H. Liu and W. Q. Zhu, Time-delay stochastic optimal control and stabilization of quasi-integrable Hamiltonian systems, *Probabilistic Engineering Mechanics*, **27**, 29–34 (2012).
30. R. H. Huan, Z. G. Ying, W. L. Jin, and W. Q. Zhu, Minimax optimal control of uncertain quasi-integrable Hamiltonian systems with time-delayed bounded feedback, *Probabilistic Engineering Mechanics*, **25**, 271–278 (2010).
31. W. Q. Zhu, Z. L. Huang, and Y. Q. Yang. Stochastic averaging of quasi integrable Hamiltonian systems, *ASME Journal of Applied Mechanics*, **64**, 975–984 (1997).

32. Y. Wang, Z. G. Ying, and W. Q. Zhu, A minimax optimal control strategy for stochastic uncertain quasi Hamiltonian systems, *Journal of Zhejiang University Science A*, **9**, 950–954 (2008).

Chapter 6

Delay Induced Strong and Weak Resonances in Delayed Differential Systems

Jian Xu, Wanyong Wang

School of Aerospace Engineering and Applied Mechanics
Tongji University, Shanghai 200092, *China*
xujian@tongji.edu.cn

In this chapter, a general and systematic scheme is provided to research delay-induced various double Hopf bifurcations with co-dimensions three and two, corresponding to strong and weak resonances physically derived from delay-induced double Hopf bifurcations in a general n-dimensional delayed differential systems. A extended method of multiple scales is proposed to obtain a common complex amplitude equation when the double Hopf bifurcations with frequency ratio $k_1 : k_2$ occurs in the systems under consideration. By analyzing the complex amplitude equation, we give the conditions of the strong and weak resonances respectively in some analytical expressions. The weak resonances correspond to the codimension-two double Hopf bifurcations since the amplitudes and the phases may be decoupled, but the strong resonances to the codimension-three double Hopf bifurcations. It is seen that the weak resonances happen in the system even for a lower-order ratio, i.e., $k_1 + k_2 \leq 4$. As applications, two examples are displayed. Four cases of the delay-induced resonance with $1 : 2$, $1 : 3$, $1 : 5$, and $1 : \sqrt{2}$ are discussed in detail and the corresponding normals are represented. Thus, the relative dynamical behaviors can be easily classified in the physical parameter space in terms of nonlinear dynamics. The dynamical behavior around the resonant double Hopf bifurcation point is analyzed in terms of three control parameters. The validity of analytical results is shown by their consistency with numerical simulations. The results show the provided conditions may be used to determine that a resonance is strong or weak.

1. Introduction

It is only a necessary condition that the nature or mode frequencies of two or more oscillators have an appropriate integer ratio for an internal resonance in a system with nonlinearly coupling oscillators. Basically, some sufficient conditions of the internal resonance are closely related to those coupling nonlinear terms in the system. Moreover, various internal resonances shows different forms in energy exchange among the coupling oscillators[1]. For this reason, the researches for various internal resonances in nonlinearly coupling systems modeled by ordinary or delayed differential equations (ODEs or DDEs) have been interesting for many years.

For nonlinear coupling systems molded by ODEs, in presence of internal resonance, there exists a mechanism of coupling and exchange of energy between the linear modes, and then the system will exhibit some special dynamical phenomena. These phenomena do not exist in linear systems with internal resonance and nonlinear systems without internal resonance. In fact, the linearizing system of a nonlinear system can determine the relationship of internal resonance between two linear modes. Then nonlinear terms will determine the strength of coupling between the two linear modes. Therefore, an appropriate integer ratio between the frequencies is only a necessary but not sufficient condition for these modes to be nonlinearly coupled. Many previous works, for examples, those of Lacarbonara et al.[2], Shaw and Peirre[3,4], Nayfeh and Nayfeh[5], King and Vakakis[6], and Li et al.[7,8], have discussed the interactions between the modes with internal resonance. In Ref. 1, Lacarbonara and Rega. investigate the activation/orthogonality conditions of nonlinear normal modes in the cases of internal resonance. They find that nonlinear modal energy exchanges when the nonlinear interaction coefficients are different from zero, and non-activation between the modes with internal resonances corresponds to a global nonlinear orthogonality. In Ref. 9, Lee and Perkins investigate a 2 : 1 internal resonance occurring in a nonlinear oscillations of suspended cables. When the natural frequencies satisfy 2 : 1 internal resonance, the modes of in-plane and out-of-plane are coupled through quadratic and cubic nonlinearities. Therefore, nonlinearity play an important role in nonlinear systems with internal resonance. The coefficients of nonlinear terms will determine the coupling strength between the modes with internal resonance. However, the relationship between the coefficients of nonlinear terms and the coupling strength has not been determined.

Mathematically, an internal resonance is derived from a resonant double Hopf bifurcation. The interactions between modes can be interpreted in the framework of the double Hopf bifurcation[10]. According to whether the ratio of two frequencies is rational or not, the double Hopf bifurcation can be divided into two cases: resonance and non-resonance. Then there is an important subdivision of resonance where $\omega_1 : \omega_2 = k_1 : k_2$ is rational into strong resonance ($k_1 + k_2 \leq 4$) and weak resonance ($k_1 + k_2 > 4$). Very recently, many authors have investigated these interactions by studying the unfoldings of Hopf-Hopf bifurcations and many methods and techniques from the geometric theory of dynamical systems on ordinary differential equations have been extended to study non-resonant and resonant double Hopf bifurcation including the center manifold reduction[11], the method of multiple scales[12,13], the perturbation-incremental scheme[14] and the frequency domain approach[15] and so on. Steen and Davis[16] obtained the lowest order normal form equations of 1 : 1 resonance in which resonant phase interactions arisen in the linear order as well as the nonlinear orders. By using the Poincaré Birkhoff normal form and Arnold normal form, Van Gils et al.[17] analyzed the behavior of a general four coordinate state space system near 1 : 1 resonance. An important previous work about 1 : 2 resonance is that of Knobloch and Proctor[18]. They reduced the original differential equation to a three-dimensional truncated differential equation which include two amplitude equations and one phase equation by using a center manifold and normal form approach. Then LeBlanc[19] gave the completely classification and unfold of the generic case of 1 : 2 resonance by using Liapunov-Schmidt reduction and singularity theory. Using the method of multiple scales, Luongo et al. study the cases of 1 : 1 resonance[20,21], 1 : 2 and 1 : 3 resonance[22]. Although these previous work have solved some questions, the relationship between the coefficients of nonlinear terms and resonance has not been determined. Some authors think a resonance is strong resonance only because $k_1 + k_2 \leq 4$. They obviously ignore the effects of nonlinearity.

For systems with nonlinearity and delay coupling, their characteristic equations are transcendental equations which have infinite eigenvalues. Therefore, a double Hopf bifurcation can easily occur even in systems governed by low-dimensional DDEs. For example, in Ref. 23, Campbell and Bélair obtained some resonant double Hopf bifurcation occurring in a delayed systems consisting of a harmonic oscillator with forcing. Recently, some works have been done to investigate resonant double Hopf bifurcation occurring in DDEs. Campbell and LeBlanc[24] used center manifold analysis to investigate a 1 : 2 resonant double Hopf bifurcation in a DDE. Xu et al.[14] developed an efficient method (PIS) for studying weak resonant double Hopf bifurcation in nonlinear systems with delayed feedbacks. Using the method of multiple scales, Luongo et al. study the cases of 1 : 1 resonance[20,21], 1 : 2 and 1 : 3 resonance[22]. Our previous work also study a 1 : 3 resonance in DDE[25]. However, the relationship between the coefficients of nonlinear terms and the strength of resonance are not clear.

In the case of resonance, because the ratio of frequencies is rational, the resonant terms will appear in the normal form at high enough order. Then the amplitude variables will do not decouple from the phase variables at this order. The normal form describing the evolution of the system is a three-dimensional system including two amplitude equations and one phase equation[18]. Here the differences between weak and strong resonances are that the resonant terms of weak resonances appear at higher order, but the resonant terms of strong resonance appear even at lower order. Thus truncated at lower order, the normal form of a weak resonance is a two-dimensional system consisting of two amplitude variables and it is a codimension-two bifurcation. However, a strong resonance is still a codimension-three bifurcation whose normal form is a three-dimensional system including two amplitude equations and one phase equation. Traditionally, people think that a low-order resonance where $k_1 + k_2 \leq 4$ is a strong resonance. However, in our recent researches, we find that even if $k_1 + k_2 \leq 4$, the double Hopf bifurcation still may be a weak resonance. When the coefficients of nonlinear terms satisfied some special conditions, for example, the coupled terms in normal form vanishing, the coupling between modes will be weak. At that moment, although the frequencies satisfy the relationship of internal resonance, the resonance is weak resonance. This case is rarely discussed except in Refs.26, 27. Nayfeh et al. find that some modes do not interact at all due to vanishing of nonlinear interaction coefficients in the normal forms when they investigate the nonlinear normal modes in clamped-clamped buckled beams with three-to-one internal resonance[26]. Then this phenomenon is again discussed by Xu and Chung and a 1 : 2 resonant double Hopf bifurcation is investigated in two cases including strong resonance and weak resonance[27]. In these previous papers, the authors find the phenomenon that a low-order resonance may not be a strong resonance, but they do not give a further investigation. In this paper, inspired by this phenomenon, we attempt to develop a general and systematic approach to determine some priori conditions which can determine whether a resonance is a strong resonance or not.

Summarily, the researches for the strong and weak resonances in ordinary and delayed differential systems have been not completed. One of the problems there is not a criterion to determine whether a resonance is strong or weak. This constructs our present motivation. This paper tries to provide a systematic scheme which can produce some conditions to determine that the resonance is strong or weak. Moreover, mechanism of the strong and weak resonances is also displayed by bifurcation analysis.

The outline of this paper is as follows. In Section 2, based on the method of multiple scales, a method to investigate double Hopf bifurcation is developed and the normal forms of all the double Hopf bifurcations are obtained. In Section 3, the conditions which determine a bifurcation is a strong or weak resonance are discussed and the real normal forms of all the weak and strong resonances are given. In Section 4, a simple application is given and three resonant cases are studied. In Section 5, another application is given, and the effects of nonlinearity on resonances are investigated. In Section 6, conclusions are given.

2. Analysis for Double Hopf Bifurcation

An n-dimensional delayed differential equation governed by the following equation

$$\dot{\mathbf{x}} = \mathbf{F}(\mathbf{x}, \mathbf{x}_\tau, \mu), \tag{1}$$

where $\mathbf{x} \in \mathbb{R}^n$ is the state variable, $\mu \in \mathbb{R}^m$ is a control parameter vector and $\mathbf{x}_\tau = \mathbf{x}(t-\tau)$ is the variable before time delay τ. Without loss of generality, we assume the trivial equilibrium solution $\mathbf{x} = 0$ is always a solution of (1) for any value of μ. Here we choose $\mu = (\mu_1, \mu_2, \tau) \in \mathbb{R}^3$. The conditions under which a $k_1 : k_2$ double Hopf bifurcation occurs for such a system is $\mathrm{Re}\ \lambda_1 = \mathrm{Re}\ \lambda_2 = 0$ and $\mathrm{Im}\ \lambda_1 : \mathrm{Im}\ \lambda_2 = k_1 : k_2$ (here we do not emphasize whether $k_1 : k_2$ is irrational or not, so the bifurcation may be non-resonance or resonance). Thus we can get three critical parameter values $\mu_{\mathbf{c}} = (\mu_{1c}, \mu_{2c}, \tau_c)$ at which a $k_1 : k_2$ double Hopf bifurcation occurs and two pairs of purely eigenvalues are $\lambda_{1,3} = \pm i\omega_1$ and $\lambda_{2,4} = \pm i\omega_2$ with $\omega_1 : \omega_2 = k_1 : k_2$.

In the following, the method of multiple scales will be used to investigate the dynamical behaviors around the double Hopf bifurcation point.

The solution of (1) is first expressed as

$$\mathbf{x} = \mathbf{x}(\epsilon, T_k, \ldots), \tag{2}$$

where $T_k = \epsilon^k t$ $(k = 0, 1, 2, \ldots)$ and $\epsilon \ll 1$. We assume a multiple scales expansion of the solution of (1) as

$$\begin{aligned}\mathbf{x}(t) &= \sum_{j=1}^{\infty} \epsilon^j \mathbf{x}_j(T_0, T_1, T_2, \ldots)\\ &= \epsilon \mathbf{x}_1(T_0, T_1, T_2, \ldots) + \epsilon^2 \mathbf{x}_2(T_0, T_1, T_2, \ldots) + \epsilon^3 \mathbf{x}_3(T_0, T_1, T_2, \ldots) + O(\epsilon^4), \end{aligned}\tag{3}$$

while the parameters are ordered as

$$\mu_1 = \mu_{1c} + \mu_{1\epsilon}, \quad \mu_2 = \mu_{2c} + \mu_{2\epsilon}, \quad \tau = \tau_c + \tau_\epsilon, \tag{4}$$

where the perturb parameters often have two cases: $\mu_\epsilon = \epsilon\hat{\mu}_\epsilon$ and $\mu_\epsilon = \epsilon^2\hat{\mu}_\epsilon$ $(\mu_\epsilon = (\mu_{1\epsilon}, \mu_{2\epsilon}, \tau_\epsilon))$ for different ratios of two frequencies ω_1 and ω_2. For the cases of $\omega_1 : \omega_2 = 1 : 1$ or $1 : 2$, the resonant terms appear at one or two order, therefore $\mu_\epsilon = \epsilon\hat{\mu}_\epsilon$; in other cases, we usually let $\mu_\epsilon = \epsilon^2\hat{\mu}_\epsilon$. In the following, the two cases are dealt with separately.

2.1. *The Case* $\mu_\epsilon = \epsilon\hat{\mu}_\epsilon$

In this case, the perturbation parameters are ordered as

$$\mu_{1\epsilon} = \epsilon\hat{\mu}_{1\epsilon}, \quad \mu_{2\epsilon} = \epsilon\hat{\mu}_{2\epsilon}, \quad \tau_\epsilon = \epsilon\hat{\tau}_\epsilon. \tag{5}$$

The delay term in (1) can be further expanded as

$$\begin{aligned}\mathbf{x}(t-\tau) &= \epsilon \mathbf{x}_1(T_0 - \tau_c - \epsilon\hat{\tau}_\epsilon, T_1 - \epsilon\tau_c - \epsilon^2\hat{\tau}_\epsilon, T_2 - \epsilon^2\tau_c - \epsilon^3\hat{\tau}_\epsilon, \ldots)\\ &\quad + \epsilon^2 \mathbf{x}_2(T_0 - \tau_c - \epsilon\hat{\tau}_\epsilon, T_1 - \epsilon\tau_c - \epsilon^2\hat{\tau}_\epsilon, T_2 - \epsilon^2\tau_c - \epsilon^3\hat{\tau}_\epsilon, \ldots)\\ &\quad + \epsilon^3 \mathbf{x}_3(T_0 - \tau_c - \epsilon\hat{\tau}_\epsilon, T_1 - \epsilon\tau_c - \epsilon^2\hat{\tau}_\epsilon, T_2 - \epsilon^2\tau_c - \epsilon^3\hat{\tau}_\epsilon, \ldots)\\ &\quad + \cdots\\ &= \epsilon \mathbf{x}_{1\tau} + \epsilon^2(\mathbf{x}_{2\tau} - \tau_c D_1 \mathbf{x}_{1\tau} - \hat{\tau}_\epsilon D_0 \mathbf{x}_{1\tau})\\ &\quad + \epsilon^3\Big(\tau_c\hat{\tau}_\epsilon D_1 D_0 \mathbf{x}_{1\tau} + \frac{1}{2}\tau_c^2 D_1^2 \mathbf{x}_{1\tau} - \tau_c D_2 \mathbf{x}_{1\tau}\\ &\quad - \tau_c D_1 \mathbf{x}_{2\tau} + \frac{1}{2}\hat{\tau}_\epsilon^2 D_0^2 \mathbf{x}_{1\tau} - \hat{\tau}_\epsilon D_0 \mathbf{x}_{2\tau}\\ &\quad - \hat{\tau}_\epsilon D_1 \mathbf{x}_{1\tau} + \mathbf{x}_{3\tau}\Big) + O(\epsilon^4),\end{aligned}\tag{6}$$

where $D_k =$ tial/tialT_k and $\mathbf{x}_{j\tau} = \mathbf{x}_j(T_0 - \tau_c, T_1, T_2, \ldots)$. Substituting (3), (5), and (6) into (1), expanding $\mathbf{F}$ as well as equating separately coefficients of like powers of ϵ yield the following equations, given by

$$D_0\mathbf{x}_1 - \mathbf{F}^{\mathbf{0}}_{\mathbf{x}}\mathbf{x}_1 - \mathbf{F}^{\mathbf{0}}_{\mathbf{x}_\tau}\mathbf{x}_{1\tau} = 0, \tag{7}$$

$$\begin{aligned} D_0\mathbf{x}_2 - \mathbf{F}^{\mathbf{0}}_{\mathbf{x}}\mathbf{x}_2 - \mathbf{F}^{\mathbf{0}}_{\mathbf{x}_\tau}\mathbf{x}_{2\tau} &= -D_1\mathbf{x}_1 + \mathbf{F}^{\mathbf{0}}_{\mathbf{x}\mu_\epsilon}\mathbf{x}_1\hat{\mu}_\epsilon + \mathbf{F}^{\mathbf{0}}_{\mathbf{x}_\tau\mu_\epsilon}\mathbf{x}_{1\tau}\hat{\mu}_\epsilon \\ &\quad + \frac{1}{2}\mathbf{F}^{\mathbf{0}}_{\mathbf{xx}}\mathbf{x}_1^2 + \mathbf{F}^{\mathbf{0}}_{\mathbf{xx}_\tau}\mathbf{x}_1\mathbf{x}_{1\tau} + \frac{1}{2}\mathbf{F}^{\mathbf{0}}_{\mathbf{x}_\tau\mathbf{x}_\tau}\mathbf{x}_{1\tau}^2 \\ &\quad + \mathbf{F}^{\mathbf{0}}_{\mathbf{x}_\tau}(-\hat{\tau}_\epsilon D_0\mathbf{x}_{1\tau} - \tau_c D_1\mathbf{x}_{1\tau}), \end{aligned} \tag{8}$$

$$\begin{aligned} D_0\mathbf{x}_3 - \mathbf{F}^{\mathbf{0}}_{\mathbf{x}}\mathbf{x}_3 - \mathbf{F}^{\mathbf{0}}_{\mathbf{x}_\tau}\mathbf{x}_{3\tau} &= -D_2\mathbf{x}_1 - D_1\mathbf{x}_2 + \frac{1}{2}\mathbf{F}^{\mathbf{0}}_{\mathbf{xx}\mu_\epsilon}\mathbf{x}_1^2\hat{\mu}_\epsilon + \frac{1}{6}\mathbf{F}^{\mathbf{0}}_{\mathbf{xxx}}\mathbf{x}_1^3 \\ &\quad + \mathbf{F}^{\mathbf{0}}_{\mathbf{x}\mu_\epsilon}\mathbf{x}_2\hat{\mu}_\epsilon + \mathbf{F}^{\mathbf{0}}_{\mathbf{xx}}\mathbf{x}_1\mathbf{x}_2 + \mathbf{F}^{\mathbf{0}}_{\mathbf{xx}_\tau\mu_\epsilon}\mathbf{x}_1\mathbf{x}_{1\tau}\hat{\mu}_\epsilon \\ &\quad + \frac{1}{2}\mathbf{F}^{\mathbf{0}}_{\mathbf{xxx}_\tau}\mathbf{x}_1^2\mathbf{x}_{1\tau} + \mathbf{F}^{\mathbf{0}}_{\mathbf{xx}_\tau}\mathbf{x}_{1\tau}\mathbf{x}_2 + \frac{1}{2}\mathbf{F}^{\mathbf{0}}_{\mathbf{x}_\tau\mathbf{x}_\tau\mu_\epsilon}\mathbf{x}_{1\tau}^2\hat{\mu}_\epsilon \\ &\quad + \frac{1}{2}\mathbf{F}^{\mathbf{0}}_{\mathbf{xx}_\tau\mathbf{x}_\tau}\mathbf{x}_1\mathbf{x}_{1\tau}^2 + \frac{1}{6}\mathbf{F}^{\mathbf{0}}_{\mathbf{x}_\tau\mathbf{x}_\tau\mathbf{x}_\tau}\mathbf{x}_{1\tau}^3 \\ &\quad + \mathbf{F}^{\mathbf{0}}_{\mathbf{x}_\tau\mu_\epsilon}\hat{\mu}_\epsilon(\mathbf{x}_{2\tau} - \tau_c D_1\mathbf{x}_{1\tau} - \hat{\tau}_\epsilon D_0\mathbf{x}_{1\tau}) \\ &\quad + \mathbf{F}^{\mathbf{0}}_{\mathbf{xx}_\tau}\mathbf{x}_1(\mathbf{x}_{2\tau} - \tau_c D_1\mathbf{x}_{1\tau} - \hat{\tau}_\epsilon D_0\mathbf{x}_{1\tau}) \\ &\quad + \mathbf{F}^{\mathbf{0}}_{\mathbf{x}_\tau\mathbf{x}_\tau}\mathbf{x}_{1\tau}(\mathbf{x}_{2\tau} - \tau_c D_1\mathbf{x}_{1\tau} - \hat{\tau}_\epsilon D_0\mathbf{x}_{1\tau}) \\ &\quad + \mathbf{F}^{\mathbf{0}}_{\mathbf{x}_\tau}(-\tau_c D_2\mathbf{x}_{1\tau} - D_1\mathbf{x}_{2\tau} + \frac{1}{2}D_1^2\mathbf{x}_{1\tau}\tau_c^2 - \hat{\tau}_\epsilon D_1\mathbf{x}_{1\tau} \\ &\quad - \hat{\tau}_\epsilon D_0\mathbf{x}_{2\tau} + \tau_c\hat{\tau}_\epsilon D_0 D_1\mathbf{x}_{1\tau} - \frac{1}{2}\hat{\tau}_\epsilon^2 D_0^2\mathbf{x}_{1\tau}), \end{aligned} \tag{9}$$

$$D_0\mathbf{x}_4 - \mathbf{F}^{\mathbf{0}}_{\mathbf{x}}\mathbf{x}_4 - \mathbf{F}^{\mathbf{0}}_{\mathbf{x}_\tau}\mathbf{x}_{4\tau} = -D_3\mathbf{x}_1 - D_2\mathbf{x}_2 + \cdots, \tag{10}$$

$$\cdots\cdots\cdots\cdots$$

where $\mathbf{F}^{\mathbf{0}}_{\mathbf{x}\mu_\epsilon} = \dfrac{\mathrm{tial}^2\mathbf{F}(\mathbf{0},\mathbf{0},0)}{\mathrm{tial}\mathbf{x}\ \mathrm{tial}\mu_\epsilon}$, $\mathbf{F}^{\mathbf{0}}_{\mathbf{x}_\tau\mu_\epsilon} = \dfrac{\mathrm{tial}^2\mathbf{F}(\mathbf{0},\mathbf{0},0)}{\mathrm{tial}\mathbf{x}_\tau\ \mathrm{tial}\mu_\epsilon}$ and similarly for higher-order derivatives. $\mathbf{x}_{j\tau} = \mathbf{x}_j(T_0 - \tau_c, T_1, T_2, \ldots)$, $j = 1, 2, 3, \ldots$. It follows from (7) that one has

$$\mathbf{x}_1 = A_1(T_1, T_2, \ldots)\mathbf{p}_1 e^{i\omega_1 T_0} + A_2(T_1, T_2, \ldots)\mathbf{p}_2 e^{i\omega_2 T_0} + c.c., \tag{11}$$

where $A_j (j = 1, 2)$ are complex constants, $\mathbf{p}_j$ are the right eigenvectors of $\mathbf{F}^{\mathbf{0}}_{\mathbf{x}} + \mathbf{F}^{\mathbf{0}}_{\mathbf{x}_\tau} e^{-i\omega_j\tau_c}$ associated with the eigenvalues $i\omega_j$, given by

$$(\mathbf{F}^{\mathbf{0}}_{\mathbf{x}} + \mathbf{F}^{\mathbf{0}}_{\mathbf{x}_\tau} e^{-i\omega_j\tau} - i\mathbf{E}\omega_j)\mathbf{p}_j = 0, \tag{12}$$

and *c.c.* stands for the complex conjugate of the preceding terms. Substituting (11) into (8), we obtain

$$
\begin{aligned}
&D_0\mathbf{x}_2 - \mathbf{F}^{\mathbf{0}}_{\mathbf{x}}\mathbf{x}_2 - \mathbf{F}^{\mathbf{0}}_{\mathbf{x}_\tau}\mathbf{x}_{2\tau}\\
&= [(\mathbf{F}^{\mathbf{0}}_{\mathbf{x}\mu_\epsilon}\mathbf{p}_1\hat{\mu}_\epsilon + \mathbf{F}^{\mathbf{0}}_{\mathbf{x}_\tau\mu_\epsilon}\mathbf{p}_1\hat{\mu}_\epsilon e^{-i\omega_1\tau_c} - i\omega_1\hat{\tau}_\epsilon\mathbf{F}^{\mathbf{0}}_{\mathbf{x}_\tau}\mathbf{p}_1 e^{-i\omega_1\tau_c})A_1 \qquad (13)\\
&-(\tau_c\mathbf{F}^{\mathbf{0}}_{\mathbf{x}_\tau}\mathbf{p}_1 e^{-i\omega_1\tau_c} + \mathbf{p}_1)D_1A_1]e^{i\omega_1T_0}\\
&+[(\mathbf{F}^{\mathbf{0}}_{\mathbf{x}\mu_\epsilon}\mathbf{p}_2\hat{\mu}_\epsilon + \mathbf{F}^{\mathbf{0}}_{\mathbf{x}_\tau\mu_\epsilon}\mathbf{p}_2\hat{\mu}_\epsilon e^{-i\omega_2\tau_c} - i\omega_2\hat{\tau}_\epsilon\mathbf{F}^{\mathbf{0}}_{\mathbf{x}_\tau}\mathbf{p}_2 e^{-i\omega_2\tau_c})A_2\\
&-(\tau_c\mathbf{F}^{\mathbf{0}}_{\mathbf{x}_\tau}\mathbf{p}_2 e^{-i\omega_2\tau_c} + \mathbf{p}_2)D_1A_2]e^{i\omega_2T_0}\\
&+\left(\frac{1}{2}\mathbf{F}^{\mathbf{0}}_{\mathbf{xx}}\mathbf{p}_1^2 + \mathbf{F}^{\mathbf{0}}_{\mathbf{xx}_\tau}\mathbf{p}_1^2 e^{-i\omega_1\tau_c} + \frac{1}{2}\mathbf{F}^{\mathbf{0}}_{\mathbf{x}_\tau\mathbf{x}_\tau}\mathbf{p}_1^2 e^{-2i\omega_1\tau_c}\right)A_1^2e^{2i\omega_1T_0}\\
&+\left(\frac{1}{2}\mathbf{F}^{\mathbf{0}}_{\mathbf{xx}}\mathbf{p}_2^2 + \mathbf{F}^{\mathbf{0}}_{\mathbf{xx}_\tau}\mathbf{p}_2^2 e^{-i\omega_2\tau_c} + \frac{1}{2}\mathbf{F}^{\mathbf{0}}_{\mathbf{x}_\tau\mathbf{x}_\tau}\mathbf{p}_2^2 e^{-2i\omega_2\tau_c}\right)A_2^2e^{2i\omega_2T_0}\\
&+\left(\frac{1}{2}\mathbf{F}^{\mathbf{0}}_{\mathbf{xx}}\mathbf{p}_1\bar{\mathbf{p}}_1 + \mathbf{F}^{\mathbf{0}}_{\mathbf{xx}_\tau}\mathbf{p}_1\bar{\mathbf{p}}_1 e^{i\omega_1\tau_c} + \frac{1}{2}\mathbf{F}^{\mathbf{0}}_{\mathbf{x}_\tau\mathbf{x}_\tau}\mathbf{p}_1\bar{\mathbf{p}}_1\right)A_1\bar{A}_1\\
&+\left(\frac{1}{2}\mathbf{F}^{\mathbf{0}}_{\mathbf{xx}}\mathbf{p}_2\bar{\mathbf{p}}_2 + \mathbf{F}^{\mathbf{0}}_{\mathbf{xx}_\tau}\mathbf{p}_2\bar{\mathbf{p}}_2 e^{i\omega_2\tau_c} + \frac{1}{2}\mathbf{F}^{\mathbf{0}}_{\mathbf{x}_\tau\mathbf{x}_\tau}\mathbf{p}_2\bar{\mathbf{p}}_2\right)A_2\bar{A}_2\\
&+[\mathbf{F}^{\mathbf{0}}_{\mathbf{xx}}\mathbf{p}_1\mathbf{p}_2 + \mathbf{F}^{\mathbf{0}}_{\mathbf{xx}_\tau}(\mathbf{p}_1\mathbf{p}_2e^{-i\omega_1\tau_c} + \mathbf{p}_1\mathbf{p}_2e^{-i\omega_2\tau_c})\\
&+\mathbf{F}^{\mathbf{0}}_{\mathbf{x}_\tau\mathbf{x}_\tau}\mathbf{p}_1\mathbf{p}_2e^{-i(\omega_1+\omega_2)\tau_c}]A_1A_2e^{i(\omega_1+\omega_2)T_0}\\
&+[\mathbf{F}^{\mathbf{0}}_{\mathbf{xx}}\bar{\mathbf{p}}_1\mathbf{p}_2 + \mathbf{F}^{\mathbf{0}}_{\mathbf{xx}_\tau}(\bar{\mathbf{p}}_1\mathbf{p}_2e^{i\omega_1\tau_c} + \bar{\mathbf{p}}_1\mathbf{p}_2e^{-i\omega_2\tau_c})\\
&+\mathbf{F}^{\mathbf{0}}_{\mathbf{x}_\tau\mathbf{x}_\tau}\bar{\mathbf{p}}_1\mathbf{p}_2e^{i(\omega_1-\omega_2)\tau_c}]\bar{A}_1A_2e^{i(\omega_2-\omega_1)T_0} + c.c.,
\end{aligned}
$$

where the resonant condition $\omega_1 : \omega_2$ is not considered. In some special cases, such as $k_1 : k_2 = 1 : 1$ and $k_1 : k_2 = 1 : 2$, some terms in (13) can be combined. For example, if $k_1 : k_2 = 1 : 2$, the terms including $e^{i(\omega_2-\omega_1)T_0}$ are also resonant terms and can be combined with the terms including $e^{i\omega_1T_0}$, and the terms including $e^{2i\omega_1T_0}$ can be combined with the terms including $e^{i\omega_2T_0}$ There are secular terms in (13). To eliminate these secular terms, we let the coefficients of secular terms vanish. Then we have the following solvability conditions

$$
\begin{aligned}
D_1A_1 &= \mathscr{L}_1(A_1, \bar{A}_1^{k_2-1}A_2^{k_1}),\\
D_1A_2 &= \mathscr{L}_1(A_2, A_1^{k_2}\bar{A}_2^{k_1-1}).
\end{aligned} \qquad (14)
$$

The solution of (13) has the type as

$$
\begin{aligned}
\mathbf{x}_2 &= \mathbf{z}_1A_1e^{i\omega_1T_0} + \mathbf{z}_2A_2e^{i\omega_2T_0} + \mathbf{z}_{11}A_1^2e^{2i\omega_1T_0} \qquad (15)\\
&+\mathbf{z}_{22}A_2^2e^{2i\omega_2T_0} + \mathbf{z}_{\bar{1}1}\bar{A}_1A_1 + \mathbf{z}_{\bar{2}2}\bar{A}_2A_2\\
&+\mathbf{z}_{12}A_1A_2e^{i(\omega_1+\omega_2)T_0} + \mathbf{z}_{\bar{1}2}\bar{A}_1A_2e^{i(\omega_2-\omega_1)T_0} + c.c.\ ,
\end{aligned}
$$

where the coefficients z_i's and z_{ij}'s can be obtained by solving linear algebraic equations which can be obtained by substituting (15) into (13) and comparing the coefficients of the two sides. Substituting (11) and (15) into (9) and eliminating the secular terms, then we can get

$$\begin{aligned} D_2A_1 &= \mathscr{L}_2(A_1, \bar{A}_1A_1, \bar{A}_2A_2, \bar{A}_1^{k_2-1}A_2^{k_1}), \\ D_2A_2 &= \mathscr{L}_2(A_2, \bar{A}_1A_1, \bar{A}_2A_2, A_1^{k_2}\bar{A}_2^{k_1-1}). \end{aligned} \tag{16}$$

The right part of (9) is determined, so the solution of (9) can be obtained. Then substituting $\mathbf{x}_1$, $\mathbf{x}_2$ and $\mathbf{x}_3$ into (10), we can get

$$\begin{aligned} D_3A_1 &= \mathscr{L}_3(A_1, \bar{A}_1A_1, \bar{A}_2A_2, \bar{A}_1^{k_2-1}A_2^{k_1}, A_1^{k_2+1}\bar{A}_2^{k_1}), \\ D_3A_2 &= \mathscr{L}_3(A_2, \bar{A}_1A_1, \bar{A}_2A_2, A_1^{k_2}\bar{A}_2^{k_1-1}, \bar{A}_1^{k_2}A_2^{k_1+1}). \end{aligned} \tag{17}$$

By taking account that

$$\frac{dA}{dt} = \epsilon D_1A + \epsilon^2 D_2A + \epsilon^3 D_3A + \cdots, \tag{18}$$

then coming back to the true time t, all the solvable conditions can be combined in a single equation[30]

$$\begin{aligned} \dot{A}_1 &= \mathscr{L}(A_1, \bar{A}_1A_1, \bar{A}_2A_2, \bar{A}_1^{k_2-1}A_2^{k_1}, A_1^{k_2+1}\bar{A}_2^{k_1}), \\ \dot{A}_2 &= \mathscr{L}(A_2, \bar{A}_1A_1, \bar{A}_2A_2, A_1^{k_2}\bar{A}_2^{k_1-1}, \bar{A}_1^{k_2}A_2^{k_1+1}), \end{aligned} \tag{19}$$

where the terms $\bar{A}_1^{k_2-1}A_2^{k_1}$, $A_1^{k_2+1}\bar{A}_2^{k_1}$, $A_1^{k_2}\bar{A}_2^{k_1-1}$, and $\bar{A}_1^{k_2}A_2^{k_1+1}$ exist due to the resonant condition $\omega_1 : \omega_2 = k_1 : k_2$. In the first one of (19), if one of the terms can be divided into $g_1(\bar{A}_1A_1, \bar{A}_2A_2)\bar{A}_1^{k_2-1}A_2^{k_1}$ or $g_2(\bar{A}_1A_1, \bar{A}_2A_2)A_1^{k_2+1}\bar{A}_2^{k_1}$, this term is a resonant term. Similarly, in the second one of (19), if one of the terms can be divided into $f_1(\bar{A}_1A_1, \bar{A}_2A_2)A_1^{k_2}\bar{A}_2^{k_1-1}$ or $f_2(\bar{A}_1A_1, \bar{A}_2A_2)\bar{A}_1^{k_2}A_2^{k_1+1}$, this term is also a resonant term.

2.2. *The Case* $\mu_\epsilon = \epsilon^2\hat{\mu}_\epsilon$

In this case, the perturbation parameters are rescaled as

$$\mu_{1\epsilon} = \epsilon^2\hat{\mu}_{1\epsilon}, \quad \mu_{2\epsilon} = \epsilon^2\hat{\mu}_{2\epsilon}, \quad \tau_\epsilon = \epsilon^2\hat{\tau}_\epsilon. \tag{20}$$

The delay term in (1) can be further expanded as

$$\begin{aligned} \mathbf{x}(t-\tau) &= \epsilon\mathbf{x}_1(T_0 - \tau_c - \epsilon^2\hat{\tau}_\epsilon, T_1 - \epsilon\tau_c - \epsilon^3\hat{\tau}_\epsilon, T_2 - \epsilon^2\tau_c - \epsilon^4\hat{\tau}_\epsilon, \ldots) \\ &\quad + \epsilon^2\mathbf{x}_2(T_0 - \tau_c - \epsilon^2\hat{\tau}_\epsilon, T_1 - \epsilon\tau_c - \epsilon^3\hat{\tau}_\epsilon, T_2 - \epsilon^2\tau_c - \epsilon^4\hat{\tau}_\epsilon, \ldots) \\ &\quad + \epsilon^3\mathbf{x}_3(T_0 - \tau_c - \epsilon^2\hat{\tau}_\epsilon, T_1 - \epsilon\tau_c - \epsilon^3\hat{\tau}_\epsilon, T_2 - \epsilon^2\tau_c - \epsilon^4\hat{\tau}_\epsilon, \ldots) \\ &\quad + \cdots \\ &= \epsilon\mathbf{x}_{1\tau} + \epsilon^2(\mathbf{x}_{2\tau} - \tau_c D_1\mathbf{x}_{1\tau}) + \epsilon^3\bigg(\frac{1}{2}\tau_c^2 D_1^2\mathbf{x}_{1\tau} - \tau_c D_2\mathbf{x}_{1\tau} \\ &\quad - \tau_c D_1\mathbf{x}_{2\tau} - \hat{\tau}_\epsilon D_0\mathbf{x}_{1\tau} + \mathbf{x}_{3\tau}\bigg) + O(\epsilon^4), \end{aligned} \tag{21}$$

where $D_k = \text{tial}/\text{tial}T_k$ and $\mathbf{x}_{j\tau} = \mathbf{x}_j(T_0 - \tau_c, T_1, T_2, \ldots)$. Substituting Equations (3), (20), and (21) into (1), expanding $\mathbf{F}$ as well as equating separately coefficients of like powers of ϵ yield the following equations, given by

$$D_0\mathbf{x}_1 - \mathbf{F}^\mathbf{0}_\mathbf{x}\mathbf{x}_1 - \mathbf{F}^\mathbf{0}_{\mathbf{x}_\tau}\mathbf{x}_{1\tau} = 0, \tag{22}$$

$$\begin{aligned} &D_0\mathbf{x}_2 - \mathbf{F}^\mathbf{0}_\mathbf{x}\mathbf{x}_2 - \mathbf{F}^\mathbf{0}_{\mathbf{x}_\tau}\mathbf{x}_{2\tau} \\ &= -D_1\mathbf{x}_1 - \tau_c\mathbf{F}^\mathbf{0}_{\mathbf{x}_\tau}D_1\mathbf{x}_{1\tau} + \frac{1}{2}\mathbf{F}^\mathbf{0}_{\mathbf{xx}}\mathbf{x}_1^2 + \mathbf{F}^\mathbf{0}_{\mathbf{xx}_\tau}\mathbf{x}_1\mathbf{x}_{1\tau} + \frac{1}{2}\mathbf{F}^\mathbf{0}_{\mathbf{x}_\tau\mathbf{x}_\tau}\mathbf{x}_{1\tau}^2, \end{aligned} \tag{23}$$

$$\begin{aligned} &D_0\mathbf{x}_3 - \mathbf{F}^\mathbf{0}_\mathbf{x}\mathbf{x}_3 - \mathbf{F}^\mathbf{0}_{\mathbf{x}_\tau}\mathbf{x}_{3\tau} \\ &= \mathbf{F}^\mathbf{0}_{\mathbf{x}\mu_\epsilon}\mathbf{x}_1\hat{\mu}_\epsilon + \frac{1}{6}\mathbf{F}^\mathbf{0}_{\mathbf{xxx}}\mathbf{x}_1^3 + \mathbf{F}^\mathbf{0}_{\mathbf{xx}}\mathbf{x}_1\mathbf{x}_2 + \mathbf{F}^\mathbf{0}_{\mathbf{x}_\tau\mu_\epsilon}\mathbf{x}_{1\tau}\hat{\mu}_\epsilon \\ &+\frac{1}{2}\mathbf{F}^\mathbf{0}_{\mathbf{xxx}_\tau}\mathbf{x}_1^2\mathbf{x}_{1\tau} + \frac{1}{2}\mathbf{F}^\mathbf{0}_{\mathbf{xx}_\tau\mathbf{x}_\tau}\mathbf{x}_1\mathbf{x}_{1\tau}^2 + \mathbf{F}^\mathbf{0}_{\mathbf{xx}_\tau}\mathbf{x}_{1\tau}\mathbf{x}_2 \\ &+\frac{1}{6}\mathbf{F}^\mathbf{0}_{\mathbf{x}_\tau\mathbf{x}_\tau\mathbf{x}_\tau}\mathbf{x}_{1\tau}^3 + \mathbf{F}^\mathbf{0}_{\mathbf{xx}_\tau}\mathbf{x}_1(\mathbf{x}_{2\tau} - \tau_c D_1\mathbf{x}_{1\tau}) + \mathbf{F}^\mathbf{0}_{\mathbf{x}_\tau\mathbf{x}_\tau}\mathbf{x}_{1\tau}(\mathbf{x}_{2\tau} - \tau_c D_1\mathbf{x}_{1\tau}) \\ &+\mathbf{F}^\mathbf{0}_{\mathbf{x}_\tau}\left(-\tau_c D_2\mathbf{x}_{1\tau} - \hat{\tau}_\epsilon D_0\mathbf{x}_{1\tau} - \tau_c D_1\mathbf{x}_{2\tau} + \frac{1}{2}D_1^2\mathbf{x}_{1\tau}\tau_c^2\right) - D_1\mathbf{x}_2 - D_2\mathbf{x}_1, \end{aligned} \tag{24}$$

$$D_0\mathbf{x}_4 - \mathbf{F}^\mathbf{0}_\mathbf{x}\mathbf{x}_4 - \mathbf{F}^\mathbf{0}_{\mathbf{x}_\tau}\mathbf{x}_{4\tau} = -D_3\mathbf{x}_1 - D_2\mathbf{x}_2 + \cdots, \tag{25}$$

$$\cdots\cdots\cdots\cdots$$

where $\mathbf{F}^\mathbf{0}_{\mathbf{x}\mu_\epsilon} = \dfrac{\text{tial}^2\mathbf{F}(\mathbf{0},\mathbf{0},0)}{\text{tial}\mathbf{x}\ \text{tial}\mu_\epsilon}$, $\mathbf{F}^\mathbf{0}_{\mathbf{x}_\tau\mu_\epsilon} = \dfrac{\text{tial}^2\mathbf{F}(\mathbf{0},\mathbf{0},0)}{\text{tial}\mathbf{x}_\tau\ \text{tial}\mu_\epsilon}$ and similarly for higher-order derivatives. $\mathbf{x}_{j\tau} = \mathbf{x}_j(T_0 - \tau_c, T_1, T_2, \ldots)$, $j = 1, 2, 3, \ldots$. The general solution of (22) is

$$\mathbf{x}_1 = A_1(T_1, T_2, \ldots)\mathbf{p}_1 e^{i\omega_1 T_0} + A_2(T_1, T_2, \ldots)\mathbf{p}_2 e^{i\omega_2 T_0} + c.c., \tag{26}$$

where $A_j(j = 1, 2)$ are complex constants, and $\mathbf{p}_j$ are the right eigenvectors of $\mathbf{F}^\mathbf{0}_\mathbf{x} + \mathbf{F}^\mathbf{0}_{\mathbf{x}_\tau}e^{-i\omega_j\tau_c}$ associated with the eigenvalues $i\omega_j$, given by

$$(\mathbf{F}^\mathbf{0}_\mathbf{x} + \mathbf{F}^\mathbf{0}_{\mathbf{x}_\tau}e^{-i\omega_j\tau_c} - i\mathbf{E}\omega_j)\mathbf{p}_j = 0, \tag{27}$$

and *c.c.* stands for the complex conjugate of the preceding terms. Substituting (26) into (23), we have

$$\begin{aligned} &D_0\mathbf{x}_2 - \mathbf{F}^\mathbf{0}_\mathbf{x}\mathbf{x}_2 - \mathbf{F}^\mathbf{0}_{\mathbf{x}_\tau}\mathbf{x}_{2\tau} \\ &= [-(\tau_c\mathbf{F}^\mathbf{0}_{\mathbf{x}_\tau}\mathbf{p}_1 e^{-i\omega_1\tau_c} + \mathbf{p}_1)D_1A_1]e^{i\omega_1T_0} \\ &+[-(\tau_c\mathbf{F}^\mathbf{0}_{\mathbf{x}_\tau}\mathbf{p}_2 e^{-i\omega_2\tau_c} + \mathbf{p}_2)D_1A_2]e^{i\omega_2T_0} \\ &+\left(\frac{1}{2}\mathbf{F}^\mathbf{0}_{\mathbf{xx}}\mathbf{p}_1^2 + \mathbf{F}^\mathbf{0}_{\mathbf{xx}_\tau}\mathbf{p}_1^2 e^{-i\omega_1\tau_c} + \frac{1}{2}\mathbf{F}^\mathbf{0}_{\mathbf{x}_\tau\mathbf{x}_\tau}\mathbf{p}_1^2 e^{-2i\omega_1\tau_c}\right)A_1^2 e^{2i\omega_1T_0} \end{aligned} \tag{28}$$

$$+\left(\frac{1}{2}\mathbf{F}^{\mathbf{0}}_{\mathbf{xx}}\mathbf{p}_2^2+\mathbf{F}^{\mathbf{0}}_{\mathbf{xx}_\tau}\mathbf{p}_2^2e^{-i\omega_2\tau_c}+\frac{1}{2}\mathbf{F}^{\mathbf{0}}_{\mathbf{x}_\tau\mathbf{x}_\tau}\mathbf{p}_2^2e^{-2i\omega_2\tau_c}\right)A_2^2e^{2i\omega_2T_0}$$

$$+\left(\frac{1}{2}\mathbf{F}^{\mathbf{0}}_{\mathbf{xx}}\mathbf{p}_1\bar{\mathbf{p}}_1+\mathbf{F}^{\mathbf{0}}_{\mathbf{xx}_\tau}\mathbf{p}_1\bar{\mathbf{p}}_1e^{i\omega_1\tau_c}+\frac{1}{2}\mathbf{F}^{\mathbf{0}}_{\mathbf{x}_\tau\mathbf{x}_\tau}\mathbf{p}_1\bar{\mathbf{p}}_1\right)A_1\bar{A}_1$$

$$+\left(\frac{1}{2}\mathbf{F}^{\mathbf{0}}_{\mathbf{xx}}\mathbf{p}_2\bar{\mathbf{p}}_2+\mathbf{F}^{\mathbf{0}}_{\mathbf{xx}_\tau}\mathbf{p}_2\bar{\mathbf{p}}_2e^{i\omega_2\tau_c}+\frac{1}{2}\mathbf{F}^{\mathbf{0}}_{\mathbf{x}_\tau\mathbf{x}_\tau}\mathbf{p}_2\bar{\mathbf{p}}_2\right)A_2\bar{A}_2$$

$$+[\mathbf{F}^{\mathbf{0}}_{\mathbf{xx}}\mathbf{p}_1\mathbf{p}_2+\mathbf{F}^{\mathbf{0}}_{\mathbf{xx}_\tau}(\mathbf{p}_1\mathbf{p}_2e^{-i\omega_1\tau_c}+\mathbf{p}_1\mathbf{p}_2e^{-i\omega_2\tau_c})$$

$$+\mathbf{F}^{\mathbf{0}}_{\mathbf{x}_\tau\mathbf{x}_\tau}\mathbf{p}_1\mathbf{p}_2e^{-i(\omega_1+\omega_2)\tau_c}]A_1A_2e^{i(\omega_1+\omega_2)T_0}$$

$$+[\mathbf{F}^{\mathbf{0}}_{\mathbf{xx}}\bar{\mathbf{p}}_1\mathbf{p}_2+\mathbf{F}^{\mathbf{0}}_{\mathbf{xx}_\tau}(\bar{\mathbf{p}}_1\mathbf{p}_2e^{i\omega_1\tau_c}+\bar{\mathbf{p}}_1\mathbf{p}_2e^{-i\omega_2\tau_c})$$

$$+\mathbf{F}^{\mathbf{0}}_{\mathbf{x}_\tau\mathbf{x}_\tau}\bar{\mathbf{p}}_1\mathbf{p}_2e^{i(\omega_1-\omega_2)\tau_c}]\bar{A}_1A_2e^{i(\omega_2-\omega_1)T_0}+c.c..$$

If we let the coefficients of secular terms in (28) equate zero, we obtain

$$\begin{aligned}(\tau_c\mathbf{F}^{\mathbf{0}}_{\mathbf{x}_\tau}\mathbf{p}_1e^{-i\omega_1\tau_c}+\mathbf{p}_1)D_1A_1=0,\\(\tau_c\mathbf{F}^{\mathbf{0}}_{\mathbf{x}_\tau}\mathbf{p}_2e^{-i\omega_2\tau_c}+\mathbf{p}_2)D_1A_2=0.\end{aligned}\tag{29}$$

From (29), we can get $D_1A_1=0$ and $D_1A_2=0$. Then the complex amplitudes A_1 and A_2 are not associated with the time scale T_1. We can get the solution of (28)

$$\mathbf{x}_2=\mathbf{z}_{11}A_1^2e^{2i\omega_1T_0}+\mathbf{z}_{22}A_2^2e^{2i\omega_2T_0}+\mathbf{z}_{\bar{1}1}\bar{A}_1A_1+\mathbf{z}_{\bar{2}2}\bar{A}_2A_2\tag{30}$$

$$+\mathbf{z}_{12}A_1A_2e^{i(\omega_1+\omega_2)T_0}+\mathbf{z}_{\bar{1}2}\bar{A}_1A_2e^{i(\omega_2-\omega_1)T_0}+c.c..$$

Substituting (26) and (30) into (24), we can get

$$\begin{aligned}D_2A_1=\mathscr{L}_2(A_1,\bar{A}_1A_1,\bar{A}_2A_2,\bar{A}_1^{k_2-1}A_2^{k_1}),\\D_2A_2=\mathscr{L}_2(A_2,\bar{A}_1A_1,\bar{A}_2A_2,A_1^{k_2}\bar{A}_2^{k_1-1}).\end{aligned}\tag{31}$$

Then by solving (28), the solution can be obtained. We substitute these solutions into next order perturbation equation and another solvability conditions can be obtained. Repeating this process up to the highest order accounted for in the analysis and coming back to the true time[30] t, we can get the complex amplitude equations

$$\begin{aligned}\dot{A}_1=\mathscr{L}(A_1,\bar{A}_1A_1,\bar{A}_2A_2,\bar{A}_1^{k_2-1}A_2^{k_1},A_1^{k_2+1}\bar{A}_2^{k_1}),\\\dot{A}_2=\mathscr{L}(A_2,\bar{A}_1A_1,\bar{A}_2A_2,A_1^{k_2}\bar{A}_2^{k_1-1},\bar{A}_1^{k_2}A_2^{k_1+1}).\end{aligned}\tag{32}$$

In this section, by using the method of multiple scales, we get the complex amplitude equations of double Hopf bifurcation occurring in a general delayed differential equation. The complex amplitude equations may be considered as one of normal forms[29] applied for bifurcation analysis. Since the ratio of two frequencies is not given in this process, the double Hopf bifurcation may be a resonance or non-resonance. In the case of non-resonances, all the solvable conditions do not contain resonant terms. Therefore we can conclude the following theorem.

Theorem 1. *If a double Hopf bifurcation is a non-resonance, then the normal forms are expressed*

$$\begin{aligned} \dot{A}_1 &= \mathscr{L}(A_1, \bar{A}_1 A_1, \bar{A}_2 A_2), \\ \dot{A}_2 &= \mathscr{L}(A_2, \bar{A}_1 A_1, \bar{A}_2 A_2); \end{aligned} \tag{33}$$

if a double Hopf bifurcation is a resonance, then the normal forms are expressed

$$\begin{aligned} \dot{A}_1 &= \mathscr{L}(A_1, \bar{A}_1 A_1, \bar{A}_2 A_2, \bar{A}_1^{k_2-1} A_2^{k_1}, A_1^{k_2+1} \bar{A}_2^{k_1}), \\ \dot{A}_2 &= \mathscr{L}(A_2, \bar{A}_1 A_1, \bar{A}_2 A_2, A_1^{k_2} \bar{A}_2^{k_1-1}, \bar{A}_1^{k_2} A_2^{k_1+1}), \end{aligned} \tag{34}$$

and at least one of the coefficients of resonant terms does not vanish.

3. Conditions for Strong Resonances and Weak Resonances

In previous section, we obtain the normal forms of a double Hopf bifurcation including non-resonances and resonances. The methods for studying non-resonances are mature, thus we do not give more discusses. Here we focus on the study on resonant double Hopf bifurcation. We discuss the normal form by dividing resonant double Hopf bifurcation into two cases: high-order resonances ($m + n > 4$) and low-order resonances ($m + n \leq 4$). High-order resonances are weak resonances, because the resonant terms appear at high-order terms (higher than three-order). However, it does not imply that all the low-order resonances are strong resonances. If all the coefficients of resonant terms vanish, the bifurcation is still a weak resonance. In the case of strong resonances, resonant terms exist even at lower order, thus the interaction between the two modes associated with strong resonance are strong. This interaction can result in the exchange of energy between the two modes. This phenomenon can not only induce rich dynamical behaviors but also offer some direct applications in engineering and science researches. In the following, we will give the real normal forms of weak resonances and strong resonances and the conditions which determine whether a resonance is a strong resonance or not.

3.1. *High-order Resonances*

The strong resonances are characterized by the fact that the amplitude variables do not decouple from the phase variables even at lower order[18]. Namely, a resonance is called a strong resonance, when the resonant terms appear in the normal form truncated at three order. Otherwise the resonance is a weak resonance. For a high-order resonance where $k_1+k_2 > 4$, the resonant terms appear first at $k_1+k_2-1(> 3)$-order, thus the high-order resonance is a weak resonance. The normal forms of a higher-order resonance truncated at three order are

$$\begin{aligned} \dot{A}_1 &= C_{1\mu}\hat{\mu}_\epsilon A_1 + C_{11\bar{1}}A_1^2\bar{A}_1 + C_{12\bar{2}}A_1A_2\bar{A}_2, \\ \dot{A}_2 &= C_{2\mu}\hat{\mu}_\epsilon A_2 + C_{1\bar{1}2}A_1\bar{A}_1A_2 + C_{22\bar{2}}A_2^2\bar{A}_2. \end{aligned} \tag{35}$$

Then the complex amplitudes can be written as

$$A_1 = r_1 e^{i\theta_1}, \quad A_2 = r_2 e^{i\theta_2}, \tag{36}$$

Substituting (36) into (35) and separating the real and imaginary parts, we can obtained the real normal form of higher-order resonance

$$\begin{aligned} \dot{r}_1 &= \mathrm{Re}[C_{1\mu}\hat{\mu}_\epsilon]r_1 + \frac{1}{4}\mathrm{Re}[C_{11\bar{1}}]r_1^3 + \frac{1}{4}\mathrm{Re}[C_{12\bar{2}}]r_1r_2^2, \\ \dot{r}_2 &= \mathrm{Re}[C_{2\mu}\hat{\mu}_\epsilon]r_2 + \frac{1}{4}\mathrm{Re}[C_{1\bar{1}2}]r_1^2r_2 + \frac{1}{4}\mathrm{Re}[C_{22\bar{2}}]r_2^3, \end{aligned} \tag{37}$$

where the amplitude variables decouple from the phase variables.

3.2. *Low-order Resonances*

Low-order resonances include three cases: 1 : 1, 1 : 2, and 1 : 3 resonances. In these three cases, their resonant terms appear first at one-order, two-order, and three-order terms. Even if we truncate the normal forms at lower-order (three-order), the resonant terms still appear at the normal forms. Therefore, low-order resonances are possibly strong resonances. However, it does not imply all the low-order resonances are strong resonances. If all the coefficients of resonant terms in normal forms vanish, the resonance is still a weak resonance. In the following, we will discuss these two cases and give their normal forms.

3.2.1. 1 : 3 *Resonance*

The normal forms of a 1 : 3 resonance occurring in a system with time delay are obtained by using the method of multiple scales in our previous work[25]. The normal forms are as follows

$$\begin{aligned} \dot{A}_1 &= C_{1\mu}\mu_\epsilon A_1 + C_{11\bar{1}}A_1^2\bar{A}_1 + C_{\bar{1}\bar{1}2}\bar{A}_1^2A_2 + C_{12\bar{2}}A_1A_2\bar{A}_2, \\ \dot{A}_2 &= C_{2\mu}\mu_\epsilon A_2 + C_{111}A_1^3 + C_{1\bar{1}2}A_1\bar{A}_1A_2 + C_{22\bar{2}}A_2^2\bar{A}_2, \end{aligned} \tag{38}$$

where A_1 and A_2 are the complex amplitudes. Owing to the two frequencies satisfying $\omega_1 : \omega_2 = 1 : 3$, there are two resonant terms $C_{\bar{1}\bar{1}2}\bar{A}_1^2A_2$ and $C_{111}A_1^3$ appearing in the normal forms. These two terms result in the differences between strong resonances and weak resonances. The complex amplitudes can be written as

$$A_1 = r_1 e^{i\theta_1}, \quad A_2 = r_2 e^{i\theta_2}, \tag{39}$$

Substituting (39) into (38) and separating the real and imaginary parts, we can obtained

$$
\begin{aligned}
\dot{r}_1 &= \mathrm{Re}[C_{1\mu}\hat{\mu}_\epsilon]r_1 + \frac{1}{4}\mathrm{Re}[C_{11\bar{1}}]r_1^3 + \frac{1}{4}\mathrm{Re}[C_{12\bar{2}}]r_1r_2^2 \\
&\quad + \frac{1}{4}\mathrm{Re}[C_{\bar{1}\bar{1}2}]r_1^2r_2\cos(\theta_2 - 3\theta_1) \\
&\quad - \frac{1}{4}\mathrm{Im}[C_{\bar{1}\bar{1}2}]r_1^2r_2\sin(\theta_2 - 3\theta_1) + O(5), \\
\dot{r}_2 &= \mathrm{Re}[C_{2\mu}\hat{\mu}_\epsilon]r_2 + \frac{1}{4}\mathrm{Re}[C_{1\bar{1}2}]r_1^2r_2 + \frac{1}{4}\mathrm{Re}[C_{22\bar{2}}]r_2^3 \\
&\quad + \frac{1}{4}\mathrm{Re}[C_{111}]\cos(\theta_2 - 3\theta_1)r_1^3 \\
&\quad + \frac{1}{4}\mathrm{Im}[C_{111}]\sin(\theta_2 - 3\theta_1)r_1^3 + O(5), \\
\dot{\theta}_1 &= \mathrm{Im}[C_{1\mu}\hat{\mu}_\epsilon] + \frac{1}{4}\mathrm{Im}[C_{11\bar{1}}]r_1^2 + \frac{1}{4}\mathrm{Im}[C_{12\bar{2}}]r_2^2 \\
&\quad + \frac{1}{4}\mathrm{Re}[C_{\bar{1}\bar{1}2}]r_1r_2\sin(\theta_2 - 3\theta_1) \\
&\quad + \frac{1}{4}\mathrm{Im}[C_{\bar{1}\bar{1}2}]r_1r_2\cos(\theta_2 - 3\theta_1) + O(4), \\
\dot{\theta}_2 &= \mathrm{Im}[C_{2\mu}\hat{\mu}_\epsilon] + \frac{1}{4}\mathrm{Im}[C_{1\bar{1}2}]r_1^2 + \frac{1}{4}\mathrm{Im}[C_{22\bar{2}}]r_2^2 \\
&\quad - \frac{1}{4}\mathrm{Re}[C_{111}]\sin(\theta_2 - 3\theta_1)\frac{r_1^3}{r_2} \\
&\quad + \frac{1}{4}\mathrm{Im}[C_{111}]\cos(\theta_2 - 3\theta_1)\frac{r_1^3}{r_2} + O(4).
\end{aligned} \tag{40}
$$

Assuming $\phi = \theta_2 - 3\theta_1$, we have

$$
\begin{aligned}
\dot{r}_1 &= \mathrm{Re}[C_{1\mu}\hat{\mu}_\epsilon]r_1 + \frac{1}{4}\mathrm{Re}[C_{11\bar{1}}]r_1^3 + \frac{1}{4}\mathrm{Re}[C_{12\bar{2}}]r_1r_2^2 \\
&\quad + \frac{1}{4}\mathrm{Re}[C_{\bar{1}\bar{1}2}]r_1^2r_2\cos\phi - \frac{1}{4}\mathrm{Im}[C_{\bar{1}\bar{1}2}]r_1^2r_2\sin\phi + O(5), \\
\dot{r}_2 &= \mathrm{Re}[C_{2\mu}\hat{\mu}_\epsilon]r_2 + \frac{1}{4}\mathrm{Re}[C_{1\bar{1}2}]r_1^2r_2 + \frac{1}{4}\mathrm{Re}[C_{22\bar{2}}]r_2^3 \\
&\quad + \frac{1}{4}\mathrm{Re}[C_{111}]r_1^3\cos\phi + \frac{1}{4}\mathrm{Im}[C_{111}]r_1^3\sin\phi + O(5), \\
\dot{\phi} &= \mathrm{Im}[C_{2\mu}\hat{\mu}_\epsilon] - 3\mathrm{Im}[C_{1\mu}\hat{\mu}_\epsilon] + \frac{1}{4}\mathrm{Im}[C_{1\bar{1}2}]r_1^2 \\
&\quad + \frac{1}{4}\mathrm{Im}[C_{22\bar{2}}]r_2^2 - \frac{3}{4}\mathrm{Im}[C_{11\bar{1}}]r_1^2 - \frac{3}{4}\mathrm{Im}[C_{12\bar{2}}]r_2^2 \\
&\quad - \frac{1}{4}\mathrm{Re}[C_{111}]\frac{r_1^3}{r_2}\sin\phi + \frac{1}{4}\mathrm{Im}[C_{111}]\frac{r_1^3}{r_2}\cos\phi \\
&\quad - \frac{3}{4}\mathrm{Re}[C_{\bar{1}\bar{1}2}]r_1r_2\sin\phi - \frac{3}{4}\mathrm{Im}[C_{\bar{1}\bar{1}2}]r_1r_2\cos\phi + O(4).
\end{aligned} \tag{41}
$$

If the coefficients C_{111} and $C_{\bar{1}\bar{1}2}$ in normal forms (38) do not vanish, (41) is the lowest dimension of the normal form of 1 : 3 resonance. However, if $C_{111} = 0$ and $C_{\bar{1}\bar{1}2} = 0$, (40) can be reduced further as follows

$$\begin{aligned}
\dot{r}_1 &= \mathrm{Re}[C_{1\mu}\hat{\mu}_\epsilon]r_1 + \frac{1}{4}\mathrm{Re}[C_{11\bar{1}}]r_1^3 + \frac{1}{4}\mathrm{Re}[C_{12\bar{2}}]r_1 r_2^2 + O(5),\\
\dot{r}_2 &= \mathrm{Re}[C_{2\mu}\hat{\mu}_\epsilon]r_2 + \frac{1}{4}\mathrm{Re}[C_{1\bar{1}2}]r_1^2 r_2 + \frac{1}{4}\mathrm{Re}[C_{22\bar{2}}]r_2^3 + O(5),\\
\dot{\theta}_1 &= \mathrm{Im}[C_{1\mu}\hat{\mu}_\epsilon] + \frac{1}{4}\mathrm{Im}[C_{11\bar{1}}]r_1^2 + \frac{1}{4}\mathrm{Im}[C_{12\bar{2}}]r_2^2 + O(4),\\
\dot{\theta}_2 &= \mathrm{Im}[C_{2\mu}\hat{\mu}_\epsilon] + \frac{1}{4}\mathrm{Im}[C_{1\bar{1}2}]r_1^2 + \frac{1}{4}\mathrm{Im}[C_{22\bar{2}}]r_2^2 + O(4).
\end{aligned} \tag{42}$$

In (42), two amplitude variables decouple from the phases, then the normal from can be reduced to a two-dimensional system. This is the same as the normal form of a weak resonance. Thus we give the following theorem.

Theorem 2. *The normal forms of a* 1 : 3 *resonant double Hopf bifurcation can be obtained as* (38). *If the coefficients of resonant terms in normal forms satisfy* $C_{111} \neq 0$ *and/or* $C_{\bar{1}\bar{1}2} \neq 0$, *then the* 1 : 3 *resonant double Hopf bifurcation is a strong resonance and the real normal form is a three-dimensional system given by* (41); *if the coefficients of resonant terms in normal forms satisfy* $C_{111} = 0$ *and* $C_{\bar{1}\bar{1}2} = 0$, *then the* 1 : 3 *resonant double Hopf bifurcation is a weak resonance and the real normal form is a two-dimensional system given by the first two of* (42).

3.2.2. 1 : 2 *Resonance*

The resonant terms of a 1 : 2 resonance appear first in quadratic terms. The normal forms of a 1 : 2 resonance are obtained by using the method of multiple scales as follows

$$\begin{aligned}
\dot{A}_1 &= (C_{1\mu}\hat{\mu}_\epsilon + C_{1\mu\mu}\hat{\mu}_\epsilon^2)A_1 + (C_{\bar{1}2} + C_{\bar{1}2\mu}\hat{\mu}_\epsilon)\bar{A}_1 A_2\\
&\quad + C_{11\bar{1}}A_1^2\bar{A}_1 + C_{12\bar{2}}A_1 A_2 \bar{A}_2,\\
\dot{A}_2 &= (C_{2\mu}\hat{\mu}_\epsilon + C_{2\mu\mu}\hat{\mu}_\epsilon^2)A_2 + (C_{11} + C_{11\mu}\hat{\mu}_\epsilon)A_1^2\\
&\quad + C_{1\bar{1}2}A_1\bar{A}_1 A_2 + C_{22\bar{2}}A_2^2\bar{A}_2,
\end{aligned} \tag{43}$$

where $C_{j\mu}\hat{\mu}_\epsilon$, $C_{j\mu\mu}\hat{\mu}_\epsilon^2$, $C_{\bar{1}2\mu}\hat{\mu}_\epsilon$, and $C_{11\mu}\hat{\mu}_\epsilon$ are the complex coefficients including perturbation parameters, and C_{ijk} are complex coefficients. The quadratic terms in (43) are the resonant terms which appear owing to the two frequencies satisfying $\omega_1 : \omega_2 = 1 : 2$. The coefficients of resonant terms $C_{1\mu\mu}\hat{\mu}_\epsilon^2$ and $C_{2\mu\mu}\hat{\mu}_\epsilon^2$ are the functions of perturbation parameters. But the coefficients $C_{\bar{1}2}$ and C_{11} are complex numbers which are not related with perturbation parameters $\hat{\mu}_\epsilon$ and $\hat{\tau}_\epsilon$, and they are determined by not only linear terms but also quadratic terms. One or two of the coefficients of resonant terms are not equal to zero, then the resonant terms will exist in the amplitude equations. Thus the amplitudes do not decoupled from the phases, and this is an important character of strong resonance. On the contrary, if the coefficients of resonant terms are all equal to zero, in this case, the bifurcation is a weak resonance. Then we will get the following theorem.

Theorem 3. *The normal forms of a* 1 : 2 *resonant double Hopf bifurcation can be obtained as* (43). *If the coefficients of resonant terms in normal forms satisfy* $C_{\bar{1}2} + C_{\bar{1}2\mu}\mu_\epsilon \neq 0$ *and/or* $C_{11} + C_{11\mu}\mu_\epsilon \neq 0$, *then the* 1 : 2 *resonant double Hopf bifurcation is a strong resonance and the real normal form is a three-dimensional system given by* (44)

$$
\begin{aligned}
\dot{r}_1 =\ & \mathrm{Re}[C_{1\mu}\hat{\mu}_\epsilon + C_{1\mu\mu}\hat{\mu}_\epsilon^2]r_1 + \frac{1}{4}\mathrm{Re}[C_{11\bar{1}}]r_1^3 \\
& + \frac{1}{4}\mathrm{Re}[C_{12\bar{2}}]r_1 r_2^2 + \frac{1}{2}\mathrm{Re}[C_{\bar{1}2} + C_{\bar{1}2\mu}\hat{\mu}_\epsilon]r_1 r_2 \cos\phi \\
& - \frac{1}{2}\mathrm{Im}[C_{\bar{1}2} + C_{\bar{1}2\mu}\hat{\mu}_\epsilon]r_1 r_2 \sin\phi + O(5), \\
\dot{r}_2 =\ & \mathrm{Re}[C_{2\mu}\hat{\mu}_\epsilon + C_{2\mu\mu}\hat{\mu}_\epsilon^2]r_2 + \frac{1}{4}\mathrm{Re}[C_{1\bar{1}2}]r_1^2 r_2 \\
& + \frac{1}{4}\mathrm{Re}[C_{22\bar{2}}]r_2^3 + \frac{1}{2}\mathrm{Re}[C_{11} + C_{11\mu}\hat{\mu}_\epsilon]r_1^2 \cos\phi \\
& + \frac{1}{2}\mathrm{Im}[C_{11} + C_{11\mu}\hat{\mu}_\epsilon]r_1^2 \sin\phi + O(5), \\
\dot{\phi} =\ & \mathrm{Im}[C_{2\mu}\hat{\mu}_\epsilon + C_{2\mu\mu}\hat{\mu}_\epsilon^2] - 2\mathrm{Im}[C_{1\mu}\hat{\mu}_\epsilon + C_{1\mu\mu}\hat{\mu}_\epsilon^2] \\
& + \frac{1}{4}\mathrm{Im}[C_{1\bar{1}2}]r_1^2 + \frac{1}{4}\mathrm{Im}[C_{22\bar{2}}]r_2^2 - \frac{1}{2}\mathrm{Im}[C_{11\bar{1}}]r_1^2 \\
& - \frac{1}{2}\mathrm{Im}[C_{12\bar{2}}]r_2^2 - \frac{1}{2}\mathrm{Re}[C_{11} + C_{11\mu}\hat{\mu}_\epsilon]\frac{r_1^2}{r_2}\sin\phi \\
& + \frac{1}{2}\mathrm{Im}[C_{11} + C_{11\mu}\hat{\mu}_\epsilon]\frac{r_1^2}{r_2}\cos\phi \\
& - \mathrm{Re}[C_{\bar{1}2} + C_{\bar{1}2\mu}\hat{\mu}_\epsilon]r_2 \sin\phi \\
& - \mathrm{Im}[C_{\bar{1}2} + C_{\bar{1}2\mu}\hat{\mu}_\epsilon]r_2 \cos\phi + O(4);
\end{aligned}
\tag{44}
$$

if the coefficients of resonant terms in normal forms satisfy $C_{\bar{1}2} + C_{\bar{1}2\mu}\hat{\mu}_\epsilon = 0$ *and* $C_{11} + C_{11\mu}\hat{\mu}_\epsilon = 0$, *then the* 1 : 2 *resonant double Hopf bifurcation is a weak resonance and the real normal form is a two-dimensional system given by* (45)

$$
\begin{aligned}
\dot{r}_1 =\ & \mathrm{Re}[C_{1\mu}\hat{\mu}_\epsilon + C_{1\mu\mu}\hat{\mu}_\epsilon^2]r_1 + \frac{1}{4}\mathrm{Re}[C_{11\bar{1}}]r_1^3 \\
& + \frac{1}{4}\mathrm{Re}[C_{12\bar{2}}]r_1 r_2^2 + O(5), \\
\dot{r}_2 =\ & \mathrm{Re}[C_{2\mu}\hat{\mu}_\epsilon + C_{2\mu\mu}\hat{\mu}_\epsilon^2]r_2 + \frac{1}{4}\mathrm{Re}[C_{1\bar{1}2}]r_1^2 r_2 \\
& + \frac{1}{4}\mathrm{Re}[C_{22\bar{2}}]r_2^3 + O(5).
\end{aligned}
\tag{45}
$$

3.2.3. 1 : 1 *Resonance*

Using the method of of multiple scales, we can get the normal forms of a 1 : 1 resonance as follows

$$\begin{aligned}\dot{A}_1 &= C_{1\mu}\hat{\mu}_\epsilon A_1 + C_{2\mu}\hat{\mu}_\epsilon A_2 + C_{11\bar{1}}A_1^2\bar{A}_1 + C_{12\bar{2}}A_1A_2\bar{A}_2\\ &\quad + C_{22\bar{2}}A_2^2\bar{A}_2 + C_{1\bar{1}2}A_1\bar{A}_1A_2 + C_{22\bar{1}}A_2^2\bar{A}_1 + C_{11\bar{2}}A_1^2\bar{A}_2,\\ \dot{A}_2 &= C'_{1\mu}\hat{\mu}_\epsilon A_1 + C'_{2\mu}\hat{\mu}_\epsilon A_2 + C'_{11\bar{1}}A_1^2\bar{A}_1 + C'_{12\bar{2}}A_1A_2\bar{A}_2\\ &\quad + C'_{22\bar{2}}A_2^2\bar{A}_2 + C'_{1\bar{1}2}A_1\bar{A}_1A_2 + C'_{22\bar{1}}A_2^2\bar{A}_1 + C'_{11\bar{2}}A_1^2\bar{A}_2,\end{aligned} \tag{46}$$

where $C_{j\mu}\hat{\mu}_\epsilon$ and C_{ijk} are complex coefficients. In (46), there are ten resonant terms which are $C_{2\mu}\hat{\mu}_\epsilon A_2$, $C_{22\bar{2}}A_2^2\bar{A}_2$, $C_{1\bar{1}2}A_1\bar{A}_1A_2$, $C_{22\bar{2}}A_2^2\bar{A}_1$, $C_{11\bar{2}}A_1^2\bar{A}_2$,$C'_{1\mu}\hat{\mu}_\epsilon A_1$, $C'_{11\bar{1}}A_1^2\bar{A}_1$, $C'_{12\bar{2}}A_1A_2\bar{A}_2$, $C'_{22\bar{1}}A_2^2\bar{A}_1$, $C'_{11\bar{2}}A_1^2\bar{A}_2$. The resonant terms appear in the linear orders as well as the nonlinear orders. Similarly, when the coefficients of all the resonant terms are equal to zero, the resonance is a weak resonance. We can get the following theorem.

Theorem 4. *The normal forms of a* 1 : 1 *resonant double Hopf bifurcation can be obtained as* (46). *If at least one of the coefficients* $C_{2\mu}\hat{\mu}_\epsilon$, $C_{22\bar{2}}$, $C_{1\bar{1}2}$, $C_{22\bar{2}}$, $C_{11\bar{2}}$, $C'_{1\mu}\hat{\mu}_\epsilon$, $C'_{11\bar{1}}$, $C'_{12\bar{2}}$, $C'_{22\bar{1}}$, $C'_{11\bar{2}}$ *in normal forms* (46) *do not vanish, then the* 1 : 1 *resonant double Hopf bifurcation is a strong resonance and the real normal form is a three-dimensional system given by* (47)

$$\begin{aligned}\dot{r}_1 &= \mathrm{Re}[C_{1\mu}\hat{\mu}_\epsilon]r_1 + (\mathrm{Re}[C_{2\mu}\hat{\mu}_\epsilon]\cos\phi - \mathrm{Im}[C_{2\mu}\hat{\mu}_\epsilon]\sin\phi)r_2\\ &\quad + \frac{1}{4}\mathrm{Re}[C_{11\bar{1}}]r_1^3 + \frac{1}{4}\mathrm{Re}[C_{12\bar{2}}]r_1r_2^2\\ &\quad + \frac{1}{4}(\mathrm{Re}[C_{22\bar{2}}]\cos\phi - \mathrm{Im}[C_{22\bar{2}}]\sin\phi)r_2^3\\ &\quad + \frac{1}{4}(\mathrm{Re}[C_{1\bar{1}2}]\cos\phi - \mathrm{Im}[C_{1\bar{1}2}]\sin\phi)r_1^2r_2\\ &\quad + \frac{1}{4}(\mathrm{Re}[C_{22\bar{1}}]\cos 2\phi - \mathrm{Im}[C_{22\bar{1}}]\sin 2\phi)r_1r_2^2\\ &\quad + \frac{1}{4}(\mathrm{Re}[C_{11\bar{2}}]\cos\phi + \mathrm{Im}[C_{11\bar{2}}]\sin\phi)r_1^2r_2 + O(5),\\ \dot{r}_2 &= (\mathrm{Re}[C'_{1\mu}\hat{\mu}_\epsilon]\cos\phi + \mathrm{Im}[C'_{1\mu}\hat{\mu}_\epsilon]\sin\phi)r_1 + \mathrm{Re}[C'_{2\mu}\hat{\mu}_\epsilon]r_2\\ &\quad + \frac{1}{4}(\mathrm{Re}[C'_{11\bar{1}}]\cos\phi + \mathrm{Im}[C'_{11\bar{1}}]\sin\phi)r_1^3\\ &\quad + \frac{1}{4}(\mathrm{Re}[C'_{12\bar{2}}]\cos\phi + \mathrm{Im}[C'_{12\bar{2}}]\sin\phi)r_1r_2^2\\ &\quad + \frac{1}{4}\mathrm{Re}[C'_{22\bar{2}}]r_2^3 + \frac{1}{4}\mathrm{Re}[C'_{1\bar{1}2}]r_1^2r_2\\ &\quad + \frac{1}{4}(\mathrm{Re}[C'_{22\bar{1}}]\cos\phi - \mathrm{Im}[C'_{22\bar{1}}]\sin\phi)r_1r_2^2\\ &\quad + \frac{1}{4}(\mathrm{Re}[C'_{11\bar{2}}]\cos 2\phi + \mathrm{Im}[C'_{11\bar{2}}]\sin 2\phi)r_1^2r_2 + O(5),\\ \dot{\phi} &= \mathrm{Im}[C'_{2\mu}\hat{\mu}_\epsilon] + (-\mathrm{Re}[C'_{1\mu}\hat{\mu}_\epsilon]\sin\phi + \mathrm{Im}[C'_{1\mu}\hat{\mu}_\epsilon]\cos\phi)\frac{r_1}{r_2}\end{aligned} \tag{47}$$

$$
\begin{aligned}
&+\frac{1}{4}(-\mathrm{Re}[C'_{11\bar{1}}]\sin\phi+\mathrm{Im}[C'_{11\bar{1}}]\cos\phi)r_1^3\\
&+\frac{1}{4}(-\mathrm{Re}[C'_{12\bar{2}}]\sin\phi+\mathrm{Im}[C'_{12\bar{2}}]\cos\phi)r_1r_2^2\\
&+\frac{1}{4}\mathrm{Im}[C'_{22\bar{2}}]r_2^3+\tfrac{1}{4}\mathrm{Im}[C'_{1\bar{1}2}]r_1^2r_2\\
&+\frac{1}{4}(\mathrm{Re}[C'_{22\bar{1}}]\sin\phi-\mathrm{Im}[C'_{22\bar{1}}]\cos\phi)r_1r_2^2\\
&+\frac{1}{4}(-\mathrm{Re}[C'_{11\bar{2}}]\sin 2\phi+\mathrm{Im}[C'_{11\bar{2}}]\cos 2\phi)r_1^2r_2\\
&-\mathrm{Im}[C_{1\mu}\hat{\mu}_\epsilon]-(\mathrm{Re}[C_{2\mu}\hat{\mu}_\epsilon]\sin\phi+\mathrm{Im}[C_{2\mu}\hat{\mu}_\epsilon]\cos\phi)\frac{r_2}{r_1}\\
&-\frac{1}{4}(\mathrm{Re}[C_{22\bar{2}}]\sin\phi+\mathrm{Im}[C_{22\bar{2}}]\cos\phi)r_2^3\\
&-\frac{1}{4}(\mathrm{Re}[C_{1\bar{1}2}]\sin\phi+\mathrm{Im}[C_{1\bar{1}2}]\cos\phi)r_1^2r_2\\
&-\frac{1}{4}\mathrm{Im}[C_{11\bar{1}}]r_1^3+\frac{1}{4}\mathrm{Im}[C_{12\bar{2}}]r_1r_1^2\\
&-\frac{1}{4}(\mathrm{Re}[C_{22\bar{1}}]\sin 2\phi+\mathrm{Im}[C_{22\bar{1}}]\cos 2\phi)r_1r_2^2\\
&-\frac{1}{4}(-\mathrm{Re}[C_{11\bar{2}}]\sin\phi+\mathrm{Im}[C_{11\bar{2}}]\cos\phi)r_1^2r_2+O(4);
\end{aligned}
$$

if the coefficients $C_{2\mu}\hat{\mu}_\epsilon$, $C_{22\bar{2}}$, $C_{1\bar{1}2}$, $C_{22\bar{2}}$, $C_{11\bar{2}}$, $C'_{1\mu}\hat{\mu}_\epsilon$, $C'_{11\bar{1}}$, $C'_{12\bar{2}}$, $C'_{22\bar{1}}$, $C'_{11\bar{2}}$ *in normal forms* (46) *all vanish, then the* $1:1$ *resonant double Hopf bifurcation is a weak resonance and the real normal form is a two-dimensional system given by* (48)

$$
\begin{aligned}
\dot{r}_1&=\mathrm{Re}[C_{1\mu}\mu_\epsilon]r_1+\frac{1}{4}\mathrm{Re}[C_{11\bar{1}}]r_1^3+\frac{1}{4}\mathrm{Re}[C_{12\bar{2}}]r_1r_2^2+O(5),\\
\dot{r}_2&=\mathrm{Re}[C'_{2\mu}\mu_\epsilon]r_2+\frac{1}{4}\mathrm{Re}[C'_{22\bar{2}}]r_2^3+\frac{1}{4}\mathrm{Re}[C'_{1\bar{1}2}]r_1^2r_2+O(5).
\end{aligned}
\tag{48}
$$

In this section, we obtain the normal forms of weak and strong resonances. In their normal forms, there are three control parameters, but not all of them are bifurcation parameters. Strong resonance is a codimension-three bifurcation, hence the three parameters are bifurcation parameters which are independent physical parameters. In the case of weak resonance, only two of them are bifurcation parameters.

4. Weak and Strong Resonances in Delayed Feedback Systems

As an application, we give a simple example to investigate several cases of resonant double Hopf bifurcations. We consider a system with delayed feedback control, given by[23]

$$
\ddot{x}+b\dot{x}(t)+ax(t)=f(\dot{x}(t-\tau)),
\tag{49}
$$

where f may be nonlinear. Letting $x_1 = x$ and $x_2 = \dot{x}$, the full system can be rewritten as

$$\begin{aligned} \dot{x}_1(t) &= x_2(t), \\ \dot{x}_2(t) &= -ax_1(t) - bx_2(t) + f(x_2(t-\tau)). \end{aligned} \tag{50}$$

Without lost of generally, we let the point (0, 0) is a equilibrium of system (50), then system (50) can be written as

$$\begin{aligned} \dot{x}_1(t) &= x_2(t), \\ \dot{x}_2(t) &= -ax_1(t) - bx_2(t) + gx_2(t-\tau) + g_2x_2^2(t-\tau) \\ &\quad + g_3x_2^3(t-\tau) + O(x_2^3(t-\tau)), \end{aligned} \tag{51}$$

where $g = f'(0)$, $g_2 = f''(0)$, and $g_3 = f'''(0)$. By solving the characteristic equation of the linearized system of (51) at point (0,0), we can get the critical values of parameters at which some resonant double Hopf bifurcations occur. As shown in Ref.23, some resonant double Hopf bifurcation points are listed as following

$$\begin{aligned} &\frac{\omega_1}{\omega_2} = \frac{1}{2}, && (b_c, g_c, \tau_c) = \left(\frac{1}{\sqrt{6}}, -\frac{\sqrt{2}}{\sqrt{3}}, \frac{4}{3}\sqrt{2}\pi\right), && \text{with} \quad \omega_1 = \frac{1}{\sqrt{2}}, \\ &\frac{\omega_1}{\omega_2} = \frac{1}{3}, && (b_c, g_c, \tau_c) = \left(0, \frac{2}{\sqrt{3}}, \frac{\sqrt{3}}{2}\pi\right), && \text{with} \quad \omega_1 = \frac{1}{\sqrt{3}}, \\ &\frac{\omega_1}{\omega_2} = \frac{1}{5}, && (b_c, g_c, \tau_c) = \left(\frac{4}{\sqrt{15}}, \frac{8}{\sqrt{15}}, \frac{\sqrt{5}}{3}\pi\right), && \text{with} \quad \omega_1 = \frac{1}{\sqrt{5}}. \end{aligned} \tag{52}$$

If we perturb the parameters as $b = b_c + b_\epsilon$, $g = g_c + g_\epsilon$, and $\tau = \tau_c + \tau_\epsilon$, (51) is written as

$$\begin{aligned} \dot{x}_1(t) &= x_2(t), \\ \dot{x}_2(t) &= -ax_1(t) - b_cx_2(t) - b_\epsilon x_2(t) + g_cx_2(t-\tau) + g_\epsilon x_2(t-\tau) \\ &\quad + g_2x_2^2(t-\tau) + g_3x_2^3(t-\tau) + O(x_2^3(t-\tau)), \end{aligned} \tag{53}$$

then we have

$$\mathbf{F}^{\mathbf{0}}_{\mathbf{x}} = \begin{pmatrix} 0 & 1 \\ -a & -b_c \end{pmatrix}, \quad \mathbf{F}^{\mathbf{0}}_{\mathbf{x}_\tau} = \begin{pmatrix} 0 & 0 \\ 0 & g_c \end{pmatrix}, \tag{54}$$

$$\mathbf{F}^{\mathbf{0}}_{\mathbf{x}}\mathbf{x} = \begin{pmatrix} x_2 \\ -ax_1 - b_cx_2 \end{pmatrix}, \quad \mathbf{F}^{\mathbf{0}}_{\mathbf{x}_\tau}\mathbf{x}_\tau = \begin{pmatrix} 0 \\ g_cx_2(t-\tau) \end{pmatrix}, \tag{55}$$

$$\mathbf{F}^{\mathbf{0}}_{\mathbf{x}\mu}\mathbf{x}\mu = \begin{pmatrix} 0 \\ -b_\epsilon x_2 \end{pmatrix}, \quad \mathbf{F}^{\mathbf{0}}_{\mathbf{x}_\tau}\mathbf{x}_\tau\mu = \begin{pmatrix} 0 \\ g_\epsilon x_2(t-\tau) \end{pmatrix}, \tag{56}$$

$$\mathbf{F}^{\mathbf{0}}_{\mathbf{x}_\tau\mathbf{x}_\tau}\mathbf{x}_\tau^2 = \begin{pmatrix} 0 \\ g_2 x_2(t-\tau)x_2(t-\tau) \end{pmatrix},$$
$$\mathbf{F}^{\mathbf{0}}_{\mathbf{x}_\tau\mathbf{x}_\tau\mathbf{x}_\tau}\mathbf{x}_\tau^3 = \begin{pmatrix} 0 \\ g_3 x_2(t-\tau)x_2(t-\tau)x_2(t-\tau) \end{pmatrix}. \quad (57)$$

In the following, we will apply the method introduced in previous section to compute the normal forms of 1 : 2 resonance, 1 : 3 resonance, and 1 : 5 resonance for system (51).

4.1. 1 : 2 *Resonance*

For system (51), a 1 : 2 resonance occurs at point $(b_c, g_c, \tau_c) = \left(\frac{1}{\sqrt{6}}, -\frac{\sqrt{2}}{\sqrt{3}}, \frac{4}{3}\sqrt{2}\pi\right)$ with two frequencies $\omega_1 = \frac{1}{\sqrt{2}}$ and $\omega_2 = \sqrt{2}$ for $a = 1$. The perturbation parameters are ordered as $b = b_c + \epsilon\hat{b}_\epsilon$, $g = g_c + \epsilon\hat{g}_\epsilon$, and $\tau = \tau_c + \epsilon\hat{\tau}_\epsilon$. Then by using the method introduced in Section 2, we can get that

$$D_1A_1 = C_{1\mu}\hat{\mu}_\epsilon A_1 + C_{\bar{1}2}\bar{A}_1A_2,$$
$$D_1A_2 = C_{2\mu}\hat{\mu}_\epsilon A_2 + C_{11}A_1^2, \quad (58)$$

where

$$C_{1\mu}\hat{\mu}_\epsilon = -(0.135099 - 0.0553902i)\hat{g}_\epsilon - (0.0319795 + 0.0779992i)\hat{\tau}_\epsilon$$
$$+(-0.115519 - 0.0893037i)\hat{b}_\epsilon,$$
$$C_{2\mu}\hat{\mu}_\epsilon = -(0.169812 + 0.0394847i)\hat{g}_\epsilon + (0.045593 - 0.196083i)\hat{\tau}_\epsilon \quad (59)$$
$$+(-0.119101 + 0.12732i)\hat{b}_\epsilon,$$
$$C_{\bar{1}2} = 0.0783335 + 0.191058i, \quad C_{11} = 0.00697997 - 0.0300189i,$$

and

$$D_2A_1 = C_{1\mu\mu}\hat{\mu}_\epsilon^2A_1 + C_{\bar{1}2\mu}\hat{\mu}_\epsilon\bar{A}_1A_2 + C_{11\bar{1}}A_1^2\bar{A}_1 + C_{12\bar{2}}A_1A_2\bar{A}_2,$$
$$D_2A_2 = C_{2\mu\mu}\hat{\mu}_\epsilon^2A_2 + C_{11\mu}\hat{\mu}_\epsilon A_1^2 + C_{1\bar{1}2}A_1\bar{A}_1A_2 + C_{22\bar{2}}A_2^2\bar{A}_2, \quad (60)$$

where

$$C_{1\mu\mu}\hat{\mu}_\epsilon^2 = -(0.0191158 + 0.0548195i)\hat{b}_\epsilon\hat{g}_\epsilon + (0.0484315 - 0.00641005i)\hat{b}_\epsilon\hat{\tau}_\epsilon$$
$$+(0.0348623 + 0.0391295i)\hat{b}_\epsilon^2 + (0.00664463 - 0.00154854i)\hat{g}_\epsilon\hat{\tau}_\epsilon$$
$$-(0.0734889 - 0.0389014i)\hat{g}_\epsilon^2 + (0.00397461 + 0.00549693i)\hat{\tau}_\epsilon^2,$$
$$C_{\bar{1}2\mu}\hat{\mu}_\epsilon = (0.212089 + 0.0766746i)\hat{b}_\epsilon + (0.0393196 + 0.375243i)\hat{g}_\epsilon$$
$$-(0.0449632 + 0.0616477i)\hat{\tau}_\epsilon,$$
$$C_{12\bar{2}} = -0.464576 + 0.281133i, \quad C_{11\bar{1}} = -0.0607052 + 0.0124769i,$$
$$C_{2\mu\mu}\hat{\mu}_\epsilon^2 = -(0.0112828 - 0.045869i)\hat{b}_\epsilon\hat{g}_\epsilon - (0.0323468 + 0.0268416i)\hat{b}_\epsilon\hat{\tau}_\epsilon$$
$$+(0.0110426 - 0.0734756i)\hat{b}_\epsilon^2 + (0.0429602 + 0.0121337i)\hat{g}_\epsilon\hat{\tau}_\epsilon \quad (61)$$

$$-(0.103476+0.0331377i)\hat{g}_\epsilon^2-(0.0133269-0.0376617i)\hat{\tau}_\epsilon^2,$$

$$C_{11\mu}\hat{\mu}_\epsilon=(0.0516144-0.0146284i)\hat{b}_\epsilon+(0.00600782-0.0722578i)\hat{g}_\epsilon$$

$$+(0.010471-0.00433283i)\hat{\tau}_\epsilon,$$

$$C_{1\bar{1}2}=-0.191222+0.00842471i, \qquad C_{22\bar{2}}=-0.209825-0.0769145i.$$

Then coming back to the real time t, we have

$$\begin{aligned}\dot{A}_1&=(C_{1\mu}\hat{\mu}_\epsilon+C_{1\mu\mu}\hat{\mu}_\epsilon^2)A_1+(C_{\bar{1}2}+C_{\bar{1}2\mu}\hat{\mu}_\epsilon)\overline{A}_1A_2\\&+C_{11\bar{1}}A_1^2\bar{A}_1+C_{12\bar{2}}A_1A_2\bar{A}_2,\\\dot{A}_2&=(C_{2\mu}\hat{\mu}_\epsilon+C_{2\mu\mu}\hat{\mu}_\epsilon^2)A_2+(C_{11}+C_{11\mu}\hat{\mu}_\epsilon)A_1^2\\&+C_{1\bar{1}2}A_1\bar{A}_1A_2+C_{22\bar{2}}A_2^2\bar{A}_2.\end{aligned} \tag{62}$$

To express the normal form in real form, a mixed form representation for complex amplitudes is introduced

$$A_1=\frac{1}{2}ae^{i\theta}, \qquad A_2=\frac{1}{2}(u+iv)e^{2i\theta}, \tag{63}$$

where we use a different representation for complex amplitudes from (39). The presentation in (39) make us understand the meaning of the normal form easily, but it is difficult to analyze the dynamics. However, the mixed representation in (63) can lead to a normal form without triangle function and it is easier to analyze.

Substituting (63) into (62) and separating the real and imaginary parts, the real normal form for the 1 : 2 resonance is drawn

$$\begin{aligned}\dot{a}&=-0.0151763a^3-0.0191158a\hat{b}_\epsilon\hat{g}_\epsilon+0.106044au\hat{b}_\epsilon-0.0383373av\hat{b}_\epsilon\\&+0.0484315a\hat{b}_\epsilon\hat{\tau}_\epsilon+0.0348623a\hat{b}_\epsilon^2-0.115519a\hat{b}_\epsilon+0.0196598au\hat{g}_\epsilon\\&-0.187622av\hat{g}_\epsilon+0.00664463a\hat{g}_\epsilon\hat{\tau}_\epsilon-0.0734889a\hat{g}_\epsilon^2-0.135099a\hat{g}_\epsilon\\&-0.116144au^2-0.0224816au\hat{\tau}_\epsilon+0.0391668au-0.116144av^2\\&+0.0308238av\hat{\tau}_\epsilon-0.0955291av+0.00397461a\hat{\tau}_\epsilon^2-0.0319795a\hat{\tau}_\epsilon,\\\dot{u}&=0.0258072a^2\hat{b}_\epsilon+0.00300391a^2\hat{g}_\epsilon-0.0478056a^2u+0.00413228a^2v\\&+0.00523552a^2\hat{\tau}_\epsilon+0.00348998a^2-0.0112828u\hat{b}_\epsilon\hat{g}_\epsilon-0.155508v\hat{b}_\epsilon\hat{g}_\epsilon\\&+0.0766746uv\hat{b}_\epsilon-0.0323468u\hat{b}_\epsilon\hat{\tau}_\epsilon+0.0110426u\hat{b}_\epsilon^2-0.119101u\hat{b}_\epsilon\\&+0.212089v^2\hat{b}_\epsilon+0.0140215v\hat{b}_\epsilon\hat{\tau}_\epsilon+0.151735v\hat{b}_\epsilon^2-0.305927v\hat{b}_\epsilon\\&+0.375243uv\hat{g}_\epsilon+0.0429602u\hat{g}_\epsilon\hat{\tau}_\epsilon-0.103476u\hat{g}_\epsilon^2-0.169812u\hat{g}_\epsilon\\&+0.0393196v^2\hat{g}_\epsilon-0.0152308v\hat{g}_\epsilon\hat{\tau}_\epsilon+0.11094v\hat{g}_\epsilon^2+0.150265v\hat{g}_\epsilon\\&-0.0524563u^3+0.159795u^2v-0.0524563uv^2-0.0616477uv\hat{\tau}_\epsilon\\&+0.191058uv-0.0133269u\hat{\tau}_\epsilon^2+0.045593u\hat{\tau}_\epsilon+0.159795v^3\end{aligned} \tag{64}$$

$$-0.0449632v^2\hat{\tau}_\epsilon + 0.0783335v^2 - 0.0266678v\hat{\tau}_\epsilon^2 + 0.0400842v\hat{\tau}_\epsilon,$$

$$\begin{aligned}\dot{v} = &-0.00731421a^2\hat{b}_\epsilon - 0.0361289a^2\hat{g}_\epsilon - 0.00413228a^2u - 0.0478056a^2v\\ &-0.00216641a^2\hat{\tau}_\epsilon - 0.0150094a^2 + 0.155508u\hat{b}_\epsilon\hat{g}_\epsilon - 0.0112828v\hat{b}_\epsilon\hat{g}_\epsilon\\ &-0.0766746u^2\hat{b}_\epsilon - 0.212089uv\hat{b}_\epsilon - 0.0140215u\hat{b}_\epsilon\hat{\tau}_\epsilon - 0.151735u\hat{b}_\epsilon^2\\ &+0.305927u\hat{b}_\epsilon - 0.0323468v\hat{b}_\epsilon\hat{\tau}_\epsilon + 0.0110426v\hat{b}_\epsilon^2 - 0.119101v\hat{b}_\epsilon\\ &-0.375243u^2\hat{g}_\epsilon - 0.0393196uv\hat{g}_\epsilon + 0.0152308u\hat{g}_\epsilon\hat{\tau}_\epsilon - 0.11094u\hat{g}_\epsilon^2\\ &-0.150265u\hat{g}_\epsilon + 0.0429602v\hat{g}_\epsilon\hat{\tau}_\epsilon - 0.103476v\hat{g}_\epsilon^2 - 0.169812v\hat{g}_\epsilon\\ &-0.159795u^3 - 0.0524563u^2v + 0.0616477u^2\hat{\tau}_\epsilon - 0.191058u^2\\ &-0.159795uv^2 + 0.0449632uv\hat{\tau}_\epsilon - 0.0783335uv + 0.0266678u\hat{\tau}_\epsilon^2\\ &-0.0400842u\hat{\tau}_\epsilon - 0.0524563v^3 - 0.0133269v\hat{\tau}_\epsilon^2 + 0.045593v\hat{\tau}_\epsilon.\end{aligned}$$

In some special cases, for example, $g_2 = 0$, the values of $C_{\overline{1}2} + C_{\overline{1}2\mu}\hat{\mu}_\epsilon$ and $C_{11} + C_{11\mu}\hat{\mu}_\epsilon$ are zero, then the bifurcation is a codimension-two weak resonance. Letting $a_1 = |A_1|$ and $a_2 = |A_2| = \sqrt{u^2+v^2}$, the real normal form is

$$\begin{aligned}\dot{a}_1 = &-0.00844366a_1^3 - 0.0675493a_1a_2^2 - 0.0191158a_1\hat{b}_\epsilon\hat{g}_\epsilon + 0.0484315a_1\hat{b}_\epsilon\hat{\tau}_\epsilon\\ &+0.0348623a_1\hat{b}_\epsilon^2 - 0.115519a_1\hat{b}_\epsilon + 0.00664463a_1\hat{g}_\epsilon\hat{\tau}_\epsilon - 0.0734889a_1\hat{g}_\epsilon^2\\ &-0.135099a_1\hat{g}_\epsilon + 0.00397461a_1\hat{\tau}_\epsilon^2 - 0.0319795a_1\hat{\tau}_\epsilon,\\ \dot{a}_2 = &-0.0212266a_1^2a_2 - 0.0424531a_2^3 + 0.0429602a_2\hat{g}_\epsilon\hat{\tau}_\epsilon - 0.103476a_2\hat{g}_\epsilon^2\\ &-0.0112828a_2\hat{b}_\epsilon\hat{g}_\epsilon - 0.169812a_2\hat{g}_\epsilon - 0.0133269a_2\hat{\tau}_\epsilon^2 - 0.0323468a_2\hat{b}_\epsilon\hat{\tau}_\epsilon\\ &+0.045593a_2\hat{\tau}_\epsilon + 0.0110426a_2\hat{b}_\epsilon^2 - 0.119101a_2\hat{b}_\epsilon.\end{aligned} \tag{65}$$

Then the equilibrium $E_0(0,0)$ is corresponding to the trivial solution of original system; the mono-modal equilibrium $E_1(a_{10},0)$ and $E_2(0,a_{20})$ are corresponding to the periodic solutions of original system; the bimodal equilibrium $E_3(a_{31},a_{32})$ is the quasi periodic or periodic-two solution of original system.

4.2. 1 : 3 *Resonance*

As shown in (52), a 1 : 3 resonant double Hopf bifurcation at point $(b_c, g_c, \tau_c) = \left(0, \frac{2}{\sqrt{3}}, \frac{\sqrt{3}}{2}\pi\right)$ with $\omega_1 = \frac{1}{\sqrt{3}}$ and $\omega_3 = \sqrt{3}$. The perturbation parameters are ordered as $b = b_c + \epsilon^2\hat{b}_\epsilon$, $g = g_c + \epsilon^2\hat{g}_\epsilon$, and $\tau = \tau_c + \epsilon^2\hat{\tau}_\epsilon$. Then the real normal form is

$$\begin{aligned}\dot{a} &= 0.00224848a^3 + 0.0180004a^2u + 0.0310266a^2v - 0.154622a\hat{b}_\epsilon \\ &\quad +0.12144a\hat{g}_\epsilon - 0.00678831au^2 - 0.00678831av^2 - 0.103081a\hat{\tau}_\epsilon, \\ \dot{u} &= 0.0021287a^3 + 0.00387179a^2u - 0.0177828a^2v - 0.0930797auv \\ &\quad +0.0540011av^2 - 0.114475u\hat{b}_\epsilon - 0.634044v\hat{b}_\epsilon + 0.269725ug_\epsilon \\ &\quad -0.57834v\hat{g}_\epsilon + 0.0789768u^3 - 0.18673u^2v + 0.0789768uv^2 \\ &\quad +0.22895u\hat{\tau}_\epsilon - 0.18673v^3 + 0.296571v\hat{\tau}_\epsilon, \\ \dot{v} &= 0.000002585287a^3 + 0.0177828a^2u + 0.00387179a^2v + 0.0930797au^2 \\ &\quad -0.0540011auv + 0.634044u\hat{b}_\epsilon - 0.114475v\hat{b}_\epsilon + 0.57834u\hat{g}_\epsilon \\ &\quad +0.269725v\hat{g}_\epsilon + 0.18673u^3 + 0.0789768u^2v + 0.18673uv^2 \\ &\quad -0.296571u\hat{\tau}_\epsilon + 0.0789768v^3 + 0.22895v\hat{\tau}_\epsilon.\end{aligned} \tag{66}$$

4.3. 1 : 5 *Resonance*

At a 1 : 5 resonant double Hopf bifurcation point $(b_c, g_c, \tau_c) = \left(\frac{4}{\sqrt{15}}, \frac{8}{\sqrt{15}}, \frac{\sqrt{5}}{3}\pi\right)$, the two frequencies are $\omega_1 = \frac{1}{\sqrt{5}}$ and $\omega_5 = \sqrt{5}$. The parameters are perturbed as $b = b_c + \epsilon^2\hat{b}_\epsilon$, $g = g_c + \epsilon^2\hat{g}_\epsilon$, and $\tau = \tau_c + \epsilon^2\hat{\tau}_\epsilon$. By using the method we developed in Section 2, we can obtain the real normal form of the 1 : 5 resonance as following

$$\begin{aligned}\dot{a}_1 &= 0.00120785a_1^3 + 0.263087a_1a_2^2 - 0.0952142a_1\hat{b}_\epsilon + 0.0886361a_1\hat{g}_\epsilon \\ &\quad -0.0542892a_1\hat{\tau}_\epsilon, \\ \dot{a}_2 &= 0.0150186a_1^2a_2 + 0.111129a_2^3 + 0.177448a_2\hat{g}_\epsilon + 0.156664a_2\hat{\tau}_\epsilon \\ &\quad -0.118099a_2\hat{b}_\epsilon.\end{aligned} \tag{67}$$

5. Weak and Strong Resonances in Van der Pol Systems with Delay Coupling

As the second application, we consider two Van der Pol-Duffing oscillators with delay coupling governed by[31]

$$\begin{aligned}\ddot{u}_1 - (\alpha - \gamma_1 u_1^2)\dot{u}_1 + u_1 + (\beta_{11} + \beta_{21}u_1)u_1^2 &= A\dot{u}_2(t-\tau), \\ \ddot{u}_2 - (\alpha - \gamma_2 u_2^2)\dot{u}_2 + u_2 + (\beta_{12} + \beta_{22}u_2)u_2^2 &= A\dot{u}_1(t-\tau),\end{aligned} \tag{68}$$

where we assume that the two oscillators have the same linear parts, then we investigate the effects of nonlinear terms on resonances. Letting $x_1 = u_1$, $x_2 = \dot{u}_1$,

$x_3 = u_2$, and $x_4 = \dot{u}_2$, the system can be rewritten as

$$\begin{aligned}
\dot{x}_1 &= x_2,\\
\dot{x}_2 &= -x_1 + \alpha x_2 + Ax_4(t-\tau) - \beta_{11}x_1^2 - \gamma_1 x_1^2 x_2 - \beta_{21}x_1^3,\\
\dot{x}_3 &= x_4,\\
\dot{x}_4 &= -x_3 + \alpha x_4 + Ax_2(t-\tau) - \beta_{12}x_3^2 - \gamma_2 x_3^2 x_4 - \beta_{22}x_3^3.
\end{aligned} \tag{69}$$

Then by solving the characteristic equation of the linearization at equilibrium (0, 0, 0, 0), we can get some double Hopf bifurcation points which are listed as following

$$\begin{aligned}
&\frac{\omega_1}{\omega_2} = \frac{1}{2}, \quad (\alpha_c, A_c, \tau_c) = \left(-\frac{\sqrt{6}}{6}, \frac{\sqrt{6}}{3}, \frac{\sqrt{2}}{3}\pi\right), \quad \text{with } \omega_1 = \frac{1}{\sqrt{2}},\\
&\frac{\omega_1}{\omega_2} = \frac{1}{3}, \quad (\alpha_c, A_c, \tau_c) = \left(-\frac{2\sqrt{3}}{3}, \frac{2\sqrt{6}}{3}, \frac{\sqrt{3}}{4}\pi\right), \quad \text{with } \omega_1 = \frac{1}{\sqrt{3}},\\
&\frac{\omega_1}{\omega_2} = \frac{1}{5}, \quad (\alpha_c, A_c, \tau_c) = \left(-\frac{4\sqrt{15}}{5}, \frac{8\sqrt{5}}{5}, \frac{\sqrt{5}}{6}\pi\right), \quad \text{with } \omega_1 = \frac{1}{\sqrt{5}},\\
&\frac{\omega_1}{\omega_2} = \frac{1}{\sqrt{2}}, (\alpha_c, A_c, \tau_c) = (-0.0962126, 0.361355, 1.51751), \text{ with } \omega_1 = 0.840894.
\end{aligned} \tag{70}$$

Proceeding as before, we investigate the effects of nonlinear terms on the conditions of resonance separately.

5.1. 1 : 2 *Resonance*

From the previous results, we know that the coefficients of resonant terms will determine whether a 1 : 2 resonance is strong resonance or not. Therefore, by ignoring some unnecessary proceedings, we compute the coefficients of resonant terms directly. We have

$$\begin{aligned}
c_{\bar{1}2} + c_{\bar{1}2\mu}\hat{\mu}_\epsilon = {} & [(-0.0525545 - 0.0427655\hat{A}_\epsilon + 0.0123333\hat{\alpha}_\epsilon - 0.0893822\hat{\tau}_\epsilon)\beta_{11}\\
& +(0.0525545 + 0.0427655\hat{A}_\epsilon - 0.0123333\hat{\alpha}_\epsilon + 0.0893822\hat{\tau}_\epsilon)\beta_{12}]\\
& +i[(0.1809 - 0.0541291\hat{A}_\epsilon + 0.096941\hat{\alpha}_\epsilon + 0.0104913\hat{\tau}_\epsilon)\beta_{11}\\
& +(-0.1809 + 0.0541291\hat{A}_\epsilon - 0.096941\hat{\alpha}_\epsilon - 0.0104913\hat{\tau}_\epsilon)\beta_{12}],\\
c_{11} + c_{11\mu}\hat{\mu}_\epsilon = {} & [(0.0335001 + 0.00129954\hat{A}_\epsilon - 0.0241556\hat{\alpha}_\epsilon - 0.0232223\hat{\tau}_\epsilon)\beta_{11}\\
& +(-0.0335001 - 0.00129954\hat{A}_\epsilon + 0.0241556\hat{\alpha}_\epsilon + 0.0232223\hat{\tau}_\epsilon)\beta_{12}]\\
& +i[(0.0673266 - 0.00644569\hat{A}_\epsilon + 0.0222268\hat{\alpha}_\epsilon - 0.0280865\hat{\tau}_\epsilon)\beta_{11}\\
& +(-0.0673266 + 0.00644569\hat{A}_\epsilon - 0.0222268\hat{\alpha}_\epsilon + 0.0280865\hat{\tau}_\epsilon)\beta_{12}].
\end{aligned} \tag{71}$$

Obviously, if $\beta_{11} = \beta_{12}$ which includes the case of $\beta_{11} = 0$ and $\beta_{12} = 0$, the coefficients of resonant terms will vanish and the resonance will be weak resonance. In other cases, the resonance will be strong resonance. What's more, we can find that 1 : 2 strong resonance is associated with quadratic nonlinearity and is not associated with other nonlinearities. Then fixing the coefficients of full system at $\gamma_1 = \gamma_2 = 1$, $\beta_{21} = \beta_{22} = 1$, $\beta_{11} = 1$, and $\beta_{12} = 0.5$, we can get the real normal form of the 1:2 strong resonance as following

$$\begin{aligned}
\dot{a} = & -0.0160701a^3 - 0.0662828a\hat{A}_\epsilon^2 - 0.0106914a\hat{A}_\epsilon u + 0.0135323a\hat{A}_\epsilon v \\
& -0.103492a\hat{A}_\epsilon\hat{\alpha}_\epsilon - 0.045908a\hat{A}_\epsilon\hat{\tau}_\epsilon + 0.192281a\hat{A}_\epsilon + 0.00112578au^2 \\
& +0.00308332au\hat{\alpha}_\epsilon - 0.0223455au\hat{\tau}_\epsilon - 0.0131386au + 0.00112578av^2 \\
& -0.0242352av\hat{\alpha}_\epsilon - 0.00262283av\hat{\tau}_\epsilon - 0.045225av + 0.0567107a\hat{\alpha}_\epsilon^2 \\
& -0.102301a\hat{\alpha}_\epsilon\hat{\tau}_\epsilon + 0.255831a\hat{\alpha}_\epsilon + 0.0162729a\hat{\tau}_\epsilon^2 - 0.10646a\hat{\tau}_\epsilon, \\
\dot{u} = & \, 0.000324886a^2\hat{A}_\epsilon - 0.0429838a^2u - 0.0432348a^2v - 0.00603889a^2\hat{\alpha}_\epsilon \\
& -0.00580559a^2\hat{\tau}_\epsilon + 0.00837502a^2 - 0.0993173\hat{A}_\epsilon^2u + 0.350296\hat{A}_\epsilon^2v \\
& -0.0270646\hat{A}_\epsilon uv - 0.179386\hat{A}_\epsilon u\hat{\alpha}_\epsilon - 0.186419\hat{A}_\epsilon u\hat{\tau}_\epsilon + 0.354544\hat{A}_\epsilon u \\
& -0.0213828\hat{A}_\epsilon v^2 - 0.181823\hat{A}_\epsilon v\hat{\alpha}_\epsilon + 0.405086\hat{A}_\epsilon v\hat{\tau}_\epsilon - 0.603868\hat{A}_\epsilon v \\
& -0.0198288u^3 + 0.017529u^2v - 0.0198288uv^2 + 0.0484705uv\hat{\alpha}_\epsilon \\
& +0.00524565uv\hat{\tau}_\epsilon + 0.09045uv + 0.0382811u\hat{\alpha}_\epsilon^2 - 0.140461u\hat{\alpha}_\epsilon\hat{\tau}_\epsilon \\
& +0.380857u\hat{\alpha}_\epsilon - 0.31422u\hat{\tau}_\epsilon^2 + 0.271446u\hat{\tau}_\epsilon + 0.017529v^3 \qquad (72) \\
& +0.00616664v^2\hat{\alpha}_\epsilon - 0.0446911v^2\hat{\tau}_\epsilon - 0.0262773v^2 - 0.0495531v\hat{\alpha}_\epsilon^2 \\
& -0.0156933v\hat{\alpha}_\epsilon\hat{\tau}_\epsilon + 0.338152v\hat{\alpha}_\epsilon - 0.112929v\hat{\tau}_\epsilon^2 + 0.187365v\hat{\tau}_\epsilon, \\
\dot{v} = & -0.00161142a^2\hat{A}_\epsilon + 0.0432348a^2u - 0.0429838a^2v + 0.00555669a^2\hat{\alpha}_\epsilon \\
& -0.00702162a^2\hat{\tau}_\epsilon + 0.0168316a^2 - 0.350296\hat{A}_\epsilon^2u - 0.0993173\hat{A}_\epsilon^2v \\
& +0.0270646\hat{A}_\epsilon u^2 + 0.0213828\hat{A}_\epsilon uv + 0.181823\hat{A}_\epsilon u\hat{\alpha}_\epsilon - 0.405086\hat{A}_\epsilon u\hat{\tau}_\epsilon \\
& +0.603868\hat{A}_\epsilon u - 0.179386\hat{A}_\epsilon v\hat{\alpha}_\epsilon - 0.186419\hat{A}_\epsilon v\hat{\tau}_\epsilon + 0.354544\hat{A}_\epsilon v \\
& -0.017529u^3 - 0.0198288u^2v - 0.0484705u^2\hat{\alpha}_\epsilon - 0.00524565u^2\hat{\tau}_\epsilon \\
& -0.09045u^2 - 0.017529uv^2 - 0.00616664uv\hat{\alpha}_\epsilon + 0.0446911uv\hat{\tau}_\epsilon \\
& +0.0262773uv + 0.0495531u\hat{\alpha}_\epsilon^2 + 0.0156933u\hat{\alpha}_\epsilon\hat{\tau}_\epsilon - 0.338152u\hat{\alpha}_\epsilon \\
& +0.112929u\hat{\tau}_\epsilon^2 - 0.187365u\hat{\tau}_\epsilon - 0.0198288v^3 + 0.0382811v\hat{\alpha}_\epsilon^2 \\
& -0.140461v\hat{\alpha}_\epsilon\hat{\tau}_\epsilon + 0.380857v\hat{\alpha}_\epsilon - 0.31422v\hat{\tau}_\epsilon^2 + 0.271446v\hat{\tau}_\epsilon.
\end{aligned}$$

The amplitude equations of the 1 : 2 weak resonance where $\beta_{11} = \beta_{12} = 1$ as following

$$\begin{aligned}
\dot{a}_1 = & -0.0122395a_1^3 - 0.0662828a_1\hat{A}_\epsilon^2 - 0.103492a_1\hat{A}_\epsilon\hat{\alpha}_\epsilon - 0.045908a_1\hat{A}_\epsilon\hat{\tau}_\epsilon \\
& +0.192281a_1\hat{A}_\epsilon + 0.0365056a_1a_2^2 + 0.0567107a_1\hat{\alpha}_\epsilon^2 - 0.102301a_1\hat{\alpha}_\epsilon\hat{\tau}_\epsilon \\
& +0.255831a_1\hat{\alpha}_\epsilon + 0.0162729a_1\hat{\tau}_\epsilon^2 - 0.10646a_1\hat{\tau}_\epsilon, \qquad (73) \\
\dot{a}_2 = & -0.072512a_1^2a_2 - 0.0993173\hat{A}_\epsilon^2a_2 - 0.179386\hat{A}_\epsilon a_2\hat{\alpha}_\epsilon - 0.186419\hat{A}_\epsilon a_2\hat{\tau}_\epsilon
\end{aligned}$$

$$+0.354544\hat{A}_\epsilon a_2 - 0.032421a_2^3 + 0.0382811a_2\hat{\alpha}_\epsilon^2 - 0.140461a_2\hat{\alpha}_\epsilon\hat{\tau}_\epsilon$$
$$+0.380857a_2\hat{\alpha}_\epsilon - 0.31422a_2\hat{\tau}_\epsilon^2 + 0.271446a_2\hat{\tau}_\epsilon.$$

5.2. 1 : 3 *Resonance*

The resonant terms of a 1 : 3 resonance appear at cubic terms in normal form. Then we have

$$\begin{aligned} c_{\bar{1}\bar{1}2} &= [0.0174208(\beta_{12}^2 - \beta_{11}^2) + 0.020303(\beta_{22} - \beta_{21}) \\ &\quad +0.0138572(\gamma_2 - \gamma_1)] + i[0.0095287\,(\beta_{12}^2 - \beta_{11}^2) \\ &\quad +0.0720042\,(\beta_{21} - \beta_{22}) + 0.00390732\,(\gamma_2 - \gamma_1)], \\ c_{111} &= [0.0000513014\,(\beta_{21} - \beta_{22}) + 0.00740001\,(\gamma_2 - \gamma_1) \\ &\quad +0.0103587\,(\beta_{12}^2 - \beta_{11}^2)] + i[0.00461892\,(\beta_{11}^2 - \beta_{12}^2) \\ &\quad +0.0128172\,(\beta_{21} - \beta_{22}) + 0.00400255\,(\gamma_1 - \gamma_2)]. \end{aligned} \tag{74}$$

If the coefficients of resonant terms satisfy $c_{\bar{1}\bar{1}2} = 0$ and $c_{111} = 0$, we can get the following equations

$$\begin{aligned} &0.0174208(\beta_{12}^2 - \beta_{11}^2) + 0.020303(\beta_{22} - \beta_{21}) + 0.0138572(\gamma_2 - \gamma_1) = 0, \\ &0.0095287\,(\beta_{12}^2 - \beta_{11}^2) + 0.0720042\,(\beta_{21} - \beta_{22}) + 0.00390732\,(\gamma_2 - \gamma_1) = 0, \\ &0.0103587\,(\beta_{12}^2 - \beta_{11}^2) + 0.0000513014\,(\beta_{21} - \beta_{22}) + 0.00740001\,(\gamma_2 - \gamma_1) = 0, \\ &0.00461892\,(\beta_{11}^2 - \beta_{12}^2) + 0.0128172\,(\beta_{21} - \beta_{22}) + 0.00400255\,(\gamma_1 - \gamma_2) = 0. \end{aligned} \tag{75}$$

Because the above algebraic equations include six variables and only four equations, there are infinite solutions. Then there are infinite cases in which this 1 : 3 resonance is a weak resonance. For example, when $\beta_{12} = \pm\beta_{11}$, $\beta_{22} = \beta_{21}$, and $\gamma_2 = \gamma_1$, this 1 : 3 resonance will be weak resonance. However, when the coefficients of nonlinear terms do not satisfy the above equations, this 1 : 3 resonance will be a strong resonance. Letting $\gamma_1 = 1$, $\gamma_2 = 0.5$, $\beta_{21} = 1$, $\beta_{22} = 0.5$, $\beta_{11} = 1$, and $\beta_{12} = 0.5$, we can get the real normal form of this 1 : 3 strong resonance as following

$$\begin{aligned} \dot{a} &= -0.00652187a^3 - 0.00753644a^2u - 0.00672548a^2v + 0.150737a\hat{A}_\epsilon \\ &\quad +0.0038648au^2 + 0.0038648av^2 + 0.166287a\hat{\alpha}_\epsilon - 0.0795992a\hat{\tau}_\epsilon, \\ \dot{u} &= -0.00200069a^3 - 0.0217556a^2u - 0.0597256a^2v + 0.0201764auv \\ &\quad -0.0226093av^2 + 0.290262\hat{A}_\epsilon u - 0.339768\hat{A}_\epsilon v - 0.0137734u^3 \\ &\quad +0.0213007u^2v - 0.0137734uv^2 + 0.2664u\hat{\alpha}_\epsilon + 0.244618u\hat{\tau}_\epsilon \\ &\quad +0.0213007v^3 + 0.284755v\hat{\alpha}_\epsilon + 0.394635v\hat{\tau}_\epsilon, \end{aligned} \tag{76}$$

$$\dot{v} = 0.00296852a^3 + 0.0597256a^2u - 0.0217556a^2v - 0.0201764au^2 \\ +0.0226093auv + 0.339768\hat{A}_\epsilon u + 0.290262\hat{A}_\epsilon v - 0.0213007u^3 \\ -0.0137734u^2v - 0.0213007uv^2 - 0.284755u\hat{\alpha}_\epsilon - 0.394635u\hat{\tau}_\epsilon \\ -0.0137734v^3 + 0.2664v\hat{\alpha}_\epsilon + 0.244618v\hat{\tau}_\epsilon.$$

If we let $\gamma_1 = \gamma_2 = 1$, $\beta_{21} = \beta_{22} = 1$, and $\beta_{11} = \beta_{12} = 1$, the resonance is a weak resonance. It's real normal form is

$$\begin{aligned} \dot{a}_1 &= -0.00730135a_1^3 + 0.0161743a_1a_2^2 + 0.166287a_1\hat{\alpha}_\epsilon + 0.150737a_1\hat{A}_\epsilon \\ &\quad -0.0795992a_1\hat{\tau}_\epsilon, \\ \dot{a}_2 &= -0.0355231a_1^2a_2 - 0.021978a_2^3 + 0.2664a_2\hat{\alpha}_\epsilon + 0.290262\hat{A}_\epsilon a_2 \\ &\quad +0.244618a_2\hat{\tau}_\epsilon. \end{aligned} \tag{77}$$

5.3. 1 : 5 *Resonance*

As we know, 1 : 5 resonance is a weak resonance and there are not resonant terms in its normal form truncated at lower order. Therefore the amplitude equation of 1 : 5 resonance is a two dimensional system consisting of two amplitude variables. In this paper, assuming the coefficients of nonlinear terms as $\gamma_1 = \gamma_2 = 1$, $\beta_{21} = \beta_{22} = 1$, and $\beta_{11} = \beta_{12} = 1$, we can get the real normal form as following

$$\begin{aligned} \dot{a}_1 &= -0.0041112a_1^3 + 0.00186391a_1a_2^2 + 0.0991747a_1\hat{\alpha}_\epsilon + 0.0966751a_1\hat{A}_\epsilon \\ &\quad -0.0494452a_1\hat{\tau}_\epsilon, \\ \dot{a}_2 &= -0.0170318a_1^2a_2 - 0.0120732a_2^3 + 0.174331a_2\hat{\alpha}_\epsilon + 0.18879\hat{A}_\epsilon a_2 \\ &\quad +0.173334a_2\hat{\tau}_\epsilon. \end{aligned} \tag{78}$$

5.4. 1 : $\sqrt{2}$ *Resonance*

The method proposed in this paper can also be used to investigate non-resonance. In the case of non-resonance, resonant terms do not appear in its normal form even truncated at any order. Therefore its normal form is a two dimensional system. As an example, we study a 1 : $\sqrt{2}$ non-resonance. Letting the coefficients of nonlinear terms as $\gamma_1 = \gamma_2 = 1$, $\beta_{21} = \beta_{22} = 1$, and $\beta_{11} = \beta_{12} = 1$, we have

$$\begin{aligned} \dot{a}_1 &= -0.0163586a_1^3 + 0.00905499a_1a_2^2 + 0.373628a_1\hat{\alpha}_\epsilon + 0.175217a_1\hat{A}_\epsilon \\ &\quad -0.103076a_1\hat{\tau}_\epsilon, \\ \dot{a}_2 &= -0.0626163a_1^2a_2 - 0.0295614a_2^3 + 0.437811a_2\hat{\alpha}_\epsilon + 0.222423\hat{A}_\epsilon a_2 \\ &\quad +0.168782a_2\hat{\tau}_\epsilon. \end{aligned} \tag{79}$$

6. Conclusions

In this paper, we have proposed an analytical method to investigate the strong and weak resonance with frequencies in all ratios $k_1 : k_2$, $(k_1, k_2 \in \mathbb{Z})$ occurring in a general n-dimensional delayed differential equation. Mechanism of the weak and strong resonances is displayed by analyzing codimension-two and codimension-three double Hopf bifurcation. The main results are represented as follows:

(a) The method of multiple scales may be extended to obtain the complex amplitude equations which may be considered as one of normal forms applied for a high co-dimensional bifurcation analysis .
(b) We analytically deduce the conditions which determine whether the resonances are strong or weak. It follows from the conditions that all of the high-order resonances, i.e., $k_1 + k_2 > 4$, are weak, but not all the low-order resonances, $k_1 + k_2 \leq 4$, are strong.
(c) The strong resonances with two frequencies correspond to the codimension-three double Hopf bifurcations. Thus, three independent physical parameters can be chosen as the unfolding parameters to classify the dynamical behaviors in the neighborhood of the resonant points. Similarly, two independent physical parameters can be used to classify the dynamical behaviors in the neighborhood of the weakly resonant points.
(d) As applications for the conditions obtained, we perform that the resonance with frequency ratio $1 : 2$ or $1 : 3$ may be strong or weak, which depends the values of the physical parameters. Similarly, the analysis for the weak resonance with ratio $1 : 5$ and the non-resonance with ratio $1 : \sqrt{2}$ is also achieved.

It should be noted that the present paper is focused on proposing the conditions to determine a low-order resonance is strong or weak. However, the obtained results are also applied for a high order resonance. In addition, the results can be also applied for ordinary differential systems when $\tau = 0$ in system (1). It is easy to obtain a normal form with strong or weak resonances by using the criterion given in this paper so that one can classify the dynamical behaviors in term of nonlinear dynamics. Therefore, this paper may provide a mathematical foundation for various applications.

Acknowledgments. This research was supported by the State Key Program of the National Natural Science Foundation of China under Grant No. 11032009.

References

1. W. Lacarbonara and G. Rega, Resonant non-linear normal modes. Part II: activation/orthogonality conditions for shallow structural systems, *Int. J. Non-Linear Mech.*, **38** (6), 873–887 (2003).
2. W. Lacarbonara, G. Rega, and A. H. Nayfeh, Resonant non-linear normal modes. Part

I: analytical treatment for structural one-dimensional system, *Int. J. Nonlin. Mech.*, **38** (6), 851–872 (2003).
3. S. W. Shaw and C. Peirre, Normal modes of vibration for non-linear continuous systems, *J. Sound Vib.*, **169** (3), 319–347 (1994).
4. S. W. Shaw and C. Peirre, Nonlinear normal modes and invariant manifolds, *J. Sound Vib.*, **150** (1), 170–173 (1991).
5. A. H. Nayfeh and S. A. Nayfeh, On nonlinear modes of continuous systems, *J. Vibr. Acoust.*, **116** (1), 129–136 (1994).
6. M. E. King and A. F. Vakakis, An energy-based approach to computing resonant nonlinear normal modes, *J. Appl. Mech.*, **63** (3), 810–819 (1996).
7. X. Y. Li, Y. S. Chen and Z. Q. Wu, Bifurcation of nonlinear internal resonant normal modes of a class of multi-degree-of-freedom systems, *Mech. Research Comm.*, **29** (5), 299–306 (2002).
8. X. Y. Li, Y. S. Chen and Z. Q. Wu, Singular analysis of bifurcation of nonlinear normal normal modes for a class of systems with dual internal resonances, *Appl. Math. Mech.*, **23** (10), 997–1007 (2002).
9. C. L. Lee and N. C. Perkins, Nonlinear oscillations of suspended cables containing a two-to-one internal resonance, *Nonlinear Dyna.*, **3** (6), 465–490 (1992).
10. G. Revel, D. M. Alonso, and J. L. Moiola, Interactions between oscillatory modes near 2 : 3 resonant Hopf-Hopf bifurcation, *Chaos*, **20** (4), 043106 (2010).
11. J. K. Hale, *Theory of Functional Differential Equations* (Springer-verlag, New York, 1977).
12. S. L. Das and A. Chatterjee, Multiple scale without center manifold reductions for delat differential equations near Hopf bifurcations, *Non. Dyna.*, **30** (4), 323–335 (2002).
13. V. Gattulli, F. Di Fabio, and A. Luongo, Simple and double Hopf bifurcations in aeroelastic oscillators with tuned mass dampers, *J. Franklin Inst.*, **338** (2–3), 201–217 (2001).
14. J. Xu, K. W. Chung and C. L. Chen, An efficient method for studying weak resonant double Hopf bifurcation in nonlinear systems with delayed feedbacks, *SIAM J. Appl. Dyn. Sys.*, **6** (1), 29–60 (2007).
15. J. L. Moiola and G. R. Chen, *Hopf Bifurcation Analysis: A Frequency Domain Approach* (World Scientific, Singapore, 1997).
16. P. Steen and S. H. Davis, Quasi-periodic bifurcations in nonlinearly coupled oscillators near a point of strong resonance, *SIAM J. Appl. Math.*, **42** (6), 1345–1368 (1982).
17. S. A. Van Gils, M. Krupa, and W. F. Langford, Hopf bifurcation with non-semisimple 1 : 1 resonance, *Nonlinearity*, **3** (3), 825–850 (1990).
18. E. Knobloch and M. R. E. Proctor, The double Hopf bifurcation with 2 : 1 resonance, *Proc. R. Soc. Lond.*, **A 415** (1848), 61–90 (1988).
19. V. LeBlanc, *On The* 1 : 2 *Resonant Hopf Bifurcation* (Ph.D. thesis, University of Waterloo, Waterloo, 1995).
20. A. Luongo, A. Di Egidio, and A. Paolone, Multiscale analysis of defective multiple-Hopf bifurcations, *Comput. Struct.*, **82** (31), 2705–2722 (2004).
21. V. Gattulli, F. Di Fabio, and A. Luongo, One to one resonant double Hopf bifurcation in aeroelastic oscillators with tuned mass dampers, *J. Sound Vib.*, **262** (2), 201–217 (2003).
22. A. Luongo, A. Paolone, and A. Di Egidio, Multiple timescales analysis for 1 : 2 and 1 : 3 resonant Hopf bifurcations, *Nonlinear Dyna.*, **34** (3–4), 269–291 (2003).
23. S. A. Campbell and J. Bélair, Resonant codimension two bifurcation in the harmonic oscillator with delayed forcing, *Canadian Appl. Math. Quart.*, **17** (3), 217–238 (1999).
24. S. A. Campbell and V. B. LeBlanc, Resonant Double Hopf-Hopf interaction in delay

differential equations, *J. Dynam. Differential Equations*, **10** (2), 327–346 (1998).
25. W. Y. Wang and J. Xu, Multiple scales analysis for double Hopf bifurcation with 1 : 3 resonance, *Nonlinear Dynam.*, **66** (1–2), 39–51 (2011).
26. A. H. Nayfeh, W. Lacarbonara, and C. M. Chin, Nollinear normal modes of buckled beams: three-to-one and one-to-one internal resonance, *Nonlinear Dynam.*, **18** (3), 253–273 (1999).
27. J. Xu and K. W. Chung, Double Hopf bifurcation with strong resonances in a delayed system with nonlinearities, *Math. Probl. Eng.*, 759363 (2009).
28. A. H. Nayfeh and D. T. Mook, *Nonlinear Oscillations* (Wiley, New York, 1979).
29. A. H. Nayfeh, *Method of Normal Forms* (Wiley, New York, 1993).
30. A. Luongo and A. Paolone, On the reconstitution problem in the multiple time scale method, *Nonlinear Dynam.*, **19** (2), 133–158 (1999).
31. W. Stephen and R. Richard, The dynamics of two coupled Van der Pol oscillators with delay coupling, *Nonlinear Dynam.*, **30** (3), 205–221 (2002).

Chapter 7

Stability and Hopf Bifurcation of Time-delayed Systems with Complex Coefficients

Zaihua Wang

State Key Laboratory of Mechanics and Control of Mechanical Structures
Nanjing University of Aeronautics and Astronautics, Nanjing 210016, *China*
zhwang@nuaa.edu.cn

Junyu Li

School of Mathematics, Physics and Information Science
Zhejiang Ocean University, Zhoushan 316000, *China*

This chapter surveys recent advances in studying the stability and bifurcation of time-delayed dynamic systems with real or complex coefficients. It begins with some results about the crossing direction of the characteristic roots passing through the imaginary axis of the complex plan, by using the sign of the real part of the first order derivative of the characteristic function with respect to the delay or some concerned parameter at the critical points. Serval special cases are addressed, such as time-delayed systems with a single delay or commensurate delays, with delay-independent coefficients or delay-dependent coefficients, with real coefficients or complex coefficients. In addition, the degenerate case is also considered, for which high order derivative of the characteristic function with respect to the delay or some concerned parameter at the critical points is required in determining the crossing direction. Next, two algorithms are presented for checking the stability of time-delayed systems by calculating the rightmost characteristic root(s) numerically or by determining the number of stability switches graphically. Then, the pseudo-oscillator analysis is introduced for the Hopf bifurcation-induced periodic solution of scalar time-delayed systems with real or complex coefficients respectively. By means of the pseudo-oscillator analysis, the amplitude of the bifurcated periodic solution and its stability can be determined in a very simple way and with high accuracy. Several examples are given to demonstrate the effectiveness of the main results. Finally, some concluding remarks are drawn from the discussion.

1. Introduction

Over the past few decades, dynamics and control of time-delayed systems have been the major topics in many engineering applications, due to the rapid development of active control theories and technologies. Usually, the state variables of a time-delayed system are displacement, or velocity, or acceleration, or density, and so on, so the state variables must be real. That is to say, a time-delayed system arising from many practical applications should be modeled by a differential equation (or a set of differential equations) with real state variables and real coefficients. However, we do have many dynamical systems in real applications that can be modeled by differential equations with complex coefficients. In laser dynamics, for example, under certain conditions, the amplitude of the field and the carrier density are governed by two coupled delay differential equations with complex coefficients, and other examples include the famous Lang-Kobayashi equation[1]. In rotor dynamics, rotor motion is measured in orthogonal directions at the stations of interest, thus the system equation can be described by a differential equation with complex coefficients if the two orthogonal displacements is integrated as a complex variable[2]. When a delayed feedback control is performed, one has a time-delayed system with complex coefficients. In addition, such a form also appears in nonlinear dynamics, say, as in the applications of the method of multiple time scales to time-delayed systems where the complex amplitude is governed by a delay differential equation with complex coefficients, or in controlling Hopf bifurcation of a dynamical system in complex normal form via delayed feedback control. In this way, the delay differential equations with complex coefficients (complex-DDEs) have a more compact form than the ones with real coefficients (real-DDEs).

In the study of system dynamics, however, the differential equations with complex coefficients are usually transformed back to the coupled ones with real coefficients. This transformation increases the computational complexity in analysis. For example, in the stability analysis of complex-DDEs through root location of the characteristic equations, the order of a polynomial derived from the coupled equations with real coefficients is 4 times that of the corresponding polynomial deduced directly from the system equation with complex coefficients[3]. Thus, dynamics analysis that uses the properties of the complex-DDEs directly is preferable.

The main objective of this chapter is to present some recent advances on the stability and Hopf bifurcation of complex-DDEs. Emphasis is placed on three problems. The first one is the determination of the crossing direction of the two branches of the characteristic roots passing through the imaginary axis of the complex plan, a key problem both in the applications of the method of stability switches and in the existence of a Hopf bifurcation. Extension of some available results in the literature is introduced[3–5]. The second one is about stability test via numerical method or graphical method. Algorithms based on the integration form of the Mikhailov criterion are introduced for the calculation of the rightmost characteristic roots of time-delayed systems and for the determination of the number of stability switches in given interval of a parameter[6]. The third one is the estimation of the periodic solution induced by Hopf bifurcation and its stability analysis. A new proposed method named pseudo-oscillator analysis is introduced[7,8] and it has found a number of successful applications in the Hopf bifurcation of the laser dynamics model, metal cutting model and so on. All the results are simple in form and easy for implementation in applications.

2. The Crossing Direction for Stability Analysis

It is well-recognized that time delay is frequently the key factor that leads to instability and complicated nonlinear dynamics. Thus, the delay effect on the stability of time-delayed systems has been widely studied in the literature. The stability of a time-delayed system may changes from stable to unstable, or vice versa, as the delay increases from zero to infinity. This is the phenomenon of stability switches[9]. It has been shown that under some mild conditions, the stability changes only if a branch of characteristic roots passes through the imaginary axis of the complex plan. This is true also for some other parameters under certain conditions. Thus, when the constant delay takes value in some given interval, it does not require to check the stability at each parameter value, but to find the critical values of the delay at first and then to determine the crossing direction of the characteristic roots.

More precisely, to make the form of a time-delayed system with commensurate delays as general as possible, assume that the characteristic equation is in the form

$$p(\lambda) = p_0(\lambda, \tau) + \sum_{k=1}^{m} p_k(\lambda, \tau) e^{-k\lambda\tau} \tag{1}$$

where $\tau > 0$ is the delay, $p_k(\lambda, \tau)$ are real polynomials and the orders with respect to λ satisfying $\deg p_0(\lambda, \tau) > \deg p_k(\lambda, \tau)$ for $k = 1, 2, \ldots, m$. At $\tau = 0$, $p(\lambda)$ is a polynomial, and when $\tau > 0$ it is not difficult to find out, from $p(\mathrm{i}\omega) = 0$, all the critical values of ω and the corresponding delay values $0 < \tau_0 < \tau_1 < \cdots$ for each critical value of ω. Once a critical point $(\omega, \tau) = (\omega_*, \tau_*)$ with $\omega_* > 0$ is in hand (obtained analytically or numerically), the crossing direction of the corresponding two branches of characteristic roots can be determined directly from the sign of

$$S \stackrel{\text{def.}}{=} \Re \left[\frac{d\lambda}{d\tau} \right]_{(\lambda, \tau) = (\mathrm{i}\omega_*, \tau_*)} . \tag{2}$$

The time-delayed system increases (or decreases) one pair of conjugate characteristic roots with positive real part as τ passes through τ_* from the left to the right (or from the right to the left), namely, the crossing direction of the corresponding two branches of characteristic roots passing through the imaginary axis is from the left-half (right-half) complex plan to the right-half (left-half) complex plan, if $S > 0$ (or $S < 0$), see Fig.1. Then, the stability of the time-delayed system in each subinterval of the delay can be determined if the number of roots with non-negative real parts for $\tau = 0$ is known. This is the key idea of the method of stability switches.

If $p_k(\lambda, \tau)$ are polynomials with complex coefficients, the roots of $p(\lambda)$ do not appear in pairs conjugately. In this case, it is required to find out all the roots of ω (positive or non-positive) from $p(\mathrm{i}\omega) = 0$. Once a critical point $(\omega, \tau) = (\omega_*, \tau_*)$ is in hand (obtained analytically or numerically), we calculate S defined in (2). The time-delayed system increases (or decreases) one characteristic root with positive real

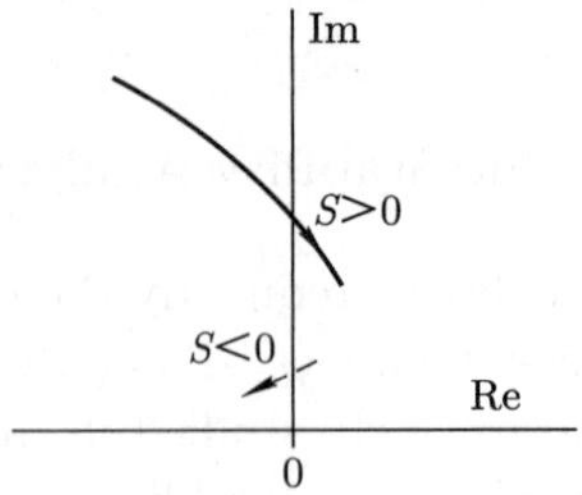

Fig. 1. The branch of the characteristic root passing through the imaginary axis from the left-half complex plan to the right-half complex plan or from the right-half complex plan to the left-half complex plan, determined by the sign of S.

part as τ passes through τ_* from the left to the right (or from the right to the left), namely, the crossing direction of the corresponding two branches of characteristic roots passing through the imaginary axis is from the left-half (right-half) complex plan to the right-half (left-half) complex plan, if $S > 0$ (or $S < 0$).

For the degenerate case with $S = 0$, it is required to use higher order derivative to determine the crossing direction.

In this section, we want to show that for some special classes of the characteristic functions, the crossing direction of the characteristic roots, namely the sign of S, can be determined in a much more straightforward way.

2.1. *The Case with a Single Delay*

Let us begin with a very simple case. When $p_0(\lambda, \tau) = p_0(\lambda)$, $p_1(\lambda, \tau) = p_1(\lambda)$ are delay-independent, we consider a time-delayed system with the characteristic function

$$p(\lambda) = p_0(\lambda) + p_1(\lambda)e^{-\lambda\tau}. \tag{3}$$

The critical values of ω satisfying $p(\mathrm{i}\omega) = 0$ must be the roots of $F(\omega) = 0$, where

$$F(\omega) = |p_0(\mathrm{i}\omega)|^2 - |p_1(\mathrm{i}\omega)|^2. \tag{4}$$

And the critical values of delay can be found from

$$\begin{cases} \cos(\omega\tau) = \dfrac{-(p_0^i(\omega)\, p_1^i(\omega) + p_1^r(\omega)\, p_0^r(\omega))}{(p_1^i(\omega))^2 + (p_1^r(\omega))^2}, \\[2ex] \sin(\omega\tau) = \dfrac{-(p_1^i(\omega)\, p_0^r(\omega) - p_1^r(\omega)\, p_0^i(\omega))}{(p_1^i(\omega))^2 + (p_1^r(\omega))^2}, \end{cases} \tag{5}$$

where $p_0(\mathrm{i}\omega) = p_0^r(\omega) + \mathrm{i}p_0^i(\omega)$, $p_1(\mathrm{i}\omega) = p_1^r(\omega) + \mathrm{i}p_1^i(\omega)$. Once a critical point (ω_*, τ_*) is in hand, and assume that $F(\omega)$ has no repeated roots, namely $F(\omega) = 0$ and $F'(\omega) = 0$ do not hold simultaneously, then at $(\omega, \tau) = (\omega_*, \tau_*)$, one has[9]

$$\mathrm{sgn}(S) = \mathrm{sgn}\, F'(\omega_*). \tag{6}$$

In fact, by differentiating $p(\lambda) = 0$ with respect to τ, we have

$$\tau'(\lambda) = (\lambda'(\tau))^{-1} = -\frac{p_0'(\lambda)}{\lambda p_0(\lambda)} + \frac{p_1'(\lambda)}{\lambda p_1(\lambda)} - \frac{\tau}{\lambda}. \tag{7}$$

Let $\bar{z}$ be the conjugate of complex z, $\Re(z)$ and $\Im(z)$ stand for the real and imaginary parts of z. Because $\mathrm{sgn}(\Re(z)) = \mathrm{sgn}(\Re(z^{-1}))$ and $p_0(\mathrm{i}\omega_*) = p_1(\mathrm{i}\omega_*)$, one has

$$\begin{aligned}
\mathrm{sgn}(S) &= \mathrm{sgn}\ \Re\left[-\frac{p_0'(\mathrm{i}\omega_*)}{\mathrm{i}\omega_* p_0(\mathrm{i}\omega_*)} + \frac{p_1'(\mathrm{i}\omega_*)}{\mathrm{i}\omega_* p_1(\mathrm{i}\omega_*)}\right] \\
&= -\mathrm{sgn}\ \Im[p_0'(\mathrm{i}\omega_*)\overline{p_0(\mathrm{i}\omega_*)} - p_1'(\mathrm{i}\omega_*)\overline{p_1(\mathrm{i}\omega_*)}] \\
&= \mathrm{sgn}[\Re p_0(\mathrm{i}\omega_*)\Re p_0'(\mathrm{i}\omega_*) + \Im p_0(\mathrm{i}\omega_*)\Im p_0'(\mathrm{i}\omega_*) \\
&\quad -\Re p_1(\mathrm{i}\omega_*)\Re p_1'(\mathrm{i}\omega_*) - \Im p_1(\mathrm{i}\omega_*)\Im p_1'(\mathrm{i}\omega_*)] \\
&= \mathrm{sgn}\ F'(\omega_*).
\end{aligned}$$

In particular, if $p(\lambda) = \lambda + a + be^{-\lambda\tau}$ and the corresponding $F(\omega)$ has exactly one positive simple root ω_*, then $F'(\omega_*) > 0$. It means that the crossing direction is always from the left-half complex plan to the right-half complex plan, namely, as the delay increases, with the critical delay values denoted by $\tau_0 < \tau_1 < \cdots$, the time-delayed system is unstable for any $\tau > 0$ if $a + b < 0$, or alternatively when $a + b > 0$, the system is stable if $\tau \in [0, \tau_0)$ and it is unstable if $\tau > \tau_0$.

Next, we consider time-delayed systems whose characteristic functions are given by the following form

$$p(\lambda) = p_0(\lambda, \tau) + p_1(\lambda, \tau)e^{-\lambda\tau}, \tag{8}$$

where $\tau > 0$ is the delay, $p_0(\lambda, \tau)$ and $p_1(\lambda, \tau)$ are delay-dependent real polynomials with respect to λ, and $\deg(p_0) > \deg(p_1)$. Such time-delayed systems can be found in population dynamics[11], laser dynamics[12], TCP/AQM networks[13,14], and neuron science[15]. This case occurs if the time delay is scaled to be 1 by proper time transformation, and it is encountered also in the study of α-stability, for which all the characteristic roots are required to satisfy $\Re(\lambda) < -\alpha$ with $\alpha > 0$.

Assume that $p_0(\mathrm{i}\omega, \tau) = 0$ and $p_1(\mathrm{i}\omega, \tau) = 0$ do not hold simultaneously for any $\omega \geq 0$, and let

$$p_0(\mathrm{i}\omega, \tau) = p_0^r(\omega, \tau) + \mathrm{i}p_0^i(\omega, \tau), \qquad p_1(\mathrm{i}\omega, \tau) = p_1^r(\omega, \tau) + \mathrm{i}p_1^i(\omega, \tau),$$

then from $p(\mathrm{i}\omega) = 0$ one sees that $\omega > 0$ must satisfy $F(\omega, \tau) = 0$, where

$$F(\omega, \tau) = (p_0^i(\omega, \tau))^2 + (p_0^r(\omega, \tau))^2 - (p_1^i(\omega, \tau))^2 - (p_1^r(\omega, \tau))^2 \tag{9}$$

and

$$\begin{cases} \cos(\omega\tau) = \dfrac{-(p_0^i(\omega, \tau)\, p_1^i(\omega, \tau) + p_1^r(\omega, \tau)\, p_0^r(\omega, \tau))}{(p_1^i(\omega, \tau))^2 + (p_1^r(\omega, \tau))^2}, \\[2ex] \sin(\omega\tau) = \dfrac{-(p_1^i(\omega, \tau)\, p_0^r(\omega, \tau) - p_1^r(\omega, \tau)\, p_0^i(\omega, \tau))}{(p_1^i(\omega, \tau))^2 + (p_1^r(\omega, \tau))^2}. \end{cases} \tag{10}$$

Similarly, for the argument of $p(\mathrm{i}\omega) = 0$, there is an integer k such that $2k\pi + G(\omega,\tau) = 0$, where

$$G(\omega,\tau) = \omega\tau + \arctan\frac{p_0^i(\omega,\tau)}{p_0^r(\omega,\tau)} - \arctan\frac{p_1^i(\omega,\tau)}{p_1^r(\omega,\tau)}. \tag{11}$$

In addition, let Δ denote the determinant of the Jacobian matrix of F, G with respect to ω, τ, described by

$$\Delta(\omega,\tau) \overset{\text{def.}}{=} \frac{\partial F}{\partial\omega}\frac{\partial G}{\partial\tau} - \frac{\partial F}{\partial\tau}\frac{\partial G}{\partial\omega} = \begin{vmatrix} F'_\omega(\omega,\tau) & F'_\tau(\omega,\tau) \\ G'_\omega(\omega,\tau) & G'_\tau(\omega,\tau)\end{vmatrix}. \tag{12}$$

Then the following statement holds[4].

Theorem 1. *For the characteristic function $p(\lambda)$ defined by (8), then at the critical point for which $p(\mathrm{i}\omega_*) = 0$ for some $\tau = \tau_*$, one has simply*

$$\operatorname{sgn}(S) = \operatorname{sgn}(\Delta(\omega_*,\tau_*)). \tag{13}$$

In fact, differentiating $p(\lambda) = 0$ with respect to τ gives

$$\frac{d\lambda}{d\tau} = -\frac{\lambda + \frac{\partial p_0}{\partial\tau}(\lambda,\tau)/p_0(\lambda,\tau) - \frac{\partial p_1}{\partial\tau}(\lambda,\tau)/p_1(\lambda,\tau)}{\tau + \frac{\partial p_0}{\partial\lambda}(\lambda,\tau)/p_0(\lambda,\tau) - \frac{\partial p_1}{\partial\lambda}(\lambda,\tau)/p_1(\lambda,\tau)}.$$

At the critical point $(\omega,\tau) = (\omega_*,\tau_*)$, one has $|p_0(\mathrm{i}\omega_*,\tau_*)|^2 = |p_1(\mathrm{i}\omega_*,\tau_*)|^2$, and

$$\begin{aligned}
&\left.\frac{d\lambda}{d\tau}\right|_{(\lambda,\tau)=(\mathrm{i}\omega_*,\tau_*)} \\
&= -\left.\frac{\lambda|p_0(\lambda,\tau)|^2 + \frac{\partial p_0}{\partial\tau}(\lambda,\tau)\overline{p_0(\lambda,\tau)} - \frac{\partial p_1}{\partial\tau}(\lambda,\tau)\overline{p_1(\lambda,\tau)}}{\tau|p_0(\lambda,\tau)|^2 + \frac{\partial p_0}{\partial\lambda}(\lambda,\tau)\overline{p_0(\lambda,\tau)} - \frac{\partial p_1}{\partial\lambda}(\lambda,\tau)\overline{p_1(\lambda,\tau)}}\right|_{(\lambda,\tau)=(\mathrm{i}\omega_*,\tau_*)} \\
&= -\frac{A + B\mathrm{i}}{C + D\mathrm{i}}
\end{aligned}$$

where

$$\begin{aligned}
A &= \frac{\partial p_0^r}{\partial\tau}(\omega_*,\tau_*)p_0^r(\omega_*,\tau_*) + \frac{\partial p_0^i}{\partial\tau}(\omega_*,\tau_*)p_0^i(\omega_*,\tau_*) \\
&\quad - \left(\frac{\partial p_1^r}{\partial\tau}(\omega_*,\tau_*)p_1^r(\omega_*,\tau_*) + \frac{\partial p_1^i}{\partial\tau}(\omega_*,\tau_*)p_1^i(\omega_*,\tau_*)\right) \\
&= \frac{1}{2}\frac{\partial F}{\partial\tau}(\omega_*,\tau_*),
\end{aligned}$$

$$\begin{aligned}
B &= \omega|p_0(\mathrm{i}\omega_*,\tau_*)|^2 + \left(\frac{\partial p_0^i}{\partial\tau}(\omega_*,\tau_*)p_0^r(\omega_*,\tau_*) - \frac{\partial p_0^r}{\partial\tau}(\omega_*,\tau_*)p_0^i(\omega_*,\tau_*)\right)\\
&\quad - \left(\frac{\partial p_1^i}{\partial\tau}(\omega_*,\tau_*)p_1^r(\omega_*,\tau_*) - \frac{\partial p_1^r}{\partial\tau}(\omega_*,\tau_*)p_1^i(\omega_*,\tau_*)\right)\\
&= |p_0(\mathrm{i}\omega_*,\tau_*)|^2\left(\omega + \left(\frac{\partial}{\partial\tau}\arctan\frac{p_0^i}{p_0^r}\right)(\omega_*,\tau_*) - \left(\frac{\partial}{\partial\tau}\arctan\frac{p_1^i}{p_1^r}\right)(\omega_*,\tau_*)\right)\\
&= |p_0(\mathrm{i}\omega_*,\tau_*)|^2\cdot\frac{\partial G}{\partial\tau}(\omega_*,\tau_*),
\end{aligned}$$

and

$$\begin{aligned}
C &= \tau|p_0(\mathrm{i}\omega_*,\tau_*)|^2 + \left(\frac{\partial p_0^i}{\partial\omega}(\omega_*,\tau_*)p_0^r(\omega_*,\tau_*) - \frac{\partial p_0^r}{\partial\omega}(\omega_*,\tau_*)p_0^i(\omega_*,\tau_*)\right)\\
&\quad - \left(\frac{\partial p_1^i}{\partial\omega}(\omega_*,\tau_*)p_1^r(\omega_*,\tau_*) - \frac{\partial p_1^r}{\partial\omega}(\omega_*,\tau_*)p_1^i(\omega_*,\tau_*)\right)\\
&= |p_0(\mathrm{i}\omega_*,\tau_*)|^2\left(\tau + \left(\frac{\partial}{\partial\omega}\arctan\frac{p_0^i}{p_0^r}\right)(\omega_*,\tau_*) - \left(\frac{\partial}{\partial\omega}\arctan\frac{p_1^i}{p_1^r}\right)(\omega_*,\tau_*)\right)\\
&= |p_0(\mathrm{i}\omega_*,\tau_*)|^2\cdot\frac{\partial G}{\partial\omega}(\omega_*,\tau_*),\\
D &= -\frac{\partial p_0^r}{\partial\omega}(\omega_*,\tau_*)p_0^r(\omega_*,\tau_*) - \frac{\partial p_0^i}{\partial\omega}(\omega_*,\tau_*)p_0^i(\omega_*,\tau_*)\\
&\quad + \frac{\partial p_1^r}{\partial\omega}(\omega_*,\tau_*)p_1^r(\omega_*,\tau_*) + \frac{\partial p_1^i}{\partial\omega}(\omega_*,\tau_*)p_1^i(\omega_*,\tau_*)\\
&= -\frac{1}{2}\frac{\partial F}{\partial\omega}(\omega_*,\tau_*).
\end{aligned}$$

It follows that

$$AC + BD = -\frac{1}{2}|p_0(\mathrm{i}\omega_*,\tau_*)|^2\Delta$$

and

$$\mathrm{sgn}(S) = -\mathrm{sgn}\,\Re\frac{A+B\mathrm{i}}{C+D\mathrm{i}} = -\mathrm{sgn}(AC+BD) = \mathrm{sgn}(\Delta).$$

This completes Theorem 1.

In particular, if $p_0(\lambda,\tau)$ and $p_1(\lambda,\tau)$ are independent of τ, the case introduced in the last subsection, one has

$$F'_\tau(\omega,\tau) = 0, \qquad G'_\tau(\omega,\tau) = \omega, \qquad \Delta = \omega F'(\omega).$$

Therefore, one has $\mathrm{sgn}(S) = \mathrm{sgn}(\Delta(\omega_*,\tau_*)) = \mathrm{sgn}\,F'(\omega_*)$, namely it holds (6).

Moreover, a delay-independent dynamical system can be transformed into a delay-dependent system with fixed delay 1 by using proper time transformation, thus Theorem 1 is expected true for delay equations with a certain parameter but with fixed time delay. Actually, let $\rho_0(\lambda,\xi)$ and $\rho_1(\lambda,\xi)$ be two polynomials satisfying $\deg(\rho_0) > \deg(\rho_1)$ with respect to λ, and assume that the characteristic function of a time-delayed system is in the form of

$$\rho(\lambda) = \rho_0(\lambda, \xi) + \rho_1(\lambda, \xi)e^{-\lambda\tau} \tag{14}$$

with ξ being a parameter other than τ. Accordingly, let

$$F(\omega, \xi) = (\rho_0^i(\omega, \xi))^2 + (\rho_0^r(\omega, \xi))^2 - (\rho_1^i(\omega, \xi))^2 - (\rho_1^r(\omega, \xi))^2, \tag{15}$$

$$G(\omega, \xi) = \omega\tau + \arctan\frac{\rho_0^i(\omega, \xi)}{\rho_0^r(\omega, \xi)} - \arctan\frac{\rho_1^i(\omega, \xi)}{\rho_1^r(\omega, \xi)}. \tag{16}$$

For fixed $\tau > 0$, assume that $\rho(i\omega_*) = 0$ for some $\xi = \xi_*$, then the crossing direction can be determined by the following theorem.

Theorem 2. *For the characteristic function $\rho(\lambda)$ defined in (14) with fixed delay τ, then at the critical point (ω_*, ξ_*), one has*

$$\mathrm{sgn}\left(\Re\left[\frac{d\lambda}{d\xi}\right]_{(\lambda,\xi)=(i\omega_*,\xi_*)}\right) = \mathrm{sgn}\left(\begin{vmatrix} F'_\omega(\omega_*, \xi_*) & F'_\xi(\omega_*, \xi_*) \\ G'_\omega(\omega_*, \xi_*) & G'_\xi(\omega_*, \xi_*)\end{vmatrix}\right). \tag{17}$$

Example 1. Consider a first-order time-delayed system described by

$$\dot{x}(t) = ax(t) + bx(t - \tau) \quad (a, b \in \mathbb{R}). \tag{18}$$

The characteristic function for the trivial solution $x = 0$ reads $p(\lambda) = \lambda - a - be^{-\lambda\tau}$. For a given $\alpha > 0$, the delay system is said to have the α-stability, if the roots of $p(\lambda)$ satisfy $\Re(\lambda) < -\alpha$. It is equivalent to that all the roots of $q(\lambda) = p(\lambda - \alpha)$ satisfy $\Re(\lambda) < 0$. Obviously, the coefficients of $q(\lambda)$ is delay-dependent, and

$$F = \omega^2 + (\alpha + a)^2 - b^2e^{2\alpha\tau}, \qquad G = \omega\tau + \arctan\frac{\omega}{-\alpha - a}.$$

If $(\alpha + a)^2 - b^2e^{2\alpha\tau} > 0$, then $F = 0$ has no real roots, and the α-stability cannot be changed as τ varies. If $(\alpha + a)^2 - b^2e^{2\alpha\tau} < 0$, then $F = 0$ has exactly one positive root $\omega_* = \sqrt{b^2e^{2\alpha\tau} - (\alpha + a)^2}$, for which the critical delay values can be found from

$$-\alpha - a - b\,e^{\tau\,\alpha}\cos(\tau\,\omega_*) = 0, \qquad \omega_* + b\,e^{\tau\,\alpha}\sin(\tau\omega_*) = 0.$$

For each critical delay value τ_*, after simplification by using the condition $F = 0$, the crossing direction can be determined simply by

$$\mathrm{sgn}(S) = \mathrm{sgn}(\Delta) = \mathrm{sgn}((-b^2a^2 - b^2\alpha\,a) + (b^4\alpha\,\tau_* + b^4)\,e^{2\,\tau_*\,\alpha}).$$

When τ_* is large enough, one has $\mathrm{sgn}(S) > 0$. It means that the crossing direction must be from the left-half complex plan to the right-half complex plan. Thus, the α-stability does not hold for large delay τ if the α-stability switch does occur.

2.2. *The Degenerate Case with Real or Complex Coefficients*

Based on the above analysis, the crossing direction can be determined only if $F'(\omega_*) \neq 0$ for the delay-independent case or $\Delta(\omega_*, \tau_*) \neq 0$ for the delay-dependent case, where the coefficients are real. Similar results hold true for complex-DDEs. For example, for the case with a single delay, the crossing direction can be determined by the sign[3] of $\omega_* F'(\omega_*)$. The point emphasized in this case is that not only the positive critical values of ω but also the non-positive critical values of ω should be taken into consideration. The characteristic roots do not appear in pairs conjugately. In addition, as τ passes through a critical point increasingly, the number of zeros with non-negative real parts of $p(\lambda)$ increases (or decreases) by only one, not by two for the cases with real coefficients.

In applications, $F'(\omega_*) = 0$ or $\Delta(\omega_*, \tau_*) = 0$ may occur. This is the so-called "degenerate case", high order derivative of $F(\omega)$ or $\Delta(\omega, \tau)$ is required for the determination of the crossing direction. To make the exposition as general as possible, let us consider a time-delayed system whose characteristic function is given by

$$p(\lambda) = p_0(\lambda, \tau) + p_1(\lambda, \tau)e^{-\lambda\tau}, \tag{19}$$

where $\tau > 0$ is the delay, $p_0(\lambda, \tau)$ and $p_1(\lambda, \tau)$ are delay-dependent complex polynomials with respect to λ, and $\deg(p_0) > \deg(p_1)$. At the critical point $(\omega, \tau) = (\omega_*, \tau_*)$, the Taylor expansion gives

$$\lambda - \mathrm{i}\omega_* = \sum_{k=1}^{\infty} \frac{\lambda^{(k)}(\tau_*)}{k!}(\tau - \tau_*)^k.$$

If there is a positive integer M such that $\Re(\lambda^{(M)}(\tau_*)) \neq 0$ but $\Re(\lambda^{(k)}(\tau_*)) = 0$ for all $k = 0, 1, 2, \ldots, M-1$, then

$$\Re(\lambda) = \Re\left(\frac{\lambda^{(M)}(\tau_*)}{M!}\right)(\tau - \tau_*)^M + o((\tau - \tau_*)^M). \tag{20}$$

In addition, let $\Delta(\omega, \tau)$ be the Jacobian determinant defined in (12), and $\Delta_\tau^{(k)}(\omega, \tau)$ be the k th order derivative of Δ with respect to τ, then straightforward calculation gives a relationship of higher-order derivatives[5]

$$\operatorname{sgn} \Re[\lambda^{(k+1)}(\tau_*)] = \operatorname{sgn}\left(\Delta_\tau^{(k)}(\omega_*, \tau_*)\right). \tag{21}$$

If $S_M = \operatorname{sgn} \Delta_\tau^{(M-1)}(\omega_*, \tau_*)$, $(M \geq 1)$, then, it holds

Theorem 3. *For a time-delayed system with characteristic function given in* (19) *with complex coefficients, let $N(\tau)$ be the number of zeros with non-negative real parts of $p(\lambda)$, and let M be the integer such that* (21) *with $S_M \neq 0$ is true, then as τ passes through τ_* increasingly,*

(I) *The integer $N(\tau)$ keeps unchanged if M is even;*
(II) *When M is odd, $N(\tau)$ increases (decreases) by one if $S_M > 0$ ($S_M < 0$).*

Example 2. Consider the following simple complex-DDE

$$\dot{Y} = (1 + \mathrm{i}\alpha)PY + ke^{-\mathrm{i}v\tau}Y(t-\tau) \quad (\tau > 0). \tag{22}$$

The trivial solution is asymptotically stable if and only if

$$\lambda - (1 + \mathrm{i}\alpha)P - ke^{-\mathrm{i}v\tau} \cdot e^{-\lambda\tau} = 0 \tag{23}$$

has roots with negative real parts only. In terms of the principal branch $W_0(z)$ of the Lambert W function[16], the rightmost characteristic root of the complex-DDE, regarding as a function with respect to the delay τ, can be written in the form

$$\lambda_0(\tau) = \frac{1}{\tau}\left((1 + \mathrm{i}\alpha)P\tau + W_0\left(k\tau e^{-\mathrm{i}v\tau - (1+\mathrm{i}\alpha)P\tau}\right)\right). \tag{24}$$

Thus, the trivial solution is asymptotically stable if and only if $\Re(\lambda_0(\tau)) < 0$. The critical delay values can be found out simply from $\Re(\lambda_0(\tau)) = 0$ with the help of Maple (a computer algebra) function `LambertW`.

In the application of the above results, the two auxiliary functions $F(\omega, \tau)$, $G(\omega, \tau)$ associated with the characteristic function can be found to be

$$F = (\omega - \alpha P)^2 + P^2 - k^2, \qquad G = \omega\tau + \arctan\frac{\omega - \alpha P}{-P} + v\tau$$

and the corresponding Jacobian determinant $\Delta(\omega, \tau)$ is given by

$$\Delta = 2(\omega + v)(\omega - \alpha P).$$

For $k = 0.13$, $P = -0.13$, $\alpha = 2$, $v = 0.5$, the condition $F = 0$ gives the repeated critical value $\omega_* = -0.26$, and the Jacobian determinant vanishes: $\Delta = 0$. Hence higher-order derivatives of the Jacobian determinant with respect to the delay is required to determine the crossing direction of the characteristic branch.

Now, straightforward calculation gives

$$\begin{aligned}\frac{d\Delta}{d\tau} &= 2(2\omega + v - \alpha P)\frac{d\omega}{d\tau} \\ &= -2(2\omega + v - \alpha P)\frac{(P^2 + (\omega - \alpha P)^2)(\omega + v)}{(P^2 + (\omega - \alpha P)^2)\tau - P},\end{aligned}$$

where $\dfrac{d\omega}{d\tau}$ is determined from $G(\omega, \tau) = 0$ by using implicit differentiation. It is easy to know that when $k = 0.13, P = -0.13, \alpha = 2$, and $v = 0.5$, one has $\mathrm{sgn}\left(\dfrac{d\Delta}{d\tau}\right) \neq 0$ and $m = 2$ at each critical point $(\omega, \tau) = (\omega_*, \tau_*)$. It means that except for the critical delay values, the stability of the complex-DDE keeps unchanged as the delay increases from zero to the infinity. This is in agreement with graphical test given in Fig.2.

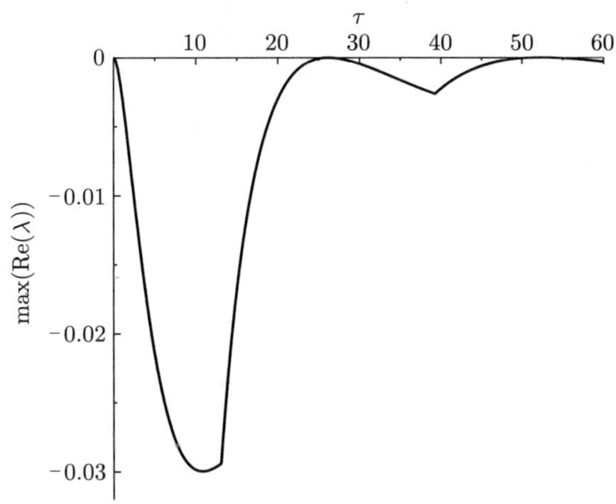

Fig. 2. The real part of the rightmost characteristic root with respect to the delay τ for Eq. (22) with $k = 0.13, P = -0.13, \alpha = 2$, and $v = 0.5$.

2.3. *The Case with Commensurate Delays*

Some above results for the non-degenerate case can be extended to the time-delayed systems with commensurate delays. Consider the case with delay-independent coefficients, namely the characteristic function has the following form

$$p(\lambda) = p_0(\lambda) + \sum_{k=1}^{m} p_k(\lambda)e^{-k\lambda\tau} \quad (m > 1). \tag{25}$$

As done for real-DDEs[10], the key procedure for the determination of the crossing direction is to define $p^{(j)}(\lambda)$ recursively as follows

$$p^{(1)}(\mathrm{i}\omega) = \overline{p_0(\mathrm{i}\omega)}p(\mathrm{i}\omega) - p_m(\mathrm{i}\omega)e^{-m\mathrm{i}\omega\tau}\overline{p(\mathrm{i}\omega)} = \sum_{k=0}^{m-1} q_k^{(1)}(\mathrm{i}\omega)e^{-k\mathrm{i}\omega\tau},$$

$$\cdots\cdots\cdots\cdots$$

$$p^{(j)}(\mathrm{i}\omega) = \overline{q_0^{(j-1)}(\mathrm{i}\omega)}p^{(j-1)}(\mathrm{i}\omega) - q_{m-j+1}^{(j-1)}(\mathrm{i}\omega)e^{-(m-j+1)\mathrm{i}\omega\tau}\overline{p^{(j-1)}(\mathrm{i}\omega)}$$

$$= \sum_{k=0}^{m-j} q_k^{(j)}(\mathrm{i}\omega)e^{-k\mathrm{i}\omega\tau},$$

$$\cdots\cdots\cdots\cdots$$

$$p^{(m-1)}(\mathrm{i}\omega) = q_0^{(m-1)}(\mathrm{i}\omega) + q_1^{(m-1)}(\mathrm{i}\omega)e^{-\mathrm{i}\omega\tau}.$$

Here $p^{(m-1)}(\lambda)$ has the same form of Eq. (3). Moreover, let

$$q_0^{(j+1)}(\omega) = |q_0^{(j)}(\omega)|^2 - |q_{m-j}^{(j)}(\omega)|^2 \quad (j = 0, 1, 2, \ldots, m-2),$$

$$F(\omega) = |q_0^{(m-1)}(\mathrm{i}\omega)|^2 - |q_1^{(m-1)}(\mathrm{i}\omega)|^2,$$

$$S_k(\omega) = \mathrm{sgn}\left[\omega F^{(k)}(\omega) \prod_{j=1}^{m-1} q_0^{(j)}(\omega)\right],$$

and assume that equation (25) has no pure imaginary zeros satisfying $q_0^{(j)}(\omega) = 0$ $(j = 1, 2, \ldots, m-1)$ simultaneously, then one has[3]

Theorem 4. *Let $N(\tau)$ be the number of zeros with non-negative real parts of $p(\lambda)$, and let M be the integer such that $S_M(\omega_*) \neq 0$, and $S_m(\omega_*) = 0$ for all $m < M$ at the critical point for which $p(\mathrm{i}\omega_*) = 0$ at some $\tau = \tau_*$, then, it holds*

(I) *The number $N(\tau)$ keeps unchanged as τ passes through τ_* increasingly if M is an even.*

(II) *If M is an odd, then as τ increases and passes through τ_*, $N(\tau)$ increases by 1 if $S_M(\omega_*) > 0$ and decreases by 1 if $S_M(\omega_*) < 0$.*

Example 3. Let us consider the following complex-DDE

$$\dot{z}(t) = (a + \mathrm{i}b)z(t) - U(t) \tag{26}$$

with $U(t) = k\sum_{n=0}^{N} R^n\{z(t - n\tau) - z[t - (n+1)\tau]\}$. This equation comes from the problem of controlling Hopf bifurcation via N time delay auto-synchronization (NTDAS) control[17]. For simplicity, let $N = 2$, then the corresponding characteristic equation reads

$$\lambda - (a - k + \mathrm{i}b) - k(1 - R)e^{-\lambda\tau} - kRe^{-2\lambda\tau} = 0. \tag{27}$$

When $a = 0.3, k = 0.75, R = 0.8$, and $b = \pi$, by following the above-introduced procedure, one has

$$\begin{aligned} q_0^{(1)}(\omega) &= (\omega - \pi)^2 - 0.1575, \\ F(\omega) &= (\omega - 2.5606)(\omega - \pi)^2(\omega - 3.7225). \end{aligned} \tag{28}$$

Obviously, $F(\omega)$ has two simple roots $\omega_1 = 2.5606, \omega_2 = 3.7225$, and a repeated root $\omega_3 = \pi$ with multiplicity two. Equation (27) is equivalent to

$$(\lambda - (a - k + \mathrm{i}b))e^{\lambda\tau} - k(1 - R) - kRe^{-\lambda\tau} = 0. \tag{29}$$

Substituting $\lambda = \mathrm{i}\omega$ into Eq. (29) and separating the real and imaginary part give

$$\begin{cases} (b - \omega)\sin(\omega\tau) + (-a + k - kR)\cos(\omega\tau) - k + kR = 0, \\ (-a + k + kR)\sin(\omega\tau) + (\omega - b)\cos(\omega\tau) = 0. \end{cases} \tag{30}$$

It follows that

$$\begin{cases} \cos(\omega\tau) = \dfrac{k(b - \omega - Rb + R\omega)}{(-a + k - kR)^2 + (\omega - b)^2}, \\ \sin(\omega\tau) = \dfrac{(-a + k + kR)(1 - R)k}{(-a + k - kR)^2 + (\omega - b)^2}. \end{cases}$$

Thus the critical delay values corresponding to $\omega_1, \omega_2, \omega_3$ can be found to be

$$\tau_{n,1} = \frac{2n\pi + \arccos 0.8750}{\omega_1}, \quad \tau_{n,2} = \frac{2(n+1)\pi - \arccos 0.8750}{\omega_2}$$

and

$$\tau_{n,3} = \frac{(2n+1)\pi}{\omega_3},$$

for $n = 0, 1, 2, \ldots$. Due to $q_0^{(1)}(\omega_j) > 0$ $(j = 1, 2)$ and $q_0^{(1)}(\omega_3) < 0$, one has

$$S_1(\omega) = \mathrm{sgn}[\omega F'(\omega) q_0^{(1)}(\omega)] = \mathrm{sgn}[F'(\omega)] = \begin{cases} -1, & \tau = \tau_{n,1}, \\ 1, & \tau = \tau_{n,2}, \\ 0, & \tau = \tau_{n,3}. \end{cases}$$

Because $S_2(\omega_3)|_{\tau=\tau_{n,3}} = 1 > 0$, no stability switch occurs as τ passes through $\tau_{n,3}$ according to Theorem 4. Moreover, as seen in Table 1 with τ in $[0, 6]$, the complex-DDE is asymptotically stable when $\tau \subset (0.1974, 1.5521) \cup (2.6511, 3.2340)$ except for $\tau = 1$ and $\tau = 3$ that correspond to $\omega_3 = \pi$, the system is critical stable. This is consistent with the graphical test shown in Fig.3 by using DDE-BIFTOOL.

Table 1. The sign of $S_j(\omega_i)$ at each critical values for Eq.(27) when $a = 0.3, k = 0.75, R = 0.8$, and $b = \pi$.

Critical delay	0.1974 ($\tau_{0,1}$)	1 ($\tau_{0,3}$)	1.5521 ($\tau_{0,2}$)	2.6511 ($\tau_{1,1}$)	3 ($\tau_{1,3}$)	3.2340 ($\tau_{1,2}$)	4.9279 ($\tau_{2,2}$)	5 ($\tau_{2,3}$)	5.1049 ($\tau_{2,1}$)
Sign of $S_1(\omega_i)$	−1	0	1	−1	0	1	1	0	−1
Sign of $S_2(\omega_3)$		1			1			1	

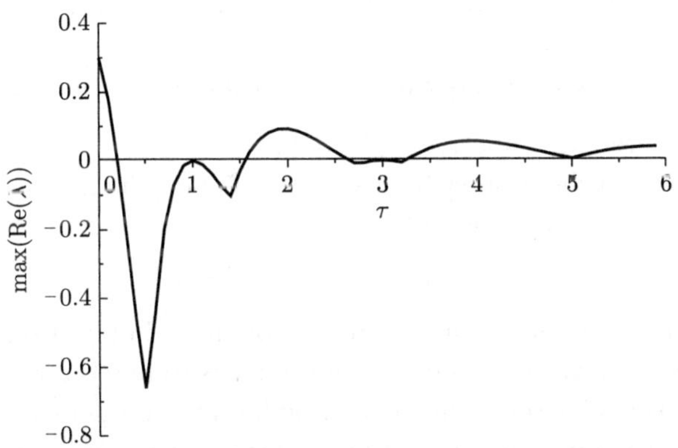

Fig. 3. The real part of the right most root with respect to the delay τ for Eq. (27) when $a = 0.3, k = 0.75, R = 0.8$, and $b = \pi$.

3. Numerical and Graphical Stability Test

If the system parameters are all fixed, then the asymptotical stability of time-delayed systems of retarded-type can be checked very conveniently by using the following theorem, an equivalent form of Mikhailov criterion[18]:

Theorem 5. *Let $p(\lambda)$ be the characteristic function of a real time-delayed system of retarded type as following*

$$p(\lambda) = \lambda^n + \sum_{i=1}^{n} p_i(e^{-\lambda\tau_1}, e^{-\lambda\tau_2}, \ldots, e^{-\lambda\tau_s})\lambda^{n-i}, \tag{31}$$

where $\tau_1, \tau_2, \ldots, \tau_s > 0$ are the delays, $p_i(\phi_1, \phi_2, \ldots, \phi_s)$ are real polynomials with respect to $\phi_1, \phi_2, \ldots, \phi_s$ for $i = 1, 2, \ldots, n$. If $p(\mathrm{i}\omega) \neq 0$ holds for all real ω, then the time-delayed system has characteristic roots with negative real parts only, if and only if

$$\int_0^\infty \Re\left(\frac{p'(\mathrm{i}\omega)}{p(\mathrm{i}\omega)}\right) d\omega \equiv \int_0^\infty \frac{S'(\omega)R(\omega) - S(\omega)R'(\omega)}{R^2(\omega) + S^2(\omega)} d\omega = \frac{n\pi}{2} \tag{32}$$

where $R(\omega) = \Re(p(\mathrm{i}\omega)),\ S(\omega) = \Im(p(\mathrm{i}\omega))$.

Because the integral in the left-hand side of Eq.(32) is some times of $\pi/2$, equation (32) holds if there is a $T > 0$ such that

$$\int_0^T \frac{S'(\omega)R(\omega) - S(\omega)R'(\omega)}{R^2(\omega) + S^2(\omega)} d\omega > \frac{(n-1)\pi}{2}. \tag{33}$$

This criterion has be generalized to time-delayed systems with fractional-order derivatives[6]. It can be used not only to check whether the characteristic functions have roots with negative real parts only, but also to calculate the rightmost characteristic root (or the pair of conjugate rightmost characteristic roots), as well as to find out the stable delay intervals and unstable delay intervals graphically.

3.1. *Calculation of the Rightmost Characteristic Root (s)*

For a given characteristic function $p(\lambda)$ defined in (31), let α_0 be the abscissa, namely the real part of the root with maximal real part (the rightmost characteristic root, or for short, the rightmost root):

$$\alpha_0 = \max\{\Re(\lambda) : \ \ p(\lambda) = 0\}. \tag{34}$$

The time-delayed system is asymptotically stable if and only if $\alpha_0 < 0$. Hence, the most straightforward way in checking the stability is to find out α_0 or the rightmost root itself. Algorithms for this problem can be found in a number of papers[19–22]. In the application of the numerical algorithms based on the Lambert W function[16,20], the domain of the principal-branch's value of the Lambert W function must be taken into consideration when the auxiliary function is constructed. A number of stability criteria[23–25] can be used to calculate α_0 as well as the rightmost root(s). Below is the algorithm based on[6] Theorem 5.

We firstly chose a sufficiently large number $T > 0$ and calculate the integral

$$\int_0^T \frac{S'(\omega)R(\omega) - S(\omega)R'(\omega)}{R^2(\omega) + S^2(\omega)} d\omega,$$

(i) if the integral value is larger than $(n-1)\pi/2$, then $\alpha_0 < 0$;
(ii) if the integral value is smaller than $(n-1)\pi/2$, then $\alpha_0 \geq 0$.

In Case (i), we chose a negative number $\beta < 0$ such that the testing integral over $[0, T]$ for $q(\lambda) = p(\lambda + \beta)$ is less than $(n-1)\pi/2$, which means that $q(\lambda)$ has at least one root with non-negative real part. It follows that $\beta \leq \alpha_0 \leq 0$. Similarly in Case (ii), chose a positive number β such that $0 \leq \alpha_0 \leq \beta$.

Next, with β obtained above, calculate the integral over $[0, T]$ for $q(\lambda) = p(\lambda + \beta/2)$. If the integral value is larger than $(n-1)\pi/2$, then the interval containing α can be shortened as $\alpha_0 \in [\beta, \beta/2]$ (Case (i)), or $\alpha_0 \in [0, \beta/2]$ (Case (ii)). If the integral value is smaller than $(n-1)\pi/2$, then the interval containing α can be shortened as $\alpha_0 \in [\beta/2, 0]$ (Case (i)), or $\alpha_0 \in [\beta/2, \beta]$ (Case (ii)). Continue this procedure repeatedly, one has a series of closed interval containing α_0:

$$[a_1, b_1] \supset [a_2, b_2] \supset \cdots \supset [a_k, b_k] \supset \cdots, \tag{35}$$

with $b_k - a_k \to 0$ as $k \to \infty$. The intersect of these closed intervals determines α_0 exactly as follows

$$\{\alpha_0\} = \bigcap_{k=1}^{\infty} [a_k, b_k]. \tag{36}$$

Finally, with this α_0, solving ω from $p(\alpha_0 + i\omega) = 0$ gives the real rightmost characteristic root $\lambda = \alpha_0$ if $p(\alpha_0 + i\omega) = 0$ holds only if $\omega = 0$, or the conjugate pair of complex rightmost characteristic roots $\lambda = \alpha_0 \pm i\omega_0$, if $p(\alpha_0 + i\omega) = 0$ holds for some $\omega = \omega_0 > 0$.

In real applications, it is more effective if the above algorithm is combined with the Newton-Raphson iteration method. One can firstly find a good estimation $\hat{\alpha}_0 + i\hat{\omega}_0$ of the rightmost root after a few number of iterations involved in the above algorithm. Then with $\hat{\alpha}_0 + i\hat{\omega}_0$ as the initial guess and $\varepsilon > 0$ as the tolerance, the rightmost root can be refined by using the Newton-Raphson iteration scheme

$$\lambda_{k+1} = \lambda_k - \frac{p(\lambda_k)}{p'(\lambda_k)} \quad (k \geq 0). \tag{37}$$

The iteration is stopped when $|\lambda_{k+1} - \lambda_k| < \varepsilon$.

3.2. *Calculation of the Number of Stability Switches Graphically*

To investigate the stability switches of time-delayed systems graphically by using Theorem 5, with τ as the parameter taking values in a given interval $[\underline{\tau}, \overline{\tau}]$, we divide this interval with points τ_i's satisfying

$$\underline{\tau} = \tau_0 < \tau_1 < \tau_2 < \cdots < \tau_{s-1} < \tau_s = \overline{\tau}$$

where s is sufficiently large such that the length of each subinterval is sufficiently small. For each fixed τ_i and sufficiently large $T > 0$, calculate the testing integral

$$q_i = \int_0^T \frac{S'(\omega)R(\omega) - S(\omega)R'(\omega)}{R^2(\omega) + S^2(\omega)} d\omega.$$

Then the sub-intervals where $q_i > (n-1)\pi/2$ holds are the stable intervals, and the endpoints of these stable sub-intervals where there is a leap of the integral value correspond to the switching points of stability. In this way, the number of stability switches can be easily figured out.

Of course, this algorithm works also for the case when some other parameter, rather than the delay τ, is addressed.

Example 4. Let the characteristic function be in the form

$$p(\lambda) = \lambda^3 + 15\lambda + 1 + b\,e^{-\lambda}, \quad b \in [-3, 3]. \tag{38}$$

At $b = 2$, straightforward numerical calculation gives the following values of the testing integral

$$\int_0^{500} \Re\left(\frac{p'(\mathrm{i}\omega)}{p(\mathrm{i}\omega)}\right) d\omega = 4.7124 > \frac{(3-1)\pi}{2},$$

$$\int_0^{500} \Re\left(\frac{p'(-0.1+\mathrm{i}\omega)}{p(-0.1+\mathrm{i}\omega)}\right) d\omega = -1.5702 < \frac{(3-1)\pi}{2},$$

$$\int_0^{500} \Re\left(\frac{p'(-0.01+\mathrm{i}\omega)}{p(-0.01+\mathrm{i}\omega)}\right) d\omega = 4.7124 > \frac{(3-1)\pi}{2},$$

$$\int_0^{500} \Re\left(\frac{p'(-0.02+\mathrm{i}\omega)}{p(-0.02+\mathrm{i}\omega)}\right) d\omega = -1.5702 < \frac{(3-1)\pi}{2}.$$

It means that the abscissa α_0 belongs to $(-0.02, -0.01) \subset (-0.1, 0)$. Solving ω from $p(-0.015 + \mathrm{i}\omega) = 0$, we obtain $\omega \approx \pm 3.9082$. With $-0.015 \pm 3.9082\mathrm{i}$ as the initial guess of the rightmost roots, then the first iteration of Newton-Raphson's method gives the rightmost roots: $-0.1431 \times 10^{-1} \pm 0.3920 \times 10\ \mathrm{i}$.

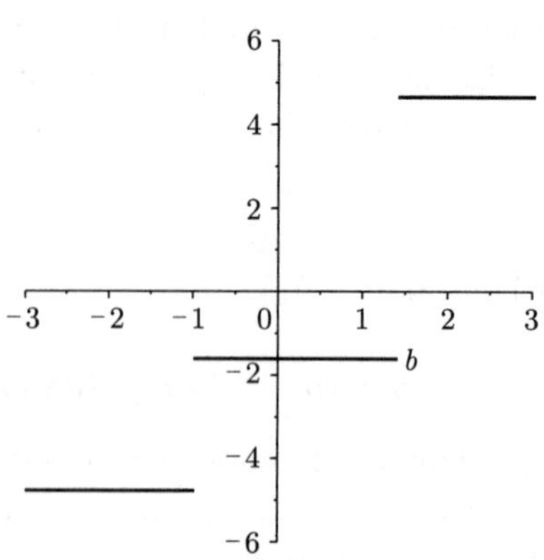

Fig. 4. The plot of the testing integral defined in the left-hand side of (38): $\int_0^{500} \Re\left(\frac{p'(\mathrm{i}\omega)}{p(\mathrm{i}\omega)}\right) d\omega$, for $b \in [-3, 3]$. For $b \in (1.3834, 3]$, the time-delayed system is asymptotically stable because the value of the testing integral is larger than $\frac{(3-1)\pi}{2} = \pi$.

Moreover, with fixed $T = 500$, the values of the testing integral in the left-hand side of (38) for different $b \in [-3, 3]$ are shown in Fig.4, which indicates that the corresponding time-delayed system is asymptotically stable only if $b \in (1.3834, 3]$. The system undergoes 1 stability switch as the parameter b varies in $[-3, 3]$.

4. Pseudo-oscillator Analysis for Hopf Bifurcation

Hopf bifurcation analysis involves two main steps: (1) the existence of the bifurcation, a case that can be carried out from linear stability analysis as done in the last section, (2) the calculation of the bifurcated periodic solution and its stability analysis, a case that will be discussed below by using the pseudo-oscillator analysis[7]. Compared with the methods such as manifold reduction, normal form theory and perturbation methods, that have been widely used in the literature, the pseudo-oscillator analysis can be used for the Hopf bifurcation analysis of scalar time-delayed systems with real or complex coefficients in a much more straightforward way.

4.1. *Scalar Time-delayed Systems with Real Coefficients*

Let us consider the scalar time-delayed system with real coefficients described by

$$F(x(t), \dot{x}(t), \ldots, x^{(n-1)}(t), x^{(n)}(t), x(t-\tau), \dot{x}(t-\tau), \ldots, x^{(n-1)}(t-\tau), p) = 0 \quad (39)$$

where p is the bifurcation parameter, $\tau > 0$ is the delay. In some applications, p does not appear and τ is the bifurcation parameter. In addition, the function F has fourth-order partial derivatives with respect to the state variables as well as the delayed state variables appeared in F, and $F(0, \ldots, 0, p) = 0$ for all p. Moreover, assume that equation (39) undergoes a Hopf bifurcation at $p = p_c$, namely, let $\varepsilon = p - p_c$, then one has

- For small $|\varepsilon|$, the characteristic equation of the linearized time-delayed system round $x = 0$ has a pair of conjugate roots $\lambda = \alpha(\varepsilon) \pm \mathrm{i}\beta(\varepsilon)$.
- At $p = p_c$, $\alpha(0) = 0$, $\beta(0) = \omega_c > 0$, and all the other characteristic roots have negative real parts.
- $\alpha'(0) = \Re\left(\dfrac{d\lambda}{d\varepsilon}\right)\Big|_{\varepsilon=0} \neq 0$.

Then for small ε, the bifurcated periodic solution can be expressed as

$$x(t) = a\cos(\omega_c t + \theta) + O(\varepsilon) \quad (40)$$

due to Fourier expansion for periodic function, where a, θ are determined by the initial condition. Let us recast (39) into the following form

$$\ddot{x} + \omega_c^2 x + (\pm F - \ddot{x} - \omega_c^2 x) = 0 \quad (41)$$

where $(\pm F - \ddot{x} - \omega_c^2 x) \approx 0$ if ε is small. It means that equation (41) can be regarded as a perturbed system of the undamped oscillator $\ddot{x} + \omega_c^2 x = 0$. Thus, it is reasonable to define the "energy function" of Eq. (41) as follows

$$E = \frac{\dot{x}^2}{2} + \omega_c^2 \frac{x^2}{2} \tag{42}$$

which is also called the pseudo-energy of Eq. (39). It follows that

$$\frac{dE}{dt} = -(\pm F - \ddot{x} - \omega_c^2 x) \cdot \dot{x} \approx -(\pm F - \ddot{x} - \omega_c^2 x) \cdot \dot{x}\Big|_{x=a\cos(\omega_c t+\theta)}. \tag{43}$$

Hence, $\frac{dE}{dt}$ is approximately periodic with period $T = 2\pi/\omega_c$, and it varies slowly. Therefore, the local dynamics can be determined by the averaged power function $h(a)$, given by

$$\begin{aligned} h(a) &= -\frac{1}{T}\int_0^T \left((\pm F - \ddot{x} - \omega_c^2 x) \cdot \dot{x}\Big|_{x=a\cos(\omega_c t+\theta)} \right) dt \\ &= -\frac{1}{T}\int_0^T \left(\pm F \cdot \dot{x}\Big|_{x=a\cos(\omega_c t+\theta)} \right) dt. \end{aligned} \tag{44}$$

With the use of Taylor approximation of the function F, straightforward calculation shows that $h(a)$ has the following form

$$h(a) = c_2 a^2 + c_4 a^4 + O(a^6) \tag{45}$$

where the coefficients c_2, c_4 are independent of a. If $c_2 < 0$ (or equivalently $h''(0) < 0$), then $h(a) < 0$ for all small a, namely the energy E decreases to zero as the time t goes to infinity. The sign before F in the averaged power function $h(a)$ should be chosen so as to make the stability condition $h''(0) < 0$ in agreement with the linear stability analysis. If there is a positive number a_0 satisfying $h(a_0) = 0$, then the bifurcated periodic solution can be approximated by $x(t) \approx a_0 \cos(\omega_c t + \theta)$, which is stable if $h'(a_0) < 0$ and is unstable if $h'(a_0) > 0$. This is the key idea of the pseudo-oscillator analysis[7].

Example 5. Consider a time-delayed system arising from machinery dynamics[26]

$$\ddot{x}(t) + 2\xi\dot{x}(t) + x(t) = p\left[-\Delta x + \frac{3}{10}\left((\Delta x)^2 - (\Delta x)^3\right)\right] \tag{46}$$

where $\xi > 0$, $\Delta x = x(t) - x(t-\tau)$. Corresponding to the trivial solution $x = 0$, the characteristic function reads

$$P(\lambda) = \lambda^2 + 2\,\xi\,\lambda + (1+p) - p\,e^{-\lambda\tau}. \tag{47}$$

A linear stability analysis tells that at $p = p_c := 2\xi(\xi+1)$, the critical values of ω, found from $P(\mathrm{i}\omega) = 0$, is $\omega = \omega_c := \sqrt{2\xi+1}$, and the corresponding minimal critical delay value is

$$\tau = \tau_c := \frac{2}{\sqrt{2\xi+1}}\left(\pi - \arctan\frac{1}{\sqrt{2\xi+1}}\right).$$

Except for the conjugate pair of characteristic roots $\lambda = \pm i\omega_c$, all the other characteristic roots have negatively real parts. In addition, for fixed $\tau = \tau_c$, one has

$$\Re\left[\frac{d\lambda}{dp}\right]_{p=p_c} = \frac{1}{2(1+\xi)^2(1+\tau_c\xi)} \neq 0. \tag{48}$$

It means that as p passes through p_c, the trivial solution $x = 0$ losses its stability and a Hopf bifurcation occurs at $p = p_c$.

To use the pseudo-oscillator analysis, we rewrite Eq.(46) in the following form

$$\ddot{x}(t) + \omega_c^2 x(t) = -\left(2\xi\dot{x}(t) + x(t) - p\left[-\Delta x + \frac{3}{10}((\Delta x)^2 - (\Delta x)^3)\right] - \omega_c^2 x(t)\right). \tag{49}$$

With the main part of the bifurcated periodic solution $x(t) \approx x_*(t) = a\cos(\omega_c t)$ up to a constant phase angle, the averaged power function $h(a)$ is found to be[27]

$$\begin{aligned} h(a) &= \frac{\omega_c}{40}a^2[9p\sin(\omega_c\tau_c)(\cos(\omega_c\tau_c) - 1)a^2 - 20(2\xi\omega_c - p\sin(\omega_c\tau_c))] \\ &= -\frac{1}{20}\,\frac{a^2\omega_c^2\xi(-9pa^2\omega_c^2 + 9pa^2 + 20p_c^2 - 20pp_c)}{p_c^2} \end{aligned} \tag{50}$$

where the condition $P(i\omega_c) = 0$ is used for simplifying $h(a)$. From $h(a) = 0$, one obtain an approximation of the amplitude of the bifurcated solution as follows

$$a_T = \frac{\sqrt{20\xi + 20}}{3}\sqrt{\left(1 - \frac{p}{p_c}\right)\Big/\frac{p}{p_c}}.$$

Moreover, it is easy to show that $h'(a_T) > 0$, which means that the bifurcated periodic solution is unstable and the Hopf bifurcation is subcritical. Hence, up to a phase angle, the periodic solution induced by the subcritical Hopf bifurcation has the following form

$$x(t) \approx \frac{\sqrt{20\xi + 20}}{3}\sqrt{\left(1 - \frac{p}{p_c}\right)\Big/\frac{p}{p_c}}\cos\left(\sqrt{2\xi + 1}\,t\right). \tag{51}$$

Table 2. The amplitude and period of the Hopf bifurcation-induced periodic solution of Eq.(46) with[27] $\xi = 0.1$, by using different methods.

Parameter p/p_c	Amplitude a_T	Amplitude a_N	Period $2\pi/\omega_c$	Period $2\pi/\omega$
0.95	0.3587	0.3615	5.7381	5.7357
0.96	0.3191	0.3210	5.7381	5.7357
0.97	0.2750	0.2760	5.7381	5.7357
0.98	0.2234	0.2239	5.7381	5.7357
0.99	0.1571	0.1581	5.7381	5.7358

To show the efficiency of the pseudo-oscillator analysis, let $\xi = 0.1$, then[26] $p_c = 0.22$, $\omega_c = 1.095$, $\tau_c = 4.385$. From Table 2, one can see that the results $(a_T,\ 2\pi/\omega_c)$ obtained by using the pseudo-oscillator analysis are in very good agreement with the ones $(a_N,\ 2\pi/\omega)$ calculated by using the shooting method. The unstable periodic solution cannot be obtained by using numerical integration.

The pseudo-oscillator analysis has been successfully applied for the Hopf bifurcation of different dynamical systems, including laser dynamics model[7], population dynamics model[28], biological model[29], maglev dynamics model[30], neural network model[31] and economy dynamics model[32]. Moreover, the pseudo-oscillator analysis has a close relation with the iteration method. In fact, let $x_0(t) = a\cos(\omega_c t + \theta)$ be the initial guess of the bifurcated periodic solution, and for $k \geq 0$, define an iteration sequence as follows[33]

$$\begin{aligned} x_{k+1}(t) = x_k(t) - F(&x_k(t), \dot{x}_k(t), \ldots, x_k^{(n-1)}(t), x_k^{(n)}(t), \\ &x_k(t-\tau), \dot{x}_k(t-\tau), \ldots, x_k^{(n-1)}(t-\tau), p). \end{aligned} \tag{52}$$

Then the amplitude a can be determined from the Galerkin-type condition

$$\int_0^{2\pi/\omega_c} ((\ddot{x}_1(t) - \ddot{x}_0(t)) \cdot \dot{x}_0(t))\, dt = 0, \tag{53}$$

a condition that is equivalent to $h(a) = 0$. The iteration method works for the Hopf bifurcation of more general time-delayed systems. In this sense, the idea of the pseudo-oscillator analysis works for the Hopf bifurcation analysis of some special non-scalar time-delayed systems.

Example 6. Consider the following time-delayed system with two delays[34]

$$\ddot{x}(t) + \dot{x}(t-\tau) + \frac{9}{2}x^2(t-\tau)\dot{x}(t-\tau) + x(t-2\tau) = 0. \tag{54}$$

The characteristic function of the trivial solution is

$$p(\lambda) = \lambda^2 + \lambda e^{-\lambda\tau} + e^{-2\lambda\tau}.$$

At $\tau = \tau_c := \pi/6 \approx 0.5236$, a Hopf bifurcation occurs, because $p(\lambda)$ has roots with negative real parts only, except for the exact one pair of conjugate roots $\lambda = \pm\omega_c \mathrm{i}$ $(\omega_c = 1)$, on the imaginary axis, and the branches passes through the imaginary axis satisfy $\Re\left[\dfrac{d\lambda}{d\tau}\right]_{\tau=\pi/6} \neq 0$. Close to the bifurcation point, with $x(t) \approx a\cos(t+\theta)$, one finds

$$\begin{aligned} h(a) &= -\frac{\omega_c}{2\pi}\int_0^{\frac{2\pi}{\omega_c}} \delta\left[\ddot{x}(t) + \dot{x}(t-\tau) + \frac{9}{2}x^2(t-\tau)\dot{x}(t-\tau) + x(t-2\tau)\right]\dot{x}(t)dt \\ &= \frac{\delta}{2}\left[a^2\cos\tau(2\sin\tau - 1) - \frac{9}{8}a^4\cos\tau\right]. \end{aligned}$$

The constant δ should be taken as 1, because the system is asymptotically stable if $0 < \tau < \tau_c$. When $\tau > \tau_c$, there is exactly one positive root of $h(a) = 0$:

$a_T = 2\sqrt{2(2\sin\tau - 1)/3}$. Hence, the bifurcation-induced periodic solution has a form $x(t) \approx a_T \cos t$ up to a phase angle, and it is stable due to $h'(a_T) < 0$. Again as shown in Table 3, the estimation of the amplitude by using the pseudo-oscillator analysis is in very good agreement with the numerical calculation.

Table 3. The amplitude of the Hopf bifurcation-induced periodic solution of Eq.(54), by using different methods[34].

Parameter τ	Amplitude a_T	Amplitude a_N
0.10	0. 6813	0. 6568
0.15	0. 7289	0. 7443
0.20	0. 7647	0. 7948
0.25	0. 7937	0. 8281
0.30	0. 8182	0. 8574
0.35	0. 8395	0. 8831

The pseudo-oscillator analysis has been proved rigorously and extended to a more general case[34].

Theorem 6. *For the scalar time-delayed system described by* (39)*, then near the Hopf bifurcation point* p_c*, let* $\varepsilon = p - p_c$*, the following statements hold true*

- *For small* $|\varepsilon|$*, in order that equation* (39) *has a periodic solution around* $x = 0$*, it is necessary that* $h(a)$ *has a positive root.*
- *If* $h(a)$ *has a positive root* a_0 *and* $h(a_0)$ *is not an extreme, then the bifurcated periodic solution has the following approximate form*

$$x \approx a_0 \cos(\omega_c t + \theta), \qquad \dot{x} \approx -a_0\omega_c \sin(\omega_c t + \theta).$$

- *If there is a* $a_0 > 0$ *and an integer* m *such that* $h(a_0) = h'(a_0) = \cdots = h^{(2m)}(a_0) = 0$, $h^{(2m+1)}(a_0) < 0$*, then the bifurcated periodic solution is stable, and if there is a* $a_0 > 0$ *and an integer* m *such that* $h(a_0) = h'(a_0) = \cdots = h^{(2m)}(a_0) = 0$, $h^{(2m+1)}(a_0) > 0$*, then the periodic solution is unstable.*

4.2. *Scalar Time-delayed Systems with Complex Coefficients*

Now, let us consider the scalar time-delayed systems in the form of Eq. (39) with complex coefficients, where p is the real bifurcation parameter. Since the characteristic roots do not come in pairs conjugately, the first two conditions that govern the existence of a Hopf bifurcation for Eq. (39) at $p = p_c$ should be changed a little bit as follows:

- For small $|\varepsilon|$, the characteristic equation of the linearized time-delayed system round $x = 0$ has a root $\lambda = \alpha(\varepsilon) + \mathrm{i}\beta(\varepsilon)$, where $\varepsilon = p - p_c$.
- At $p = p_c$, $\alpha(0) = 0$, $|\beta(0)| = \omega_c > 0$, and all the other characteristic roots have negative real parts.

- $\alpha'(0) = \Re\left(\frac{d\lambda}{d\varepsilon}\right)\Big|_{\varepsilon=0} \neq 0.$

For small ε, the Hopf bifurcation-induced periodic solution can be expressed as

$$x(t) = a\, e^{\mathrm{i}(\omega_c t+\theta)} + O(\varepsilon). \tag{55}$$

Then similar to the case with real coefficients, we define the "pseudo-energy" of Eq. (39) in the following form

$$E = \frac{1}{2}\dot{x}\dot{\bar{x}} + \frac{1}{2}\omega_c^2 x\bar{x}$$

where $\bar{x}$ stands for the conjugate of x. The "pseudo-power function" is

$$\frac{dE}{dt} = -\frac{1}{2}[\overline{(\pm F - \ddot{x} - \omega_c^2 x)}\dot{x} + (\pm F - \ddot{x} - \omega_c^2 x)\dot{\bar{x}}]\Big|_{x=a\, e^{\mathrm{i}(\omega_c t+\theta)}}.$$

By using the averaging technique over $[0, T]$ with $T = 2\pi/|\omega_c|$, the "pseudo-power function" $h(a)$ is simplified to

$$\begin{aligned} h(a) &= -\frac{1}{2T}\int_0^T \left[\overline{(\pm F - \ddot{x} - \omega_c^2 x)}\dot{x} + (\pm F - \ddot{x} - \omega_c^2 x)\dot{\bar{x}}]\Big|_{x=ae^{\mathrm{i}(\omega_c t+\theta)}}\right] dt \\ &= c_2 a^2 + c_4 a^4 + O(a^6). \end{aligned} \tag{56}$$

The sign "+" or "−" before F in the "pseudo-power function" should be selected to match the linear stability analysis. At this stage, the Hopf bifurcation-induced periodic solution of scalar time-delayed systems with complex coefficients can be analyzed as done in Section 4.1.

Example 7. Let us consider the simplified Lang-Kobayashi equation arising from laser dynamics[1]

$$\frac{dF}{dt} = (1 + \mathrm{i}\alpha)(p - |F|^2)F + \eta\, e^{-\mathrm{i}\Omega}F(t-1) \tag{57}$$

where $\alpha, p, \eta, \Omega \in \mathbb{R}$. The characteristic function of the trivial solution $F = 0$ reads

$$p(\lambda) = \lambda - (1 + \mathrm{i}\alpha)p - \eta e^{-\mathrm{i}\Omega}e^{-\lambda}. \tag{58}$$

In terms of the principal branch $W_0(z)$ of the Lambert W function, $p(\lambda)$ has roots with negative real parts only, if and only if $\Re[W_0(\eta\, e^{-\mathrm{i}\Omega-(1+\mathrm{i}\alpha)p})] + p < 0$, and the bifurcation point can be found from

$$\Re[W_0(\eta\, e^{-\mathrm{i}\Omega-(1+\mathrm{i}\alpha)p})] + p = 0. \tag{59}$$

Now the pseudo-oscillator associated with Eq. (57) near a Hopf bifurcation is

$$\ddot{F} + \omega_c^2 F + \delta(\dot{F} - (1 + \mathrm{i}\alpha)(p - |F|^2)F - \eta e^{-\mathrm{i}\Omega}F(t-1)) = 0 \tag{60}$$

where ω_c is vibration frequency at the bifurcation point solved from $p(\mathrm{i}\omega_c) = 0$, or

$$\begin{cases} -p - \eta\cos(\Omega + \omega_c) = 0, \\ \omega_c - \alpha p + \eta\sin(\Omega + \omega_c) = 0. \end{cases}$$

Then the averaged pseudo-power function $h(a)$ is found to be[8]

$$h(a) = -\delta\omega_c\alpha a^4 + \delta\omega_c\, a^2[-\omega_c + \alpha p - \eta\sin(\omega_c + \Omega)]. \qquad (61)$$

In order that the stability condition $h''(0) < 0$ for the trivial solution $F = 0$ is consistent with the linear stability analysis, it requires that $\delta = \mathrm{sgn}(\alpha\omega_c)$.

Case 1. Let $p = p_c$ be the bifurcation point, then with small $\varepsilon = p - p_c$, the averaged pseudo-power function $h(a)$ can be simplified as

$$h_p(a) = -\delta\omega_c\alpha a^2(a^2 - \varepsilon).$$

When $\varepsilon > 0$, $F = 0$ is unstable and the amplitude of the bifurcation-induced periodic solution is about $a_T = \sqrt{\varepsilon}$. Since $h_p'(a_T) = -\delta\omega_c\alpha\varepsilon^{3/2} < 0$, the periodic solution is stable and the Hopf bifurcation is supercritical.

Case 2. Assume that the system undergoes a Hopf bifurcation at $\eta = \eta_c$, then for small $\varepsilon = \eta - \eta_c$, the function $h(a)$ is simplified to

$$h_\eta(a) = -\delta\omega_c\alpha a^2\left[a^2 - \frac{\varepsilon(\omega_c - \alpha p)}{\alpha\eta_c}\right].$$

Hence, $a_T = \sqrt{\varepsilon\,(\omega_c - \alpha\, p)/(\alpha\,\eta_c)}$ is the unique positive root of $h_\eta(a) = 0$, and the bifurcation is supercritical due to $h_\eta'(a_T) < 0$.

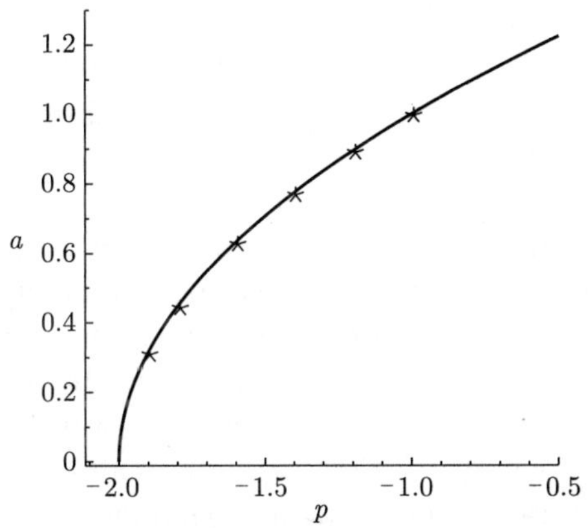

Fig. 5. The amplitude of the bifurcated periodic solution for $\alpha = \pi/12$ (≈ 0.2618), $\eta = 2$, $\Omega = 2k\pi + \pi/6$ ($k \in \mathbb{Z}$), where equation (57) undergoes a Hopf bifurcation at $p = p_c := -2$ with $\omega_c = -\pi/6$.

The pseudo-oscillator analysis offers an estimation of the amplitude of the bifurcation-induced periodic solution with high accuracy. For example, when $\alpha = \pi/12$, $\Omega = \pi/6$, and $\eta = 2$, then a Hopf bifurcation occurs at $p = p_c = -2$, because at $p = -2$, $\lambda = -\dfrac{\pi}{6}\mathrm{i}$ is the unique simple characteristic root on the imaginary axis, and all the other characteristic roots stay in the open left-half complex plane. In addition, for small $\varepsilon = p - (-2)$, let $\lambda_0(\varepsilon) = W_0(2e^{-\mathrm{i}\frac{\pi}{6} - (1+\mathrm{i}\frac{\pi}{12})p}) + \left(1 + \mathrm{i}\dfrac{\pi}{12}\right)p$, then $\lambda_0(0) := \omega_c = -\pi/6$ and $\Re\lambda_0'(0) \neq 0$. As shown in Fig.5, the amplitude obtained by the pseudo-oscillator analysis is in very good agreement with the numerical simulation, even when the parameter p is not close to the bifurcation point, where the solid curve is the plot of the amplitude derived by the pseudo-oscillator analysis, and the dots are obtained by numerical simulation. More details and illustrative examples are available for the applications of the proposed method[5,8].

5. Conclusions

Compared with the intensive study on the stability and Hopf bifurcation of real-DDEs, results that can be used directly for complex-DDEs are considerable few in the literature. Though some engineering applications use complex-DDEs to model dynamical systems such as in rotor dynamics, laser dynamics, the complex-DDEs are usually transformed into coupled real-DDEs in dynamics analysis. This procedure greatly increases the computational complexity. In this chapter, some new results on the stability and Hopf bifurcation of time-delayed systems with real or complex coefficients are presented. The proposed stability criteria and numerical algorithms generalize some available results in the literature, and they can be numerically implemented in a simple way. In particular, a new auxiliary function derived from the argument of the critical stability condition and a Jacobian determinant are introduced to determine the crossing direction of systems with delay-dependent coefficients. The results are extended to the degenerated case for which higher order derivatives are required to determine the crossing direction of the characteristic function. The pseudo-oscillator analysis is a combination of the energy analysis and the averaging technique for slightly perturbed systems. It works for the scalar time-delayed systems near a Hopf bifurcation with high accuracy, because the main part of the corresponding bifurcation-induced periodic solution is harmonic so that the time-delayed systems near a Hopf bifurcation can be considered as a small perturbation of an undamped oscillator. The idea is expected to be applied directly for the Hopf bifurcation of general time-delayed systems or discrete dynamical systems. It should be mentioned that the pseudo-oscillator analysis may fail to work in some applications due to the use of the averaging technique.

Acknowledgments. The authors thank the financial support of the National Natural Science Foundation of China under Grant Nos. 10825207 and 11032009. Zaihua Wang thanks Alexander von Humboldt Foundation, Germany, for supporting his renewed visit to Professor Tassilo Küpper at the Institute of Mathematics, University of Cologne, Germany, and thanks Professor Küpper for the kind hospitality.

References

1. T. Erneux, *Applied Delay Differential Equations* (Springer Verlag, Berlin, 2009).
2. A. Muszynska, Transition to fluid-induced limit cycle seld-excited vibrations of a rotor

and instability threshold "Hysterisis", *International Journal of Rotating Machinery*, **5**(2), 123–133 (1999).
3. J. Y. Li, L. Zhang, and Z. H. Wang, Two effective stability criteria for linear time-delay systems with complex coefficients, *Journal of System Science and Complexity*, **24**(5), 835–849 (2011).
4. Z. H. Wang, A very simple criterion for characterizing the crossing direction of time-delay systems with delay-dependent parameters, *International Journal of Bifurcation and Chaos*, **22**(3), 1250048-8 (2012).
5. J. Y. Li, *Stability Analysis and Dynamic Response of Time-delay Systems with Complex Coefficients*, Dissertation (Nanjing University of Aeronautics and Astronautics, Nanjing, 2010) (in Chinese).
6. Z. H. Wang, M. L. Du, and M. Shi, Stability test of fractional-delay systems via integration, *Science China: Physics, Mechanics & Astronomy*, **54**(10), 1839–1846 (2011).
7. Z. H. Wang and H. Y. Hu, Pseudo-oscillator analysis of scalar nonlinear time-delay systems near a Hopf bifurcation, *International Journal of Bifurcation and Chaos*, **17**(8), 2805–2814 (2007).
8. J. Y. Li and Z. H. Wang, Local Hopf bifurcation of complex nonlinear systems with time-delay, *International Journal of Bifurcation and Chaos*, **19**(3), 1069–1079 (2009).
9. Y. Kuang, *Delay Differential Equation with Application to Population Dynamics* (Academic Press, San Diego, 1993).
10. J. E. Marshall, H. Gorecki, A. Korytowski, and K. Walton, *Time-Delay Systems: Stability and Performance Criteria with Applications* (Prentice Hall, New York, 1992).
11. E. Beretta and Y. Kuang, Geometric stability switch criteria in delay differential systems with delay dependent parameters, *SIAM Journal of Mathematical Analysis*, **33**(5), 1144–1165 (2002).
12. P. Tass, J. Kurths, M. G. Rosenblum, G. Guasti and H. Hefter, Delay-induced transitions in visually guided movements, *Physical Review E*, **54**(3), R2224–R2227 (1996).
13. C. V. Hollot, V. Misra, D. Towsley, and W. B. Gong, Analysis and design of controllers for AQM routers supporting TCP flows, *IEEE Transactions on Automatic Control*, **47**(6), 945–959 (2002).
14. Y. G. Zheng and Z. H. Wang, Stability and Hopf bifurcation of a class of TCP/AQM networks, *Nonlinear Analysis: Real World Applications*, **11**(3), 1552–1559 (2010).
15. X. Xu, H. Y. Hu, and H. L. Wang, Stability switches, Hopf bifurcation and chaos of a neuro model with delay-dependent parameter, *Physics Letters A*, **354**(1–2), 126–136 (2006).
16. H. Shinozaki and T. Mori, Robust Stability Analysis of linear time-delay systems by Lambert W function: some extreme point results, *Automatica*, **42**(6), 1791–1799 (2006).
17. J. E. S. Socolar and D. J. Gauthier, Analysis and comparison of multiple-delay schemes for controlling unstable fixed points of discrete maps, *Physical Review E*, **57**, 6589–6595 (1998).
18. V. Kolmanovskii and A. Myshkis, *Introduction to The Theory and Applications of Functional Differential Equations* (Kluwer, Dordrecht, 1999).
19. S. Arunsawtwong, Stability of retarded differential equations, *International Journal of Control*, **65**(2), 347–364 (1996).
20. Z. H. Wang and H. Y. Hu, Calculation of the rightmost characteristic root of retarded time-delay systems via Lambert W function, *Journal of Sound and Vibration*, **318** (4–5), 757–767 (2008).
21. K. Verheyden, T. Luzyanina, and D. Rose, Efficient computation of characteristic

roots of delay differential equations using LMS methods, *Journal of Computational and Applied Mathematics*, **214**(1), 209–226 (2008).
22. Z. Wu and W. Michiels, Reliably computing all characteristic roots of delay differential equations in a given right half plane using a spectral method, *Journal of Computational and Applied Mathematics*, **236**(9), 2499–2514 (2011).
23. G. Stepán, *Retarded Dynamical Systems: Stability and Characteristic Functions* (Longman Scientific and Technical, Essex, 1989).
24. M. Buslowicz, Stability of linear continuous-time fractional order systems with delays of the retarded type, *Bulletin of The Polish Academy of Sciences: Technical Sciences*, **56**(4), 319–324 (2008).
25. M. Shi and Z. H. Wang, An effective analytical criterion for stability testing of fractional-delay systems, *Automatica*, **47**(9), 2001–2005 (2011).
26. T. Kalmár-Nagy, G. Stepán, and F. C. Moon, Subcritical Hopf bifurcation in the delay equation model for machine tool vibrations, *Nonlinear Dynamics*, **26**(2), 121–142 (2001).
27. F. Gao, H. L. Wang, and Z. H. Wang, Hopf bifurcation of a nonlinear delayed system of machine tool vibration via pseudo-oscillator analysis, *Nonlinear Analysis, Series B: Real World Applications*, **8**(5), 1561–1568 (2007).
28. J. Li and Z. H. Wang, Hopf bifurcation of a nonlinear Lasota-Wazewska-type population model with maturation delay, *Dynamics of Continuous Discrete and Impulsive Systems: Series B-Applications and Algorithms*, **14**(5), 611–623 (2007).
29. J. Y. Li, Hopf bifurcation of the sunflower equation, *Nonlinear Analysis: Real World Applications*, **10**(4), 2574–2580 (2009).
30. L. L. Zhang, L. H. Huang, and Z. Z. Zhang, Hopf bifurcation of the maglev time-delay feedback system via pseudo-oscillator analysis, *Mathematical and Computer Modeling*, **52**(5-6), 667–673 (2010).
31. M. J. Sun, *Stability and Hopf Bifurcation of a Delayed Model of Neural Network*, Master Thesis (Nanjing University of Aeronautics and Astronautics, Nanjing, 2007) (in Chinese).
32. J. Li, Hopf bifurcation of an economic model, *Journal of Dynamics and Control*, **8**(4), 380–384 (2010) (in Chinese).
33. Z. H. Wang, An iteration method for calculating the periodic solution of time-delay systems after a Hopf bifurcation, *Nonlinear Dynamics*, **53**(1–2), 1–11 (2008).
34. Y. J. Yu and Z. H. Wang, Proof of the pseudo-oscillator analysis and its application to higher-order Hopf bifurcation, *Journal of Dynamics and Control*, **10**(3) (2012) (in Chinese).

Chapter 8

Estimation and Control in Time-delayed Dynamical Systems Using the Chebyshev Spectral Continuous Time Approximation and Reduced Liapunov-Floquet Transformation

Eric A. Butcher, Oleg Bobrenkov, Morad Nazari, Shahab Torkamani

New Mexico State University, Las Cruces, NM, 88003, USA

eab@nmsu.edu

In this chapter, a detailed description is provided of the authors' recent work on estimation and control of retarded time-delayed systems with (possibly) time-periodic coefficients using the Chebyshev spectral continuous time approximation (CSCTA) and reduced Liapunov-Floquet transformation (RLFT). First, these specialized computational tools are outlined and illustrated with several examples including various forms of the delayed Mathieu equation with single and multiple discrete delays as well as discontinuous distributed delay. Specifically, it is demonstrated how a time-periodic linear delayed system may be transformed into a set of constant-coefficient ODEs of low dimension in which the eigenstructure (including the dominant characteristic exponents of the delayed system) is exactly preserved in the transformed domain. Next, the control problem of (possibly time-periodic) time-delayed systems is explored, in which three strategies are suggested for designing closed-loop linear feedback controllers. Specifically, it is shown that using CSCTA and RLFT enables the control design to take place in the transformed coordinate system in which both the delay and periodic coefficients are effectively removed, thereby enabling the use of traditional control design tools for linear time-invariant systems. Next, the problem of estimation of states and parameters (including the delay) for nonlinear delayed systems using optimal stochastic filtering with noise-corrupted, possibly incomplete measurements is explored using CSCTA. After discretizing the delayed system with a set of ordinary differential equations, the stochastic estimation problem is implemented via the use of state augmentation and an extended Kalman-Bucy filter. This estimator can also be employed in an observer-based feedback controller, and for an illustrative example in a practical engineering problem, the problem of spacecraft attitude estimation and multi-actuator regulation control in the presence of time delay in one actuator is discussed and illustrated.

1. Introduction

Time-delayed dynamical systems, due to the vast variety of fields in science and engineering they can be relevant to, have attracted an increasing interest during past few decades. Papers on the stability analysis of delay differential equations (DDEs) started to appear in the 1940s. Unlike in the case of ODEs, dependence of the current state derivative on the past state at prior times brings additional complexity to the stability analysis of DDEs. One of the first approaches was Krasovskii's generalization of the second method of Liapunov[1]. Various other methods were developed and various problems associated with the stability analysis of DDEs were posed. Some fundamental results in the frequency domain were obtained by Pontryagin[2], while Chebotarev and Meiman[3] and Hayes[4] studied the Routh-Hurwitz problems of quasi polynomials. Similar to the stability of linear time-invariant (LTI) ODEs which are usually analyzed through the computation of characteristic values of the state matrix, or equivalently the computation of the roots of the characteristic equation, for LTI DDEs this characteristic equation is represented by an exponential polynomial with an infinite number of complex roots. Different frequency domain stability criteria which take into account robustness and computational considerations appeared in 1990's and were closely connected to the rapid development of computer technologies. The theory of analysis and control of DDEs is covered in several modern texts, e.g., Refs.5–10.

Most stability and control studies of time-delayed systems assume single or multiple discrete delays. However, whenever the past effect of the system is distributed over an interval, distributed delay terms appear in the system model. Analysis of a system modeled by a DDE having distributed delay, or as it is called in some references a "Delayed Integro-Differential Equation" (DIDE), is more complicated than that for a system with discrete delays, e.g., Ref.11, and can include the case in which the distributed delay is discontinuous[12]. Distributed delay has been used to model the chip formation process in milling dynamics[13] and also can be used in delayed feedback control[14].

Two of the first studies on time-periodic systems with time delay were due to Hahn[15] and Stokes[16]. Recently, much attention has been devoted to the topic, e.g., Refs.7, 17 and references therein. Most of this work has been concerned with the stability problem in machining dynamics, in particular the problem of chatter vibrations in milling, e.g., Ref.18. In such problems, frequently the time-periodic coefficients of the DDE are discontinuous[19], which is a fundamentally different problem than that when the distributed delay is discontinuous in a DIDE. When either ODEs or DDEs have time-periodic coefficients, finding solutions analytically is rarely possible. Stability analysis for periodic ODEs requires computing the characteristic values (Floquet multipliers) of the finite-dimensional monodromy matrix[20], which is obtained by evaluating the state transition matrix at the principal parametric period. For periodic DDEs, however, the Floquet multipliers are eigenvalues of an infinite-dimensional monodromy operator. The positions of the dominant Floquet multipliers with respect to the unit circle in the complex plane should be analyzed. According to the unit circle stability criteria, if all eigenvalues lie within the unit circle then the system is asymptotically stable.

There have been a few research studies concerning state estimation and parameter identification for time delay systems, e.g., Refs.21–30. While in most of these studies the critical role of measurement noise and model uncertainties is ignored, in Refs.22, 23 the optimal joint filtering and identification of states and parameters in a linear stochastic time delay system is studied through designing an optimal finite-dimensional filter. Parameter identification or state estimation of time-varying DDEs, however, has been treated in Refs.29, 30. There has also been much research in the literature concerning controller design for time-delayed systems, e.g., Refs.8, 9, 14, 31–41, including the case of delayed feedback control in which there is no choice but to use delayed states for feedback because the current states are simply not available. Unlike the case with the estimation problem, however, the control of periodic delayed systems has lately received significant attention, e.g., Refs.7, 17, 42–51.

Recently, new computational tools have been demonstrated that enable many of the traditional tools for time-invariant non-delayed systems to be successfully applied to various kinds of DDEs, including DIDEs and DDEs with periodic coefficients. These include the Chebyshev Spectral Continuous Time Approximation (CSCTA)[52], and the Reduced Liapunov-Floquet Transformation (RLFT)[53]. These specialized computational tools can be used in conjunction with estimation and control problems in time-delayed systems. This work explores the application of CSCTA and RLFT to the problems of estimation of parameters, states, and time delay as well as delayed feedback control of nonlinear delay differential equations (DDEs) having time-varying coefficients and discrete or distributed delay. In particular, a novel approach in state, parameter, and delay estimation of DDEs is proposed through exploiting optimal stochastic filtering in conjunction with CSCTA. The nonlinear delay differential equation is first discretized with a set of ordinary differential equations for which the stochastic estimation problem of the resulting ODE system is represented as an optimal filtering problem using a state augmentation technique. The extended Kalman-Bucy filter is used to obtain the estimated states, parameters, and delay from a noise-corrupted, possibly incomplete measurement of the states.

In addition, the control problem of time-delayed systems is also explored, in which both CSCTA and RLFT can be used for strategic advantage in designing delayed feedback controllers. For linear periodic time-delayed systems, this enables the control design to take place in a reduced order transformed coordinate system in which both the delay and periodic coefficients are effectively removed, thereby enabling the use of traditional tools for non-delayed LTI systems, while for nonlinear systems the linear part is transformed thus enabling the use of many additional tools for analysis and control of nonlinear systems. The estimator described above can also be employed to provide estimated states to be used in the feedback controller using an observer-based controller.

For an illustrative example of the above concepts in a practical engineering problem, the problem of attitude estimation and multi-actuator regulation control of a rigid spacecraft in the presence of time delays in one actuator is discussed.

A nonlinear feedback controller is designed using both delayed and non-delayed states such that the controlled system has the desired linear closed-loop dynamics which contains a time-delayed term using an inverse dynamics approach. First, the closed-loop stability is shown to be approximated by a second order linear DDE for the attitude coordinate for which the Hsu-Bhatt-Vyshnegradskii stability chart can be used to choose the control gains that result in a stable closed-loop response. It is shown that the suggested control methodology which was recently proposed in Ref.40 is an effective strategy for the problem of spacecraft attitude control using delayed feedback. Finally, the optimal estimator is combined with the controller in order to implement observer-based delayed feedback control of the spacecraft attitude with unknown time delay using optimally estimated states and delay obtained from noise-corrupted measurements.

2. Chebyshev Spectral Continuous Time Approximation

2.1. *Formulation*

In Refs.54–56, it was demonstrated that for a LTI DDE, an equivalent abstract LTI ODE can be introduced in which the infinite-dimensional system matrix is the infinitesimal generator of the solution operator. Previously, other strategies discretized the infinitesimal generator for constant DDEs via approximation of the derivatives using a finite difference-based differentiation[37,57] and pseudospectral differencing[58–60]. While it is predictable that the (linear) convergence of finite difference-based differentiation is slower than the (exponential) convergence of pseudospectral differencing[61], the efficiencies of both approaches were directly compared in the context of the stability analysis of specific DDEs in Ref.52. In addition, by approximating the equivalent abstract ODE a large set of tools for analysis, estimation, and control of ODEs becomes available to use with DDEs. It is noted that other techniques such as the Galerkin method (e.g., Ref.62) have also been presented to obtain a set of ODEs from DDEs.

In this section, a new CSCTA technique which was recently introduced in Ref.52 will be reviewed. This technique allows for a DDE system to be represented as a larger system of ODEs. The advantages of this technique are that the nonlinearities and external excitation can be simply incorporated in the formulation. Also, this technique is applicable to constant and periodic DDEs with multiple delays, as well as DIDEs with discontinuous distributed delay.

Consider a system of nonlinear retarded autonomous DDEs with a single discrete time delay described by

$$\begin{aligned} \dot{\mathbf{x}}(t) &= \mathbf{A}_1\mathbf{x}(t) + \mathbf{A}_2\mathbf{x}(t-\tau) + \mathbf{f}(\mathbf{x}(t), \mathbf{x}(t-\tau)), \\ \mathbf{x}(t) &= \phi(\theta), \quad -\tau \le t = \theta \le 0, \end{aligned} \tag{1}$$

where $\mathbf{x} \in \mathbb{R}^q$ is a q-dimensional state vector, $\mathbf{A}_1$ and $\mathbf{A}_2$ are constant $q \times q$ matrices, $\mathbf{f}(\cdot)$ is a nonlinear q-dimensional vector function of the current and delayed states, and $\phi(t)$ is the initial vector function. As explained in Refs.54–56, the abstract representation of Eq.(1) is the evolution of an initial function in a Banach space, i.e.,

$$\dot{\mathcal{Y}}(t) = \mathcal{A}\mathcal{Y}(t) + \mathcal{F}(\mathcal{Y}(t)), \quad \mathcal{Y}(0) = \phi(\theta) \tag{2}$$

where $\mathcal{Y}(t)$ and $\mathcal{F}(\cdot)$ are infinite-dimensional vectors and $\mathcal{A}$ is the infinitesimal generator of the solution operator corresponding to the linearized system, i.e.,

$$\mathcal{T}(t) : \mathcal{C}\left([-\tau, 0], \mathbb{R}^q\right) \rightarrow \mathcal{C}\left([-\tau, 0], \mathbb{R}^q\right) = \frac{\partial \mathcal{Y}(t)}{\partial \phi} \tag{3}$$

that maps the initial function deviations $\delta\phi$ forward in time. This operator is a strongly continuous semigroup, and it satisfies all the semigroup properties including $\mathcal{T}(0) = I$, $\mathcal{T}(t)\mathcal{T}(s) = \mathcal{T}(t+s)$. The infinitesimal generator $\mathcal{A}$ is defined as

$$\begin{aligned}
&\mathcal{A}\phi = \frac{d\phi}{d\theta} = \lim_{t\to 0}\frac{1}{t}\left(\mathcal{T}(t)\phi - \phi\right), \quad -\tau \le \theta < 0, \\
&\mathcal{A}\phi = \mathbf{A}_1\mathbf{x}(0) + \mathbf{A}_2\mathbf{x}(-\tau), \quad \theta = 0, \\
&\mathcal{A}\phi + \mathcal{F}(\phi) = \mathbf{A}_1\mathbf{x}(0) + \mathbf{A}_2\mathbf{x}(-\tau) + \mathbf{f}(\mathbf{x}(0), \mathbf{x}(-\tau)), \quad \theta = 0
\end{aligned} \tag{4}$$

and can be thought of as an infinite-dimensional square matrix. Approximation of the infinite dimensional vectors $\mathcal{Y}(t)$ and $\mathcal{F}(\cdot)$ and operator $\mathcal{A}$ with finite dimensional ones is the main idea of CSCTA.

Accordingly, the main steps are setting up an uneven grid of Chebyshev collocation points, fitting an interpolating polynomial through the values of the function evaluated at the grid points, and then differentiating the polynomial and evaluating the result at each grid point as a linear combination of the nodal function values[63]. The Chebyshev collocation points are unevenly spaced points in the domain $[-1, 1]$ corresponding to the extremum points of the Chebyshev polynomial of the first kind[64] of degree N. As seen in Fig. 1(a), we can also define these points as the projections of equispaced points on the upper half of the unit circle as $t_j = \cos(j\pi/N)$, $j = 0, 1, \ldots, N$ where the number of collocation points used is $m = N + 1$. The main advantage of this strategy lies in its "spectrally accurate" exponential convergence characteristics, which are significantly more desirable than those of finite difference, finite elements, cubic splines, or wavelets on analytic functions. In addition, the unequally spaced Chebyshev grid are nearly optimal interpolation points for minimizing uniform error compared with the zeros or extremum points of other orthogonal polynomials and avoid the Gibbs phenomenon of interpolating functions which are typically associated with an equally spaced grid.

Associated with the Chebyshev grid, we can find the $m \times m$ Chebyshev differ-

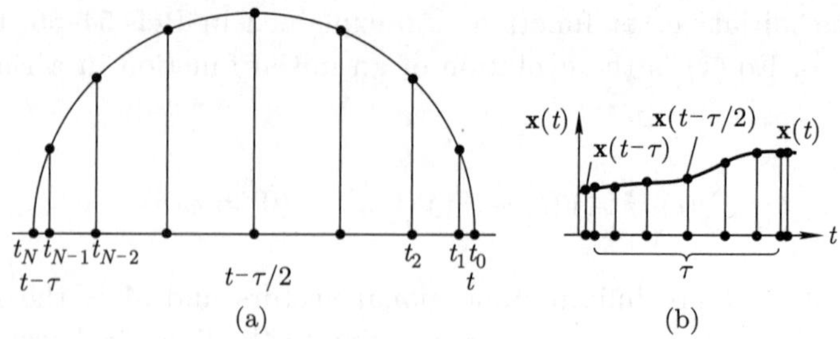

Fig. 1. Diagrams of (a) Chebyshev collocation points as defined by projections from the unit circle and (b) solution of $\mathbf{x}(t)$ discretized by CSCTA.

entiation matrix $\mathbf{D}_N$ for any order N using indices from 0 to N as Ref.61

$$\begin{aligned} D_{00} &= \frac{2N^2+1}{6}, \quad D_{NN} = -\frac{2N^2+1}{6}, \\ D_{jj} &= \frac{-t_j}{2(1-t_j^2)}, \quad j = 1, \ldots, N-1, \\ D_{ij} &= \frac{c_i(-1)^{i+j}}{c_j(t_i - t_j)}, \quad i \neq j,\ i,j = 0, \ldots, N, \\ c_i &= \begin{cases} 2, & i = 0, N, \\ 1, & \text{otherwise.} \end{cases} \end{aligned} \tag{5}$$

Also, let the $mq \times mq$ differential operator $\mathbb{D}$ (corresponding to q first order DDEs) be defined as $\mathbb{D}_\mathbf{N} = \mathbf{D_N} \otimes I_q$, where I_q is a $q \times q$ identity matrix.

A finite dimensional approximation to $\mathcal{Y}(t)$ is now defined as

$$\begin{aligned} \mathbf{y}(t) &= \left[\mathbf{x}^T(t), \ldots, \mathbf{x}^T(t-\tau)\right]^T \\ &= \left[\mathbf{x}^T(t_0), \mathbf{x}^T(t_1), \mathbf{x}^T(t_2), \ldots, \mathbf{x}^T(t_N)\right]^T \\ &= \left[\mathbf{y}_1^T(t), \mathbf{y}_2^T(t), \mathbf{y}_3^T(t), \ldots, \mathbf{y}_m^T(t)\right]^T \end{aligned} \tag{6}$$

where $\mathbf{x}(t_0) = \mathbf{x}(t)$ and $\mathbf{x}(t_N) = \mathbf{x}(t-\tau)$ (see Fig. 1(b)). Equation (2) is now approximated as

$$\dot{\mathbf{y}}(t) = \hat{\mathbf{A}}\mathbf{y}(t) + \hat{\mathbf{f}}(\mathbf{y}(t)) \tag{7}$$

where $\hat{\mathbf{A}}$ is obtained from $\mathbb{D}_\mathbf{N}$ by (1) replacing the first q rows by zeros and (2) replacing the $q \times q$ left upper corner by the $\mathbf{A}_1$ matrix and the $q \times q$ right upper corner by the $\mathbf{A}_2$ matrix. The nonlinear vector $\hat{\mathbf{f}}(\cdot)$ is nonzero only on the first q elements which take the form $\mathbf{f}(\mathbf{y}_1(t), \mathbf{y}_m(t))$. Thus, equation (7) is as follows:

$$\begin{bmatrix} \dot{\mathbf{y}}_1(t) \\ \dot{\mathbf{y}}_2(t) \\ \vdots \\ \dot{\mathbf{y}}_m(t) \end{bmatrix} = \begin{bmatrix} \mathbf{A}_1 \quad \mathbf{0}_q \quad \cdots \quad \mathbf{0}_q \quad \mathbf{A}_2 \\ \frac{2}{\tau}\left[\mathbb{D}_\mathbf{N}\right]^{(q+1,mq)} \end{bmatrix} \begin{bmatrix} \mathbf{y}_1(t) \\ \mathbf{y}_2(t) \\ \vdots \\ \mathbf{y}_m(t) \end{bmatrix} + \begin{bmatrix} \mathbf{f}(\mathbf{y}_1(t), \mathbf{y}_m(t)) \\ 0 \\ \vdots \\ 0 \end{bmatrix} \tag{8}$$

where the superscript $(q+1, mq)$ on $\mathbb{D}_\mathbf{N}$ refers to the fact that only rows of $\mathbb{D}_\mathbf{N}$ between $q+1$ and mq are written into the remaining $Nq \times mq$ part of matrix $\hat{\mathbf{A}}$. Note that the $2/\tau$ factor in front of the portion of $\mathbb{D}_\mathbf{N}$ above accounts for the fact of rescaling the standard collocation expansion interval $[-1, 1]$ to $[0, \tau]$.

In Ref.12 a new extended CSCTA-based technique was introduced for the response and stability analysis of DDEs with the simultaneous effects of the periodic time variation of coefficients and of discontinuous distributed delay. To illustrate the extended formulation, consider a q-dimensional linear system of periodic DDEs with discontinuous distributed delay as

$$\begin{aligned}\dot{\mathbf{x}}(t) &= \mathbf{A}(t)\mathbf{x}(t) + \sum_{i=1}^{k} \int_{-\tau_i}^{-\tau_{i-1}} \boldsymbol{\eta}_i(\theta, t)\mathbf{x}(t+\theta)d\theta, \\ \mathbf{x}(t) &= \boldsymbol{\phi}(t), \quad -\max(\tau_i) \leq t \leq 0\end{aligned} \tag{9}$$

where $\mathbf{A}(t+T) = \mathbf{A}(t)$ is a time-periodic matrix. Each $\boldsymbol{\eta}_i(\theta, t) = \boldsymbol{\eta}_i(\theta, t+T)$ is a matrix of periodic continuous functions valid in some time delay interval $\theta \in [-\tau_i, -\tau_{i-1}]$, $i = 1, 2, \ldots, k$. Equivalently, this sum of integrals can be expressed as $\int_{-\max(\tau_i)}^{0} \boldsymbol{\eta}(\theta, t)\mathbf{x}(t+\theta)d\theta$, where $\boldsymbol{\eta}(\theta, t)$ is periodic in t with period T and discontinuous in θ. Distributed delays in Eq.(9) can become discrete in the case when a $\boldsymbol{\eta}_i(\theta, t)$ matrix includes a Dirac delta function. The functions $\boldsymbol{\eta}_i(\theta, t)$ may also vanish in various intervals if the state derivative does not depend on the states in that interval.

Let the extended state vector $\mathbf{y}(t)$ be introduced as

$$\mathbf{y}(t) = \left[\mathbf{x}^T(t),\ \cdots,\ \mathbf{x}^T(t-\tau_1),\ \cdots,\ \mathbf{x}^T(t-\tau_{k-1}),\ \cdots,\ \mathbf{x}^T(t-\tau_k)\right]^T \tag{10}$$

where k is the number of discontinuous distributed delay terms and k separate Chebyshev grids each with m points are defined in each of the intervals $[t-\tau_i, t-\tau_{i-1}]$ of length h_i, $i = 1, 2, \cdots, k$. Therefore, as shown in Ref.12, equation (9) can be discretized as

$$\dot{\mathbf{y}}(t) = \hat{\mathbf{A}}(t)\mathbf{y}(t), \tag{11}$$

where

$$\hat{\mathbf{A}}(t) = \begin{bmatrix} \mathbf{A}(t)\ \mathbf{0}_2\ \cdots\ \mathbf{0}_2\ \mathbf{B}_{21}(t)\ \cdots\ \mathbf{B}_{2,m_2}(t)\ \cdots\ \mathbf{B}_{k1}(t)\ \cdots\ \mathbf{B}_{k,m_k}(t) \\ \left[\ \frac{2}{h_1}\mathbb{D}_{N_1}^{(q+1,m_1 q)}\ \right] \\ \qquad\qquad \left[\ \frac{2}{h_2}\mathbb{D}_{N_2}^{(q+1,m_2 q)}\ \right] \\ \qquad\qquad\qquad\qquad \ddots \\ \qquad\qquad\qquad\qquad\qquad \left[\ \frac{2}{h_k}\mathbb{D}_{N_k}^{(q+1,m_k q)}\ \right] \end{bmatrix}, \tag{12}$$

$\mathbf{B}_{ij}(t)$ ($i = 1, 2, \ldots, k$, $j = 1, 2, \ldots, m_i$) represent the matrix functions $\boldsymbol{\eta}_i(\theta, t)$ evaluated at the Chebyshev points of each interval multiplied by weighting coefficients based on Clenshaw-Curtis quadrature[61,65] and rescaled by a factor $h_i/2$ which allows a definite integral to be computed as a linear combination of the integrand values at the Chebyshev collocation points within the interval of integration.

The blocks $\frac{2}{h_i}\mathbb{D}_{N_i}^{(q+1,m_i q)}$ ($i = 1, 2, \ldots, k$) in Eq.(12) are differential operators $\mathbb{D}_{N_i} = \mathbf{D}_{N_i} \otimes I_q$ (rescaled to account for intervals $[-\tau_i, -\tau_{i-1}]$) with the first q rows omitted. The number of Chebyshev collocation points for the ith interval is $m_i = N_i + 1$. The overlapping occurrences of the contiguous $\mathbb{D}_{N_i}^{(q+1,m_i)}$ blocks are due to the common nodes of the adjacent collocation intervals. The T-periodic matrix $\hat{\mathbf{A}}(t)$ introduced in Eq.(12) is a square matrix of order $\left(q\sum_{i=1}^{k} m_i\right)$. However, as was noted in Ref.12 this matrix is not an approximation to the infinitesimal generator but is simply the discretized matrix that results from rewriting the DDE as a boundary value problem, a strategy which is also valid for any linear or nonlinear DDE.

2.2. *Examples*

2.2.1. *First Order Scalar Linear DDE*

Consider the following scalar linear DDE with constant coefficients

$$\dot{x}(t) = -ax(t) - bx(t - \tau). \tag{13}$$

Assuming the solution in the form $x(t) = e^{\lambda t}$ yields the characteristic equation

$$\lambda + a + be^{-\lambda\tau} = 0. \tag{14}$$

The stability boundaries for the scalar DDE can be obtained analytically[9] by assuming $\lambda = 0$ as

$$a + b = 0 \tag{15}$$

and also by assuming $\lambda = i\omega$ and separating the real and imaginary parts to obtain the delay dependent boundary

$$\tau\sqrt{b^2 - a^2} = \arccos\left(-\frac{a}{b}\right). \tag{16}$$

The number of first order equations equals one in this case, so matrices $\mathbf{A}_1 = -a$ and $\mathbf{A}_2 = -b$ in Eq.(8) are constant scalar parameters. The matrix $\hat{\mathbf{A}}$ for this example with $m = 5$ collocation points and $\tau = 1$ is given by

$$\hat{\mathbf{A}} = \begin{bmatrix} -a & 0 & 0 & 0 & -b \\ 3.4142 & -1.4142 & -2.8284 & 1.4142 & -0.5858 \\ -1 & 2.8284 & 0 & -2.8284 & 1 \\ 0.5858 & -1.4142 & 2.8284 & 1.4142 & -3.4142 \\ -1 & 2.3431 & -4 & 13.6569 & -11 \end{bmatrix}. \tag{17}$$

Stability of the equilibrium solution of Eq.(13) is determined by the eigenvalues of $\hat{\mathbf{A}}$ via the left half plane stability criterion, i.e., the spectral abscissa $\alpha(\hat{\mathbf{A}})$ must be less than zero for asymptotic stability. Four stability boundaries are shown in Fig.2. For comparison purposes, forward and centered finite difference methods with equally-spaced grid points are also used to generate approximations to the infinitesimal generator. The dots indicate the plot of the curves in Eqs.(15) and (16), the thin solid line represents the stability chart obtained by CSCTA, the thick dashed line is the stability boundary obtained using centered finite differences as described in Ref.57, and the dash-dotted line is used for the forward finite difference boundary. It can be seen that 5 collocation points give sufficient accuracy for both stability boundaries, whereas the matrices resulting from finite differences yield an inaccurate Hopf stability boundary. (Note $\hat{\mathbf{A}}$ is 5×5 in all three cases.)

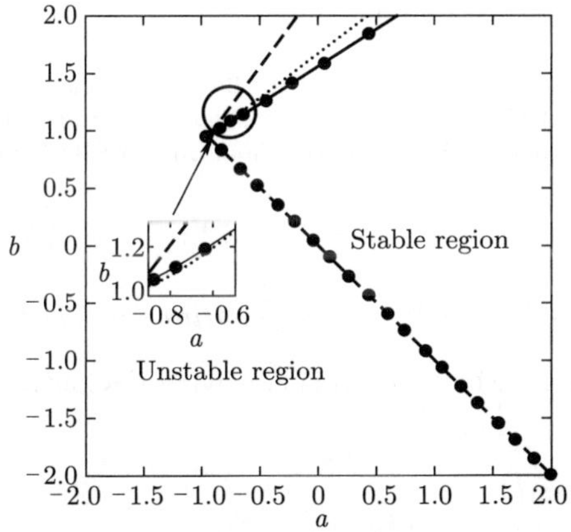

Fig. 2. Diagrams of stability of the scalar DDE (13) for $\tau = 1$ obtained analytically (dotted), by CSCTA (solid), CTA involving centered finite differences from Ref.57 (dashed), and forward finite differences (dash-dotted). Five collocation points and four finite differences were used to keep the same dimension of all approximate $\hat{\mathbf{A}}$ matrices.

Finally, in Fig. 3 the eigenvalues of the 20×20 and 40×40 $\hat{\mathbf{A}}$ matrices from the three approximation methods are compared with the exact roots of the characteristic equation. It is seen that CSCTA more accurately captures the true eigenvalue spectrum compared with the method of finite differences which often produces spurious eigenvalues.

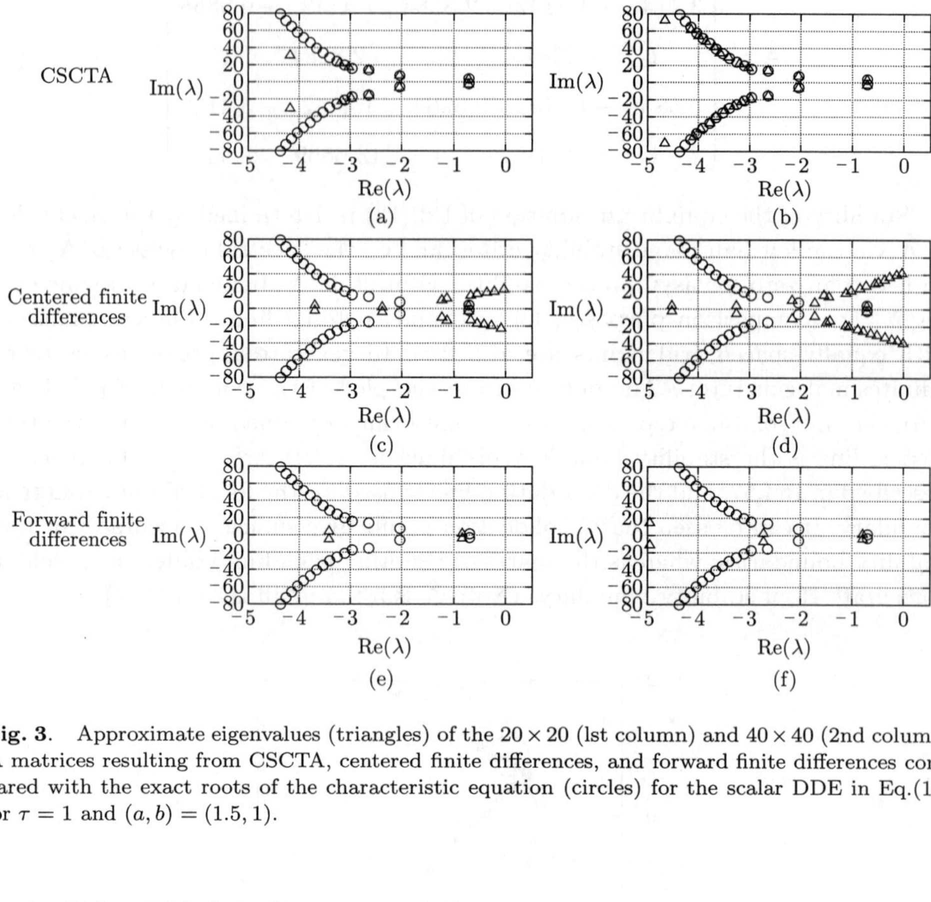

Fig. 3. Approximate eigenvalues (triangles) of the 20×20 (lst column) and 40×40 (2nd column) $\hat{\mathbf{A}}$ matrices resulting from CSCTA, centered finite differences, and forward finite differences compared with the exact roots of the characteristic equation (circles) for the scalar DDE in Eq.(13) for $\tau = 1$ and $(a, b) = (1.5, 1)$.

2.2.2. *Delayed Mathieu Equation with Discontinuous Distributed Delays*

Now consider the delayed Mathieu equation with discontinuous distributed delays given by

$$\ddot{x}(t) + (a + b\cos t)x(t)$$
$$= c\,x(t-2\pi) + \cos 2t\left(c_1\int_{-0.6\pi}^{-0.2\pi} w_1(\theta)x(t+\theta)d\theta + c_2\int_{-2\pi}^{-\pi} w_2(\theta)x(t+\theta)d\theta\right). \tag{18}$$

Note that the parametric period $T = \tau_{\max} = 2\pi$. To obtain the CSCTA form of Eq.(18) the submatrices in Eq. (12) are obtained as

$$
\begin{aligned}
\mathbf{A}(t) &= \begin{bmatrix} 0 & 1 \\ -(a + b\cos t) & 0 \end{bmatrix}, \\
\mathbf{B}_{ji}(t) &= c_k W_{ji} \begin{bmatrix} 0 & 0 \\ w_k(\theta_{ji}) & 0 \end{bmatrix} \cos 2t, \quad k = 1, 2; \quad j = 2, 4, \\
& \qquad i = \begin{cases} 1, 2, \cdots, m, & j = 2, \\ 1, 2, \cdots, m-1, & j = 4, \end{cases} \\
\mathbf{B}_{4,m}(t) &= c_2 W_{4,m} \begin{bmatrix} 0 & 0 \\ w_2(\theta_{4,m}) & 0 \end{bmatrix} \cos 2t + \begin{bmatrix} 0 & 0 \\ c & 0 \end{bmatrix}
\end{aligned}
\tag{19}
$$

where $w_i(\theta_{ji})$ are Clenshaw-Curtis weights for the collocation points θ_{ji}. The monodromy matrices corresponding to Eq.(11) were subsequently obtained for various parameter sets and the stability determined from their eigenvalues (Floquet multipliers) by the unit circle criteria. Stability charts for Eq.(18) are shown in Fig. 4. The values of discontinuous distributed delay coefficients are changed simultaneously, and the stability charts in the (a, c) plane are shown. The first column of plots can be recognized as those of the standard delayed Mathieu equation (for $c_1 = c_2 = 0$)[66], which is turn are similar to both the Hsu-Bhatt-Vyshnegradskii stability chart for the constant second order time-delayed system[67] and the Strutt-Ince stability chart for the standard Mathieu equation[68].

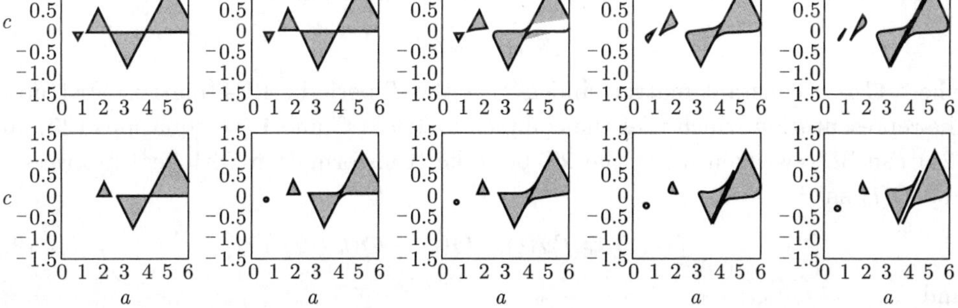

Fig. 4. Stability charts for Eq.(18) in the (a, c) plane for different values of b and $c_1 = c_2$. The parameter values are: b = 0.1, 1, and 2 in the first, second, and third row, respectively, and $c_1 = c_2$ = 0, 0.1, 0.2, 0.3, and 0.4 in the respective columns from 1 to 5. The shaded regions are stable.

3. Reduced Liapunov-Floquet Transformation

3.1. *Formulation*

As was discussed from a theoretical perspective in Ref.20 and from a practical perspective in Ref.69, application of the Liapunov-Floquet Transformation (LFT) enables the analysis of time-periodic ODEs by employing theoretical and computational tools for ODEs with time-invariant linear part. This approach was successfully implemented for various applications in linear and nonlinear time-periodic systems analysis such as order reduction[70,71], chaos and bifurcation control[72,73], construction of time-invariant forms[74], and Hamiltonian perturbations[75], among others.

In Ref.53, LFT was applied for the first time to a time-periodic linear DDE discretized by CSCTA. The proposed combined approach allows for the stability and time response analysis of the periodic DDE by applying a $2T$-periodic LFT to the equivalent large-order periodic ODE system resulting from CSCTA. It was shown, however, that the less accurate Floquet multipliers (which are clustered about the origin in the complex plane) do not allow the computation of the full-size LFT matrix. Therefore, a linear order reduction transformation must be applied which results in a RLFT and a corresponding reduced order constant ODE system which together contain the dominant modes and thus the approximate dynamics of the DDE. In fact, combining CSCTA and RLFT results in a new technique for the analysis of time-periodic nonlinear DDEs which is an alternative to the existing techniques for dimensional reduction such as center manifold analysis for nonlincar constant[76–78] and periodic nonlinear DDEs[79] and the multiple scales method[80] applied to nonlinear constant DDEs.

Consider the application of the standard LFT as[69]

$$\mathbf{y}(t) = \mathbf{L}(t)\mathbf{z}(t), \quad \mathbf{L}(t) = \mathbf{L}(t+T) \tag{20}$$

to Eq.(11) such that it is transformed to a time-invariant system as

$$\dot{\mathbf{z}}(t) = \mathbf{C}\mathbf{z}(t) \tag{21}$$

where $\mathbf{C}$ is a constant matrix, and $\mathbf{L}(t)$ is the T-periodic LFT matrix. Instead of the representation which uses the complex matrices $\mathbf{C}$ and $\mathbf{L}(t)$, equations (20) and (21) can be rewritten using the $2T$-periodic transformation with real matrices $\mathbf{R}$ and $\mathbf{Q}(t)$ as

$$\mathbf{y}(t) = \mathbf{Q}(t)\mathbf{z}(t), \quad \mathbf{Q}(t) = \mathbf{Q}(t+2T) \tag{22}$$

and

$$\dot{\mathbf{z}}(t) = \mathbf{R}\mathbf{z}(t). \tag{23}$$

It has been shown in Ref.69 that the LFT matrix $\mathbf{L}(t)$ can be computed in closed form for a commutative system for which the state transition matrix (STM) is known. If the system is not commutative, then the STM must be approximated, for example by Chebyshev polynomial expansion[81]. In either case, the STM can be factored as

$$\mathbf{\Phi}(t) = \mathbf{L}(t)e^{\mathbf{C}t} = \mathbf{Q}(t)e^{\mathbf{R}t} \tag{24}$$

where

$$\mathbf{C} = \frac{1}{T}\ln\mathbf{\Phi}(T); \quad \mathbf{R} = \frac{1}{2T}\ln\mathbf{\Phi}(2T) \tag{25}$$

are computed via an eigenanalysis of the monodromy matrix $\mathbf{\Phi}(T)$ or its square $\mathbf{\Phi}(2T) = \mathbf{\Phi}^2(T)$. Therefore, the T- and $2T$-periodic matrices $\mathbf{L}(t)$ and $\mathbf{Q}(t)$ are solved as

$$\mathbf{L}(t) = \mathbf{\Phi}(t)e^{-\mathbf{C}t}, \quad \mathbf{Q}(t) = \mathbf{\Phi}(t)e^{-\mathbf{R}t} \tag{26}$$

and can be expressed as a real Fourier series up to some desired accuracy. Since we are interested in the real $2T$-periodic transformation, the future discussion will be around $\mathbf{Q}(t)$ rather than $\mathbf{L}(t)$.

As previously discussed, a RLFT is required in which only the r modes corresponding to r dominant Floquet multipliers are retained, since the inaccurate multipliers clustered around the origin result in a set of inaccurate eigenvalues of $\mathbf{R}$ (characteristic exponents) with highly negative real parts which in turn makes Eq.(26) impractical to compute the full LFT matrix. Often we choose $r = q$, that is, we retain the same number of modes as the size of the DDE. Therefore, consider an order reduction transformation $\mathbf{z} = \hat{\mathbf{T}}\mathbf{w}$ where

$$\hat{\mathbf{T}} = \mathbf{M}\mathbf{T}\mathbf{M}_{11}^{-1} = \begin{bmatrix} \mathbf{M}_{11} & \mathbf{M}_{12} \\ \mathbf{M}_{21} & \mathbf{M}_{22} \end{bmatrix} \begin{bmatrix} \mathbf{I}_q \\ \mathbf{0}_{(m-1)q\times q} \end{bmatrix} \mathbf{M}_{11}^{-1} = \begin{bmatrix} \mathbf{I}_q \\ \mathbf{M}_{21}\mathbf{M}_{11}^{-1} \end{bmatrix} \tag{27}$$

is a $mq \times q$ rectangular matrix. $\mathbf{M}$ is the modal matrix of $\mathbf{R}$ such that $\mathbf{R} = \mathbf{M}\mathbf{\Lambda}\mathbf{M}^{-1}$ and the matrix

$$\mathbf{\Lambda} - \begin{bmatrix} \mathbf{\Lambda}_1 & \mathbf{0} \\ \mathbf{0} & \mathbf{\Lambda}_2 \end{bmatrix} \tag{28}$$

is diagonal. Only the modal coordinates corresponding to $\mathbf{\Lambda}_1$ are retained. Also, $\mathbf{M}_{11}$ is the top left $q \times q$ partition of $\mathbf{M}$. Application to Eq.(23) yields

$$\dot{\mathbf{w}} = \bar{\mathbf{R}}\mathbf{w} \tag{29}$$

where $\bar{\mathbf{R}} = (\hat{\mathbf{T}}^T\hat{\mathbf{T}})^{-1}\hat{\mathbf{T}}^T\mathbf{R}\hat{\mathbf{T}}$ is a projection of matrix $\mathbf{R}$ onto $\mathbf{w}$ such that equation (29) exactly preserves the eigenstructure of Eq.(23). Note that in Eq.(27) $\mathbf{z}_1 = \mathbf{w}$ (master coordinates) and $\mathbf{z}_2 = \mathbf{M}_{21}\mathbf{M}_{11}^{-1}\mathbf{w}$ is the relation between master and slave coordinates where $\mathbf{z}^T = \left[\mathbf{z}_1^T, \ \mathbf{z}_2^T\right]$ is partitioned into vectors of length q and $(m-1)q$, respectively.

Because the state vector $\mathbf{x}(t)$ is actually a reduced subset of the first q elements of vector $\mathbf{y}(t)$, we define a RLFT matrix $\bar{\mathbf{Q}}(t)$ that transforms the original DDE in Eq.(11) into Eq.(29) and that satisfies

$$\mathbf{x}(t) = \mathbf{T}^T\mathbf{y}(t) = \mathbf{T}^T\mathbf{Q}(t)\mathbf{z}(t) = \bar{\mathbf{Q}}(t)\mathbf{w}(t) = \bar{\mathbf{Q}}(t)\mathbf{T}^T\mathbf{z}(t) \tag{30}$$

where $\mathbf{T}^T = [\mathbf{I}_q, \ \mathbf{0}_{q\times(m-1)q}]$. Hence,

$$\mathbf{T}^T\mathbf{Q}(t) = \bar{\mathbf{Q}}(t)\mathbf{T}^T \tag{31}$$

and the RLFT matrix may be obtained as

$$\bar{\mathbf{Q}}(t) = \mathbf{T}^T\mathbf{Q}(t)\mathbf{T} = \mathbf{Q}_{11}(t) \tag{32}$$

where $\mathbf{Q}_{11}(t)$ is the top left $q \times q$ partition of $\mathbf{Q}(t)$. Equation (32) represents a projection of the full LFT matrix. However, recall that it is impractical to compute $\mathbf{Q}(t)$ in its entirety to begin with. Therefore, we seek a simple way to obtain $\bar{\mathbf{Q}}(t)$ without computing $\mathbf{Q}(t)$. Recalling the form for $\mathbf{Q}(t)$ in Eq.(26), an analogous formula for an approximate $\bar{\mathbf{Q}}(t)$ that was used successfully in Ref.53 is

$$\bar{\mathbf{Q}}(t) = \mathbf{\Phi}_{11}(t)e^{-\mathbf{R}_{11}t} \tag{33}$$

where $\mathbf{\Phi}_{11}$ and $\mathbf{R}_{11}$ are the upper $q\times q$ partitions of $\mathbf{\Phi}(t)$ and $\mathbf{R}$, respectively. Note that the necessary initial condition at $t = 0$ for Eq.(32) is satisfied.

The transformations in Eqs.(20), (22), and (30) convert Eq.(11) into a time-invariant system while preserving the stability properties. The unit-circle stability criteria is applied for Floquet multipliers (eigenvalues of $\mathbf{\Phi}(T)$) and their squares (eigenvalues of $\mathbf{\Phi}(2T)$), while for the corresponding eigenvalues of $\mathbf{C}$, $\mathbf{R}$, and $\bar{\mathbf{R}}$ (characteristic exponents) the left half-plane stability criteria is used.

3.2. *Example*: *Delayed Mathieu Equation*

As an illustrative example of the use of LFT, consider the delayed Mathieu's equation[66]

$$\ddot{x}(t) + (a + b\cos t)x(t) = cx(t - 2\pi). \tag{34}$$

The CSCTA was applied to Eq.(34) in Ref.52 to transform the DDE into an approximate ODE representation of the system given by

$$\dot{\mathbf{y}}(t) = \begin{bmatrix} \begin{bmatrix} 0 & 1 \\ -a - b\cos t & 0 \end{bmatrix} & \mathbf{0}_q & \cdots & \mathbf{0}_q & \begin{bmatrix} 0 & 0 \\ c & 0 \end{bmatrix} \\ & \dfrac{1}{\pi}\left[\mathbb{D}^{(3,2m)}\right] & & & \end{bmatrix} \mathbf{y}(t). \tag{35}$$

Consider a specific set of parameters $(a, b, c) = (1.5, 0.1, 0.1)$ that results in a stable location shown on the stability chart in Fig.5, and assume $m = 20$ collocation points are used. CSCTA results in a pair of complex conjugate squares of Floquet multipliers $(-0.4296\pm0.2250i)$ and a number of inaccurate multipliers of the matrix $\mathbf{\Phi}(2T)$ clustered around zero within the unit circle shown in Fig.6(a). This set of inaccurate multipliers are mapped into a set of inaccurate eigenvalues of $\mathbf{R}$ (characteristic exponents) as shown in Fig.6(b). Hence, a reduced order $2T$-periodic LFT matrix is computed from Eq.(33).

We can get the expressions for the elements of the RLFT matrix as a Fourier series with 5 harmonics by evaluating the product of the reduced STM and the

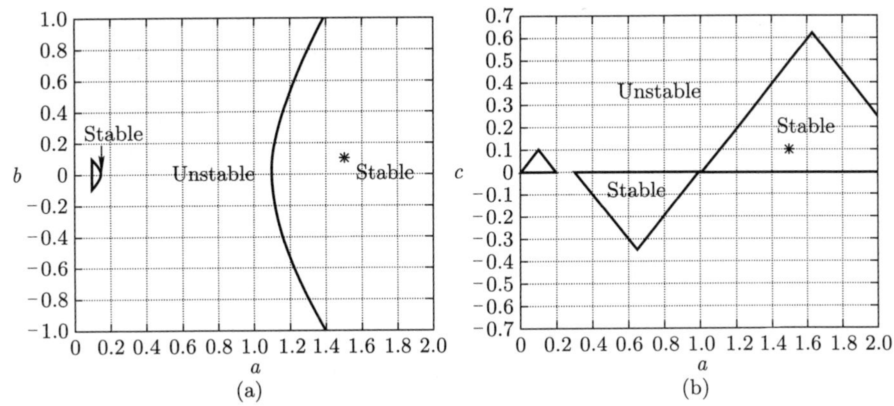

Fig. 5. The stability charts for the delayed Mathieu's equation for (a) $c = 0.1$ in the (a, b) plane also obtained in Ref.52 and (b) $b = 0.1$ in the (a, c) plane. The asterisk shows the stable point $(a, b) = (a, c) = (1.5, 0.1)$.

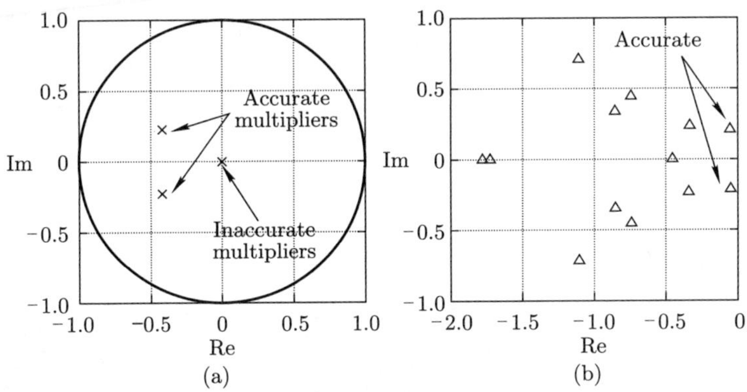

Fig. 6. In the first row, (a) the squares of Floquet multipliers and (b) the characteristic exponents of the LF-transformed system are given. Note the sets of accurate and inaccurate multipliers and exponents.

matrix exponential at the intermediate points of the interval $[0, 2T]$. These entries are also plotted in Fig.7. The constant reduced system in terms of 2-dimensional vector $\mathbf{w}(t)$ is given by

$$\dot{\mathbf{w}}(t) = \begin{bmatrix} -0.04799 & 0.1681 \\ -0.2665 & -0.06717 \end{bmatrix} \mathbf{w}(t) \tag{36}$$

whose eigenvalues are $-0.05758 \pm 0.2115i$.

In Fig.8, the responses of the dominant modes of the reduced order system transformed by the RLFT matrix $\bar{\mathbf{Q}}(t)$ (i.e., the $\mathbf{w}(t)$ vector back-transformed to $\mathbf{y}$ coordinates) are compared to the corresponding numerically obtained dominant modes of the full time-periodic ODE system defined in Eq.(35) (i.e., the first two components of the $\mathbf{y}(t)$ vector) and the two states of the DDE in Eq.(34) (i.e., $x(t)$

and $\dot{x}(t)$). To compare the three responses, it is essential to choose equivalent initial conditions as explained in Ref.53. For the time series shown in Fig.8, the initial conditions are

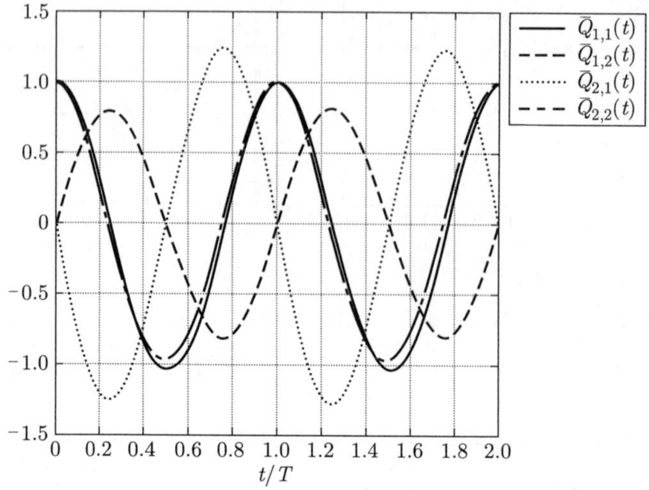

Fig. 7. Elements of the 2×2 reduced order LFT matrix for the stable point $(a, b) = (1.5, 0.1)$ on the stability chart in Fig.5.

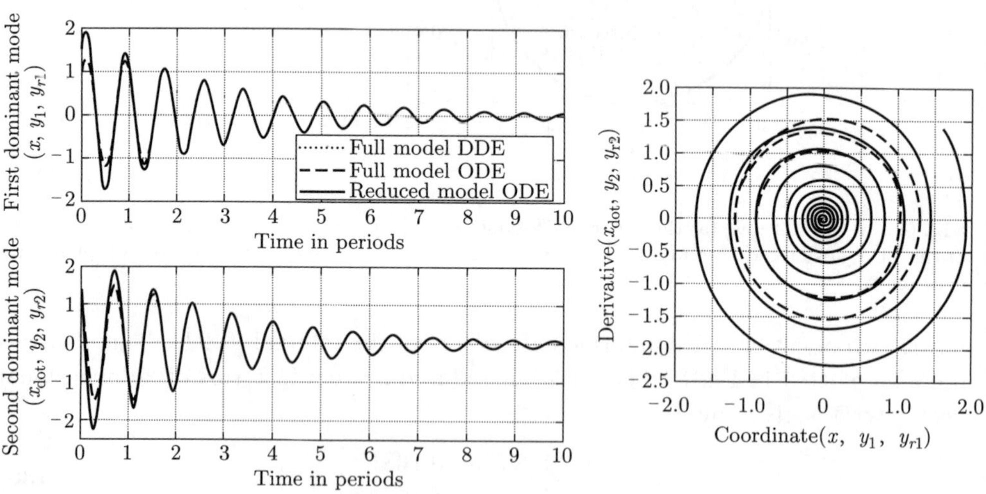

Fig. 8. The time series and the phase plane of the two dominant mode responses obtained by integrating the time-periodic DDE 34 (dotted, matches the dashed line), the full size time-periodic ODE 35 (dashed), and the reduced constant system 36 back-transformed by the reduced LFT (solid) are shown.

$$\mathbf{y}(0) = \underbrace{[1,\ 1, \cdots, 1]^T}_{2m \text{ times}},$$

$$\mathbf{w}(0) = [1.5932,\ 1.4091]^T, \tag{37}$$

$$\phi(t) = [1,\ 1]^T \text{ for } t \in [-\tau, 0].$$

It can be seen that the responses of the reduced constant system with the RLFT applied do not match those of the time-periodic DDE and the time-periodic ODE system for a short period of time in the beginning but become almost indistinguishable later in time. This short interval is the time it takes the exponential solutions corresponding to the neglected characteristic exponents to completely die out.

4. Feedback Control of Periodic Delayed Systems

Many strategies have been utilized to control time-delayed systems or to design delayed feedback control laws for non-delayed systems in the case that the current states are unavailable for feedback. For linear time-invariant control systems with delay, the stabilization problem is to design a control law such that all of the closed-loop poles (eigenvalues) lie in the left-half plane. If pole placement is desired to ensure a specified response, however, one is faced with the problem that infinitely many poles should be specified, while the number of control parameters (gains) is finite. One strategy is to assign the place of the dominant poles only or to specify only the spectral abscissa[9,38]. An alternative strategy is to use an infinite (or large) number of control gains to control all (or many) of the poles[7]. Two techniques to implement this strategy include the use of distributed delays in the controller (in which the kernel function of the distributed delay serves as an infinite-dimensional vector of control gains) or by allowing the control gains to be time-periodic in which their Fourier coefficients form a set of infinitely many control parameters. While the stabilization of non-delayed systems by means of periodic control gains has been the focus of recent papers (e.g., Refs.82, 83), the combined effect of feedback delay and time-periodic control gains results in time-periodic DDEs, for which the stability analysis requires the use of the infinite-dimensional Floquet theory. This technique was utilized in Ref.47, for instance, in which two Fourier coefficients for the periodic control gains were utilized.

Yet a third strategy for delayed feedback control is that of finite spectrum assignment, in which the resulting closed-loop system has only a finite number of poles that can be assigned arbitrarily and thus requires only a finite number of control gains to be specified. Two techniques to implement this strategy include the use of a feedback based on a prediction of the state with distributed delay, e.g., Refs.14, 32, 34, 35, and also utilizing a special case of periodic controllers call "act-and-wait control" where the feedback term is switched on and off periodically in time[42–44].

The advantage in this control strategy is that if the switch-off (waiting) period is longer than the feedback delay, then the system can be transformed to a discrete map of finite dimension thus presenting a finite spectrum assignment problem. In Ref.7, several examples showed that an unstable process can be stabilized by the application of the act-and-wait concept where the traditional constant gain controllers cannot stabilize the system.

The remainder of this section will outline several strategies for the use of the CSCTA and RLFT methods can help to reduce a (possibly time-periodic) nonlinear delayed system with linear and/or nonlinear feedback control into a form more conducive for the application of many of the techniques for control of delayed systems discussed above and allow the application of standard tools for control design of LTI systems. While control design based on finite difference discretization of the infinitesimal generator has been demonstrated in Ref.37 and the LFT has been utilized for the control of time-periodic non-delayed systems[72,84–91], the use of CSCTA and RLFT have thus far not been investigated for their application in control design for delayed periodic systems or for delayed feedback control using periodic gains of non-delayed time-invariant systems. Therefore, the remainder of this section offers a few ideas for implementation of these strategies.

4.1. *Formulation*

Consider the q-dimensional nonlinear (quasi linear) system with a single point delay given by

$$\dot{\mathbf{x}} = \mathbf{A}_1(t)\mathbf{x}(t) + \mathbf{A}_2(t)\mathbf{x}(t-\tau) + \mathbf{f}(\mathbf{x}(t), \mathbf{x}(t-\tau), t) + \mathbf{B}(t)\mathbf{u}(t) \tag{38}$$

in which $\mathbf{A}_1(t), \mathbf{A}_2(t), \mathbf{B}(t)$, and the purely nonlinear vector function $\mathbf{f}(\cdot)$ are periodic in t with period T. Application of CSCTA using m Chebyshev collocation points transforms Eq.(38) into the periodic nonlinear ODE system

$$\dot{\mathbf{y}} = \hat{\mathbf{A}}(t)\mathbf{y} + \hat{\mathbf{f}}(\mathbf{y}, t) + \hat{\mathbf{B}}(t)\mathbf{u}(t), \quad \hat{\mathbf{B}}(t) = \begin{bmatrix} \mathbf{B}(t) \\ 0 \\ \vdots \\ 0 \end{bmatrix} \tag{39}$$

where $\hat{\mathbf{A}}(t)$ and $\hat{\mathbf{f}}(\cdot)$ are time-periodic analogues of those in Eq.(7), and $\mathbf{y}$ is an mq-dimensional vector. At this point a full-state feedback control law can be designed using a linear, nonlinear, or a combined linear/nonlinear controller with periodic control gains as

$$\mathbf{u}(t) = -\mathbf{K}(t)\mathbf{y} + \mathbf{k}(\mathbf{y}, t) \tag{40}$$

where $\mathbf{K}(t)$ and the purely nonlinear function $\mathbf{k}(\cdot)$ are periodic in t with period T and can thus be expressed using finite Fourier series with coefficients to be specified.

Note that the form of the feedback control law in $\mathbf{y}(t)$ translates into a distributed delay control law in $\mathbf{x}(t)$, i.e.,

$$\mathbf{u}(t) = -\int_{-\tau}^{0} \bar{\mathbf{K}}(t,\theta)\mathbf{x}(t+\theta)d\theta + \int_{-\tau}^{0} \bar{\mathbf{k}}(\mathbf{x}(t+\theta),t,\theta)d\theta \tag{41}$$

where Clenshaw-Curtis quadrature is used to obtain the gain matrix $\mathbf{K}(t)$ and the nonlinear control term $\mathbf{k}(\cdot)$ from $\bar{\mathbf{K}}(t)$ and $\bar{\mathbf{k}}(\cdot)$. Note that the control law in Eqs.(40) and (41) includes feedback of the current state $\mathbf{x}(t)$. If sensor and/or actuator delays make the current and delayed states $\mathbf{x}(t+\theta)$, $-\tau_c < \theta \leq 0$ unavailable for feedback, however, then the appropriate weights in Eq.(41) must be set to zero. Alternatively, the control input can be changed to $\mathbf{u}(t-\tau_c)$ or output feedback can be employed.

Consider the linear versions of Eqs.(38)–(41) in which we have to design the matrix $\mathbf{K}(t)$ such that the closed-loop system is asymptotically stable. We do not require both the uncontrolled system and the control law to have delay and periodic coefficients, but we assume that each property shows up in at least one place. Note that a large number of control gains are available, both from the discretization of the distributed delay feedback control and from the Fourier series expansion (up to any desired order) of the periodic coefficients of $\mathbf{K}(t)$, to control a large number of the closed-loop poles. Our assumption of controllability applies only to $(\mathbf{A}_1(t), \mathbf{B}(t))$ in Eq.(38), because in fact the pair $(\hat{\mathbf{A}}(t), \hat{\mathbf{B}}(t))$ may not be controllable even if the pair $(\mathbf{A}_1(t), \mathbf{B}(t))$ is originally controllable as shown in Ref.37. Practically, this is not a problem since all we require is to control the current state $\mathbf{y}_1(t) = \mathbf{x}(t)$ and not $\mathbf{y}_i(t)$, $i = 2, \ldots, m$ as these are simply delayed states. Hence, for pole placement we typically cannot place all mq poles even if the matrices $\hat{\mathbf{A}}, \hat{\mathbf{B}}, \mathbf{K}$ are time-invariant. Again, this is not a concern as long as the poles being placed include the dominant ones.

We now outline three possible strategies for determining the gain matrix $\mathbf{K}(t)$ in the linear problem. First, the stability of the closed-loop response may be investigated in the parameter space of available control gains using Floquet theory, including the possibility of optimization of the control gains by minimizing the spectral radius which must be less than unity for asymptotic stability[37]. This strategy does not use either CSCTA or RLFT. We note that rather than obtaining the monodromy matrix associated with the closed-loop matrix $(\hat{\mathbf{A}}(t) - \hat{\mathbf{B}}(t)\mathbf{K}(t))$, other numerical techniques such as Chebyshev collocation[18] and semidiscretization[7] may be directly applied to the closed-loop DDE to more efficiently generate an approximation to the infinite-dimensional monodromy operator. This strategy was employed in Refs.49–51 as well as in Ref.47 in which the monodromy operator was approximated symbolically via Chebyshev polynomial expansion[17,92] in terms of constant and periodic (with unknown Fourier coefficients) control gains.

The second strategy utilizes CSCTA but not RLFT. It uses optimal (time-varying LQR) control to minimize the cost function

$$J = \frac{1}{2}\mathbf{y}^T(t_f)\mathbf{S}\mathbf{y}(t_f) + \frac{1}{2}\int_0^{t_f} \left(\mathbf{y}^T(t)\mathbf{Q}(t)\mathbf{y}(t) + \mathbf{u}^T(t)\mathbf{R}(t)\mathbf{u}(t)\right) dt, \qquad (42)$$

where $\mathbf{Q}(t)$ and $\mathbf{R}(t)$ are symmetric positive definite T-periodic matrices, by directly solving the time-periodic Riccati equation

$$\dot{\mathbf{P}}(t) = -\left(\hat{\mathbf{A}}^T(t)\mathbf{P}(t) + \mathbf{P}(t)\hat{\mathbf{A}}(t) + \mathbf{Q}(t)\right) + \mathbf{P}(t)\hat{\mathbf{B}}(t)\mathbf{R}^{-1}(t)\hat{\mathbf{B}}^T(t)\mathbf{P}(t) \qquad (43)$$

backwards in time from the final value $\mathbf{P}(t_f) = \mathbf{S}$. The control gain is then obtained as

$$\mathbf{K}(t) = \mathbf{R}^{-1}(t)\hat{\mathbf{B}}^T(t)\mathbf{P}(t). \qquad (44)$$

Note, however, that solving the large-dimensional Riccati equation is computationally intensive and may not be practical in applications. If $\hat{\mathbf{A}}$, $\hat{\mathbf{B}}$, $\mathbf{Q}$, and $\mathbf{R}$ are constant and $\mathbf{S} = \mathbf{0}$, however, then the much simpler algebraic LQR problem can be solved for the case where $t_f \to \infty$.

The third strategy for obtaining a non-delayed control law is adopted from Refs.87, 90 and involves use of both CSCTA and RLFT as well as pole-placement for LTI systems. Again, consider the control problem associated with the linear parts of Eqs.(38)–(41) where the output matrix $\mathbf{C}(t) = \mathbf{I}_{mq}$. Application of RLFT with q dominant modes to $\dot{\mathbf{y}} = \hat{\mathbf{A}}(t)\mathbf{y} + \hat{\mathbf{B}}(t)\mathbf{u}(t)$ yields

$$\dot{\mathbf{w}}(t) = \bar{\mathbf{R}}\mathbf{w}(t) + (\hat{\mathbf{T}}^T\hat{\mathbf{T}})^{-1}\hat{\mathbf{T}}^T\mathbf{Q}^{-1}(t)\hat{\mathbf{B}}(t)\mathbf{u}(t). \qquad (45)$$

Recall from Section 3 that $\mathbf{T}^T\mathbf{Q}(t) = \bar{\mathbf{Q}}(t)\mathbf{T}^T$. It can be shown by solving for $\mathbf{Q}^T(t)\mathbf{T}{=}\hat{\mathbf{T}}(\hat{\mathbf{T}}^T\hat{\mathbf{T}})^{-1}\bar{\mathbf{Q}}^T(t)$, taking the transpose, and pre- and post-multiplying the result by $\bar{\mathbf{Q}}^{-1}(t)$ and $\mathbf{Q}^{-1}(t)$, respectively, that the projection $(\hat{\mathbf{T}}^T\hat{\mathbf{T}})^{-1}\hat{\mathbf{T}}^T\mathbf{Q}^{-1}(t)$ reduces to $\bar{\mathbf{Q}}^{-1}(t)\mathbf{T}^T$. Also recalling the form of $\hat{\mathbf{B}}(t)$ in Eq.(39), equation (45) is expressed in terms of the RLFT matrix $\bar{\mathbf{Q}}(t)$ as

$$\dot{\mathbf{w}}(t) = \bar{\mathbf{R}}\mathbf{w}(t) + \bar{\mathbf{Q}}^{-1}(t)\mathbf{B}(t)\mathbf{u}(t). \qquad (46)$$

Now consider an auxiliary system of the form

$$\dot{\bar{\mathbf{w}}}(t) = \bar{\mathbf{R}}\bar{\mathbf{w}}(t) + \bar{\mathbf{B}}\mathbf{v}(t) \qquad (47)$$

where $\bar{\mathbf{B}}$ is an arbitrary constant matrix with the same order and rank as $\bar{\mathbf{Q}}^{-1}(t)\mathbf{B}(t)$ such that $(\bar{\mathbf{R}}, \bar{\mathbf{B}})$ is a controllable pair. The state feedback control law $\mathbf{v}(t) = -\bar{\mathbf{K}}\bar{\mathbf{w}}(t)$ can be designed by applying pole placement or optimal control theory such that equation (47) is asymptotically stable. Defining $\mathbf{e}(t) = \mathbf{w}(t) - \bar{\mathbf{w}}(t)$, the error dynamics from Eqs.(46) and (47) are given as

$$\dot{\mathbf{e}}(t) = (\bar{\mathbf{R}} - \bar{\mathbf{B}}\bar{\mathbf{K}})\mathbf{e}(t) + \bar{\mathbf{Q}}^{-1}(t)\mathbf{B}(t)\mathbf{u}(t) - \bar{\mathbf{B}}\mathbf{v}(t). \qquad (48)$$

Assuming $\mathbf{B}(t)$ is a rectangular matrix, the control law for $\mathbf{u}(t)$ is found by using least squares to minimize the norm of the time-varying term $\bar{\mathbf{Q}}^{-1}(t)\mathbf{B}(t)\mathbf{u}(t)-\bar{\mathbf{B}}\mathbf{v}(t)$ along with the above control law for $\mathbf{v}(t)$ as

$$\mathbf{u}(t) = -\mathbf{B}^{\dagger}(t)\bar{\mathbf{Q}}(t)\bar{\mathbf{B}}\bar{\mathbf{K}}\mathbf{w}(t) \tag{49}$$

where $\mathbf{B}^{\dagger}(t)$ is the least squares inverse of $\mathbf{B}(t)$. Substituting $\mathbf{w}(t) = \bar{\mathbf{Q}}^{-1}(t)\mathbf{x}(t)$ into Eq.(49) provides the non-delayed feedback control law $\mathbf{u}(t) = -\mathbf{K}(t)\mathbf{x}(t)$ where

$$\mathbf{K}(t) = \mathbf{B}^{\dagger}(t)\bar{\mathbf{Q}}(t)\bar{\mathbf{B}}\bar{\mathbf{K}}\bar{\mathbf{Q}}^{-1}(t). \tag{50}$$

As is discussed in Ref.88, due to use of the least squares inverse asymptotic stability of the closed-loop dynamics is not guaranteed and one must search for the optimal $\bar{\mathbf{B}}$ matrix that provides the best response. Alternatively, use of a backstepping strategy as detailed in Ref.88 does guarantee asymptotic stability for underactuated systems. Also, as is explained in Refs.72, 84, 88–90 the strategy for computing $\bar{\mathbf{Q}}^{-1}(t)$ involves finding the STM $\mathbf{\Psi}(t)$ associated with the adjoint system

$$\dot{\mathbf{s}}(t) = -\hat{\mathbf{A}}^{T}(t)\mathbf{s}(t) \tag{51}$$

and using the relationships $\mathbf{\Phi}^{-1}(t) = \mathbf{\Psi}^{T}(t)$ and

$$\mathbf{Q}^{-1}(t) = \left[\mathbf{\Phi}e^{-\mathbf{R}t}\right]^{-1} = e^{\mathbf{R}t}\mathbf{\Phi}^{-1}(t) = e^{\mathbf{R}t}\mathbf{\Psi}^{T}(t). \tag{52}$$

Hence $\bar{\mathbf{Q}}^{-1}(t)$ is a projection of $\mathbf{Q}^{-1}(t)$ that satisfies $\bar{\mathbf{Q}}^{-1}(t)\mathbf{T}^{T} = \mathbf{T}^{T}\mathbf{Q}^{-1}(t)$, and hence $\bar{\mathbf{Q}}^{-1}(t) = \mathbf{T}^{T}\mathbf{Q}^{-1}(t)\mathbf{T} = (\mathbf{Q}^{-1}(t))_{11}$. Again, however, it is impractical to compute the full $\mathbf{Q}^{-1}(t)$ in the first place. A formula for $\bar{\mathbf{Q}}^{-1}(t)$ analogous to that for $\bar{\mathbf{Q}}(t)$ in Section 3 is

$$\bar{\mathbf{Q}}^{-1}(t) = e^{\mathbf{R}_{11}t}\mathbf{\Psi}_{11}^{T}(t). \tag{53}$$

Additional strategies for control of time-periodic systems are discussed in Refs.87, 90 which could be used for control of linear periodic DDEs after employing CSCTA and RLFT include output feedback and an observer-based controller. Possible strategies for designing the full nonlinear control law in Eq.(40) also involve use of CSCTA and RLFT applied to Eq.(39) which yields (assuming $T = \tau$)

$$\begin{aligned}\dot{\mathbf{w}}(t) &= \bar{\mathbf{R}}\mathbf{w}(t) + \bar{\mathbf{Q}}^{-1}(t)\mathbf{T}^{T}\left[\hat{\mathbf{f}}(\mathbf{y}(t),t) + \hat{\mathbf{B}}(t)\mathbf{u}(t)\right] \\ &= \bar{\mathbf{R}}\mathbf{w}(t) + \bar{\mathbf{Q}}^{-1}(t)\left[\mathbf{f}\left(\bar{\mathbf{Q}}(t)\mathbf{w}(t), \bar{\mathbf{Q}}(t)\mathbf{w}(t-\tau), t\right) + \mathbf{B}(t)\mathbf{u}(t)\right].\end{aligned} \tag{54}$$

Hence, it is seen that the combined use of CSCTA and RLFT allows for the linear part of the uncontrolled periodic DDE to be expressed in an equivalent time-invariant and non-delayed form which preserves the dominant characteristic exponents, while the nonlinear part of the transformed system retains both time delay and periodic coefficients. Hence, various reduced order controllers[93], chaos controllers[73,85,86], or bifurcation controllers[72] for nonlinear periodic systems could now be applied after modification to account for possible time delays in the nonlinear part by designing both the linear and nonlinear controls in Eq.(40). We also point out that if the nonlinear DDE is uncontrolled with critical Floquet multipliers/exponents, then the strategy for invariant-manifold based order reduction used in Refs.70, 71 becomes an alternative for center manifold reduction for nonlinear constant or periodic DDEs[76–80].

4.2. *Delayed State Feedback Control of the Delayed Mathieu Equation*

Consider the controlled delayed Mathieu equation

$$\begin{gathered}\ddot{x} + (0.2 + 0.1\cos 2\pi t)x(t) = 0.5x(t-1) + u(t),\\ (x, \dot{x}) = (1, 0),\ -1 \le t \le 0.\end{gathered} \tag{55}$$

Following the first strategy described in Section 4.1, the monodromy operator of the uncontrolled system is approximated and is found to have a spectral radius larger than unity such that the uncontrolled system is unstable. In order to obtain the asymptotic stability of the controlled system, we design a delayed state feedback controller with a periodic gain matrix in which each gain component is expressed in a 3-term Fourier series with unknown coefficients to be determined as

$$\begin{aligned}u(t) = &-(k_{11} + k_{12}\cos 2\pi t + k_{13}\sin 2\pi t)x(t-1)\\ &-(k_{21} + k_{22}\sin 2\pi t + k_{23}\cos 2\pi t)\dot{x}(t-1).\end{aligned} \tag{56}$$

Note that the period and delay are the same as in Eq.(55), although this is not necessary. Any relationship between the delays in the plant and control law are acceptable when using the multi-interval version of CSCTA in Eq.(12), while the periods of the coefficients should be rationally related in order for Floquet theory to be applied using the fundamental period of the closed-loop system.

First, we set $k_{12} = k_{13} = k_{22} = k_{23} = 0$ and find the stable region in the k_{11}-k_{21} parameter space. For this purpose, the monodromy operator $\mathbf{U}$ is approximated using the Chebyshev collocation method[18]. The set of the points in the k_{11}-k_{21} parameter space for which the spectral radius of $\mathbf{U}$ lies on the circumference of the unit circle forms the stability boundary. Starting from a unity spectral radius and plotting contours for gradually decreasing spectral radii, the (k_{11}, k_{21}) location corresponding to the minimum spectral radius can be obtained. The optimum set of the first pair of the control gains is $(k_{11}, k_{21}) = (0.40, 0.35)$ which corresponds to the spectral radius $\rho = 0.54577$. Next, the optimal values obtained for k_{11} and k_{21} are held constant with $k_{13} = k_{23} = 0$ while the stable region in the k_{12}-k_{22} parameter space is found as well as the values of k_{12} and k_{22} that correspond to the minimum spectral radius for the new controlled system. The optimum set of the second pair of the control gains, i.e., $(k_{12}, k_{22}) = (-0.33, -0.06)$, corresponds to the spectral radius $\rho = 0.52154$. Finally, the optimum values obtained for k_{11}, k_{21}, k_{12}, and k_{22} are held constant while the stability region in the k_{13}-k_{23} parameter space is found as well as their optimum values. The minimum spectral radius is $\rho_{\text{opt}} = 0.51977$ which corresponds to $(k_{13}, k_{23}) = (0.03, -0.06)$. Altogether, the optimal set of the control gains is $(k_{11}, k_{21}, k_{12}, k_{22}, k_{13}, k_{23}) = (0.40, 0.35, -0.33, -0.06, 0.03, -0.06)$. These results are obtained using 150 Chebyshev collocation points and a 500×500 meshgrid for each parameter space. The optimum control gains are shown with • in the stability diagrams shown in Fig.9. Figure 10 shows the uncontrolled response of the system in Eq.(55), the controlled response using the optimal constant control law, and that using the optimal full 6-term periodic control law, along with the corresponding control efforts. The simulated results indicate that when the optimal values of the control gains are used, the closed-loop response has less settling time due to a greater effective damping, and that k_{11} and k_{21} have a more significant role in the convergence speed.

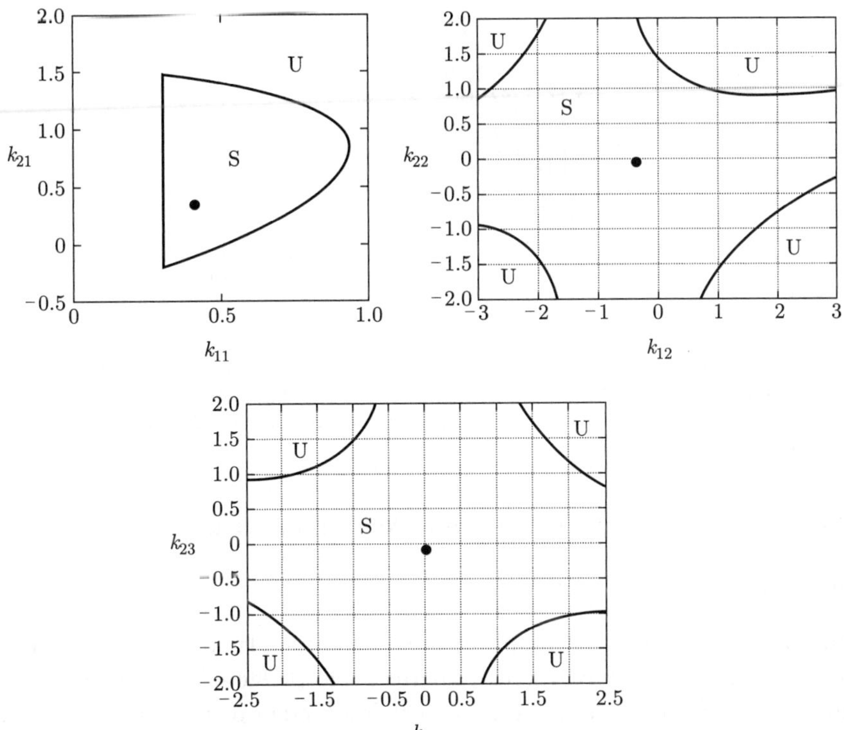

Fig. 9. The stable (labeled with S) and unstable (labeled with U) regions in different parameter spaces. The optimal control pairs are shown with •.

Note that although semi-discretization can also be utilized for the optimum control design, because of the spectral convergence in the Chebyshev collocation method, we have only shown the results obtained by the Chebyshev collocation method. We also note that the above example only illustrates the first of the three control strategies discussed in Section 4.1. For future work it is planned to directly compare the advantages and disadvantages of these techniques.

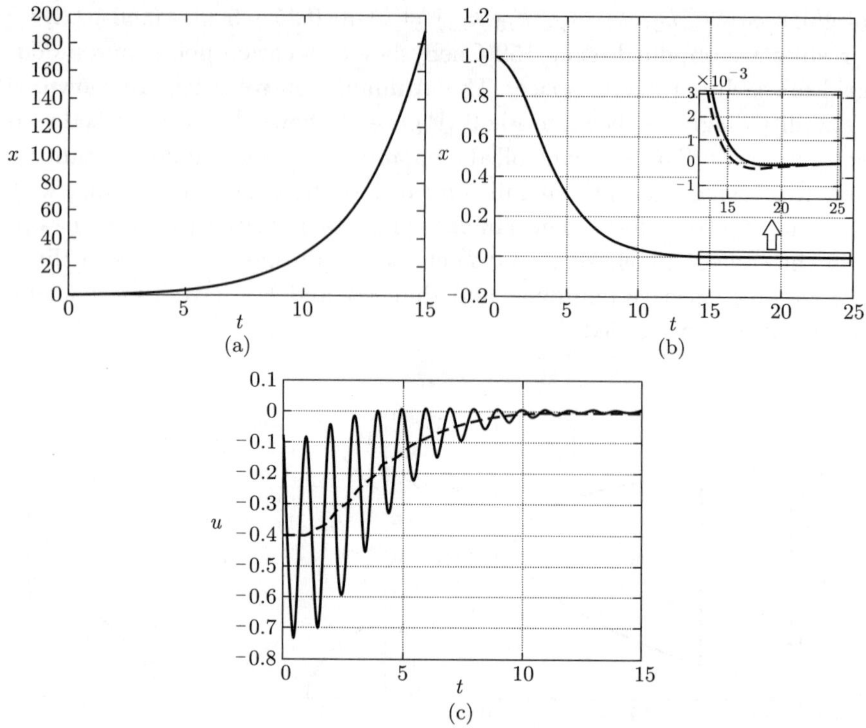

Fig. 10. (a) the uncontrolled response for Eq.(55), (b) the controlled response for Eq.(55) with the constant gain control law (thin line) versus that with the full 6-term periodic control law (thick line), and (c) the control effort corresponding to the constant gain control law (thin line) versus that corresponding to the full 6-term periodic control law (thick line).

5. Stochastic State, Parameter, and Delay Estimation

5.1. *Formulation*

In a general stochastic estimation problem it is assumed that the form of the model is known only approximately due to imperfect knowledge of the dynamical model that describes the motion and/or imperfect knowledge of parameters. The goal is to obtain the best estimate of the state as well as of model parameters based on measured data that has a random component due to observation errors[94]. In addition, a second source of stochastic excitation typically appears in the state dynamics as so-called process noise which can be either additive or multiplicative[95]. Similarly, a stochastic estimation problem in a delayed system aims to find the best estimate of the state and parameters of a stochastically-excited delayed system from measured data that contains an additive random observation error.

There has been little work reported in the literature concerning stochastic estimation or parameter identification for time delay systems. Among those reported, most are limited to an identifiability analysis of parameters[24,25,28]. Orlov et al.[25]

developed an adaptive parameter identifier for linear dynamic systems with finitely many lumped delays in the state vector and control input to simultaneously identify the system parameters and the delay. Mann et al.[29] examined the experimental data with empirical Floquet theory and principle orthogonal decomposition to estimate the parameters of a linear time-periodic delayed system from a reduced order map of the system. Tang and Guan[27] studied the problem of estimating time delay and parameters of time-delayed first-order scalar chaotic systems by first converting the problem into an optimization problem with a suitable objective function and applying a particle swarm optimization algorithm. Sun and Yang[26] exploited chaos synchronization for parameter identification of chaotic delayed systems with varying time delay through using an adaptive feedback controller based on the Razumikhin condition and the invariance principle of functional differential equations in the framework of Lyapunov-Krasovskii theory. Also, parameter estimation of nonlinear time-varying DDEs with constant delay from fully and partially available data has been studied by Deshmukh[30] wherein an ideal case of parameter linearity with no external random disturbance was considered. Chebyshev spectral collocation is used to convert the delayed dynamic system (model) into an algebraic system with unknown parameters, and subsequently a standard least-squares optimization is employed to find the solution for the unknown parameters. Other studies have investigated parameter identification in delayed chaotic systems[26,27]. While in all of the aforementioned studies, the critical role of measurement noise and model uncertainties are ignored, in Refs.22, 23 the problem of optimal filtering in stochastic time delay systems was considered by designing an optimal finite-dimensional filter.

It is clear that the problem of identifying the unknown parameters of a stochastic time-delayed system from the time series response is still open to novel efficient techniques. Here, we present a novel approach for state, delay, and parameter identification of time-varying DDEs through exploiting continuous time approximation and optimal filtering. Specifically, optimal estimation of stochastic linear and nonlinear constant and periodic DDEs having time-varying coefficients and possibly unknown delay from possibly incomplete measurements is studied. In this technique, which is outlined below, the stochastic delay differential equation is first discretized with a set of ODEs using CSCTA. Then the estimation problem in the resulting stochastic ODE system is represented as an optimal filtering problem using a state augmentation technique. Finally, using extended Kalman-Bucy (EKB) filters, the unknown state, delay, and parameters of a nonlinear DDE are estimated from a noise-corrupted, possibly incomplete measurement of the states.

The estimation problem of a delayed system in its general form can be formulated as an optimal continuous-time filtering problem in the form of a set of Itô stochastic delay differential equations as

$$\begin{aligned} d\mathbf{x}(t) &= \mathbf{f}(\mathbf{x}(t), \mathbf{x}(t-\tau), \mathbf{a}(t), t)dt + \mathbf{G}(\mathbf{x}(t), t)d\boldsymbol{\beta}(t), \\ d\mathbf{z}(t) &= \mathbf{h}(\mathbf{x}(t), \mathbf{x}(t-\tau), t)dt + \mathbf{J}(t)d\boldsymbol{\eta}(t) \end{aligned} \tag{57}$$

where $\boldsymbol{\beta}(t)$ and $\boldsymbol{\eta}(t)$ are independent Brownian motion additive stochastic processes with $E[d\boldsymbol{\beta}(t)] = E[d\boldsymbol{\eta}(t)] = 0$, $E[d\boldsymbol{\beta}(t)d\boldsymbol{\beta}^T(t)] = \mathbf{Q}dt$, $E[d\boldsymbol{\eta}(t)d\boldsymbol{\eta}^T(t)] = \mathbf{R}dt$, $E[\cdot]$ represents the expectation operator, and $\mathbf{a}(t)$ represents a vector of parameters to be estimated. Note that the process noise is multiplicative if $\mathbf{G}(\cdot)$ depends on $\mathbf{x}(t)$ and additive if it does not depend on $\mathbf{x}(t)$. Alternatively, the estimation problem for additive noise problems may be formulated in terms of the stationary zero-mean Gaussian white noise processes formally defined as $\mathbf{v}(t) = d\boldsymbol{\beta}(t)/dt$, $\mathbf{w}(t) = d\boldsymbol{\eta}(t)/dt$ and measurement process $\mathbf{y}(t) = d\mathbf{z}(t)/dt$ as

$$\begin{aligned} \dot{\mathbf{x}}(t) &= \mathbf{A}_1(\mathbf{a}(t),t)\mathbf{x}(t) + \mathbf{A}_2(\mathbf{a}(t),t)\mathbf{x}(t-\tau) \\ &\quad + \mathbf{g}(\mathbf{x}(t),\mathbf{x}(t-\tau),\mathbf{a}(t),t) + \mathbf{G}(t)\mathbf{v}(t), \\ \mathbf{y}(t) &= \mathbf{h}(\mathbf{x}(t),\mathbf{x}(t-\tau),t) + \mathbf{J}(t)\mathbf{w}(t) \end{aligned} \tag{58}$$

where $\mathbf{G}(t)$ is assumed to be independent of $\mathbf{x}(t)$, $\mathbf{v}(t)$, and $\mathbf{w}(t)$ are assumed to be both mutually independent and independent from the state and observation with constant covariance matrices of $\mathbf{Q}$ and $\mathbf{R}$, respectively, i.e., $\mathbf{v} \sim N(\mathbf{0},\mathbf{Q})$ and $\mathbf{w} \sim N(\mathbf{0},\mathbf{R})$, and the linear part of the process model is shown explicitly in the $\mathbf{A}_1$ and $\mathbf{A}_2$ matrices. Note that $\mathbf{a}(t)$ can be a constant vector and it is assumed to be time varying here only to keep the generality of the notation.

The filtering problem can be represented in the context of parameter estimation, using the so-called state augmentation method, in which the parameter vector $\mathbf{a}$ (now assumed to be constant) is included with the state vector while being constrained to have a zero rate of change, i.e.,

$$\begin{aligned} \begin{Bmatrix} \dot{\mathbf{x}}(t) \\ \dot{\mathbf{a}}(t) \end{Bmatrix} &= \begin{Bmatrix} \mathbf{A}_1(\mathbf{a},t)\mathbf{x} + \mathbf{A}_2(\mathbf{a},t)\mathbf{x}_\tau + \mathbf{g}(\mathbf{x},\mathbf{x}_\tau,\mathbf{a},t) \\ \mathbf{0} \end{Bmatrix} + \begin{Bmatrix} \mathbf{G}(t) \\ \mathbf{0} \end{Bmatrix} \mathbf{v}(t), \\ \mathbf{y}(t) &= \mathbf{h}(\mathbf{x},\mathbf{x}_\tau,t) + \mathbf{J}(t)\mathbf{w}(t) \end{aligned} \tag{59}$$

where $\mathbf{x}_\tau$ denotes $\mathbf{x}(t-\tau)$. The purpose of the optimal continuous-time filtering problem is to recursively obtain estimates of the states and parameters from the mean, median, or mode of the time-varying conditional probability density

$$P\left(\mathbf{x}(t) | \left\{\mathbf{y}(\rho) :\ 0 \le \rho \le t\right\}\right). \tag{60}$$

The parameter vector $\mathbf{a}(t)$ is assumed to initially have a Gaussian distribution with mean $\mathbf{a}_0$ and covariance $\mathbf{P}_0$. Therefore, the augmented state method along with optimal filtering problem provides a pertinent approach for simultaneous estimate of the state and parameters of a nonlinear system. However, in order for the approach to be applicable, the DDE of the system needs to be expressed as a system of ODEs using CSCTA. This results in the system of ODEs given by

$$\begin{Bmatrix} \dot{\mathbf{Y}}(t) \\ \dot{\mathbf{a}}(t) \end{Bmatrix} = \left\{ \begin{matrix} \begin{bmatrix} \mathbf{A}_1(\mathbf{a},t) & \mathbf{0}_{n\times n} & \cdots & \mathbf{0}_{n\times n} & \mathbf{A}_2(\mathbf{a},t) \\ & & \frac{2}{\tau}\left[\mathbb{D}^{(n+1:nm,:)}\right] & & \end{bmatrix} \mathbf{Y}(t) + \begin{bmatrix} \mathbf{g}(\mathbf{Y}_1(t),\mathbf{Y}_m(t),\mathbf{a},t) \\ \mathbf{0}_{n(m-1)\times 1} \end{bmatrix} \\ \mathbf{0}_{r\times 1} \end{matrix} \right\} + \left\{ \begin{matrix} \begin{bmatrix} \mathbf{G}(t) \\ \mathbf{0}_{n(m-1)\times 1} \end{bmatrix} \\ \mathbf{0}_{r\times 1} \end{matrix} \right\} \mathbf{v}(t), \tag{61}$$

$$\mathbf{y}(t) = \mathbf{h}(\mathbf{Y}_1(t),\mathbf{Y}_m(t),t) + \mathbf{J}(t)\mathbf{w}(t)$$

where $\mathbf{Y}(t)$ is now used for the expanded state vector since $\mathbf{y}(t)$ denotes the measurement.

Note that since the linear matrix in Eq.(61) is a function of the augmented states, even if the delay system and the resulting approximated ODE are linear ($\mathbf{g} = \mathbf{0}$), the optimal filtering problem is nonlinear. Therefore the filtering technique should be capable of handling nonlinearity in the process. Hence we use the EKB filter for both linear and nonlinear delay differential equations. In order to present the equations for propagation of the estimate and error covariance, we express Eq.(61) in the non-delayed form of the nonlinear optimal filtering problem described in Eq.(57). Assuming the augmented delayed state to be incorporated in a finite-dimensional state $\mathbf{X}$, i.e., $\mathbf{X}^T(t) = [\mathbf{Y}^T(t),\ \mathbf{a}^T(t)]$, equation (61) can be written as a nonlinear optimal filtering problem without delay in the Itô form

$$\begin{aligned} d\mathbf{X}(t) &= \boldsymbol{\mathcal{F}}(\mathbf{X},t)dt + \boldsymbol{\mathcal{G}}(t)d\boldsymbol{\beta}(t), \\ d\mathbf{z} &= \boldsymbol{\mathcal{H}}(\mathbf{X},t)dt + \mathbf{J}(t)d\boldsymbol{\eta}(t). \end{aligned} \tag{62}$$

In order to use the EKB filter, a linearization of both $\boldsymbol{\mathcal{F}}(\cdot)$ and $\boldsymbol{\mathcal{H}}(\cdot)$ is needed. Rather than linearizing about a reference trajectory, the EKB filter employs a linearization about the state estimate itself. We also note that the time delay may be estimated similar to any parameter. For this purpose, however, it turns out to be far easier to estimate the delay by first normalizing the time via $t = \tau t'$ (which brings τ out as a coefficient in the $\mathbf{A}_1$ and $\mathbf{A}_2$ matrices and nonlinear function $\mathbf{g}(\cdot)$) before employing CSCTA, rather than estimating τ directly from Eq.(61).

To this end, we utilize the following structure for a nonlinear observer which, as in the case of the conventional KB filter, is obtained by taking the expectation of the dynamic model and adding a feedback term consisting of the measurement residual times an (as yet) unknown gain matrix, i.e.,

$$d\hat{\mathbf{X}}(t) = E[\boldsymbol{\mathcal{F}}(\mathbf{X},t)]dt + \mathbf{L}(t)[d\mathbf{z}(t) - E[\boldsymbol{\mathcal{H}}(\mathbf{X},t)]dt] \tag{63}$$

where $\hat{\mathbf{X}} = E[\mathbf{X}]$ denotes the estimated state. The differential observation error is

$$\begin{aligned} d\mathbf{e}(t) &= d\mathbf{X}(t) - d\hat{\mathbf{X}}(t) \\ &= \boldsymbol{\mathcal{F}}(\mathbf{X},t)dt - E[\boldsymbol{\mathcal{F}}(\mathbf{X},t)]dt - \mathbf{L}(t)\left(\boldsymbol{\mathcal{H}}(\mathbf{X},t)dt - E[\boldsymbol{\mathcal{H}}(\mathbf{X},t)]dt\right) + d\boldsymbol{B}(t) \end{aligned} \tag{64}$$

where $d\boldsymbol{B}(t) = \boldsymbol{\mathcal{G}}(t)d\boldsymbol{\beta}(t) - \mathbf{L}(t)\mathbf{J}(t)d\boldsymbol{\eta}(t)$ is a differential Brownian motion process with

$$E[d\boldsymbol{B}(t)d\boldsymbol{B}^T(t)] = \left[\boldsymbol{\mathcal{G}}(t)\mathbf{Q}\boldsymbol{\mathcal{G}}^T(t) + \mathbf{L}(t)\mathbf{J}(t)\mathbf{R}\mathbf{J}^T(t)\mathbf{L}^T(t)\right] dt. \tag{65}$$

Linearizing about the current estimate yields

$$\begin{aligned} \boldsymbol{\mathcal{F}}(\mathbf{X},t) &= \boldsymbol{\mathcal{F}}(\hat{\mathbf{X}},t) + \left.\frac{\partial \boldsymbol{\mathcal{F}}}{\partial \mathbf{X}}\right|_{\mathbf{X}=\hat{\mathbf{X}}}(\mathbf{X}-\hat{\mathbf{X}}) + \cdots, \\ \boldsymbol{\mathcal{H}}(\mathbf{X},t) &= \boldsymbol{\mathcal{H}}(\hat{\mathbf{X}},t) + \left.\frac{\partial \boldsymbol{\mathcal{H}}}{\partial \mathbf{X}}\right|_{\mathbf{X}=\hat{\mathbf{X}}}(\mathbf{X}-\hat{\mathbf{X}}) + \cdots \end{aligned} \tag{66}$$

from which $E[\boldsymbol{\mathcal{F}}(\mathbf{X},t)] = \boldsymbol{\mathcal{F}}(\hat{\mathbf{X}},t) + \cdots$ and $E[\boldsymbol{\mathcal{H}}(\mathbf{X},t)] = \boldsymbol{\mathcal{H}}(\hat{\mathbf{X}},t) + \cdots$. Truncating the Taylor series after the first order terms yields the differential observation error as

$$d\mathbf{e}(t) = [\tilde{\mathbf{F}}(t) - \mathbf{L}(t)\tilde{\mathbf{H}}(t)]\mathbf{e}(t)dt + d\mathbf{B}(t) \tag{67}$$

where $\tilde{\mathbf{F}}(t)$ and $\tilde{\mathbf{H}}(t)$ are the Jacobian matrices

$$\tilde{\mathbf{F}}(t) := \left.\frac{\partial \boldsymbol{\mathcal{F}}(\mathbf{X},t)}{\partial \mathbf{X}}\right|_{\mathbf{X}=\hat{\mathbf{X}}}, \qquad \tilde{\mathbf{H}}(t) := \left.\frac{\partial \boldsymbol{\mathcal{H}}(\mathbf{X},t)}{\partial \mathbf{X}}\right|_{\mathbf{X}=\hat{\mathbf{X}}}. \tag{68}$$

Similarly, the error covariance matrix can be obtained by differentiating $E[\mathbf{e}(t)\mathbf{e}^T(t)]$ using the Itô differential rule[94,95] to obtain

$$\begin{aligned} d\mathbf{P}(t) = {} & [\tilde{\mathbf{F}}(t) - \mathbf{L}(t)\tilde{\mathbf{H}}(t)]\mathbf{P}(t)dt + \mathbf{P}(t)[\tilde{\mathbf{F}}(t) - \mathbf{L}(t)\tilde{\mathbf{H}}(t)]^T dt \\ & + \boldsymbol{\mathcal{G}}(t)\mathbf{Q}\boldsymbol{\mathcal{G}}^T(t)dt + \mathbf{L}(t)\mathbf{J}(t)\mathbf{R}\mathbf{J}^T(t)\mathbf{L}^T(t)dt. \end{aligned} \tag{69}$$

The optimal gain matrix $\mathbf{L}(t)$ which leads to a *minimum variance estimator* can be obtained by minimizing $\mathrm{tr}(d\mathbf{P}(t))$ which yields

$$\mathbf{L}(t) = \mathbf{P}(t)\tilde{\mathbf{H}}^T(t)\left(\mathbf{J}(t)\mathbf{R}\mathbf{J}^T(t)\right)^{-1}. \tag{70}$$

The optimal estimator so obtained is given by

$$d\hat{\mathbf{X}}(\mathbf{t}) = \boldsymbol{\mathcal{F}}(\hat{\mathbf{X}},t)dt + \mathbf{L}(t)[d\mathbf{z} - \boldsymbol{\mathcal{H}}(\hat{\mathbf{X}},t)dt] \tag{71}$$

where the covariance is propagated via the Riccati equation

$$d\mathbf{P}(t) = [\tilde{\mathbf{F}}(t)\mathbf{P}(t) + \mathbf{P}(t)\tilde{\mathbf{F}}^T(t) - \mathbf{L}(t)\tilde{\mathbf{H}}(t)\mathbf{P}(t) + \boldsymbol{\mathcal{G}}(t)\mathbf{Q}\boldsymbol{\mathcal{G}}^T(t)]dt. \tag{72}$$

We point out that the above technique is similar in concept (in that it uses CSCTA but not RLFT) to the second of the three strategies for control described in Section 3.1, i.e., by direct numerical solution of the time-dependent Riccati equation from which the control gain was obtained. In fact, the LQR and KBF problems are dual to each other. Other estimation strategies which are dual to the first and third control strategies are also possible to formulate, in which observer design problems may also be solved by traditional pole placement without taking into account the stochastic noise in the process and measurement. In fact, such observer designs and observer-based controllers for linear and nonlinear periodic systems of ODEs which utilize the Liapunov-Floquet transformation were proposed in Refs.87, 96, respectively. Hence, we point out that observer designs for periodic DDEs dual to the controller formulation based on CSCTA, RLFT, and the auxiliary systems in Section 4.1 are possible. However, we do not explore this concept further here since the KBF-based estimator will be used in the following example as well as in the observer-based feedback control design for attitude control in Section 6.

5.2. *Parametrically Forced Second Order Nonlinear DDE*

Consider a second order damped delayed system with cubic nonlinearity and a parametric sinusoidal excitation as

$$\ddot{x}(t) + kx(t) + \epsilon x^3(t) + c\dot{x}(t) = f_1 x(t-1) + f_2 \sin(2\pi t)x(t) + v(t) \tag{73}$$

where the the position $x(t)$ and velocity $\dot{x}(t)$ are measured directly and the parameter vector of the system is $\mathbf{a} = [k, f_2, c, f_1, \epsilon]$ with a true value of $\mathbf{a} = [7, 5, 0.05, 0.1, 1]$. After performing a convergence study the delay system is approximated with a set of ODEs using 8 Chebyshev collocation points.

First, the states and all five parameters of the system are simultaneously estimated from a noise-corrupted measurement. The above DDE is integrated using the `dde23` function in Matlab[97] with a time step of $dt = 0.0001$ sec. The response corrupted by an additive zero-mean Gaussian noise $\mathbf{w}(t)$ with a covariance of $\mathbf{R} = 0.1\mathbf{I}$ is fed into the filtering algorithm with a measurement function of $\mathbf{h}(\mathbf{Y}(t)) = \mathbf{Y}_1(t)$ as the continuous-time measured states. The estimation filtering sequence is initiated with a first guess for the unknown parameters as $\mathbf{a}_0 = [5, 3, 0.5, 0.8, 2]$. In order to account for uncertainties in the acceleration model a zero-mean Gaussian noise $v(t)$ with a variance of $Q = 0.1$ is included in the process state. The initial value of the error covariance matrix $\mathbf{P}_0$ is $10\mathbf{I}$. Note that the augmented state-space process of the Eq.(61) is comprised of 2 primary states (the second of which includes the process noise) plus 16 discretized delayed states plus 5 augmented parameter states. The results of using EKB filter for simultaneous estimation of the states and the unknown parameters are shown in Fig.11. The estimated unknown parameters are $\hat{\mathbf{a}} = [6.963, 4.984, 0.056, 0.107, 1.006]$ with an average error of 4.38%.

The quality of the measurements represented by the stochastic term $\mathbf{w}(t)$ and the extent of uncertainty of the model represented by stochastic term $v(t)$, are two determining factors in the accuracy of the estimated parameters using the current approach. To show this, we reduce the variance of process and covariance of measurement noise to $Q = 10^{-4}$ and $\mathbf{R} = 10^{-4}\mathbf{I}$. The initial value of the error covariance matrix is also reduced relatively to $\mathbf{P}_0 = 0.1\mathbf{I}$ while the initial guess for the unknown parameters $\mathbf{a}_0$ is still the same. Following the same approach

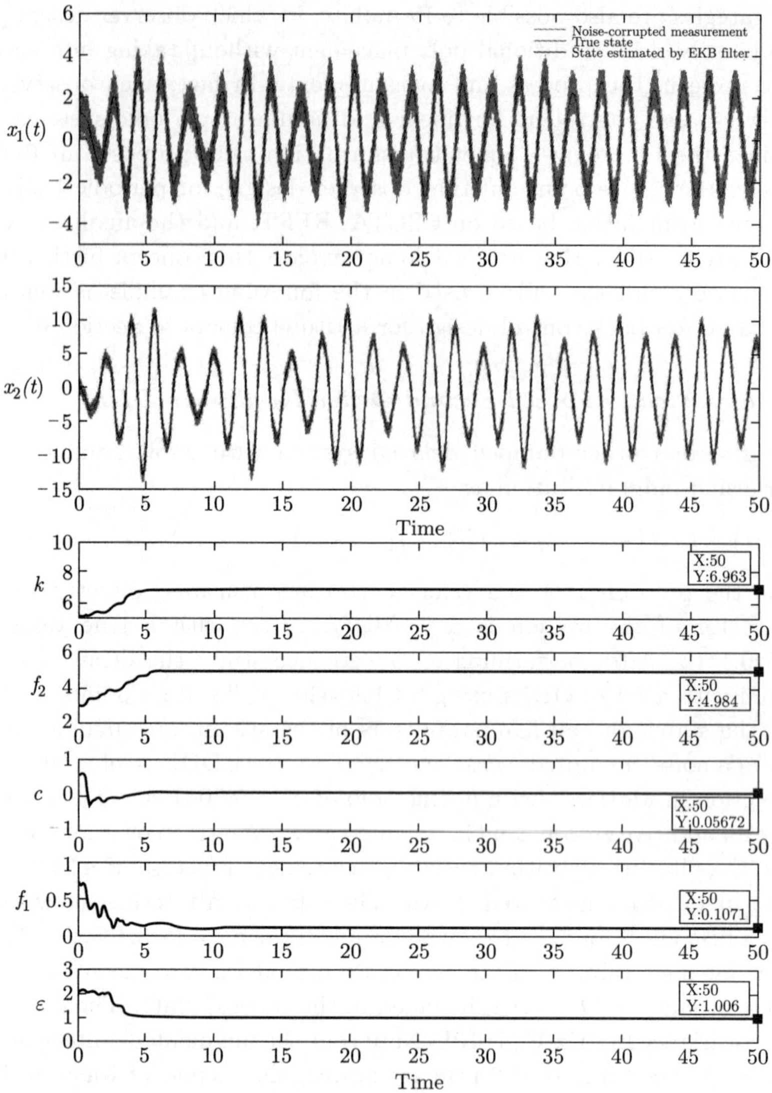

Fig. 11. States and parameters of the parametrically forced second order nonlinear DDE of Eq.(73) estimated from noise-corrupted measurement using EKB filter.

using CSCTA and the EKB filter yields the estimated vector of unknowns with an improved accuracy as $\hat{\mathbf{a}} = [6.99, 5.00, 0.05, 0.10, 1.00]$ which has only 0.02% error on average.

So far, the state observation has been assumed to be complete, i.e., there exist measurements of all state components of the system, and the time delay has been assumed to be known exactly. However in some real-world situations it frequently happens that one or more of the states cannot be measured and the time delay is not known. However, the EKB filter can still be used for estimation in case of an incomplete measurement as long as the system remains observable. In fact, the EKB filter can be used to estimate a globally-observable process by sequentially incorporating only measurements of locally unobservable processes provided that sufficient number of measurements is taken[98]. Also, the time delay may be treated as a parameter multiplying the $\mathbf{A}_1$ and $\mathbf{A}_2$ matrices and vector $\mathbf{g}(\cdot)$ in Eqs.(58) and (59) by a transformation $t = \tau t'$ to a nondimensional time t' in which the delay is normalized to unity, after which state augmentation and CSCTA may be applied as before. In this way, the time delay itself is estimated just as the parameters are estimated in Fig.11.

The nonlinear DDE of Eq.(73) is now reconsidered where only the position $x(t)$ is measured, i.e., it is assumed that the velocity $\dot{x}(t)$ is unobserved and no measurement is available from this state. In addition, the initial guess $\hat{\tau}_0 = 3$ for the unknown time delay is twice the true value ($\tau = 1.5$), and the remaining five parameters are now fixed to the same true values as before (i.e., they are not estimated). The covariances of the measurement and process noise are reduced to $\mathbf{R} = \mathbf{Q} = 0.1\mathbf{I}$. The filtering sequence is initiated with the same initial value of the error covariance matrix. The estimated time delay along with the observed and unobserved states of the system estimated using EKB filter are depicted in Fig.12.

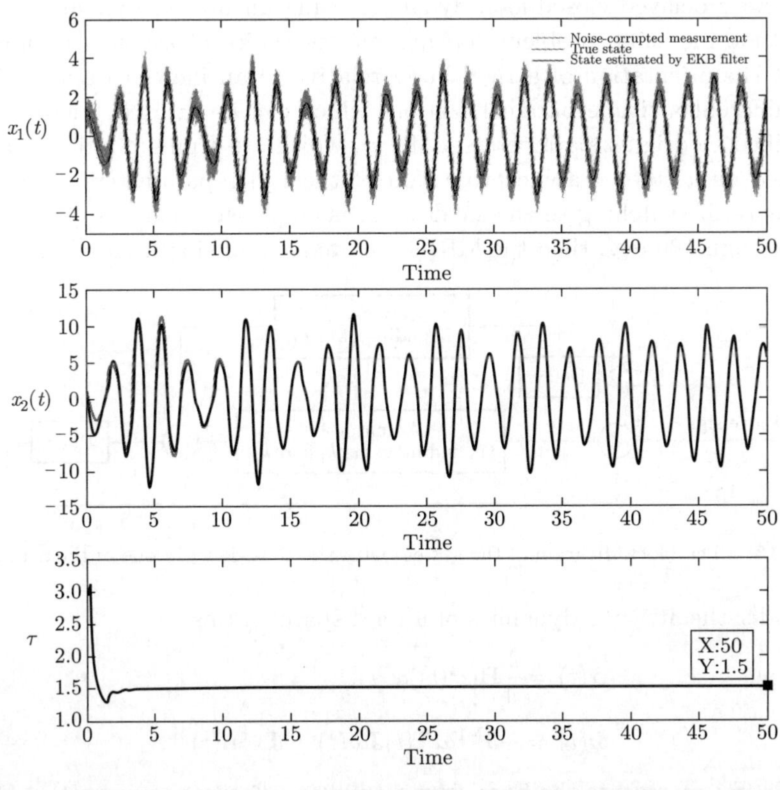

Fig. 12. States and estimated delay of the parametrically forced second order nonlinear DDE of Eq.(73) with unknown time delay estimated from noise-corrupted measurement of $x(t)$ and without any measurement of $\dot{x}(t)$ using EKB filter.

The estimate of the time delay converges to its true value in under 10 sec while the estimated states show a good accuracy. Note that the state $x_2(t) = \dot{x}(t)$ is estimated with absolutely no measurement of this state.

6. Application to Observer-based Delayed Feedback Control of Spacecraft Attitude

The observer-based attitude control of a rigid spacecraft with nonlinear delayed multi-actuator feedback control is studied in this section. It is assumed that the time delay occurs in one of the actuators (e.g., reaction control thruster) while the other actuator (e.g., reaction wheel (RW) or control moment gyro (CMG)) does not have significant time delay. This is relevant to the problem of spacecraft desaturation maneuvers in which the RW or CMG is brought to its null state while using the thruster to keep the spacecraft motionless without a resulting torque applied. Figure 13 shows the block diagram of the system with time delay in one of the actuators, where actuator 1 is nondelayed with gains k_1 and k_2, while actuator 2 has time delay with gain k_3. Therefore, a nonlinear feedback controller using both delayed and nondelayed states is sought for the controlled system to have the desired linear delayed closed-loop dynamics using an inverse dynamics approach. The attitude modeling problem depends on the choice of attitude parameters to represent the orientation of a rigid body relative to an inertial frame. There are several different attitude parametrizations which can be utilized. Here, we utilize the modified Rodriguez parameters (MRPs), which are a minimal three-parameter set defined as a stereographic projection of the Euler parameters (quaternions). Furthermore, if switching to the shadow set is employed when the principal angle passes through 180 deg, then the MRPs are unique and singularity-free[99].

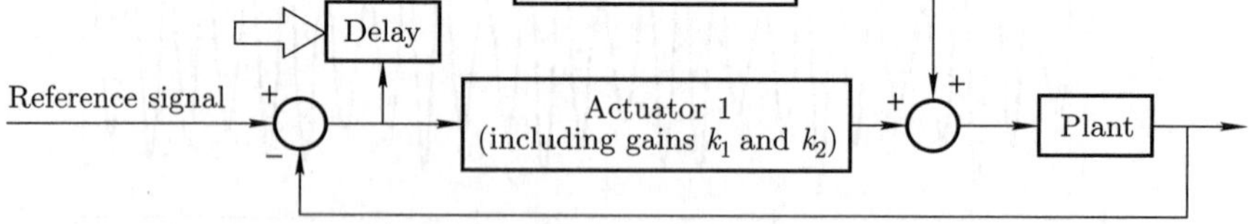

Fig. 13. The block diagram of the system with the time delay in one of the actuators

Consider the attitude dynamics of a rigid spacecraft as

$$\begin{aligned} \dot{\boldsymbol{\sigma}}(t) &= \frac{1}{4}\mathbf{B}(\boldsymbol{\sigma}(t))\boldsymbol{\omega}(t), \\ \dot{\boldsymbol{\omega}}(t) &= -\mathbf{J}^{-1}\boldsymbol{\omega}^{\times}(t)\mathbf{J}\boldsymbol{\omega}(t) + \mathbf{J}^{-1}\mathbf{u}(t) \end{aligned} \tag{74}$$

where $\boldsymbol{\omega} \in \mathbb{R}^3$ represents the body-frame angular velocity vector, $\boldsymbol{\sigma}(t) \in \mathbb{R}^3$ is the MRP set, $\mathbf{u}(t) \in \mathbb{R}^3$ is the control input, $\mathbf{J}$ is the 3×3 symmetric inertia matrix,

and

$$\mathbf{B}(\boldsymbol{\sigma}) = \left[(1 - \boldsymbol{\sigma}^T\boldsymbol{\sigma})\mathbf{I}_{3\times 3} + 2\boldsymbol{\sigma}^{\times} + 2\boldsymbol{\sigma}\boldsymbol{\sigma}^T\right]. \tag{75}$$

Note that $(\cdot)^{\times} : \mathbb{R}^3 \rightarrow \mathrm{so}(3)$ is the skew-symmetric mapping given by

$$\boldsymbol{\Gamma}^{\times} = \begin{bmatrix} 0 & -\Gamma_3 & \Gamma_2 \\ \Gamma_3 & 0 & -\Gamma_1 \\ -\Gamma_2 & \Gamma_1 & 0 \end{bmatrix} \tag{76}$$

where the space of 3×3 real skew-symmetric matrices is denoted by so(3), the Lie algebra of the Lie group SO(3).

6.1. *Inverse Dynamics Approach for Feedback Control Law*

There are different approaches for controlling the attitude dynamics of a rigid body. One method is to assume a linear control law which results in a nonlinear model for the closed-loop dynamics of the system[36,41]. Another method is to assume a nonlinear control law which results in a linear model for the closed-loop dynamics of the system[99]. The second approach will be utilized in this section. In particular, an inverse dynamics approach common in robotics open-loop path-planning problems is utilized here, in which the desired closed-loop response is approximated by a set of second order delay differential equations. This approach (without time delay) has been used in the attitude control problem using both MRPs[99] and quaternions[100].

Based on the desired closed-loop dynamics for the delay-free case[99], for the delayed system, we assume that the desired closed-loop system is

$$\ddot{\boldsymbol{\sigma}}(t) + k_1\dot{\boldsymbol{\sigma}}(t) + k_2\boldsymbol{\sigma}(t) = k_3\boldsymbol{\sigma}(t-\tau) \tag{77}$$

where k_1, k_2, and k_3 are scalar control gains. The body angular acceleration vector corresponding to Eq.(77) can be simplified to[40]

$$\begin{aligned}\dot{\boldsymbol{\omega}}(t) = -k_1\boldsymbol{\omega}(t) - \left[\boldsymbol{\omega}(t)\boldsymbol{\omega}^T(t) + \left(\frac{4k_2}{1+||\boldsymbol{\sigma}(t)||^2} - \frac{||\boldsymbol{\omega}(t)||^2}{2}\right)\mathbf{I}_{3\times 3}\right]\boldsymbol{\sigma}(t) \\ +4\mathbf{B}^{-1}(\boldsymbol{\sigma}(t))k_3\boldsymbol{\sigma}(t-\tau).\end{aligned} \tag{78}$$

By following the inverse dynamics approach[40], it can be shown that the nonlinear control law given by

$$\begin{aligned}\mathbf{u}(t) = {} & \boldsymbol{\omega}^{\times}(t)\mathbf{J}\boldsymbol{\omega}(t) - \mathbf{J}k_1\boldsymbol{\omega}(t) \\ & -\mathbf{J}\left[\boldsymbol{\omega}(t)\boldsymbol{\omega}(t)^T + \left(\frac{4k_2}{1+||\boldsymbol{\sigma}(t)||^2} - \frac{||\boldsymbol{\omega}(t)||^2}{2}\right)\mathbf{I}_{3\times 3}\right]\boldsymbol{\sigma}(t) \\ & +4\mathbf{J}\frac{1}{[1+||\boldsymbol{\sigma}(t-\tau)||^2]^2}\mathbf{B}^T(\boldsymbol{\sigma}(t-\tau))k_3\boldsymbol{\sigma}(t-\tau)\end{aligned} \tag{79}$$

can locally asymptotically stabilize the system in Eq.(74) if the scalar control gains k_1, k_2, and k_3 are selected such that equation (77) is stable. Hence, the closed-loop stability is given by the same second-order DDE for which the Hsu-Bhatt-Vyshnegradskii stability chart can be used to select stable control gains. In the controller given in Eq.(79), the first three nondelayed terms correspond to the control force in actuator 1 in Fig.13, while the last term depending on the delayed MRP parameter set corresponds to the control force in actuator 2.

6.2. *Observer-based Controller Design*

6.2.1. *Delayed Feedback Control from Estimated States*

In developing the feedback control law in Section 6.1, both the MRP parametrization $\boldsymbol{\sigma}(t)$ and the angular velocity $\boldsymbol{\omega}(t)$ of the rigid body are assumed to be available. A problem arises if the internal states of the system are not known, in which case we can design an observer or estimator that attempts to reconstruct the internal state vector of the plant using measurable noisy outputs in a way that is applicable in state feedback. Assuming the system is controllable and observable, the controlled model with measurements takes the form of

$$\begin{bmatrix} \dot{\boldsymbol{\sigma}}(t) \\ \dot{\boldsymbol{\omega}}(t) \end{bmatrix} = \begin{bmatrix} \frac{1}{4}\mathbf{B}(\boldsymbol{\sigma}(t))\boldsymbol{\omega}(t) \\ -\mathbf{J}^{-1}\boldsymbol{\omega}^{\times}(t)\mathbf{J}\boldsymbol{\omega}^{\times}(t) + \mathbf{J}^{-1}\hat{\mathbf{u}}(t) \end{bmatrix} = \mathbf{f}\left(\begin{bmatrix} \boldsymbol{\sigma}(t) \\ \boldsymbol{\omega}(t) \end{bmatrix}\right) + \mathbf{b}\hat{\mathbf{u}}(t) + \begin{bmatrix} 0 \\ 0 \\ 0 \\ 1 \\ 1 \\ 1 \end{bmatrix} v(t), \tag{80}$$

$$\mathbf{y} = \mathbf{h}\left(\begin{bmatrix} \boldsymbol{\sigma}(t) \\ \boldsymbol{\omega}(t) \end{bmatrix}, t\right) + \mathbf{w}(t)$$

where $v(t)$ and $\mathbf{w}(t)$ are zero-mean Gaussian noise processes, $\mathbf{b} = [\mathbf{0}_{3\times3},\ \mathbf{J}^{-1}]^T$, and $\hat{\mathbf{u}}(t)$ is the control law given by Eq.(79) wherein the states $\boldsymbol{\sigma}(t)$, $\boldsymbol{\omega}(t)$, and the delayed MRP set $\boldsymbol{\sigma}(t-\tau)$ are replaced with their estimates $\hat{\boldsymbol{\sigma}}(t), \hat{\boldsymbol{\omega}}(t), \hat{\boldsymbol{\sigma}}(t-\tau)$ obtained from the state observer. Note that the process noise is added only to the dynamics, not the kinematics.

Following Section 5.1 we employ CSCTA and define the expanded $(6(N+1)\times 1)$ vector $\mathbf{Y}(t)$ as

$$\mathbf{Y}(t) = \left[\begin{bmatrix} \boldsymbol{\sigma}(t) \\ \boldsymbol{\omega}(t) \end{bmatrix}^T, \begin{bmatrix} \boldsymbol{\sigma}(t-\tau_1) \\ \boldsymbol{\omega}(t-\tau_1) \end{bmatrix}^T, \begin{bmatrix} \boldsymbol{\sigma}(t-\tau_2) \\ \boldsymbol{\omega}(t-\tau_2) \end{bmatrix}^T, \ldots, \begin{bmatrix} \boldsymbol{\sigma}(t-\tau) \\ \boldsymbol{\omega}(t-\tau) \end{bmatrix}^T\right]^T \tag{81}$$

such that $\tau_i = \frac{\tau}{2}\left(1 - \cos\frac{i\pi}{N}\right)$, $i = 1, 2, \ldots, N$, where N is the number of collocation intervals. Then, as in Section 5 the observer is given by the set of ODEs:

$$\dot{\hat{\mathbf{Y}}}(t) = \begin{bmatrix} \mathbf{0}_{6\times6} & \mathbf{0}_{6\times6} & \cdots & \mathbf{0}_{6\times6} & \mathbf{0}_{6\times6} \\ & & \frac{2}{\tau}\mathbb{D}^{(7:6(N+1),:)} & & \end{bmatrix} \hat{\mathbf{Y}}(t)$$

$$+ \begin{bmatrix} \mathbf{f}\left(\hat{\mathbf{Y}}_1(t)\right) \\ \mathbf{0}_{6N\times1} \end{bmatrix} + \begin{bmatrix} \mathbf{b} \\ \mathbf{0}_{6N\times3} \end{bmatrix} \hat{\mathbf{u}}(t) + \mathbf{L}(t)\left[\mathbf{y}(t) - \mathbf{h}\left(\hat{\mathbf{Y}}_1(t), t\right)\right]$$

$$= \mathcal{F}\left(\hat{\mathbf{Y}}(t)\right) + \mathcal{B}\hat{\mathbf{u}}(t) + \mathbf{L}(t)\left[\mathbf{y}(t) - \mathbf{h}\left(\hat{\mathbf{Y}}_1(t), t\right)\right] \tag{82}$$

where the differential operator $\mathbb{D}$ can be obtained as described in Section 2.1. The observer above delivers the estimated states $\hat{\boldsymbol{\omega}}(t)$, $\hat{\boldsymbol{\sigma}}(t)$, and $\hat{\boldsymbol{\sigma}}(t-\tau)$ which are required to form the control input $\hat{\mathbf{u}}(t)$ in Eq.(80). Therefore, the observer-based controller using the KBF as in Section 5 can be expressed as

$$\begin{aligned}
\dot{\mathbf{Y}}(t) &= \mathcal{F}(\mathbf{Y}(t)) + \mathcal{B}\hat{\mathbf{u}}(t) + \mathcal{G}v(t), \\
\mathbf{y}(t) &= \mathbf{h}\left(\mathbf{Y}_1(t), t\right) + \mathbf{w}(t), \\
\dot{\hat{\mathbf{Y}}}(t) &= \mathcal{F}\left(\hat{\mathbf{Y}}(t)\right) + \mathcal{B}\hat{\mathbf{u}}(t) + \mathbf{L}(t)\left[\mathbf{y}(t) - \mathbf{h}\left(\hat{\mathbf{Y}}_1, t\right)\right], \\
\hat{\mathbf{u}}(t) &= \hat{\boldsymbol{\omega}}^{\times}(t)\mathbf{J}\hat{\boldsymbol{\omega}}(t) - \mathbf{J}k_1\hat{\boldsymbol{\omega}}(t) \\
&\quad - \mathbf{J}\left[\hat{\boldsymbol{\omega}}(t)\hat{\boldsymbol{\omega}}(t)^T + \left(\frac{4k_2}{1 + ||\hat{\boldsymbol{\sigma}}(t)||^2} - \frac{||\hat{\boldsymbol{\omega}}(t)||^2}{2}\right)\mathbf{I}_{3\times3}\right]\hat{\boldsymbol{\sigma}}(t) \\
&\quad + 4\mathbf{J}\frac{1}{\left[1 + ||\hat{\boldsymbol{\sigma}}(t-\tau)||^2\right]^2}\mathbf{B}^T(\hat{\boldsymbol{\sigma}}(t-\tau))k_3\hat{\boldsymbol{\sigma}}(t-\tau)
\end{aligned} \tag{83}$$

where $\mathcal{G} = [0,\ 0,\ 0,\ 1,\ 1,\ 1,\ 0,\ \ldots,\ 0]^T$, and the KBF observer gain is $\mathbf{L}(t) = \mathbf{P}(t)\tilde{\mathbf{H}}^T(t)\mathbf{R}^{-1}$ where $\mathbf{R}$ is the covariance of the measurement noise. $\mathbf{P}(t)$ is the error covariance propagated with Eq.(83) as

$$\dot{\mathbf{P}}(t) = \tilde{\mathbf{F}}(t)\mathbf{P}(t) + \mathbf{P}(t)\tilde{\mathbf{F}}^T(t) - \mathbf{L}(t)\tilde{\mathbf{H}}(t)\mathbf{P}(t) + \mathcal{G}(t)\mathbf{Q}\mathcal{G}^T(t) \tag{84}$$

where $\tilde{\mathbf{H}}(t)$ and $\tilde{\mathbf{F}}(t)$ are the Jacobians in Eq.(68).

6.2.2. *Delayed Feedback Control from Estimated Delay and State*

As mentioned in the previous section, if parameters of the system are not known the observer can be designed to estimate the parameters along with the states in a way that is applicable in the state feedback. The control law of Eq.(79) requires the current states $\boldsymbol{\sigma}(t)$ and $\boldsymbol{\omega}(t)$ and the delayed MRP set $\boldsymbol{\sigma}(t-\tau)$. Since the approach used for designing the observer in the observer-based controller was shown in Section 5 to be also capable of estimating the delay, another observer can be designed to generate estimates of the current states $\hat{\boldsymbol{\sigma}}(t)$, $\hat{\boldsymbol{\omega}}(t)$, and the time delay $\hat{\tau}$. Therefore, the estimated delayed MRP set can be obtained by first finding the estimated current states and then moving back in time to the extent of the estimated delay. Obviously, the observer error in this case is expected to be more than in the case where both the current and the delayed states are directly estimated and the fixed time delay is assumed to be known.

6.3. *Simulation Results*

Numerical simulations of the observer-based controllers described in Sections 6.2.1 and 6.2.2 are discussed in this section. In the first simulation the observer estimates the current states and the delayed MRP. The combined observer-controller of Eq.(83) is applied with an inertia matrix of $\mathbf{J} = \text{diag}([30, 20, 10])$ kg $\cdot$ m^2. The measurement function is assumed to be $\mathbf{h}(\mathbf{Y}_1) = \mathbf{Y}_1$ and corrupted with a zero-mean Gaussian noise of the covariance $\mathbf{R} = 1 \times 10^{-4}\mathbf{I}$. The initial state of the system is assumed to be $[\boldsymbol{\sigma}_0^T,\ \boldsymbol{\omega}_0^T]^T = [-0.3, -0.4, 0.2, 0.2, 0.2, 0.2]^T$ and no process noise is assumed to exist ($\mathbf{Q} = \mathbf{0}$). The error covariance propagation is initialized with $\mathbf{P}_0 = 1 \times 10^{-2}\mathbf{I}$ and the initial guess for the estimated states of the system is assumed to be 50% deviated from the true state. The controller parameters are $k_1 = 8$, $k_2 = 16$, $k_3 = 8$, the time delay is assumed to be $\tau = 0.5$ sec, and 10 Chebyshev collocation points are used in CSCTA. The combined observer-controller states are compared with those of the controller without any observer in Fig.14, in which it is seen that the combined delayed state feedback controller/observer successfully brings the states of the rigid spacecraft to zero.

In the second simulation the observer estimates only the current states and the delay. The delayed MRP is obtained by moving back in time to the extent of the estimated delay. The measurement function is assumed to be the same as before but corrupted with a zero-mean Gaussian noise of the covariance $\mathbf{R} = 1 \times 10^{-5}\mathbf{I}$. The initial state of the system is again $[\boldsymbol{\sigma}_0^T,\ \boldsymbol{\omega}_0^T]^T = [-0.3, -0.4, 0.2, 0.2, 0.2, 0.2]^T$ and the process noise has a variance of $Q = 1 \times 10^{-5}$. The error covariance propagation is initialized with $\mathbf{P}_0 = 1 \times 10^{-3}\mathbf{I}$ and the initial guess for the estimated states of the system is assumed to be 50% deviated from the true state. The controller parameters are $k_1 = 8$, $k_2 = 16$, $k_3 = 8$, the true value of the time delay is assumed to be $\tau = 0.95$ sec, and 10 Chebyshev collocation points are used in CSCTA. The estimated delay is depicted in Fig.15 as well as the comparison between the observer-based controlled states and those of the controller without any observer (assuming the true states are available for feedback). As clear from the figure, the estimated delay converges to the close vicinity of the true delay after 2 seconds. As expected, the estimation error in this case is more than in the case where the delayed states are directly estimated with the constant time delay assumed to be known.

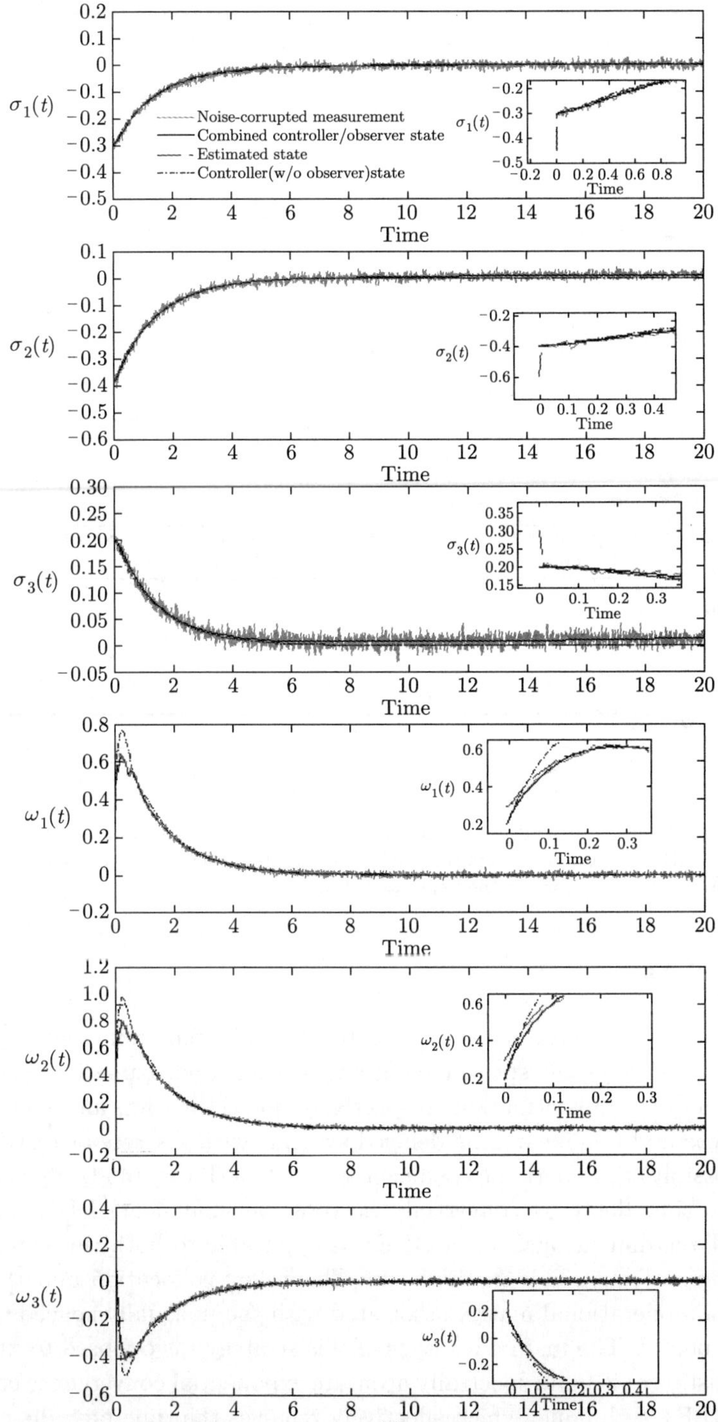

Fig. 14. The estimated attitude states $\boldsymbol{\sigma}$, and angular velocity components $\boldsymbol{\omega}$ of the controller without observer compared with those of the observer-based controller where the delayed states are directly estimated and the fixed time delay is assumed to be known.

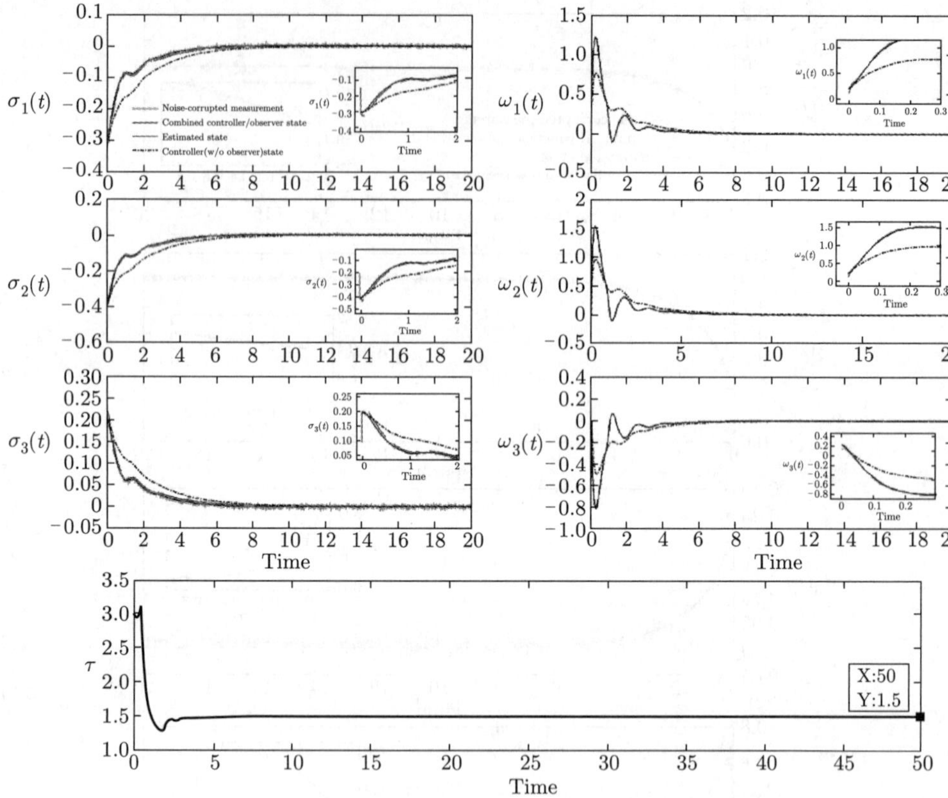

Fig. 15. The estimated attitude states $\boldsymbol{\sigma}$, angular velocity components $\boldsymbol{\omega}$, and time delay τ of the controller without observer compared with those of the observer-based controller where the delayed states are estimated using the estimated delay.

7. Conclusions

In this chapter, the authors have presented two different numerical approaches, based on the Chebyshev spectral continuous time approximation and reduced Liapunov-Floquet transformation, respectively, for estimation and control of (retarded) linear and nonlinear time-delayed systems with discrete or distributed delay and (possibly) time-periodic coefficients. The CSCTA method, which effectively removes the time delay by converting the (possibly nonlinear) DDEs into an approximate large-dimensional set of ODEs, is applicable to both time-invariant and time-varying systems. This method uses Chebyshev collocation and the spectral differentiation operational matrix associated with the unequally-spaced Chebyshev collocation points. The main advantage of this strategy (as opposed to finite differences, for instance) is in its spectrally accurate exponential convergence characteristics. The RLFT technique, which effectively removes the time-periodic coefficients of the resulting time-periodic linear ODEs after applying CSCTA to a system of time-periodic linear DDEs, is specifically used in conjunction with CSCTA for periodic delayed systems. Thus, the resulting low-dimensional set of constant-coefficient ODEs (i.e., Eq.(29)) exactly preserves the eigenstructure (including the dominant characteristic exponents) of the delayed system as represented in the transformed domain (i.e., Eq.(23)). Approximate expressions for the RLFT matrix (i.e., Eqs.(32) and (33)) and its inverse (obtained from the adjoint system) were provided. For linear control problems with delayed feedback or nonlinear delayed systems with a non-vanishing linear part, the combination of CSCTA and RLFT results in a low-dimensional system with time-invariant linear part and periodic coefficients multiplying the control or nonlinear term. Several illustrative examples were used to demonstrate the techniques, including a scalar time-invariant DDE (in which the CSCTA form was compared with the continuous time approximation obtained using forward and centered finite differences) and various forms of the delayed Mathieu equation with single and multiple discrete delays as well as discontinuous distributed delay. It should be noted that no discussion of delay or parametric resonances and their effects on the RLFT were provided, but the reader is directed to Ref.53 for further discussion of this issue.

Next, the control problem associated with time-invariant and time-periodic delayed systems was discussed, in which three strategies were suggested for designing closed-loop feedback controllers. It is shown that using the CSCTA and RLFT methods can help to reduce a (possibly time-periodic) nonlinear delayed system with linear and/or nonlinear feedback control into a form more conducive for the application of many techniques for control of delayed and non-delayed systems in the literature. Specifically, the following three strategies were suggested for designing closed-loop linear feedback controllers for linear periodic delayed systems. First, the stability of the closed-loop response may be investigated in the parameter space of available control gains using Floquet theory, and the optimal control gains may be found for a minimum spectral radius. This strategy does not use either CSCTA or RLFT but instead utilizes one of several existing techniques (e.g., Chebyshev collocation, semi-discretization, and temporal finite elements) to construct a linear map representation (which approximates the monodromy operator) of the closed-loop periodic DDE system. An example of a controlled delayed Mathieu equation illustrated this strategy in which the periodic control gains utilized a three-term Fourier series with coefficients to be obtained by sequentially selecting higher order pairs of gain coefficients that minimize the spectral radius of the closed-loop system. Second, the use of optimal control (time-varying LQR) was suggested in which a time-varying Riccati equation is solved directly from which the time-varying control gain matrix is obtained. This method utilizes CSCTA to obtain an equivalent set of ODEs for which the LQR control is designed, but does not use RLFT. Finally, the third suggested strategy involves the use of both CSCTA and RLFT as well as either pole placement or optimal control design for linear time-invariant systems. Specifically, the sequential application of both CSCTA and RLFT enables the control design to take place in a transformed coordinate system in which both the delay and periodic coefficients are effectively removed, thereby enabling the use of traditional control design tools for linear time-invariant systems. Although the stability for underactuated systems cannot be guaranteed using this approach due to use of a least squares inverse to solve for the time-periodic control gain matrix, the alternative use of a backstepping strategy for time-periodic systems[88] would guarantee asymptotic stability even for these cases. Additional linear control strategies not discussed here that could be employed in conjunction with CSCTA and RLFT include output feedback and observer-based control, while possible nonlinear control strategies to be used with CSCTA and RLFT include nonlinear reduced order controllers, bifurcation controllers, and chaos controllers. The authors are currently planning the investigation of these alternatives.

Next, the estimation of states, parameters, and delay for nonlinear delayed systems using optimal stochastic filtering from noise-corrupted, possibly incomplete measurements was explored using CSCTA. After discretizing the delayed system with a set of ordinary differential equations, the stochastic estimation problem was implemented via state augmentation (to include the unknown parameters and/or delay in the estimate) and by using an extended Kalman-Bucy filter to propagate the equations for the estimate and the state error covariance (through a time-varying Riccati equation). In contrast to the standard Kalman-Bucy filter which employs a linearization about a reference solution, in the extended version the linearization is about the estimate itself (similar to the extended Kalman filter for discrete estimation), thus providing additional accuracy for nonlinear estimation problems. An example of a second order damped system with cubic nonlinearity forced by a delayed term and parametric excitation served as an illustrative example in which two estimation scenarios were analyzed: a) The two states (position and velocity) along with five parameters were estimated from full noise-corrupted state measurements, and b) the two states and the time delay were estimated from incomplete noise-corrupted measurements of the position only. Initial errors in the states, parameters, and delay along with a process noise to help account for unmodeled dynamics were employed in each case. This optimal estimator can also be employed to provide estimated states, parameters, and delay to be used in an observer-based feedback controller. For an illustrative example in a practical engineering problem, the problem of spacecraft attitude estimation and multi-actuator regulation control in the presence of an unknown time delay in one actuator was discussed and illustrated with simulations. This problem is relevant to spacecraft desaturation maneuvers in which the reaction wheel or control moment gyro is brought to its null state while using a reaction control thruster to keep the spacecraft motionless without a resulting torque applied. A nonlinear feedback controller using both delayed and non-delayed states was employed in order for the closed-loop system to have the desired linear dynamics using an inverse dynamics approach along with modified Rodriguez parameters for the attitude parameterization. The results demonstrated that the observer-based delayed feedback regulation controller can successfully stabilize a desired attitude in the presence of noisy measurements and unknown time delay. Future planned research efforts will include the investigation of alternative control laws obtained through the use of Liapunov-Krasovskii functionals and linear matrix inequalities, more realistic attitude measurement models, and the application of these methods to the more difficult problem of delayed feedback attitude estimation and control of dual-spin spacecraft.

Acknowledgments. The first author would like to thank colleagues Brian Mann, Hanspeter Schaub, and Amit Sanyal. The support of NSF grants CMMI-0900289 and CMMI-1131646 is gratefully appreciated.

References

1. N. Krasovskii, *Stability of Motion: Applications of Lyapunov's Second Method to Differential Systems and Equations with Delay* (Stanford University Press, Stanford, 1963).
2. L. Pontryagin, On the zeros of some elementary transcendental functions (in Russian), *Izvestiya Akademii Nauk SSSR.*, **6**, 115–134 (1942). (English translation in American Mathematical Society Translations, 96–110, 1955.)
3. N. Chebotarev and N. Meiman, The Routh-Hurwitz problem for polynomials and entire functions (in Russian), *Trudy Matematicheskogo Instituta imeni V. A. Steklova.*, **26**, 3–331 (1949).
4. N. Hayes, Roots of the transcendental equations associated with certain differential-difference equations, *Journal of the London Mathematical Society*, **25**, 226–232 (1950).
5. R. Bellman and K. Cooke, *Differential-Difference Equations* (Academic, 1963).
6. K. Gu, V. Kharitonov, and J. Chen, *Stability of Time-Delay Systems* (2003).
7. T. Insperger and G. Stépán, *Semi-Discretization for Time-Delay Systems: Stability and Engineering Applications* (Springer, New York, 2011).
8. W. Michiels and S.-I. Niculescu, *Stability and Stabilization of Time-delay Systems: an Eigenvalue-based Approach* (SIAM Press, Philadelphia, 2008).
9. S.-I. Niculescu, *Delay Effects on Stability. A Robust Control Approach* (Springer-Verlag, London, 2001).
10. G. Stépán, *Retarded Dynamical Systems: Stability and Characteristic Functions* (Longman Scientific and Technical, Harlow, UK, 1989).
11. F. Khasawneh and B. Mann, Stability of delay integro-differential equations using a spectral element method, *Mathematical and Computer Modeling*, **54**(9–10), 2493–2503 (2011).
12. O. Bobrenkov, M. Nazari, and E. Butcher, Response and stability analysis of periodic delayed systems with discontinuous distributed delay, *Journal of Computational and Nonlinear Dynamics*, **7**, (2012).
13. F. Khasawneh, B. Mann, T. Insperger, and G. Stépán, Increased stability of low-speed turning through a distributed force and continuous delay model, *Journal of Computational and Nonlinear Dynamics*, **4**(4), 1–12 (2009).
14. S. Mondie, M. Dambrine, and O. Santos, Approximation of control laws with distributed delays: a necessary condition for stability, *Kybernetika*, **38**, 541–551 (2002).

15. W. Hahn, On difference differential equations with periodic coefficients, *Journal of Mathematical Analysis and Application*, **8**, 70–101 (1961).
16. A. Stokes, A floquet theory for functional differential equations, *Proceedings of the National Academy of Sciences, USA*, **48**, 1330–1334 (1962).
17. E. Butcher and B. Mann, *Stability Analysis and Control of Linear Periodic Delayed Systems Using Chebyshev and Temporal Finite Element Methods* (*Delay Differential Equations*: *Recent Advances and New Directions*, B. Balachandran, D. Gilsinn, and T. Kalmar-Nagy (Ed.), New York (2009).
18. E. Butcher, O. Bobrenkov, E. Bueler, and P. Nindujarla, Analysis of milling stability by the Chebyshev collocation method: algorithm and optimal stable immersion levels, *Journal of Computational and Nonlinear Dynamics*, **3**, 1–12 (2009).
19. F. Khasawneh, B. Mann, and E. Butcher, A multi-interval Chebyshev collocation approach for the stability of periodic delayed systems with discontinuities, *Communications in Nonlinear Science and Numerical Simulation*, **16**, 4408–4421 (2011).
20. V. Yakubovich and V. Starzhinski, *Linear Differential Equations with Periodic Coefficients, Parts* 1–2 (Wiley, New York, 1975).
21. Y. Orlov, L. Belkoura, J. Richard, and M. Dambrine, Online parameter identification of linear time delay systems, *Proceedings of the* 41*st IEEE Conference on Decision and Control, Las Vegas, NV, USA*, **1**, 630–635 (2002).
22. M. Basin, P. Shi, and D. Calderon-Alvarez, Optimal state filtering and parameter identification for linear time-delay systems, *Proceedings of the American Control Conference, Seattle, USA*, 7–12 (2008).
23. N. Ahmedova, V. Kolmanovskii, and A. Matasov, Constructive filtering algorithms for delayed systems with uncertain statistics, *Journal of Dynamic Systems, Measurement, and Control*, **125**, 229–235 (2003).
24. G. Ferretti, C. Maffezzoni, and R. Scattolini, On the identifiability of the time delay with least-squares methods, *Automatica*, **32**, 449–453 (1996).
25. Y. Orlov, L. Belkoura, J. P. Richard, and M. Dambrine, On identifiability of linear time-delay systems, *IEEE Transactions on Automatic Control*, **47**, 1319–1324 (2002).
26. Z. Sun and X. Yang, Parameters identification and synchronization of chaotic delayed systems containing uncertainties and time-varying delay, *Mathematical Problems in Engineering*, 105309 (2010).
27. Y. Tang and X. Guan, Parameter estimation for time-delay chaotic systems by particle swarm optimization, *Chaos, Solitons and Fractals*, **40**, 1391–1398 (2009).
28. J. Zhang, X. Xia, and C. H. Moog, Parameter identifiability of nonlinear systems with time-delay, *IEEE Transactions on Automatic Control*, **51**, 371–375 (2006).
29. B. Mann and K. Young, An empirical approach for delayed oscillator stability and parametric identification, *Proceedings of the Royal Society A*, **462**, 2145–2160 (2006).
30. V. Deshmukh, Parametric estimation for delayed nonlinear time-varying dynamical systems, *Journal of Computational Nonlinear Dynamics*, **6**, 041003-1 (2011).
31. D. Eller, G. Aggarwal, and H. Banks, Optimal control of linear time-delay systems, *IEEE Transactions on Automatic Control*, **AC-14**, 678–687 (1969).
32. A. Manitius and A. Olbrot, Finite spectrum assignment problem for systems with delays, *IEEE Transactions on Automatic Control*, **24**, 541–553 (1979).
33. I.-R. Horng and J. -H. Chou, Analysis, parameter estimation and optimal control of time-delay systems via Chebyshev series, *International Journal of Control*, **41**, 1221–1234 (1985).
34. K. Engelborghs, M. Dambrine, and D. Roose, Limitations of a class of stabilization methods for delay systems, *IEEE Transactions on Automatic Control*, **46**, 336–339 (2001).

35. S. Mondie and W. Michiels, Finite spectrum assignment of unstable time-delay systems with a safe implementation, *IEEE Transactions on Automatic Control*, **48**, 2207–2212 (2003).
36. A. Ailon, R. Segev, and S. Arogeti, A simple velocity-free controller for attitude regulation of a spacecraft with delayed feedback, *IEEE Transactions on Automatic Control*, **49**, 125–130 (2004).
37. J. -Q. Sun and B. Song, Control studies of time-delayed dynamical systems with the method of continuous time approximation, *Communications in Nonlinear Systems and Numerical Simulation*, **14**(11), 3933–3944 (2009).
38. W. Michiels, T. Vyhlidal, and P. Zitek, Control design for time-delay systems based on quasi-direct pole placement, *Journal of Process Control*, **20**, 337–343 (2010).
39. A. Chunodkar and M. Akella, Attitude stabilization with unknown bounded delay in feedback control implementation, *Journal of Guidance, Control, and Dynamics*, **34**, 533–542 (2011).
40. M. Nazari, E. Samiei, E. Butcher, and H. Schaub, Attitude stabilization using nonlinear delayed actuator control with an inverse dynamics approach, *22nd AIAA/AAS Space Flight Mechanics Meeting, Charleston, SC, USA* (Jan. 29–Feb. 2, 2012).
41. E. Samiei, M. Nazari, E. Butcher, and H. Schaub, Delayed feedback attitude control using neural networks and Lyapunov-Krasovskii functionals, *22nd AIAA/AAS Space Flight Mechanics Meeting, Charleston, SC, USA* (Jan. 29–Feb. 2, 2012).
42. T. Insperger, Act and wait concept for time-continuous control systems with feedback delay, *IEEE Transactions on Control Systems*, **14**, 974–977 (2006).
43. T. Insperger and G. Stepan, Act-and-wait concept for discrete-time systems with feedback delay, *IET Control Theory A*, **1**, 553–557 (2007).
44. G. Stepan and T. Insperger, Stability of time-periodic and delayed systems — A route to act-and-wait control, *Annual Reviews in Control*, **30**, 159–168 (2006).
45. V. Deshmukh, H. Ma, and E. Butcher, Optimal control of parametrically excited linear delay differential systems via Chebyshev polynomials, *Optimal Control Applications and Methods*, **27**, 123–136 (2006).
46. H. Ma, E. Butcher, and E. Bueler, Chebyshev expansion of linear and piecewise linear dynamic systems with time delay and periodic coefficients under control excitations, *Journal of Dynamic Systems, Measurement, and Control*, **125**, 236–243 (2003).
47. H. Ma, V. Deshmukh, E. Butcher, and V. Averina, Delayed state feedback and chaos control for time-periodic systems via a symbolic approach, *Communications in Nonlinear Science and Numerical Simulation*, **10**, 479–497 (2005).
48. J. Zhang and J.-Q. Sun, Robustness analysis of optimally designed feedback control of linear periodic systems with time-delay, *Proceedings of the 6th International Conference on Multibody Systems, Nonlinear Dynamics, and Control*, ASME DETC 2007, Las Vegas, NV. (Sept. 4–7, 2007).
49. O. Elbeyli and J.-Q. Sun, On the semi-discretization method for feedback control design of linear systems with time delay, *Journal of Sound and Vibration*, **273**, 429–440 (2004).
50. J. Sheng and J.-Q. Sun, Feedback controls and optimal gain design of delayed periodic linear systems, *Journal of Vibration and Control*, **11**(2), 277–294 (2005).
51. J. Sheng, O. Elbeyli, and J.-Q. Sun, Stability and optimal feedback controls for time-delayed linear periodic systems, *AIAA Journal*, **42**(5), 908–911 (2004).
52. E. Butcher and O. Bobrenkov, On the Chebyshev spectral continuous time approximation for constant and periodic delay differential equations, *Communications in Nonlinear Science and Numerical Simulation*, **16**, 1541–1554 (2011).
53. O. Bobrenkov, E. Butcher, and B. Mann, On application of the Liapunov-Floquet

transformation to differential equations with time delay and periodic coefficients, *Journal of Vibration and Control* (2012). doi:10.1177/1077546311433914.
54. O. Diekmann, S. van Gils, S. Verduyn Lunel, and H.-O. Walther, *Delay Equations: Functional, Complex, and Nonlinear Analysis* (Springer-Verlag, New York, 1995).
55. A. Halanay, *Differential Equations: Stability, Oscillations, Time Lags* (Academic Press, New York, 1966).
56. J. Hale and S. Verduyn Lunel, *Introduction to Functional Differential Equations* (Springer-Verlag, New York, 1993).
57. J.-Q. Sun, A method of continuous time approximation of delayed dynamical systems, *Communications in Nonlinear Science and Numerical Simulation*, **14**(4), 998–1007 (2009).
58. D. Breda, S. Maset, and R. Vermiglio, Pseudospectral differencing methods for characteristic roots of delay differential equations, *SIAM Journal of Scientific Computing*, **27**(2), 482–495 (2005).
59. D. Breda, S. Maset, and R. Vermiglio, Computing the characteristic roots for delay differential equations, *IMA Journal of Numerical Analysis*, **24**, 1–19 (2004).
60. D. Breda, Solution operator approximations for characteristic roots of delay differential equations, *Applied Numerical Mathematics*, **56**, 305–317 (2006).
61. L. Trefethen, *Spectral Methods in MATLAB* (SIAM Press, Philadelphia, 2000).
62. P. Wahi and A. Chatterjee, Galerkin projections for delay differential equations, *Journal of Dynamic Systems, Measurement, and Control*, **127**, 80–87 (2005).
63. E. Bueler, Error bounds for approximate eigenvalues of periodic-coefficient linear delay differential equations, *SIAM Journal of Numerical Analysis*, **45**(6), 2510–2536 (2007).
64. L. Fox and I. B. Parker, *Chebyshev Polynomials in Numerical Analysis* (Oxford University Press, London, 1968).
65. C. Clenshaw, The numerical solution of linear differential equations in chebyshev series, *Proceedings of Cambridge Philosophical Society*, **53**, 134–149 (1957).
66. T. Insperger and G. Stépán, Stability chart for the delayed Mathieu equation, *Proceedings of Royal Society of London A: Mathematical, Physical, and Engineering Sciences*, **458**, 1989–1998 (2002).
67. C. Hsu and S. Bhatt, Stability charts for second-order dynamical systems with time lag, *Journal of Applied Mechanics (ASME)*, **33E**(1), 119–124 (1966).
68. F. Van der Pol and M. J. O. Strutt, On the stability of the solutions of Mathieu's equation, *Philosophical Magazine and Journal of Science*, **5**, 18–38 (1928).
69. S. Sinha, R. Pandiyan, and J. Bibb, Liapunov-Floquet transformation: computation and applications to periodic systems, *Journal of Vibration and Acoustics*, **118**, 209–219 (1996).
70. S. Sinha, S. Redkar, and E. Butcher, Order reduction of nonlinear systems with time-periodic coefficients using invariant manifolds, *Journal of Sound and Vibration*, **284**, 985–1002 (2005).
71. S. Sinha, S. Redkar, V. Deshmukh, and E. Butcher, Order reduction of parametrically excited nonlinear systems: techniques and applications, *Nonlinear Dynamics*, **41**, 237–273 (2005).
72. A. Dávid and S. Sinha, Bifurcation control of nonlinear systems with time-periodic coefficients, *Journal of Dynamic Systems, Measurement, and Control*, **125**, 541–548 (2003).
73. S. Sinha and A. David, Control of chaos in nonlinear systems with time-periodic Coefficients, *Philosophical Transactions of the Royal Society A*, **364**, 2417–2432 (2006).
74. E. B. S. C. Sinha and A. Dávid, Construction of dynamically equivalent time-

invariant forms for time-periodic systems, *Nonlinear Dynamics*, **16**, 203–221 (1998).
75. E. Butcher and S. Sinha, Canonical perturbation of a fast time-periodic Hamiltonian via Liapunov-Floquet transformation, *Journal of Applied Mechanics*, **65**, 209–217 (1998).
76. T. Kalmár-Nagy, G. Stépán, and F. Moon, Subcritical Hopf bifurcation in the delay equation model for machine tool vibrations, *Nonlinear Dynamics*, **26**, 121–142 (2001).
77. S. Zhao and T. Kalmár-Nagy, *Center Manifold Analysis of the Delayed Liénard Equation*, 203–220 (Delay Differential Equations: Recent Advances and New Directions, B. Balachandran, T. Kalmár-Nagy, and D. Gilsinn (Eds.), Springer, 2009).
78. D. Gilsinn, Estimating critical Hopf bifurcation parameters for a second-order delay differential equation with application to machine tool chatter, *Nonlinear Dynamics*, **30**, 103–154 (2002).
79. V. Deshmukh and E. Butcher, Dimensional reduction of nonlinear delay differential equations with periodic coefficients using Chebyshev spectral collocation, *Nonlinear Dynamics*, **52**, 137–149 (2008).
80. A. Nayfeh, Order reduction of retarded nonlinear systems — the method of multiple scales versus center-manifold reduction, *Nonlinear Dynamics*, **51**, 483–500 (2008).
81. E. Butcher and S. Sinha, A hybrid formulation for the analysis of time-periodic linear systems via Chebyshev polynomials, *Journal of Sound and Vibration*, **195**, 518–527 (1996).
82. L. Moreau and D. Aeyels, Periodic output feedback stabilization of single-input single-output continuous-time systems with odd relative degree, *Systems and Control Letters*, **51**, 395–406 (2004).
83. J. Allwright, A. Astolfi, and H. Wong, A note on asymptotic stabilization of linear systems by periodic, piecewise constant output feedback, *Automatica*, **41**, 339–344 (2005).
84. A. David and S. Sinha, Some ideas on the local control of nonlinear systems with time-periodic coefficients, *Proceedings of the ASME Design Engineering Technical Conferences, Las Vegas, NV*, 1–10 (Sept. 12–15, 1999).
85. S. Sinha, E. Gowrdon, and Y. Zhang, Control of time-periodic systems via symbolic computation with application to chaos control, *Communications in Nonlinear Science and Numerical Simulation*, **10**, 835–854 (2005).
86. S. Sinha, J. Henrichs, and B. Ravindra, A general approach in the design of active controllers for nonlinear systems exhibiting chaos, *International Journal of Bifurcation and Chaos*, **10**, 165–178 (2000).
87. S. Sinha and P. Joseph, Control of general dynamic systems with periodically varying parameters via Liapunov-Floquet transformation, *Journal of Dynamic Systems, Measurement, and Control*, **116**, 650–658 (1994).
88. V. Deshmukh and S. Sinha, Control of dynamic systems with time-periodic coefficients via the Lyapunov-Floquet transformation and backstepping technique, *Journal of Vibration and Control*, **10**, 1517–1533 (2004).
89. R. Pandiyan and S. Sinha, Time-varying controller synthesis for nonlinear systems subjected to periodic parametric loadings, *Journal of Vibration and Control*, **7**, 73–90 (2001).
90. V. Deshmukh, S. Sinha, and P. Joseph, Order reduction and control of parametrically excited dynamical systems, *Journal of Vibration and Control*, **6**, 1017–1028 (2000).
91. S. Sinha and R. Pandiyan, Analysis of quazilinear systems with periodic coefficients via Liapunov-Floquet transformation, *International Journal of Non-Linear Mechanics*, **29**(5), 687–702 (1994).

92. E. A. Butcher, H. Ma, E. Bueler, V. Averina, and Z. Szabó, Stability of linear time-periodic delay-differential equations via Chebyshev polynomials, *International Journal Numerical Methods in Engineering*, **59**, 895–922 (2004).
93. A. Gabale and S. Sinha, Model reduction of nonlinear systems with external periodic excitations via construction of invariant manifolds, *Journal of Sound and Vibration*, **330**(11), 2596–2607 (2011).
94. J. Speyer and W. Chung, *Stochastic Processes, Estimation, and Control* (*Advances in Design and Control*) (SIAM, Philadelphia, PA, 2008).
95. A. Jazwinski, *Stochastic Processes and Filtering Theory* (Academic, New York, 1970).
96. S. Sinha and Y. Zhang, Observer design for nonlinear systems with time-periodic coefficients via normal form theory, *Journal of Computational and Nonlinear Dynamics*, **4**(3), 031001 (2009).
97. L. Shampine and S. Thompson, Solving delay differential equations in MATLAB, *Applied Numerical Mathematics*, **37**(4), 441–458 (2001).
98. G. Welch, *SCAAT*: *Incremental Tracking with Incomplete Information* (PhD dissertation, University of North California at Chapel Hill, 1996).
99. H. Schaub and J. Junkins, *Analytical Mechanics of Space Systems* (AIAA, Reston, VA, USA, 2009).
100. R. Paielli and R. Bach, Attitude control with realization of linear error dynamics, *Journal of Guidance, Control, and Dynamics*, **16**, 182–189 (1993).

Chapter 9

Noise-induced Dynamics of Time-delayed Stochastic Systems

Yanfei Jin, Haiyan Hu

School of Aerospace Engineering, Beijing Institute of Technology
Beijing 100081, China
jinyf@bit.edu.cn

The noise-induced dynamical behaviors and phenomena in time-delayed stochastic systems have been important topics in many scientific fields, such as biology, physics, neural networks and engineering over the past few decades. In this chapter, we first give a brief introduction to several key concepts and approaches, such as the method of multiple scales, the stochastic averaging method, the delayed Fokker-Planck equations and the two-state model, related to time-delayed stochastic systems. Then, we present a review on some of our recent theoretical and numerical results about the dynamical behaviors and the noise-induced resonances in time-delayed stochastic systems. For instance, the principal resonance and the moment stability of stochastic systems with delayed feedback control are studied under the excitations of narrow-band random noise and Gaussian white noise, respectively. The coherence resonance can be demonstrated through the power-spectrum of a delayed bistable system driven by additive and multiplicative Gaussian white noises. When a weak harmonic signal is applied to this system, the expression of linear spectrum amplification is derived to qualify the stochastic resonance. Finally, we present an overview of the established results of noise-induced dynamics in time-delayed stochastic systems, future challenges and expected prospects.

1. Introduction

Time delays, which exist in many natural and artificial systems, can cause complex dynamical behaviors even in simple systems. Some examples are the time delays in digital controllers, analogue antialiasing and reconstruction filters, signal transmissions. Dynamical systems with time delays can be described by a set of delay differential equations (DDEs). The theoretical analysis of delayed dynamical systems is usually very difficult because the state space of a DDE is infinite-dimensional. Over the past decades, a great progress has been made in studying both general theory of DDEs and practical problems associated with delayed dynamical systems in engineering. Among them, we refer the readers to the classic books by Hale[1],

Driver[2], Qin et al.[3], Stépán[4], Gopalsamy[5], Kuang[6], Hale and Lunel[7], Diekmann[8], Hu and Wang[9].

In practice, numerous physical systems are associated with the random fluctuating environment or noise, and modeled by using stochastic differential equations (SDEs). The SDE plays an important role in the mathematical description of a non-deterministic system whose solution is a random function in some sense. Einstein's first paper on Brownian motion[10] is the cornerstone of the modern theory of stochastic processes and it can be regarded as the beginning of stochastic modeling of natural phenomena. Meanwhile, Smoluchowski[11] developed a systematic explanation and an experimental verification of Brownian motion theory independently. Later, Langevin proposed a new method and established the famous Langevin's equation, which must be the first example of the SDE[12]. Then, over the past decades the theory of SDEs has been well established[13–19], providing us with an unified theory of differential equations with random fluctuation. Especially, Arnold published the first monograph on random dynamical systems[20] in 1998, which bridges the gap between stochastic analysis and dynamical systems.

Stochastic delay differential equations (SDDEs) are used to model the evolution phenomena of practical systems with both noises and time delays. They have found applications in different fields such as engineering[21], biology[22–25], optics[26–27], neural[28–30], and climatology[31], and attracted great attention from the scientists in both theoretical and experimental fields[32–41]. For instance, in Ref.39, the behaviors of a non-Markovian bistable system with noise were analyzed experimentally by using a vertical cavity surface emitting laser with time-delayed optoelectronic feedback. It is thus necessary to develop a good understanding of SDDEs. Unfortunately, it is generally much more difficult to formulate the evolution equations for the probability density with time delays than those without time delays. And, the understanding of complex dynamical behaviors in the stochastic systems with time delays is far from complete. Therefore, the goal of this chapter is both to highlight important results on the noise-induced dynamics for researchers in the domain of the stochastic time-delayed systems, and to provide others with a useful guide for understanding the dynamics of SDDEs.

2. Fundamentals for Time-delayed Stochastic Systems

In this section we introduce several important concepts and the approximate solution methods of the time-delayed stochastic systems. Sections 2.1 and 2.2 present a general overview of the approximate methods for SDDEs, such as the method of multiple scales and the stochastic averaging method. Section 2.3 is an introduction of the Fokker-Planck approach to SDDEs with a small time delay approximation. And Section 2.4 concentrates on introducing the two-state model that typifies a class of symmetric bistable system.

2.1. *The Method of Multiple Scales*

The method of multiple scales has been widely applied to the nonlinear dynamics of various systems[42–51]. Here we describe this method through the following equation

$$\ddot{x}(t) + \omega^2 x(t) = \varepsilon f(x(t), \dot{x}(t)), \tag{1}$$

where ε is a small parameter and satisfies $|\varepsilon| \ll 1$, $f(x, \dot{x})$ is a smooth function of two variables, and ω is the natural frequency of undamped vibration.

The key idea of the method is to consider the expansion representing the response to be a function of multiple independent variables. We assume that the solution of Eq.(1) can be expanded as follows

$$x(t, \varepsilon) = \sum_{j=0}^{m-1} \varepsilon^j x_j(T_0, T_1, \ldots, T_m) + O(\varepsilon T_m), \quad T_j = \varepsilon^j t, \quad \varepsilon \ll 1, \tag{2}$$

where $x(t, \varepsilon)$ is a vector and its expansion is considered to be uniformly valid for times up to $O(\varepsilon^{-m})$, $T_0 = t$ is a fast time scale and $T_1 = \varepsilon t$ is a slow time scale. The time derivatives have the expansions in terms of the partial derivatives with respect to T_j

$$\begin{cases} \dfrac{d}{dt} = D_0 + \varepsilon D_1 + \cdots, \\ \dfrac{d^2}{dt^2} = D_0^2 + 2\varepsilon D_0 D_1 + \cdots, \end{cases} \tag{3}$$

where $D_j = \partial/\partial T_j$ $(j = 0, 1, 2, \ldots)$.

Substituting Eqs. (2) and (3) into Eq.(1) and equating the coefficients with respect to the powers of ε, we obtain a set of perturbation equations that can be solved successively. Secular terms that cause resonance have to be eliminated at each step in solving the perturbation equations. For example, for the zeroth order perturbation solution, whose solution can be expressed in the form

$$x_0 = A(T_1, T_2)e^{i\omega T_0} + \overline{A}(T_1, T_2)e^{-i\omega T_0}, \tag{4}$$

where the unknown amplitude A is determined by eliminating the secular terms, and the overbar represents complex conjugacy. When this stage is reached, the perturbation solutions for each different resonance condition can be obtained. The above procedure is called the general method of multiple scales.

Later, Rajan and Davies[52], Nayfeh and Serhan[53] extended the method of multiple scales to the analysis of the nonlinear systems under random external excitations, especially to the random narrow-band excitation. Rong et al.[54] studied the principal response of a Duffing oscillator under combined deterministic and random excitation by using the method of multiple scales. With the equations of modulation of amplitude and phase, they observed some new phenomena that do not exist in determined systems. For example, the trivial steady-state solution may lose its stability as the intensity of the random excitation increases and then the system may have a nontrivial steady-state solution. Plaut and Hsieh[55] considered the nonlinear vibrations in a system with weak quadratic and cubic nonlinearities in the restoring force and a time delay in the damping. They analyzed six resonance conditions by using the method of multiple scales, including principal, subharmonic, and superharmonic resonances. Hu et al.[56] studied the principal resonance and the 1:3 subharmonic resonance of a Duffing oscillator under linear state feedback control with a time delay. They derived the first-order approximations of the resonances and proposed the concept of an equivalent damping related to the delay feedback from the viewpoint of vibration control. Based on the work done by Hu et al., Ji et al.[57] investigated the principal, subharmonic and superharmonic resonances of a Duffing system with two distinct time delays in the linear state feedback. Furthermore, the method of multiple scales also can be used to treat with the SDDEs as shown in Section 3.1.

2.2. *Stochastic Averaging Method*

Stochastic averaging, an extension of the averaging method of Bogoliubov and Mitropolsky[58] to SDEs, was proposed by Stratonovich[59] from the viewpoint of physics. Later, a rigorous mathematical proof and interpretation for stochastic averaging was provided by Khasminski[60], Papanicolaou and Kohler[61]. Ariaratnam[62] first applied this method to study the dynamic stability of a system under random loading. Since then, the stochastic averaging method has been successfully used to solve problems in the context of random vibration. And many researchers, such as Spanos[63–64], Lin[65–66], and Zhu[67–69], have made great efforts on this aspect. Recently, the stochastic averaging method has been used to analyze the SDDEs as shown in Section 3.2 and Refs.70, 71.

In the following, the stochastic averaging procedure can be introduced briefly. Considering a SDE with a small parameter ε $(0 < \varepsilon \ll 1)$ in the following form

$$\frac{d}{dt}X_i(t) = \varepsilon f_i(\mathbf{X}, t) + \varepsilon^{1/2} g_{ik}(\mathbf{X}, t)\xi_k(t), \tag{5}$$

where $X_i(t)$ is the ith component of $\mathbf{X}(t)$, f_i and g_{ik} $(i = 1, 2, \ldots, n;\ k = 1, 2, \ldots, m)$ are deterministic functional forms, and $\xi_k(t)$ are random stationary process with zero means and correlation functions

$$E[\xi_k(u)\xi_l(v)] = R_{kl}(u - v). \tag{6}$$

As a measure of memory of any two excitation processes, the correlation time of $\xi_k(t)$ can be defined as[72]

$$\tau_{kl} = \frac{1}{\sqrt{R_{kk}(0)R_{ll}(0)}} \int_{-\infty}^{0} |R_{kl}(\tau)|\, d\tau. \tag{7}$$

When the τ_{kl}s' are much smaller than the relaxation time of the dynamical system, the present state $\mathbf{X}(t)$ can be approximated by a Markov vector. Then, equation (5) can be described by the following smoothed Itô equations

$$dX_i(t) = \varepsilon m_i(\mathbf{X})dt + \varepsilon^{1/2}\sigma_{il}(\mathbf{X})dB_l(t), \tag{8}$$

where m_is' are the drift coefficients, σ_{il} satisfy $\sigma_{il}(\mathbf{X})\sigma_{jl}(\mathbf{X}) = b_{ij}(\mathbf{X})$, $B_l(t)$ are mutually independent unit Wiener processes. The time-averaged drift and diffusion coefficients can be determined by

$$m_i(\mathbf{X}) = \left\langle f_i(\mathbf{X}, t) + \int_{-\infty}^{0} \frac{\partial g_{ik}(\mathbf{X}, t)}{\partial X_j} g_{jl}(\mathbf{X}, t+\tau)R_{kl}(\tau)d\tau \right\rangle_t, \tag{9}$$

$$b_{ij}(\mathbf{X}) = \left\langle \int_{-\infty}^{\infty} g_{ik}(\mathbf{X}, t)g_{jl}(\mathbf{X}, t+\tau)R_{kl}(\tau)d\tau \right\rangle_t, \tag{10}$$

where $\langle\cdot\rangle_t = \lim_{T\to\infty} \int_{t_0}^{t_0+T} \langle\cdot\rangle\, dt/T$ is the time-averaging operation.

The above stochastic averaging refers to the smoothed procedure. This time-averaging procedure is applicable only if the right-hand sides of Eq. (5) are small. The original equations may not satisfy this condition, and then Lin et al.[66] proposed a transformation to handle such a case.

2.3. *Delayed Fokker-Planck Equations*

The Fokker-Planck equation (FPE) is a second-order parabolic partial differential equation that governs the transition probability density of a Markov process. It was firstly proposed by Fokker and Planck for investigating the Brownian motion in a radiation field, and was developed by Kolmogorov in a rigorous mathematical way. Therefore, FPE is also called Fokker-Planck-Kolmogorov (FPK) equation. In most situations, it is convenient to analyze the corresponding SDE without time delays by using FPE because the SDE can hardly be treated analytically in a direct way. In this subsection, the Fokker-Planck approach to SDDE and the delayed FPE are summarized.

Consider a SDDE with delayed drift and diffusion terms governed by the following Itô equation

$$dx(t) = f(x(t), x(t-\tau))dt + g(x(t), x(t-\tau))dB(t), \tag{11}$$

where $f(x_0, x_\tau)$ and $g(x_0, x_\tau)$ are given functions, $\tau > 0$ is the time delay, $B(t)$ is a unit Wiener process. In addition, the state variable $x \in [a, b]$, and the initial condition is assumed to be $\{x(t) = x'(t) | t \in [t' - \tau, t']\}$.

The FPE corresponding to Eq.(11), proposed by Guillouzic[73], is given as follows

$$\begin{aligned} \frac{\partial}{\partial t}p(x_0, t_0 | x', t') = &-\frac{\partial}{\partial x_0}[\bar{f}(x_0, t_0|x', t')p(x_0, t_0|x', t')] \\ &+\frac{1}{2}\frac{\partial^2}{\partial x_0^2}[\bar{g}^2(x_0, t_0|x', t')p(x_0, t_0|x', t')], \end{aligned} \tag{12}$$

where the conditional averages are given by

$$\bar{f}(x_0, t_0|x', t') = \int_a^b f(x_0, x_\tau)p(x_\tau, t_0 - \tau|x_0, t_0; x', t')dx_\tau, \tag{13}$$

$$\bar{g}^2(x_0, t_0|x', t') = \int_a^b g^2(x_0, x_\tau)p(x_\tau, t_0 - \tau|x_0, t_0; x', t')dx_\tau. \tag{14}$$

Unfortunately, the above FPE (12) is not self-sufficient. Thus, Guillouzic et al.[74] proposed a small delay approximation method and applied it to a SDDE with non-delayed diffusion. In order to clearly illustrate this technique, they used a delayed linear equation and a delayed logistic equation as examples. Frank[75] showed that FPE (12) can be derived by using the extended phase-space approach[76], which is frequently used in the theory of DDEs. The analytical expression for stationary distributions can be obtained by using the first-order perturbation. However, the closed-form expressions of the transition probability densities can be found only for some special diffusion processes, such as the Ornstein-Uhlenbeck process and the Wiener process. He[77] also revealed that the delayed FPE can be derived, due to Novikow's theorem, in a similar way for getting the ordinary FPE. The delayed FPE of Eq.(11) can be written in the following form

$$\frac{\partial}{\partial t}p(x,t) = -\frac{\partial}{\partial x}\int f(x, x_\tau)p(x, t; x_\tau, t-\tau)dx_\tau \tag{15}$$
$$+\frac{1}{2}\int \frac{\partial^2}{\partial x_0^2}[g^2(x, x_\tau)p(x, t; x_\tau, t-\tau)]dx_\tau.$$

For small time delays, the stationary distributions determined from Eq.(15) can be established. This method has been widely applied to the dynamical analysis of SDEs with time delays as shown in Refs.78–82. However, the aforementioned analytical results are only valid for small time delays. We will introduce the two-state model with delays in Section 2.4 to explore the noise-induced dynamical behaviors in SDEs with arbitrary time delay.

2.4. *Two-state Model*

The two-state model, regarded as an adiabatic approximation to any continuous bistable system, was originally proposed by Mc Namara and Wiesenfeld[83] in the study of stochastic resonance. They considered a symmetric unperturbed system with two discrete states $\pm x_m$, which can switch from one state to the other with rate W_0. Adding a periodic input signal $A(t) = A_0 \cos \Omega t$ to this bistable system, the modulation of the system potential may cause the positions of the potential minima and maximum to oscillate. Thus, the master equation for $n_\pm(t)$, the probabilities of the system variable occupies either state $\pm$ at time t, reads

$$\dot{n}_\pm(t) = -W_\mp(t)n_\pm + W_\pm(t)n_\mp, \tag{16}$$

where $W_\pm(t)$ is the transition rate out of the $\pm$ state, $n_\pm(t)$ satisfies the normalization condition as $n_+ = 1 - n_-$.

The solution to the rate equation (16) is given by

$$n_\pm(t) = g(t)\left[n_\pm(t_0) + \int_{t_0}^{t} W_\pm(t')g^{-1}(t')dt'\right], \tag{17}$$

$$g(t) = \exp\left(-\int_{t_0}^{t} [W_+(t') + W_-(t')]\,dt'\right). \tag{18}$$

Assuming that $W_\pm(t)$ takes the following form

$$W_\pm(t) = f(\mu \pm \eta_0 \cos\omega_s t), \tag{19}$$

where μ is a dimensionless parameter associated with the ratio of a potential barrier to noise, η_0 is the dimensionless strength of the modulation of μ by the periodic signal.

In the condition of small modulation amplitude, we can use the following expansion in the small parameter $\eta_0 \cos\omega_s t$

$$W_\pm(t) = \frac{1}{2}\left\{\alpha_0 \mp \alpha_1\eta_0 \cos\omega_s t + \alpha_2\eta_0^2 \cos^2\omega_s t \mp \cdots\right\}, \tag{20}$$

where

$$\frac{1}{2}\alpha_0 = f(\mu), \quad \frac{1}{2}\alpha_n = \frac{(-1)^n}{n!}\frac{d^n f(\mu)}{d\eta^n}.$$

The integral in Eq.(17) can be given to the first-order in $\eta_0 \cos\omega_s t$

$$\begin{aligned} n_+(t|x_0,t_0) = \frac{1}{2}\Bigg\{&\exp[-\alpha_0(t-t_0)] \times \left[2\delta_{x_0,x_m} - 1 - \frac{\alpha_1\eta_0\cos(\omega_s t_0 - \phi)}{\sqrt{\alpha_0^2+\omega_s^2}}\right] \\ &+1 + \frac{\alpha_1\eta_0\cos(\omega_s t_0 - \phi)}{\sqrt{\alpha_0^2+\omega_s^2}}\Bigg\}, \end{aligned} \tag{21}$$

where $\phi = \arctan(\omega_s/\alpha_0)$, the Kronecker delta δ_{x_0,x_m} is 1 when the system variable is in the state $+$, otherwise, it is 0.

In order to compute the moments of the bistable system, the probability density can be defined as

$$p(x,t) = n_+(t)\delta(x - x_+) + n_-(t)\delta(x - x_-). \tag{22}$$

Then, the first-order moment and the autocorrelation function can be obtained from the definition

$$\begin{aligned} \langle x(t)|x_0,t_0\rangle &= \int xp(x,t|x_0,t_0)dx, \\ \langle x(t+t')x(t)|x_0,t_0\rangle &= \iint xyp(x,t+t'|y,t)p(y,t|x_0,t_0)dxdy. \end{aligned} \tag{23}$$

Based on the above two-state model, Ohira and Sato[84] introduced the concept of a delayed random walk as a mathematical framework for theoretically studying SDDEs. They defined it as a random walk in which the transition probability depends on the position of the walker at some time in the past. Subsequently, they[85] proposed a stochastic binary element whose transition rate depends on its state at a fixed interval in the past. However, they assumed that the states of the dynamical systems at two consecutive iterations are completely uncoupled, which may not realistic in practice. Then, Tsimring and Pikovsky[86] developed the theory of a prototypical noise-driven bistable system with delayed feedback for small noise and small feedback gains. An improved two-state model was proposed for arbitrary time delay in which the transition rates depend on the earlier state of the system. They proved that the theoretical results are in a good agreement with the numerical simulations of the original system. Jiang et al.[87] extended the theoretical results obtained in Ref.86 to the noise-driven bistable system with multiple time-delayed feedbacks. In Section 4, we will apply the two-state model with delays to studying the noise-induced resonances in a delayed bistable system driven by additive and multiplicative noises.

3. Dynamical Behaviors of the Stochastic Systems with Time-delayed Feedback Control

The study on stochastic delayed dynamical systems is a tough problem because of the aforementioned reasons. Some researchers have investigated the dynamical systems with time delays driven by random excitations. Among them, Grigoriu[88] found statistics of the response of a linear control system with deterministic and random time delays subject to Gaussian white noise, Di Paola and Pirrotta[89] studied the effects of time delay on the controlled linear system with random excitation and obtained the approximate solution by expanding the control force in a Taylor series. Mackey et al.[90,91] presented the solution of moment stability for a linear stochastic delay differential equation applied to check the local stability of the stochastic hematopoietic stem cell regulation system. Elbeyli et al.[92] developed an extension of the semi-discretization method to the time-delayed linear stochastic systems. These studies, however, are still limited to the simple linear systems. Klosek et al.[93,94] applied the multi-scale analysis to study the effect of noise near critical delays for SDDEs, which requires the unperturbed system possessing a periodic solution.

This section is devoted to the nonlinear dynamics of the stochastic time-delayed systems for the case of external and parametric random excitations. Section 3.1 presents the principal resonance of a Duffing oscillator with delayed state feedback under narrow-band random excitation by using the method of multiple scales and numerical simulations. The moment stability for delayed dynamical systems under Gaussian white noise excitations are studied in Section 3.2 by using the stochastic averaging method. These techniques are illustrated via a van der Pol oscillator and a linear system.

3.1. *Principal Resonance of a Duffing Oscillator with Delayed Feedback Control under Narrow-band Random Excitation*

The aim of this subsection is to reveal the dynamical behaviors existing in the nonlinear stochastic time-delayed systems through an illustrative example, a Duffing oscillator under a random narrow-band excitation and a linear delayed feedback control. The motion of such a system can be described by the following equation

$$\ddot{x}(t) + \varepsilon\beta\dot{x}(t) + \omega^2 x(t) + \varepsilon\alpha x^3(t) = \varepsilon u x(t-\tau_1) + \varepsilon v\dot{x}(t-\tau_2) + \varepsilon g(x)\xi(t), \tag{24}$$

where ε is a small parameter, $\varepsilon\beta$ and ω are the damping coefficient and the natural frequency respectively, $\varepsilon\alpha$ is the density of the nonlinear term, εu and εv are the feedback gains, $\tau_i \geq 0$ $(i = 1, 2)$ are the time delays in the paths of displacement feedback and velocity feedback, $g(x)$ is generally a deterministic function, and $\xi(t)$ is a random process proposed by Wedig[05] as follows:

$$\xi(t) = h\cos(\Omega t + \bar{\gamma}W(t)), \tag{25}$$

where the standard Wiener process $W(t)$ is a diffusion process with a null drift coefficient and a unit diffusion coefficient[18]. It is easy to derive the following power spectrum $S_\xi(\omega)$ of $\xi(t)$ from Eq.(25) as

$$S_\xi(\omega) = \frac{1}{2}\frac{h^2\bar{\gamma}^2(\Omega^2 + \omega^2 + \bar{\gamma}^4/4)}{(\Omega^2 - \omega^2 + \bar{\gamma}^4/4)^2 + \omega^2\bar{\gamma}^4}. \tag{26}$$

Figure 1 shows the power spectrum $S_\xi(\omega)$, where there is a single peak at the frequency Ω. When the noise intensity $\bar{\gamma}$ increases, so does the bandwidth of the peak. Thus, equation (26) indicates that the generalized fluctuation model (25) connects two opposite extreme cases. In particular, the limiting procedure $\bar{\gamma} \to \infty$ leads to the uniformly distributed power spectrum of white noise, and if $\bar{\gamma} \to 0$, $S_\xi(\omega)$ vanishes in the entire frequency range except at the singular frequency $\omega =$

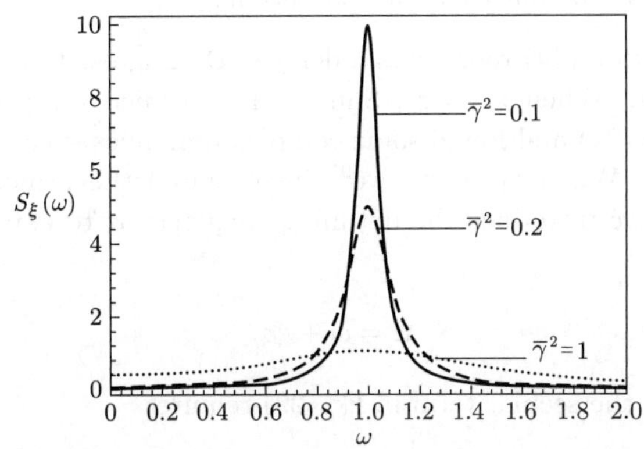

Fig. 1. The power spectrum of $\xi(t)$ with $\Omega = 1, h = 1$.

$\pm\Omega$, where $S_\xi(\pm\Omega) \to \infty$. In what follows, $\bar{\gamma}$ is assumed sufficiently small so that $\xi(t)$ is bound to a narrow-band random process.

In the following, the method of multiple scales presented in Section 2.1 is applied to Eq.(24). Substituting Eqs. (2) and (3) into Eq.(24) and equating the same power of ε, we obtain

$$D_0^2 x_0(T_0, T_1) + \omega^2 x_0(T_0, T_1) = 0, \tag{27}$$

$$\begin{aligned} &D_0^2 x_1(T_0, T_1) + \omega^2 x_1(T_0, T_1) \\ &= -2D_0 D_1 x_0(T_0, T_1) - \beta D_0 x_0(T_0, T_1) - \alpha x_0^3(T_0, T_1) \\ &\quad + u x_0(T_0 - \tau_1, T_1) + v D_0 x_0(T_0 - \tau_2, T_1) \\ &\quad + g(x_0(T_0, T_1)) \cdot h\cos(\Omega T_0 + \gamma W(T_1)). \end{aligned} \tag{28}$$

We should point out that the statistical property of the standard Wiener process $\bar{\gamma}W(t) = \bar{\gamma}W(\varepsilon t)/\sqrt{\varepsilon} = \gamma W(T_1)$ is used here to reach Eq.(28).

Substituting the general solution (4) into Eq.(28) leads to

$$\begin{aligned} D_0^2 x_1 + \omega^2 x_1 = &- 2i\omega A' \exp(i\omega T_0) - i\omega\beta A \exp(i\omega T_0) \\ &+ uA\exp[i\omega(T_0 - \tau_1)] + i\omega v A \exp[i\omega(T_0 - \tau_2)] \\ &- \alpha A^3 \exp(3i\omega T_0) - 3\alpha \bar{A} A^2 \exp(i\omega T_0) \\ &+ \frac{h}{2}\exp[i(\Omega t + \gamma W(T_1))] \cdot g(A\exp(i\omega T_0) \\ &+ \bar{A}\exp(-i\omega T_0)) + c.c., \end{aligned} \tag{29}$$

where the prime represents the derivative with respect to T_1, *c.c.* denotes the conjugate term. The amplitude $A(T_1)$ can be expressed in the polar form as

$$A(T_1) = \frac{1}{2} a(T_1)\exp(i\varphi(T_1)). \tag{30}$$

3.1.1. *Narrow-band Random External Excitation*

Let $g(x) = 1$, equation (24) represents a delayed Duffing oscillator under a random external excitation. When $\tau_1 = \tau_2$, Sun et al.[96] studied the principal resonance response of system (24) and found some complex dynamics such as bifurcation and jump phenomena. When $\tau_1 \neq \tau_2$, we[97] investigated the principal resonance of Eq.(24). At first, we introduce the detuning frequency σ to express the principal resonance as

$$\Omega = \omega + \varepsilon\sigma. \tag{31}$$

The elimination of the secular term in Eq.(29) requires

$$-\omega i(2A' + \beta A - ve^{-i\omega\tau_2}A) + uAe^{-i\omega\tau_1} - 3\alpha A^2\bar{A} + \frac{h}{2}e^{(i\sigma T_1 + i\gamma W(T_1))} = 0. \tag{32}$$

Substituting Eq.(30) into Eq.(32) and separating the real and imaginary parts of Eq.(32), we obtain

$$\begin{cases} a' = -\left(\beta + \dfrac{u}{\omega}\sin(\omega\tau_1) - v\cos(\omega\tau_2)\right)\dfrac{a}{2} + \dfrac{h}{2\omega}\sin\eta, \\ a\eta' = \left(\sigma + \dfrac{u}{2\omega}\cos(\omega\tau_1) + \dfrac{v}{2}\sin(\omega\tau_2)\right)a - \dfrac{3\alpha}{8\omega}a^3 + \dfrac{h}{2\omega}\cos\eta + a\gamma W'(T_1), \end{cases} \tag{33}$$

where $\eta(T_1) = \sigma T_1 - \varphi(T_1) + \gamma W(T_1)$.

Solving the amplitude a and the phase η from Eq.(33), the first-order uniform expansion of the solution of system can be written as

$$x(t) = a(\varepsilon t)\cos(\Omega t - \eta(\varepsilon t)) + O(\varepsilon). \tag{34}$$

(1) *Steady-state Principal Resonance and Stability Analysis*

The solution of Eq.(33) can be assumed in the form

$$a = a_0 + a_1, \quad \eta = \eta_0 + \eta_1, \tag{35}$$

where parameters a_1 and η_1 are the arbitrarily small deviations from a_0 and η_0, a_0 and η_0 are the steady-state solutions of the corresponding deterministic Eq.(33) and can be obtained by setting $a_0' = \eta_0' = 0$. That is,

$$-\beta_e a_0 + \frac{h}{\omega}\sin\eta_0 = 0, \quad \sigma_e a_0 - \frac{3\alpha}{8\omega}a_0^3 + \frac{h}{2\omega}\cos\eta_0 = 0, \tag{36}$$

where the equivalent damping ratio $\beta_e = \beta + u\sin(\omega\tau_1)/\omega - v\cos(\omega\tau_2)$, the equivalent detuning frequency $\sigma_e = \sigma + u\cos(\omega\tau_1)/2\omega + v\sin(\omega\tau_2)/2$. From Eq.(36), the frequency response equation can be derived

$$\left[\beta_e^2 + 4\left(\sigma_e - \frac{3\alpha}{8\omega}a_0^2\right)^2\right]a_0^2 = \frac{h^2}{\omega^2}. \tag{37}$$

The linearization of Eq.(33) at (a_0, η_0) for $\gamma \neq 0$ is

$$\begin{cases} a_1' = -\dfrac{\beta_e}{2}a_1 + \dfrac{h}{2\omega}\cos\eta_0 \cdot \eta_1, \\ \eta_1' = \dfrac{\sigma_e}{a_0}a_1 - \dfrac{9\alpha}{8\omega}a_0 a_1 - \dfrac{h}{2\omega a_0}\sin\eta_0 \cdot \eta_1 + \gamma W'(t). \end{cases} \tag{38}$$

According to Eq.(36), equation (38) can be rewritten as follows

$$\begin{cases} a_1' = -\dfrac{\beta_e}{2}a_1 + \left(\dfrac{3\alpha}{8\omega}a_0^2 - \sigma_e\right)a_0 \cdot \eta_1, \\ \eta_1' = \left(\dfrac{\sigma_e}{a_0} - \dfrac{9\alpha}{8\omega}a_0\right)a_1 - \dfrac{\beta_e}{2} \cdot \eta_1 + \gamma W'(t). \end{cases} \tag{39}$$

The first-order and second-order steady-state moments of a_1 and η_1 can be derived by using the moment method[98]. For the steady-state moments, the conditions

$dEa_1/dT_1 = dE\eta_1/dT_1 = 0$ hold. Here $E[\cdot]$ denotes the mathematical expectation. Taking expectation in both sides of Eq.(39), we obtain

$$Ea_1 = E\eta_1 = 0. \tag{40}$$

and

$$\begin{cases} Ea_1^2 = \dfrac{C_2^2\gamma^2}{4C_1(C_1^2 - C_2C_3)}, \\ Ea_1\eta_1 = \dfrac{C_2\gamma^2}{4(C_1^2 - C_2C_3)}, \\ E\eta_1^2 = \dfrac{(2C_1^2 - C_2C_3)\gamma^2}{4C_1(C_1^2 - C_2C_3)}, \end{cases} \tag{41}$$

where $C_1 = \beta_e/2$, $C_2 = \left(3\alpha a_0^2/8\omega - \sigma_e\right) a_0$, $C_3 = (\sigma_e/a_0 - 9\alpha a_0/8\omega)$.

Then, the expressions of the first-order and second-order steady-state moments of system read

$$\begin{cases} Ea = a_0, Ea^2 = a_0^2 + Ea_1^2, \\ E\eta = \eta_0, E\eta^2 = \eta_0^2 + E\eta_1^2. \end{cases} \tag{42}$$

To analyze the stability of the steady-state solutions (42), we give the necessary conditions for the existence of the second-order steady-state moments determined by Eq.(41). That is, the conditions $Ea_1^2 > 0$ and $E\eta_1^2 > 0$ can be satisfied only if

$$\frac{\beta_e^2}{4} + \left(\sigma_e - \frac{3\alpha}{8\omega}a_0^2\right)\left(\sigma_e - \frac{9\alpha}{8\omega}a_0^2\right) > 0. \tag{43}$$

Moreover, the necessary and sufficient condition that governs the stability of the second-order steady-state moments is that the corresponding coefficient matrix

$$\begin{bmatrix} -2C_1 & 2C_2 & 0 \\ 0 & 2C_3 & -2C_1 \\ C_3 & -2C_1 & C_2 \end{bmatrix} \tag{44}$$

must be negative defined. Based on Routh-Hurwitz criterion, the second-order steady-state moments are asymptotically stable if and only if the following two inequalities hold simultaneously

$$\begin{cases} 2C_1 - C_2 - 2C_3 > 0, \\ C_1(C_2^2 + 2C_2C_3 + 4C_3^2) - C_2C_3(C_2 + 2C_3) > 0. \end{cases} \tag{45}$$

From Eqs. (42) and (45), we know that if there are three solution branches, only the top and the bottom ones are stable and the third one between them is unstable. When the noise intensity γ is small enough, the noise is not able to change the stability of these branches.

(2) *Bifurcations of the Steady-state Solutions and Jumps*

The following analysis is to determine the effects of β_e and σ_e on the steady-state moments (42). If σ_e and h are chosen as bifurcation parameters, the variations of the steady-state moments are shown in Figs.2 and 3. In Fig.2, when σ_e increases from a small value, the variation of a_0 with σ_e moves along $A_4 \to A_1 \to A_2$. At the point A_1, i.e., when

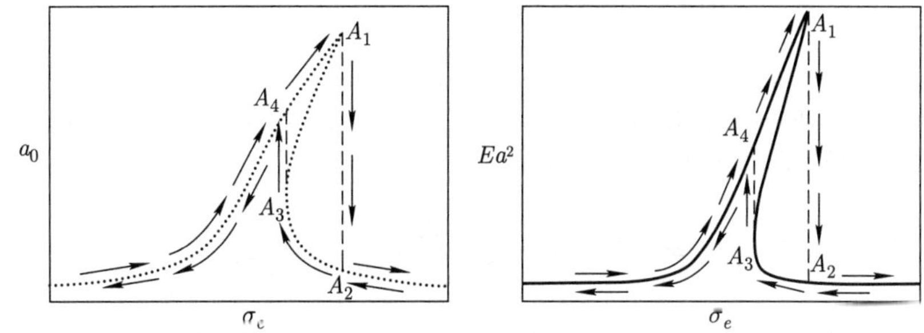

Fig. 2. The variations of the steady-state moments versus σ_e.

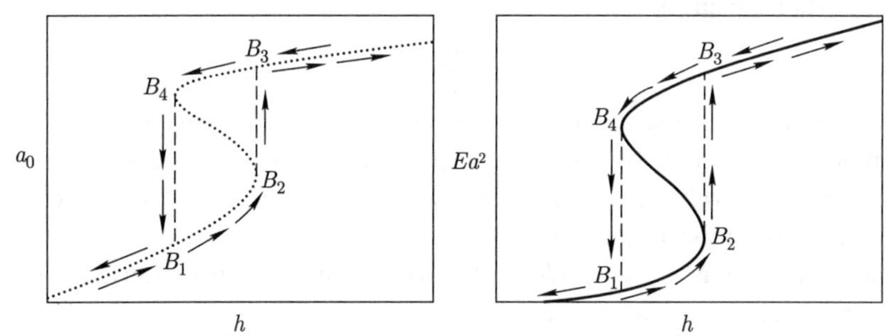

Fig. 3. The variations of the steady-state moments versus h.

$$\sigma_{e1} = \frac{3\alpha a_0^2}{4\omega} + \sqrt{\frac{9\alpha^2 a_0^4}{16\omega^2} - \beta_e^2}, \tag{46}$$

the larger stable amplitude turns to the smaller one. That is, σ_{e1} is a bifurcation point. When σ_e decreases from a large value, the variation of a_0 with σ_e moves along $A_2 \to A_3 \to A_4$. At the point A_3, i.e., when

$$\sigma_{e2} = \frac{3\alpha a_0^2}{4\omega} - \sqrt{\frac{9\alpha^2 a_0^4}{16\omega^2} - \beta_e^2}, \tag{47}$$

the steady-state solution will jump from A_3 to A_4. And, σ_{e2} is also a bifurcation point of the steady-state solution. Therefore, when $\beta_e < 3|\alpha|a_0^2/4\omega$ and $\sigma_{e2} < \sigma_e < \sigma_{e1}$ hold, the steady-state solution a_0 will have three values. The largest steady-state solution and the smallest one are stable, while the middle steady-state solution is unstable. Similarly, jumps exist in Fig.2(b). When h increases from a small value, the variation of a_0 with h moves along $B_1 \to B_2 \to B_3$ as shown in Fig.3. At point B_2, i.e., when

$$h_1 = \left[\beta_e^2 + 4\left(\sigma_{e1} - \frac{3\alpha}{8\omega}a_0^2\right)^2\right]^{1/2} \omega a_0, \tag{48}$$

the steady-state solution will jump from B_2 to B_3. Hence, h_1 is a bifurcation point. When h decreases from a large value, the variation of a_0 with h moves along $B_3 \to B_4 \to B_1$. At point B_4, i.e., when

$$h_2 = \left[\beta_e^2 + 4\left(\sigma_{e2} - \frac{3\alpha}{8\omega}a_0^2\right)^2\right]^{1/2} \omega a_0, \tag{49}$$

the larger steady-state solution turns to the smaller one. That is, h_2 is also a bifurcation point.

It is easy to see that jumps and hysteresis circle exist in Figs.2 and 3. When $\beta_e = 3|\alpha|a_0^2/4\omega$, the interval of σ_e will degenerate to one point $\sigma_e = 3\alpha a_0^2/4\omega$. The critical h is then given by

$$h_c = \sqrt{2}\beta_e \omega a_0. \tag{50}$$

When equation (50) satisfies, jumps and hysteresis phenomena will vanish. Thus, we can modify σ_e and β_e by choosing appropriate time delays and feedback gains to avoid the unfavorable phenomena appearing. Figure 4 shows the comparison between the approximate analytic results and numerical results for the variations of Ea and Ea^2 versus σ. It is clear that the approximate analytical results well coincide with the numerical results.

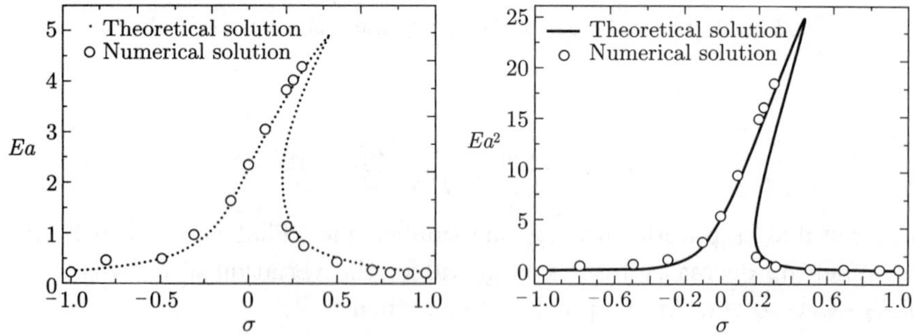

Fig. 4. Frequency response computed by the multiple scale method and the numerical method.

(3) *The Design of Feedback Control*

We give a design of delayed feedback from the viewpoint of vibration control.

When $\tau_1 \neq \tau_2$, we have two cases $0 < \tau_2 < \tau_1$ and $0 < \tau_1 < \tau_2$. For simplify, we let $\omega = 1$ and discuss only the feedback control for the case of $0 < \tau_2 < \tau_1$

here. With $\psi = |\tau_1 - \tau_2|$, assuming that $\tau_1 = \tau + \psi, \tau_2 = \tau$ and letting $d\sigma_e/d\tau = 0$, $d\beta_e/d\tau = 0$, we get $\psi = n\pi/2$ $(n = 1, 3, 5, \ldots)$. Here we only need to consider the cases of $\psi = \pi/2$ and $\psi = 3\pi/2$. When $\psi = \pi/2$, we can obtain the extreme values of σ_e and β_e as

$$\beta_e = \beta \pm |u - v|, \ \sigma_e = \sigma \quad \text{or} \quad \beta_e = \beta, \ \sigma_e = \sigma \pm |u - v|. \tag{51}$$

Substituting the first equation of Eq.(51) into Eq.(37) arrives at

$$a_{0\,\max} = \frac{h}{\beta \pm |u - v|}. \tag{52}$$

Similarly, the amplitude peak of the principal resonance for $\psi = 3\pi/2$ can be obtained as following

$$\tilde{a}_{0\,\max} = \frac{h}{\beta \pm |u + v|}. \tag{53}$$

The above expressions (52) and (53) indicate that if the time delays are appropriately chosen so that $\beta_e = \beta + |u \pm v|$, the amplitude peak can be reduced to a minimum by the feedback gains. Otherwise, the feedback gains will greatly increase the amplitude peak if the time delays make $\beta_e = \beta - |u \pm v|$. This property enables one to find the optimal control performance by implementing the condition $\beta_e > \beta$. That is, when $\psi = \pi/2$, the inequality $(u - v)\cos\tau > 0$ should be satisfied. When $\psi = 3\pi/2$, the condition $(u + v)\cos\tau < 0$ should be met. Thus, we propose the following combinations of the feedback gains and time delays in the feedback control in Table 1 for $\psi = \pi/2$ or $\psi = 3\pi/2$.

Table 1. Different parameter combinations in the design of feedback control for the case of $\tau_2 < \tau_1$.

Displacement feedback gain u	Velocity feedback gain v	Relation between u and v	Time delay τ
$u > 0$	$v < 0$	$\|u\| < \|v\|$	$[0, \pi/2] \cup [3\pi/2, 2\pi]$
$u < 0$	$v > 0$	$\|u\| < \|v\|$	$[\pi/2, 3\pi/2]$

In Fig.5, τ is chosen as $\tau = 0$, $\pi/4$ and $\pi/2$, which correspond to one time delay in the displacement feedback, and two distinct time delays in the feedback control. When $v > 0$, the variations of β_e with ψ are shown in Figs.5(a) and 5(b). It is seen that the value of β_e for $\tau = \pi/2$ is the largest one, while that for $\tau = 0$ is the smallest one. Hence, the control performance for two distinct time delays is better than that for only one time delay in displacement feedback. When $v < 0$, the variations of β_e with ψ are presented in Figs.5(c) and 5(d). It is easily found that the value of β_e for $\tau = 0$ is the largest one among three cases. This shows that the control performance for only one time delay in displacement feedback is superior to that for two distinct time delays in this case.

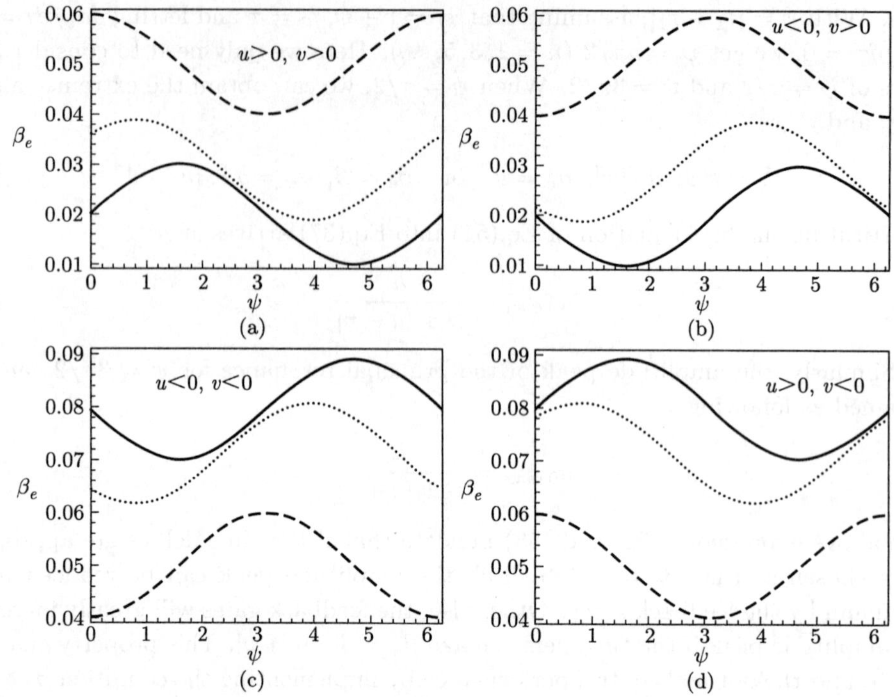

Fig. 5. The variation of β_e with ψ for $\beta = 0.05$, $|u| = 0.01$, $|v| = 0.03$ (thick curves: $\tau = 0$, dotted curves: $\tau = \pi/4$, dash curves: $\tau = \pi/2$).

Similarly, the combinations of the feedback gains and time delays for the case $0 < \tau_1 < \tau_2$ are listed in Table 2, when $\psi = \pi/2$ or $\psi = 3\pi/2$. Figure 6 displays the variations of β_e with ψ in this case. Obviously, the control performance can be better than the optimal one of the velocity feedback only, if the displacement feedback gain u is positive.

Table 2. Different parameter combinations in the design of feedback control for the case of $\tau_1 < \tau_2$.

Displacement feedback gain u	Velocity feedback gain v	Relation between u and v	Time delay τ
$u > 0$	$v < 0$	$\lvert u\rvert > \lvert v\rvert$	$[0, \pi]$
$u < 0$	$v > 0$	$\lvert u\rvert > \lvert v\rvert$	$[\pi, 2\pi]$

(4) *Numerical Simulations*

This subsection gives the numerical simulations of system (24) by using the fourth-order Runge-Kutta algorithm for external random excitation. Equation (25) can be rewritten as the following form[99]

$$\xi(t) = h\cos\phi(t), \quad \dot{\phi}(t) = \Omega + \bar{\gamma}\theta(t), \quad \theta(t) = \dot{W}(t). \tag{54}$$

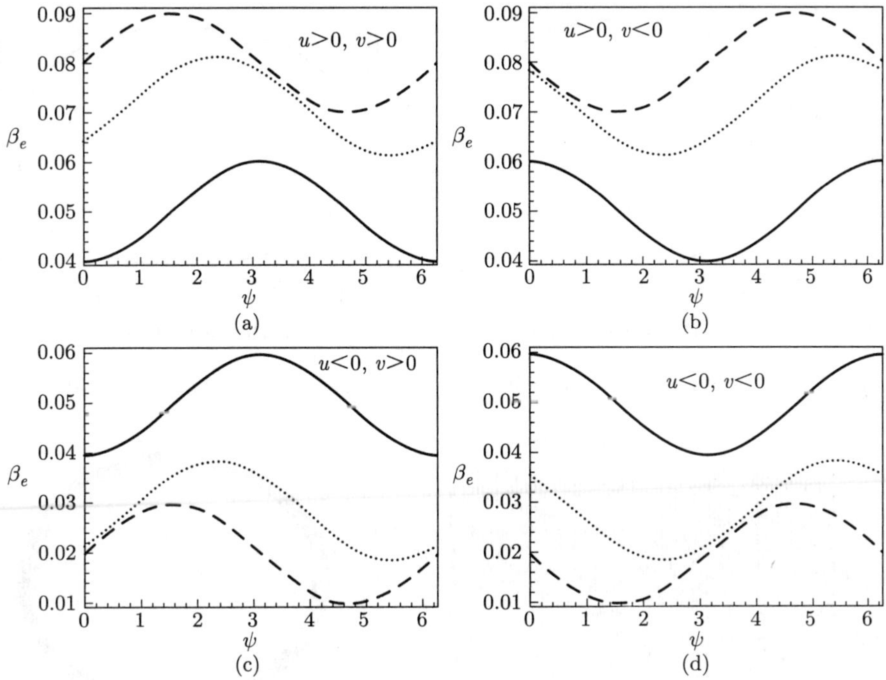

Fig. 6. The variation of β_e with ψ for $\beta = 0.05$ $|u| = 0.03$, $|v| = 0.01$ (thick curves: $\tau = 0$, dotted curves: $\tau = \pi/4$, dash curves: $\tau = \pi/2$).

The formal derivation $\theta(t)$ of the unit Wiener process is a Gaussian white noise, which has the power spectrum of a constant and is unrealized. Therefore, it is more convenient to use the pseudorandom signal to model[100] $\theta(t)$

$$\theta(t) = \sqrt{\frac{4\omega}{N}} \sum_{k=1}^{N} \cos\left(\frac{2k-1}{N}\omega t + \phi_k\right), \tag{55}$$

where ϕ_k are independent and uniformly distributed in $(0, 2\pi]$, N is a large positive integer.

In the numerical simulation, the parameters of system (24) are chosen as follows: $N = 1000$, $\varepsilon = 0.1$, $\beta = 1$, $\omega = 1$, $\alpha = 0.5$, $h = 5$, $u = 0.2$, $v = -0.2$, $\tau_1 = \pi/2$, $\tau_2 = \pi$, $\gamma = 0.1$. The numerical results with different initial values $x(t) = 4$, $\dot{x}(t) = -4$, $t \in [-\tau, 0]$ and $x(t) = -3.5$, $\dot{x}(t) = 4$, $t \in [-\tau, 0]$ and different noise intensities are shown in Figs.7 and 8. It is observed that the steady-state solution of Eq.(24) may converge to the smaller steady-state solution or the larger one as shown in Fig.7, depending on its initial condition. Meanwhile, in Fig.8, the steady-state solution is changed from a limit cycle to a diffused limit cycle when noise (25) is added.

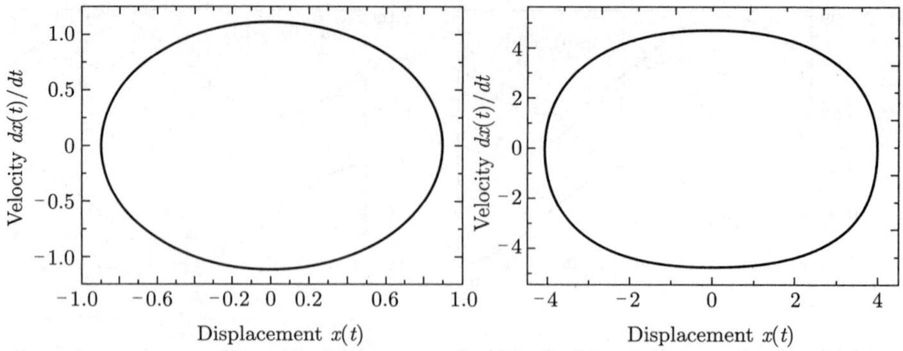

Fig. 7. Phase trajectory with different initial values for $\sigma = 0.25$ and $\gamma = 0$.

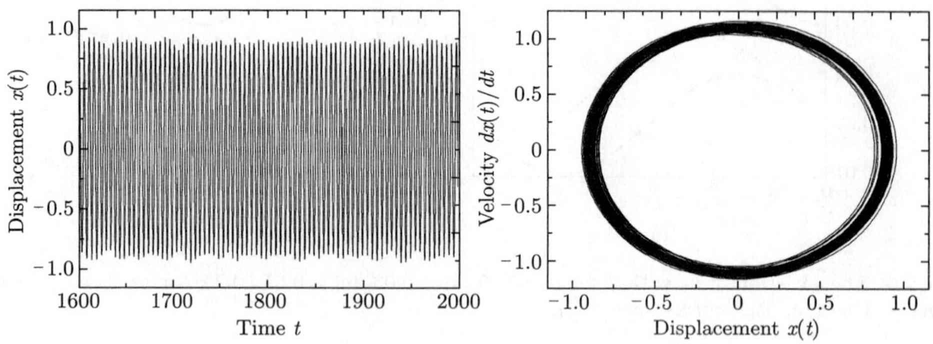

Fig. 8. Time history and phase trajectory with $\sigma = 0.25$ and $\gamma = 0.1$.

3.1.2. *Narrow-band Random Parametric Excitation*

When $g(x) = x$, equation (24) describes a delayed Duffing oscillator subject to a random parametric excitation. When $\tau_1 = \tau_2$, we[101] studied the principal resonance of Eq.(24) and introduced the detuning frequency σ to express the principal resonance as

$$\Omega = 2\omega + \varepsilon\sigma. \tag{56}$$

The elimination of the secular term in Eq.(29) requires

$$\begin{aligned} &-2i\omega A' - i\omega\beta A + uA\exp(-i\omega\tau) + i\omega vA\exp(-i\omega\tau) \\ &-3\alpha A^2\bar{A} + \frac{h\bar{A}}{2}\exp[i\sigma T_1 + i\gamma W(T_1)] = 0. \end{aligned} \tag{57}$$

Substituting Eq.(30) into Eq.(57) and separating the real and imaginary parts leads to

$$\begin{cases} a' = -\dfrac{\beta a}{2} - \dfrac{ua}{2\omega}\sin\omega\tau + \dfrac{va}{2}\cos\omega\tau + \dfrac{ha}{4\omega}\sin\eta, \\ a\eta' = \sigma a - \dfrac{3\alpha a^3}{\omega} + \dfrac{ua}{\omega}\cos\omega\tau + va\sin\omega\tau + \dfrac{ha}{2\omega}\cos\eta + \gamma a W'(T_1), \end{cases} \tag{58}$$

where $\eta(T_1) = \sigma T_1 + \gamma W(T_1) - 2\varphi(T_1)$. With the amplitude a and the phase η determined by Eq.(58), the first-order uniform expansion of the solution can be expressed in the same form as Eq.(34).

(1) *Trivial Steady-state Solution and Stability Analysis*

In order to analyze the stability of the trivial steady-state solution $a = 0$ of Eq.(58), we rewrite the linearized equation (58) at (0, 0) as the following Itô SDE

$$\begin{cases} d\rho = \left(-\dfrac{\beta}{2} - \dfrac{u}{2\omega}\sin\omega\tau + \dfrac{v}{2}\cos\omega\tau + \dfrac{h}{4\omega}\sin\eta\right) dT_1, \\ d\eta = \left(\sigma + \dfrac{u}{\omega}\cos\omega\tau + v\sin\omega\tau + \dfrac{h}{2\omega}\cos\eta\right) dT_1 + \gamma dW, \end{cases} \tag{59}$$

where $\rho = \ln a$ is assumed. Moreover, with $\bar{h} = -h/(2\omega\gamma^2)$ and $\bar{\sigma} = (\sigma + u\cos\omega\tau/\omega + v\sin\omega\tau)/\gamma^2$, the corresponding FPE for the steady-state probability density $p(\eta)$ of $\eta(T_1)$ is given as follows:

$$\frac{d^2p}{d\eta^2} - \frac{d}{d\eta}[(\bar{\sigma} - \bar{h}\cos\eta)p] = 0. \tag{60}$$

Considering the periodicity condition and the normality condition, we find that

$$p(\eta) = \frac{\exp[\bar{\sigma}(\eta + 3\pi/2) - \bar{h}\sin\eta]}{4\pi^2 I_{i\bar{\sigma}}(\bar{h}) I_{-i\bar{\sigma}}(\bar{h})} \int_{\eta}^{\eta+2\pi} \exp[-\bar{\sigma}(y + \pi/2) + \bar{h}\sin y]\, dy, \tag{61}$$

where $I_n(x)$ is the Bessel function of the first kind and n can be any real and complex number.

According to Eq.(61) and the formulas given by Gradshteyn and Ryzhik[102], the largest Lyapunov exponent $\lambda = \lambda_{\max}$ can be derived to determine the almost certain stability of the trivial steady-state solution. That is,

$$\begin{aligned} \lambda &= \lim_{T_1\to\infty} \frac{1}{T_1} \ln\left|\frac{a(T_1)}{a(0)}\right| = \lim_{T_1\to\infty} \frac{1}{T_1}(\rho(T_1) - \rho(0)) \\ &= -\frac{\beta_e}{2} + \frac{h}{8\omega}\left\{\frac{I_{1-i\bar{\sigma}}(-\bar{h})}{I_{-i\bar{\sigma}}(-\bar{h})} + \frac{I_{1+i\bar{\sigma}}(-\bar{h})}{I_{i\bar{\sigma}}(-\bar{h})}\right\}, \end{aligned} \tag{62}$$

where β_e and $I_n(x)$ are defined as above.

To check the above theoretical results, we turn to a case study, where the parameters of Eq.(62) are chosen as $\beta = 0.2$, $\alpha = 0.1$, $\omega = 1$, $h = 0.5$, $\gamma = \bar{\gamma}/\sqrt{\varepsilon} = 0.5$, $\sigma = 0.5$, $u = 0.2$, $v = -0.2$, and $\tau = \pi/2$, respectively. Figures 9–11 show the variation of the largest Lyapunov exponent λ for different parameters. In Fig.9(a), there are two different solution ranges for λ. Near $\sigma = 0$, corresponding to the parameter resonance at excitation frequency $\Omega = 2\omega$, λ reaches the maximum in the centre of the instability area. Beside the mountain, there are two approximate plane ranges where the variations of λ for σ and h are quite small. Meanwhile, λ increases with an increase of h, which is consistent with the physical intuition. Figure 9(b) illustrates the corresponding contour curves of λ. In Fig.10(a), λ decreases with an increase of the displacement feedback u. λ reaches the maximum when $v = -0.5$. When $\lambda = 0$, Figure 10(b) shows the corresponding $u - v$ curve, which is the stability boundary of the trivial steady-state solution. In Fig.11(a), there are two different resonance ranges for λ. Near $\tau = 4$ and $\gamma = 0.4$, λ reaches the peak values, respectively. Figure 11(b) shows the $\lambda - \gamma$ curve with different values of τ. Here, λ is a non-monotonic function of γ. It is clear that λ increases with an increase of τ for the case of $\tau < 4$.

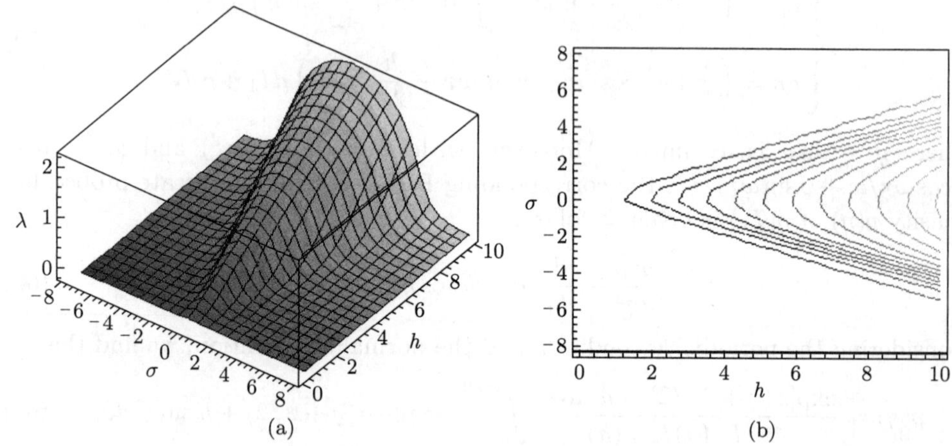

Fig. 9. The largest Lyapunov exponent of the trivial solution: (a) mesh surface of λ-(σ, h); (b) contour curves of λ.

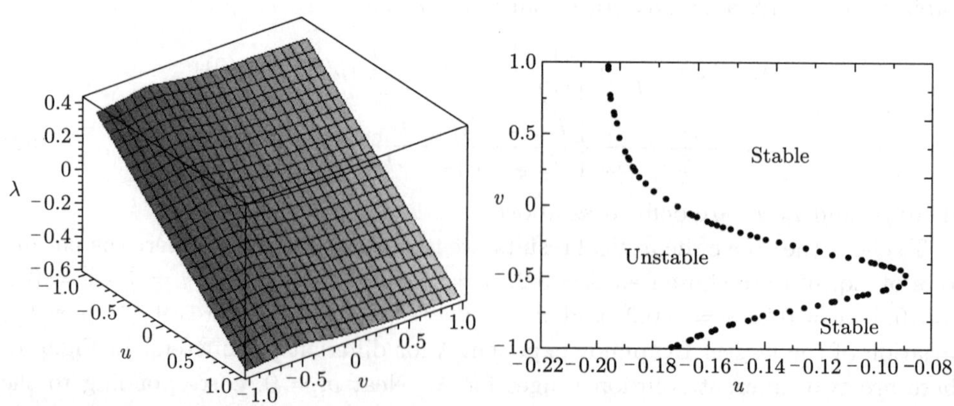

Fig. 10. The largest Lyapunov exponent of the trivial solution: (a) mesh surface of λ-(u, v); (b) stability boundary of the trivial solution.

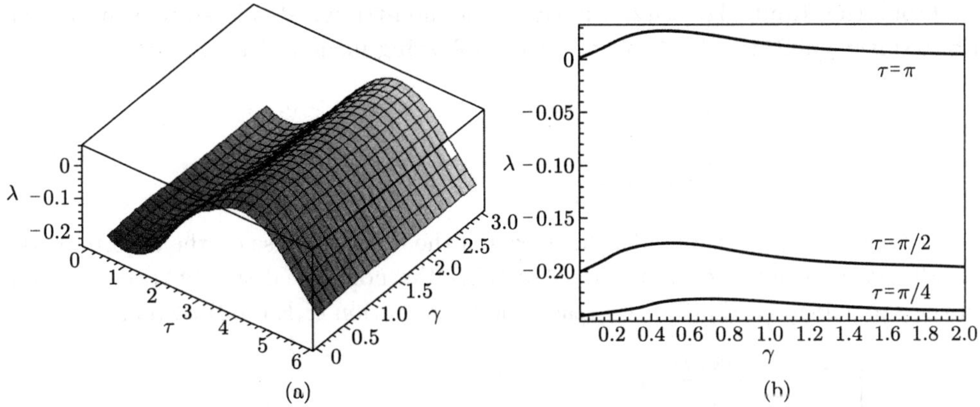

Fig. 11. The largest Lyapunov exponent of the trivial solution: (a) mesh surface of λ-(τ,γ); (b) λ versus γ with different τ.

(2) *Non-trivial Steady-state Solution and Stability Analysis*

The Itô SDE of the non-trivial steady-state solution can be found to be

$$\begin{cases} da = \left(-\dfrac{\beta a}{2} - \dfrac{ua}{2\omega}\sin\omega\tau + \dfrac{va}{2}\cos\omega\tau + \dfrac{ha}{4\omega}\sin\eta\right)dT_1, \\ d\eta = \left(\sigma - \dfrac{3\alpha a^2}{\omega} + \dfrac{u}{\omega}\cos\omega\tau + v\sin\omega\tau + \dfrac{h}{2\omega}\cos\eta\right)dT_1 + \gamma dW(T_1). \end{cases} \tag{63}$$

Firstly, the non-trivial steady-state solution $a = a_0$ and $\eta = \eta_0$ of Eq.(63) for $\gamma = 0$ is determined by the following deterministic equations

$$\begin{cases} a' = -\dfrac{\beta a}{2} - \dfrac{ua}{2\omega}\sin\omega\tau + \dfrac{va}{2}\cos\omega\tau + \dfrac{ha}{4\omega}\sin\eta, \\ \eta' = \sigma - \dfrac{3\alpha a^2}{\omega} + \dfrac{u}{\omega}\cos\omega\tau + v\sin\omega\tau + \dfrac{h}{2\omega}\cos\eta. \end{cases} \tag{64}$$

Under the conditions $a' = 0$, $\eta' = 0$, and $a \neq 0$, equation (64) gives the following relations

$$\begin{cases} a_0 = \dfrac{1}{\sqrt{3\alpha}}\left[\omega\left(\sigma + \dfrac{u\cos(\omega\tau)}{\omega} + v\sin(\omega\tau)\right) \pm \sqrt{\dfrac{h^2}{4} - \omega^2\beta_e^2}\right]^{1/2}, \\ \eta_0 = \arcsin\left(\dfrac{2}{h}\omega\beta_e\right). \end{cases} \tag{65}$$

The existence of the non-trivial steady-state solution is governed by Eq.(65) with $h > 2\omega\beta_e$. When $\alpha > 0$ and $\sigma < \sqrt{(h/2\omega)^2 - \beta_e^2} - u\cos(\omega\tau)/\omega - v\sin(\omega\tau)$ hold true, there is a trivial steady-state solution, together with a non-trivial steady-state solution whose stability needs further study. When $\alpha > 0$ and $\sigma > \sqrt{(h/2\omega)^2 - \beta_e^2} - u\cos(\omega\tau)/\omega - v\sin(\omega\tau)$ hold true, there is a trivial steady-state solution, together with two non-trivial steady-state solutions which may be stable or not.

From the Routh-Hurwitz criterion, the non-trivial steady-state solution of Eq.(64) is asymptotically stable when the following inequalities satisfy

$$\begin{cases} \beta_e = \beta + u\sin\omega\tau/\omega - v\cos\omega\tau > 0, \\ \cos\eta_0 = \dfrac{1}{h}\sqrt{h^2 - 4\omega^2\beta_e^2} > 0 \end{cases} \tag{66}$$

Secondly, when $\gamma \neq 0$ holds, the effects of the random noise on the deterministic steady-state motion are discussed. Assuming the non-trivial steady-state solution in the same form as Eq.(35), we reach the linearization of Eq.(63) at (a_0, η_0)

$$\begin{cases} a_1' = \dfrac{ha_0\cos\eta_0}{4\omega}\eta_1, \\ \eta_1' = -\dfrac{6\alpha a_0}{\omega}a_1 - \left(\beta + \dfrac{u}{\omega}\sin(\omega\tau) - v\cos(\omega\tau)\right)\eta_1 + \gamma W'(T_1). \end{cases} \tag{67}$$

From Eq.(67), the following equation holds

$$a_1'' + \left(\beta + \frac{u\sin(\omega\tau)}{\omega} - v\cos(\omega\tau)\right)a_1' + \frac{3\alpha h a_0^2\cos\eta_0}{2\omega^2}a_1 = \frac{h\gamma a_0\cos\eta_0}{4\omega}\zeta(T_1), \tag{68}$$

where $\zeta(T_1)$ is a standard Gussian white noise process.

For the Gaussian process (a_1, a_1'), the method of generalized stationary potential can be used to obtain the steady-state probability density

$$p(a_1, a_1') = C\exp\left\{-\frac{16\omega^2\bar{\beta}}{(\gamma h a_0\cos\eta_0)^2}\left[\frac{3\alpha h a_0^2\cos\eta_0}{2\omega^2}a_1^2 + (a_1')^2\right]\right\}, \tag{69}$$

where C is the normalization constant. It is clear that a_1 and a_1' are independent of each other according to Eq.(69). The steady-state probability density for a_1 reads

$$p(a_1) = C_0\exp\left\{-\frac{24\alpha(\beta + u\sin(\omega\tau)/\omega - v\cos(\omega\tau))}{\gamma^2\sqrt{h^2 - 4(\omega\beta + u\sin(\omega\tau) - v\omega\cos(\omega\tau))^2}}a_1^2\right\}, \tag{70}$$

where the normalization constant C_0 is determined by Eq.(70). The second-order moment of a_1 obtained from Eq.(70) is given by

$$Ea_1^2 = \int_0^\infty a_1^2 p(a_1)da_1 = \frac{\gamma^2\sqrt{h^2 - 4\omega^2\beta_e^2}}{48\alpha\beta_e}. \tag{71}$$

Taking the expectation on both sides of Eq.(35), we obtain the first-order and second-order steady-state moments of the non-trivial steady-state solution

$$Ea = E(a_0 + a_1) = a_0, \quad Ea^2 = a_0^2 + \frac{\gamma^2\sqrt{h^2 - 4\omega^2\beta_e^2}}{48\alpha\beta_e}. \tag{72}$$

In Fig.12, the bifurcations and jumps of the second-order steady-state moment (72) are presented. In Fig.12(a), σ is chosen to be a bifurcation parameter. The critical values σ_1' and σ_2', which form the jump zone, can be derived as follows

$$\sigma_1' = -\frac{\sqrt{h^2 - 4\omega^2\beta_e^2}}{2\omega} - \frac{u}{\omega}\cos(\omega\tau) - v\sin(\omega\tau), \tag{73}$$

$$\sigma_2' = \frac{(-\gamma^2 + 8\beta_e)\sqrt{h^2 - 4\omega^2\beta_e^2}}{16\omega\beta_e} - \frac{u}{\omega}\cos(\omega\tau) - v\sin(\omega\tau). \tag{74}$$

The second-order steady-state moment will jump at any point within the interval $[\sigma_1', \sigma_2']$ and the jumps may occur back and forth.

When σ is fixed, h is chosen as a bifurcation parameter. The variation of the second-order steady-state moment with h is shown in Fig. 12(b). The critical values of h can be determined as follows

$$h_1' = \frac{2\omega\beta_e\sqrt{64(\sigma + u\cos(\omega\tau)/\omega + v\sin(\omega\tau))^2 + (8\beta_e + \gamma^2)^2}}{8\beta_e + \gamma^2}, \tag{75}$$

$$h_2' = 2\omega\beta_e. \tag{76}$$

Obviously, as the bandwidth of the excitation increases, the jump zone decreases as shown in Fig.12. When the value of γ is greater than the upper bound, the two level responses merge into a single level one and no jump occurs. This is a feature of jump phenomena under random excitations[103].

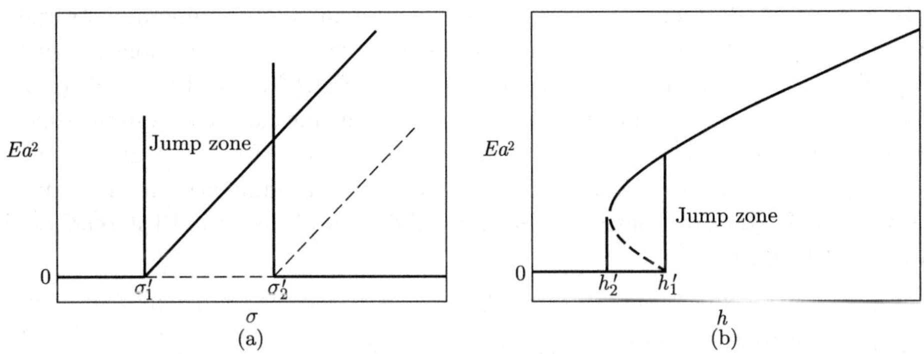

Fig. 12. The second-order steady-state moment for $\gamma \neq 0$: (a) as a function of σ; (b) as a function of h.

(3) *Numerical Simulations*

To verify the approximate results obtained above, we present the numerical solutions of Eq.(24) for the case of parametric random excitation by using the fourth-order Runge-Kutta algorithm and Eqs. (54) and (55). The parameters are chosen as $N = 1000$, $\varepsilon = 0.1$, $\beta = 0.2$, $\alpha = 0.1$ and $\omega = 1$. To demonstrate the stability conditions in Figs.9–11, Figures 13–15 show the time histories and power spectrums of the displacement of Eq.(24), starting from $x(t) = \cos(\Omega t)$, $t \in [-\tau, 0]$, $\Omega = 2.05$, corresponding to different parametric combinations.

The results in Fig.13 indicate that a change in the frequency of the narrow-band random excitation can cause a qualitative change in the response. When $\sigma = 0$, the response of system under any random excitation is oscillatory and becomes unstable as shown in Fig.13(a). When $\sigma = 1.5$, the response of system consists of two terms having two different frequencies. The response tends to the stable trivial solution after the transient response is neglected as shown in Fig.13(b).

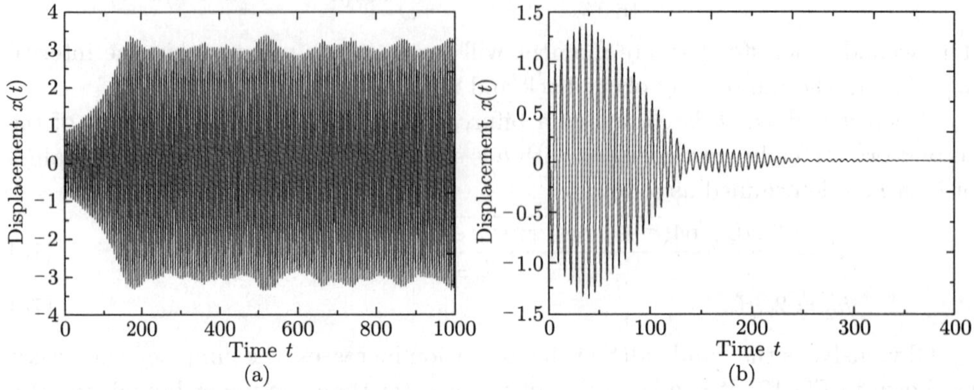

Fig. 13. Numerical results of Eq.(24): $h = 2$, $u = 0.2$, $v = -0.2$, $\tau = \pi/2, \gamma = 0.5$. (a) Time history for $\sigma = 0$; (b) time history for $\sigma = 1.5$.

In Fig.14, the different values of v make the trivial solution unstable or stable. When (u, v) falls into the unstable region shown in Fig.10(b), the response of system is unstable under random perturbation as shown in Fig.14(a). In Fig.14(b), (u, v) is chosen from the stable region shown in Fig.10(b). In this case, the system response indicates the random vibration from initial disturbance. After the transient stage, however, the random excitation does not have a strong influence on the resulting motion. The system response will oscillate slightly and decay until it reaches the stable trivial solution.

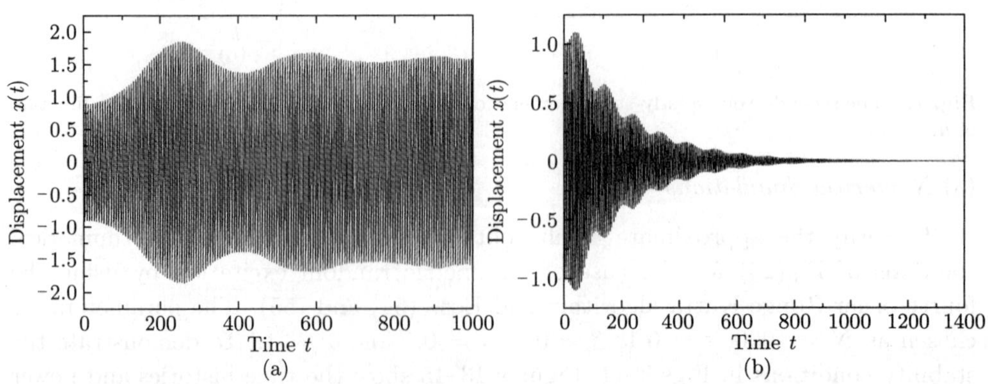

Fig. 14. Numerical results of Eq.(24): $h = 0.5$, $\sigma = 0.5$, $u = -0.1$, $\tau = \pi/2$, $\gamma = 0.5$. (a) Time history for $v = -0.5$; (b) time history for $v = 0.2$.

Figure 15 shows the effects of τ on the stability of the trivial solution. When

$\tau = \pi$, the trivial solution is unstable as shown in Fig.15(a). When $\tau = \pi/2$, the response of system decays to the stable trivial solution rapidly and looks like the free vibration of a deterministic oscillator. This behavior can be explained as follows: when the initial amplitude of response is small and very close to the stable trivial solution, the initial random disturbance does not contribute to the system motion. It is clear that the theoretical results in Figs. 9–11 get an agreement with the numerical results as shown in Figs.13–15.

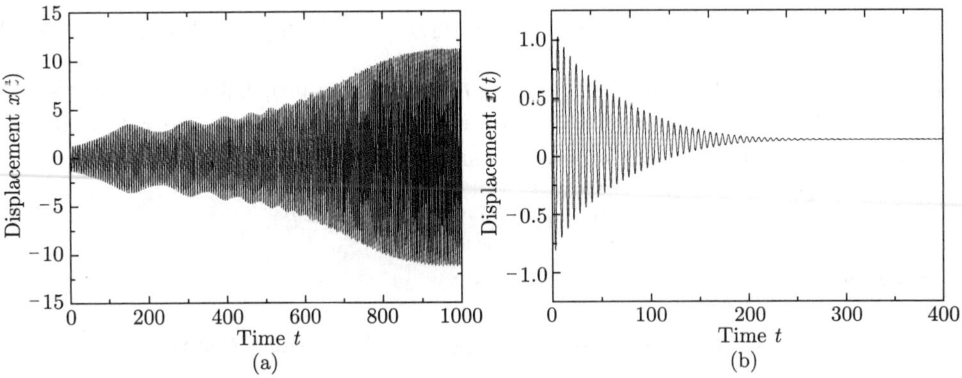

Fig. 15. Numerical results of Eq.(24): $h = 0.5$, $\sigma = 0.5$, $u = 0.2$, $v = -0.2$, $\gamma = 0.5$. (a) Time history for $\tau = \pi$; (b) time history for $\tau = \pi/2$.

Figure 16 gives the effects of noise intensity on the system response through numerical results. The noise may change the steady-state response from a periodic solution to an aperiodic one. And the corresponding phase trajectory can be changed from a limit cycle to a diffused limit cycle.

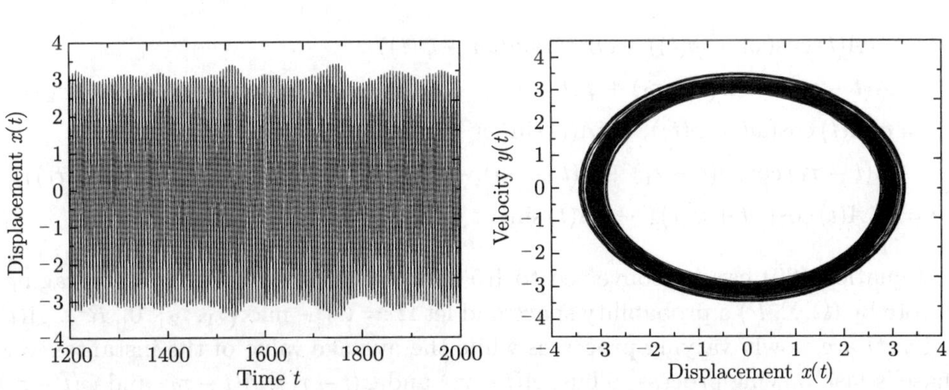

Fig. 16. Numerical results of Eq.(24): $\gamma = 0.5$. and $h = 1.5$. (a) time history; (b) phase trajectory.

3.2. *Moment Stability of Stochastic Systems with Delayed Feedback Control*

In this subsection, the asymptotic moment stability for the generalized delay system subject to external or parametric Gaussian white noise is investigated by using the stochastic averaging method presented in Section 2.2. Consider the governing equation

$$\begin{aligned}&\ddot{x}(t)+\omega^2 x(t)+\varepsilon g(x(t),\dot{x}(t),x(t-\tau_1),\dot{x}(t-\tau_1))\\&=\varepsilon f(x(t),\dot{x}(t),x(t-\tau_2),\dot{x}(t-\tau_2))+\varepsilon^{1/2}h(x(t),\dot{x}(t))\xi(t),\end{aligned} \tag{77}$$

where $0<\varepsilon\ll 1$ is the small parameter, $f(0,0,0,0)=g(0,0,0,0)=0$, $x\in\mathbb{R}$, εf is the control force, $\tau_1\geq 0$ and $\tau_2\geq 0$ are the time delays in the control plant and the controller, respectively. $\xi(t)$ is a Gaussian white noise with

$$\langle\xi(t)\rangle=0,\quad \langle\xi(t)\xi(t')\rangle=2D\delta(t-t'). \tag{78}$$

Now, let

$$x(t)=A(t)\cos\theta,\quad \theta=\omega t+\varphi(t),\quad \dot{x}(t)=-\omega A(t)\sin\theta, \tag{79}$$

and substitute Eq.(79) into Eq.(77), we have

$$\begin{cases}\dot{A}(t)=\dfrac{\varepsilon(g-f)\sin\theta(t)}{\omega}-\dfrac{\varepsilon^{1/2}h\sin\theta(t)\xi(t)}{\omega},\\[2ex]\dot{\varphi}(t)=\dfrac{\varepsilon(g-f)\cos\theta(t)}{A(t)\omega}-\dfrac{\varepsilon^{1/2}h\cos\theta(t)\xi(t)}{A(t)\omega},\end{cases} \tag{80}$$

where

$$\begin{aligned}f:=\,&f(A(t)\cos(\omega t+\varphi(t)),-\omega A(t)\sin(\omega t+\varphi(t)),\\&A(t-\tau_2)\cos(\omega(t-\tau_2)+\varphi(t-\tau_2)),-\omega A(t-\tau_2)\sin(\omega(t-\tau_2)+\varphi(t-\tau_2))),\\g:=\,&g(A(t)\cos(\omega t+\varphi(t)),-\omega A(t)\sin(\omega t+\varphi(t)),\\&A(t-\tau_1)\cos(\omega(t-\tau_1)+\varphi(t-\tau_1)),-\omega A(t-\tau_1)\sin(\omega(t-\tau_1)+\varphi(t-\tau_1))),\\h:=\,&h(A(t)\cos(\omega t+\varphi(t)),-\omega A(t)\sin(\omega t+\varphi(t))).\end{aligned}$$

Equation (80) can be converted to Itô SDE by means of stochastic averaging. Denote by (Ω,Σ,P) a probability space and let $\Omega:=C([-\max\{\tau_1,\tau_2\},0],R^2)$. $A(t)$ and $\varphi(t)$ are slowly varying processes while the average value of the instantaneous phase is fast varying process. Thus, $A(t-\tau_1)$ and $\varphi(t-\tau_1)$, $A(t-\tau_2)$ and $\varphi(t-\tau_2)$ can be replaced by $A(t)$ and $\varphi(t)$ over one period $T=2\pi/\omega$, respectively. And we obtain the Itô equation of amplitude as

$$dA=\bar{a}_1(A)dt+(\bar{b}_{11}(A))^{1/2}dB_1(t), \tag{81}$$

where

$$\bar{a}_1(A) = \frac{1}{2\pi}\int_0^{2\pi} \frac{\varepsilon(g_1 - f_1)\sin\theta}{\omega} + \varepsilon D\left[-\frac{h\sin\theta}{\omega}\cdot\frac{\partial h}{\partial \dot{x}} + \frac{h^2\cos^2\theta}{A\omega^2}\right] d\theta,$$

$$\bar{b}_{11}(A) = \frac{\varepsilon D}{\pi\omega^2}\int_0^{2\pi} h^2 \sin^2\theta d\theta,$$

$$f_1 := f(A\cos\theta, -\omega A\sin\theta, A\cos(\theta - \omega\tau_2), -\omega A\sin(\theta - \omega\tau_2)),$$

$$g_1 := g(A\cos\theta, -\omega A\sin\theta, A\cos(\theta - \omega\tau_1), -\omega A\sin(\theta - \omega\tau_1)).$$

From Eq.(81), $A(t)$ is approximately a scalar Markov process. And the corresponding FPE reads

$$\frac{\partial p}{\partial t} = -\frac{\partial}{\partial\alpha}[a_1(\alpha)p] + \frac{1}{2}\frac{\partial^2}{\partial\alpha^2}[\bar{b}_{11}(\alpha)p], \tag{82}$$

where α is the state variable for $A(t)$, and the initial condition of Eq.(82) is $p(\alpha, t|\alpha_0, t_0)$. The stationary probability density $p_s(\alpha)$ can be obtained from solving Eq.(82) if it exists. Then, the nth moment of $A(t)$ is given by

$$EA^n = \int_0^\infty \alpha^n p_s(\alpha)d\alpha, \quad n = 1, 2, \ldots. \tag{83}$$

The stability analysis is presented for the following two cases through two examples as described next.

3.2.1. *External Gaussian White Noise*

Consider a delayed van der Pol oscillator with linear delayed feedback control and external Gaussian white noise

$$\ddot{x}(t) + \omega^2 x(t) - \varepsilon[1 - x^2(t - \tau_1)]\dot{x}(t) = \varepsilon[ux(t - \tau_2) + v\dot{x}(t - \tau_2)] + \varepsilon^{1/2}\xi(t). \tag{84}$$

Upon performing the stochastic averaging, the drift and diffusion coefficients for $A(t)$ are obtained as

$$\begin{aligned}\bar{a}_1(A) &= \frac{\varepsilon}{8A\omega^2}\left[-A^4\omega^2(2 - \cos 2\omega\tau_1)\right.\\ &\quad \left.+4A^2\omega(\omega - u\sin\omega\tau_2 + \omega v\cos\omega\tau_2) + 4D\right],\\ \bar{b}_{11}(A) &= \frac{\varepsilon D}{\omega^2}.\end{aligned} \tag{85}$$

The first-order and second-order moments can be derived from Eqs. (82), (83), and (85) as follows:

$$EA = -\frac{\sqrt{2}w_2 C}{\omega w_1}\cdot\left(\frac{w_2^2}{\omega w_1^2}\right)^{1/4}\exp\left(\frac{w_2^2}{2Dw_1}\right) \cdot\left[\text{BesselK}\left(\frac{3}{4}, \frac{w_2^2}{2Dw_1}\right) - \text{BesselK}\left(\frac{1}{4}, \frac{w_2^2}{2Dw_1}\right)\right], \tag{86}$$

$$EA^2 = \frac{4\sqrt{D}C}{\omega^2 w_1^{3/2}}\cdot\left[\sqrt{w_1 D} + \sqrt{\pi}w_2\exp\left(\frac{w_2^2}{Dw_1}\right)\cdot\left(1 + \text{erf}\left(\frac{w_2}{\sqrt{Dw_1}}\right)\right)\right], \tag{87}$$

where $C = \sqrt{\omega^2 w_1/\pi D}\left[1+\mathrm{erf}\left(w_2/\sqrt{w_1 D}\right)\right]^{-1}\exp\left(-w_2^2/w_1 D\right)$, the error function is defined by $\mathrm{erf}(x) = 2\int_0^x \exp(-t^2)dt/\sqrt{\pi}$, $\mathrm{BesselK}(n,x)$ is the modified Bessel functions of the second kind.

From Eqs. (86) and (87), the stability of the first-order and second-order moments require that $w_2 < 0$. That is,

$$\omega - u\sin\omega\tau_2 + \omega v\cos\omega\tau_2 < 0. \tag{88}$$

This condition is the same as that presented in Ref.104 for the determined system. To demonstrate the validity of the theoretical results (88), we present some case studies for the parameters $\omega = 1$, $\varepsilon = 0.1$, $D = 0.05$, $\tau_1 = 0.1$, $u = 0$, $\tau_2 \in [0, 0.2]$. When $v < -1.0203$, the condition (88) holds true such that the trivial solution is stable. Otherwise, the trivial solution is unstable. When $v = -0.9$, $\tau_2 = 0.05$ and $v = -0.9$, $\tau_2 = 0.7$ are selected, corresponding to the cases of unstable solutions, the corresponding time histories are plotted in Fig.17. When $v = -1.4$, $\tau_2 = 0.1$ and $v = -1.4$, $\tau_2 = 0.2$ are chose, the time histories corresponding to the cases of stable solutions are shown in Fig.18. It is observed that the theoretical result is consistent with the numerical one.

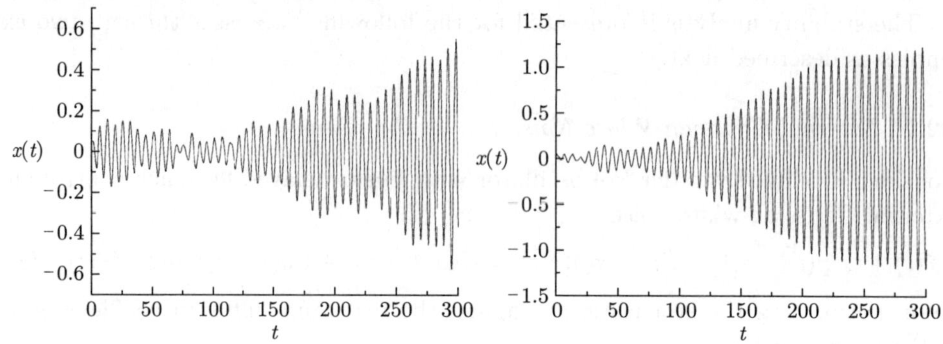

Fig. 17. Time history of Eq.(84) for $v = -0.9$, $\tau_2 = 0.05$ and 0.7.

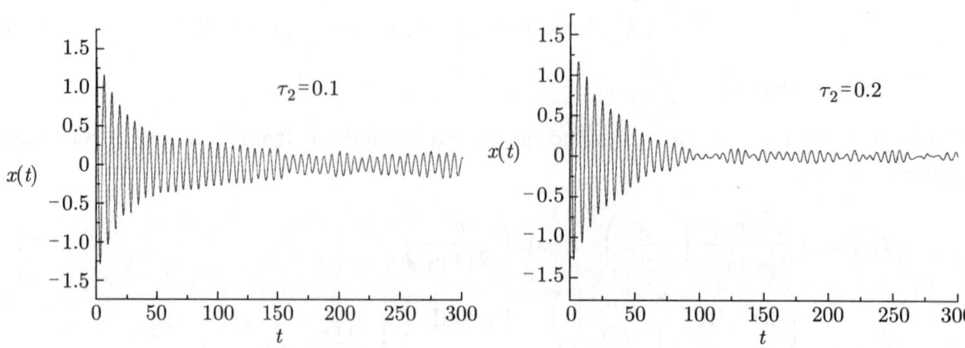

Fig. 18. Time history of Eq.(84) for $v = -1.4$, $\tau_2 = 0.1$ and 0.2.

3.2.2. *Parametric Gaussian White Noise*

Consider a delayed linear system under a parametric Gaussian white noise excitation, governed by

$$\ddot{x}(t) + 2\varepsilon\zeta\dot{x}(t) + \omega^2 x(t) = \varepsilon u x(t-\tau) + \varepsilon v \dot{x}(t-\tau) + \varepsilon^{1/2} x(t)\xi(t), \tag{89}$$

where $\zeta \geq 0$ is the damping ratio and the other system parameters are defined as above.

The stochastic averaging procedure can be used to obtain the averaged amplitude process of $A(t)$

$$dA = \varepsilon\left[-\frac{2\zeta + u\sin\omega_0\tau/\omega_0 - v\cos\omega_0\tau}{2} + \frac{3D}{16\omega^2}\right]Adt + \left[\varepsilon\frac{D}{8\omega^2}\right]^{1/2} AdB(t). \tag{90}$$

In order to find the nth moment equation of $A(t)$, we use the Itô differential rule to the function $F(A) = A^n$. The averaged Itô equation of $A^n(t)$ is

$$dA^n = \varepsilon n\left[-\frac{2\zeta + u\sin\omega_0\tau/\omega_0 - v\cos\omega_0\tau}{2} + \frac{(n+2)D}{16\omega^2}\right]A^n dt + n\left[\varepsilon\frac{D}{8\omega^2}\right]^{1/2} A^n dB(t). \tag{91}$$

Taking the ensemble average of Eq.(91), the equation of the nth moment is then obtained as

$$\frac{dEA^n}{dt} = \varepsilon n\left[-\frac{2\zeta + u\sin\omega_0\tau/\omega_0 - v\cos\omega_0\tau}{2} + \frac{(n+2)D}{16\omega^2}\right]EA^n. \tag{92}$$

From Eq.(92), the nth moment is asymptotic stable if the following inequality holds

$$\zeta > \frac{1}{2}\left[-\frac{u\sin\omega\tau}{\omega} + v\cos\omega\tau + \frac{(n+2)D}{8\omega^2}\right]. \tag{93}$$

Thus, the stability conditions for the first-order and second-order moments can be obtained by setting $n = 1$ and 2 in Eq.(93). To verify the above theoretical results,

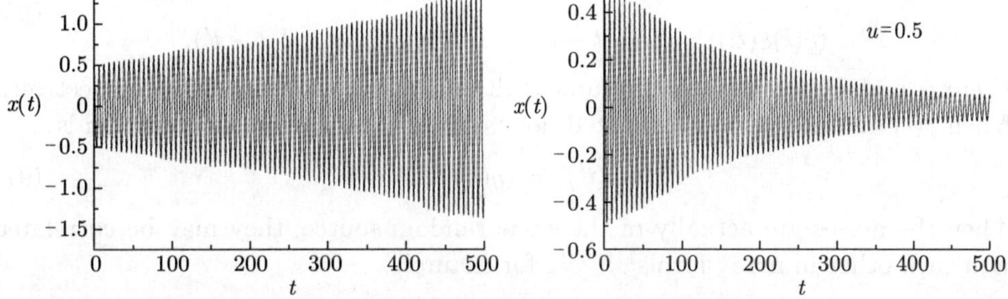

Fig. 19. Time history of Eq.(89) for different u.

we turn to a case study for $\omega = 1$, $\varepsilon = 0.1$, $\zeta = 0.2$, $\tau = 0.3$, $D = 0.05$, and $v = 0.5$. When $u > 0.347$, the condition (92) is satisfied. In Fig.19, the plots of Eq.(89) are given for the cases of $u = 0.1$ and $u = 0.5$, respectively. It is clear that the theoretical results coincide with the numerical results very well.

4. Noise-induced Resonances in Delayed Bistable Systems

Noise-induced dynamical behaviors, such as coherence resonance (CR) and stochastic resonance (SR), have attracted considerable attention over the last decades, see for example Ref.105 and references therein. SR proposed by Benzi et al.[106] and Nicolis et al.[107] is a phenomenon where the response of a dynamical system subject to an input is enhanced by the addition of an optimal amount of noise. Later, Fauve and Heslot[108] verified SR through an experiment of an ac-driven Schmitt trigger. This counterintuitive phenomenon has been applied to many fields including, but are not limited to, physics, neural network, laser, biology, and optics. CR occurs in an excitable system whose output can become quite regular at certain finite noise intensity. This effect is called as SR without external periodic excitation[109], internal[110] or autonomous SR[111], and CR[112]. In practice, time delays always exist and play an important role in many systems, such as physical and biological systems. This section focuses on the stochastic bistable system with linear time-delayed feedback, which is a prototype in describing the effects of noises and time delays. The two-state model approximation is extended to the delayed bistable system subjected to various noise sources[113,114]. The system of concern is an over-damped particle in a double-well quartic potential and with linear time-delayed feedback described by the following Langevin equation

$$\frac{dx(t)}{dt} = -\frac{\partial U(x(t), x(t-\tau))}{\partial x(t)} + x\xi(t) + \eta(t) + A_0 \cos \Omega t, \tag{94}$$

$$U(x(t), x(t-\tau)) = -\frac{x^2(t)}{2} + \frac{x^4(t)}{4} - \varepsilon x(t)x(t-\tau), \tag{95}$$

where τ is the time delay, ε is the feedback gain, A_0 and Ω are the amplitude and the frequency of the harmonic excitation, respectively. $\xi(t)$ and $\eta(t)$ are Gaussian white noises, and satisfy the following statistical properties

$$\begin{aligned} &\langle \xi(t) \rangle = \langle \eta(t) \rangle = 0, \\ &\langle \xi(t)\xi(t') \rangle = 2D\delta(t-t'), \quad \langle \eta(t)\eta(t') \rangle = 2\alpha\delta(t-t'), \end{aligned} \tag{96}$$

where α and D are the additive and multiplicative noise intensities, respectively. When $\xi(t)$ and $\eta(t)$ are uncorrelated noises, their cross-correlation function is

$$\langle \xi(t)\eta(t') \rangle = \langle \eta(t)\xi(t') \rangle = 0. \tag{97}$$

When the noises are actually of the same random source, they may be correlated with each other in many forms[115–118], for example,

$$\langle \xi(t)\eta(t') \rangle = \langle \eta(t)\xi(t') \rangle = 2\lambda\sqrt{\alpha D}\delta(t-t'), \quad |\lambda| < 1, \tag{98}$$

where λ is the correlation strength between the multiplicative and additive noises. By introducing a new noise term[119] $\zeta(t) = \eta(t) - \lambda\sqrt{\alpha/D}\xi(t)$, the new noise $\zeta(t)$ and $\xi(t)$ are no longer correlated. And $\zeta(t)$ has the following statistical properties

$$\begin{aligned} \langle\zeta(t)\rangle = 0, \quad \langle\zeta(t)\zeta(t')\rangle = 2\alpha(1-\lambda^2)\delta(t-t'), \\ \langle\xi(t)\zeta(t')\rangle = \langle\zeta(t)\xi(t')\rangle = 0. \end{aligned} \tag{99}$$

Then equation (94) can be transformed into the following form

$$\frac{dx(t)}{dt} = -\frac{\partial U(x(t), x(t-\tau))}{\partial x(t)} + (x - \lambda\sqrt{\alpha/D})\xi(t) + \zeta(t) + A_0 \cos\Omega t. \tag{100}$$

It is clear that equation (100) describes a delayed bistable system subject to uncorrelated additive and multiplicative noises and a periodic signal. In the following, the analytical expression of power spectrum, used as a measure of CR, is derived in Section 4.1. After this, Section 4.2 presents the spectrum amplification to measure the SR within a linear response approximation.

4.1. *Coherence Resonance*

When no harmonic excitation is applied to the system (94), i.e., $A_0 = 0$, equation (94) can be recast as

$$\frac{dx(t)}{dt} = -\frac{\partial U(x(t), x(t-\tau))}{\partial x(t)} + x\xi(t) + \eta(t). \tag{101}$$

The over-damped particle governed by Eq.(101) spends most of its time near the stable equilibriums $x = \pm 1$ if the delayed feedback is absent, i.e., $\varepsilon = 0$, and undergoes an occasional transition over the barrier in the center because of a moderate amount of random excitation. Hence, for small ε, D, and α, the intrawell motion can be neglected and equation (101) can be approximated with a two-state model with the dynamical variable $s(t) = \pm 1$, corresponding to $x > 0$ and $x < 0$, respectively. The delayed feedback is introduced by considering two different switching rates, depending on the state $s(t-\tau)$: p_+ if the state at the time $t-\tau$ is the same as that at the time t, and p_- otherwise.

Assume that $x_\pm = \pm\sqrt{1+\varepsilon}$ and $x_0 = 0$ are the stable and unstable equilibriums of Eq.(101). Under the adiabatic limit condition, then, we can find the transition rates out of $x_\pm$ as follows:

$$p_\pm = \frac{|U''(x_\pm)U''(x_0)|^{1/2}}{2\pi} \exp\left[\frac{\overline{U}(x_\pm) - \overline{U}(x_0)}{D}\right], \tag{102}$$

where the modified potential $\overline{U}(x(t), x(t-\tau))$ can be written in the form

$$\begin{aligned}&\overline{U}(x(t), x(t-\tau))\\&= \frac{x^2}{2} - 2\lambda\sqrt{R}x - \frac{[1-R(4\lambda^2-1)]}{2}\ln\left|D(x^2+2\lambda\sqrt{R}x+R)\right| \qquad (103)\\&+ \left\{\frac{\lambda}{\sqrt{1-\lambda^2}}\cdot[1+R(3-4\lambda^2)] - \varepsilon\frac{x(t-\tau)}{\sqrt{R(1-\lambda^2)}}\right\}\arctan\left[\frac{x+\lambda\sqrt{R}}{\sqrt{R(1-\lambda^2)}}\right].\end{aligned}$$

Here $R = \alpha/D$ is the noise intensity ratio. For a small ε, hence, it is easy to truncate Eq.(102) to the first order of ε as follows:

$$\begin{aligned}p_\pm = &\frac{\sqrt{2\pm3\varepsilon}}{2\pi}\exp\left\{-\frac{1}{D}\left[-\frac{1+\varepsilon}{2} \pm 2\lambda\sqrt{R(1+\varepsilon)}\right.\right. \qquad (104)\\&+\frac{(1-R(4\lambda^2-1))}{2}\ln\left|\frac{1+\varepsilon}{R}\pm2\lambda\sqrt{\frac{1+\varepsilon}{R}}+1\right|\\&-\frac{1}{\sqrt{1-\lambda^2}}(\lambda(1+R(3-4\lambda^2)))\left(\arctan\frac{\pm\sqrt{1+\varepsilon}+\lambda\sqrt{R}}{\sqrt{R(1-\lambda^2)}} - \arctan\frac{\lambda}{\sqrt{1-\lambda^2}}\right)\\&\left.\left.\mp\frac{\varepsilon\sqrt{1+\varepsilon}}{\sqrt{R}}\arctan\frac{\pm\sqrt{1+\varepsilon}+\lambda\sqrt{R}}{\sqrt{R(1-\lambda^2)}}\right)\right]\right\}.\end{aligned}$$

When $\lambda = 0$, equations (103) and (104) degenerate to the case of uncorrelated noises.

The master equation for the probabilities $n_\pm$ of attaining $x = \pm1$ at t is found to be

$$\dot{n}_+ = -\dot{n}_- = p_+(n_- - n_+) - (p_+ - p_-)\left[n_-(t-\tau)n_- - n_+(t-\tau)n_+\right]. \qquad (105)$$

The correlation function $C(t)$ can be expressed with the initial condition $s(0) = 1$ as

$$C(t) = \langle s(t)s(0)\rangle = \langle s(t)\rangle = n_+(t) - n_-(t). \qquad (106)$$

Therefore, the following differential equation of $C(t)$ can be obtained from Eqs. (105) and (106)

$$\frac{dC(t)}{dt} = -(p_+ + p_-)C(t) + (p_- - p_+)C(t-\tau). \qquad (107)$$

And equation (107) should be complemented with the symmetry condition $C(-t) = C(t)$ and the normalization condition $C(0) = 1$. Then, the solution of Eq.(107) on the interval $[0, \tau]$ can be assumed in the following form

$$C(t) = A\exp(-\rho|t|) + (1-A)\exp(\rho|t|), \qquad (108)$$

where

$$\rho = 2\sqrt{p_+ p_-}, \quad A = \frac{\sqrt{p_+} + \sqrt{p_-}}{\sqrt{p_+} + \sqrt{p_-} + (\sqrt{p_-} - \sqrt{p_+})e^{-\rho\tau}}. \tag{109}$$

When $t > \tau$, the correlation function $C(t)$ can be calculated from Eqs. (107)–(109) as follows

$$\begin{aligned} C(t) &= e^{-(p_+ + p_-)t} C(n\tau) \\ &+ (p_- - p_+) \cdot \int_0^t C((n-1)\tau + t') e^{(p_+ + p_-)(t'-t)} dt, \quad n = 0, 1, 2, \ldots . \end{aligned} \tag{110}$$

Performing the Fourier transform on Eq.(110) yields the power spectrum of system

$$S(\omega) = 2\mathrm{Re}\left(\int_0^\infty C(t) e^{i\omega t} dt\right). \tag{111}$$

Figure 20 shows the power spectrum for uncorrelated noises as a function in the additive noise intensity α, the frequency of oscillator ω, the time delay τ and the feedback gain ε. In Fig.20(a), the peaks at the frequency $\omega \approx 2n\pi/\tau$ attain the maximum for a certain amount of additive noise intensity. The peak height and width depend non-monotonously on α. This is a feature of the CR. In Fig.20(b), the peaks for the case of positive feedback gain appear at the frequency $\omega \approx 2n\pi/\tau$, while the peaks for the case of negative feedback gain appear at the frequency $\omega \approx (2n-1)\pi/\tau$. The peak values decrease with an increase in ω. Figure 20(c) shows the variations of $S(\omega)$ versus α and τ. It is clear that the spectral peaks reach the maxima for the time delay close to any multiples of the fundamental period ($\tau_{max} \approx 2n\pi/\omega$) of the linearized system and a certain value of additive noise intensity. This phenomenon can be explained as the resonance between the Kramers rate and time delay. When τ is between such values $\tau_{\max}$, the spectral peaks reach the minima. Thus, the time-delayed feedback has a negative effect on the enhancement of system's coherence. Figure 20(d) shows $S(\omega)$ as a function in τ with different values of ε. The spectral peaks increase with an increase in τ and in feedback gain.

In Fig.21, the output spectrum for correlated noises displays an asymmetric bimodal structure as the function of noises correlation. The peaks increase with an increase in α. And the positions of the peaks are shifted towards both directions with the increase of α. For example, the curve with $\alpha = 0.2$ shows that the position of the higher peak is shifted to the right (see Fig.21(a)) and the position of the lower peak is shifted to the left (see Fig.21(b)). The values of peaks for the case of negative correlation ($\lambda < 0$) is larger than those for the case of positive correlation ($\lambda > 0$). In other words, the coherence or the regularity of this system can attain the maximum by choosing the value of λ.

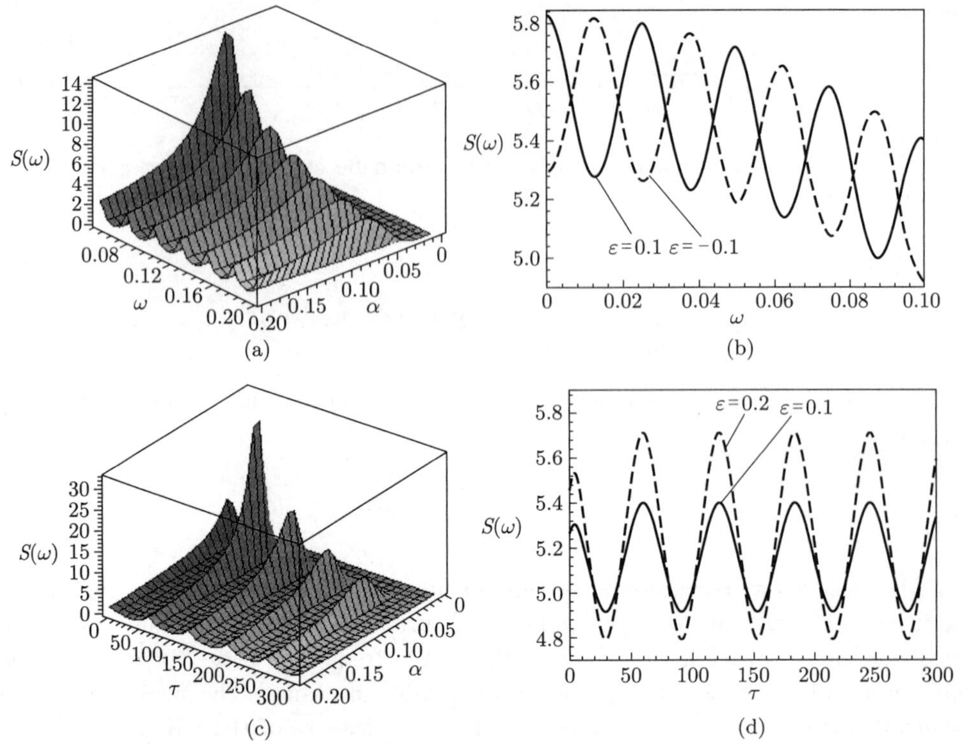

Fig. 20. Power spectrum of the delayed system (94) with uncorrelated noises as a function in ω, α and τ. (a) mesh surface of $S(\omega)$-(ω, α) with $\varepsilon = 0.1$, $\tau = 250$, $D = 0.5$; (b) $S(\omega)$ versus ω with $\tau = 250$, $D = 0.5$, $\alpha = 1$ and different ε; (c) mesh surface of $S(\omega)$-(τ, α) with $\varepsilon = 0.1$, $\omega = 0.1$, $D = 0.5$; (d) $S(\omega)$ versus τ with $\omega = 0.1$, $D = 0.5$, $\alpha = 1$ and different ε.

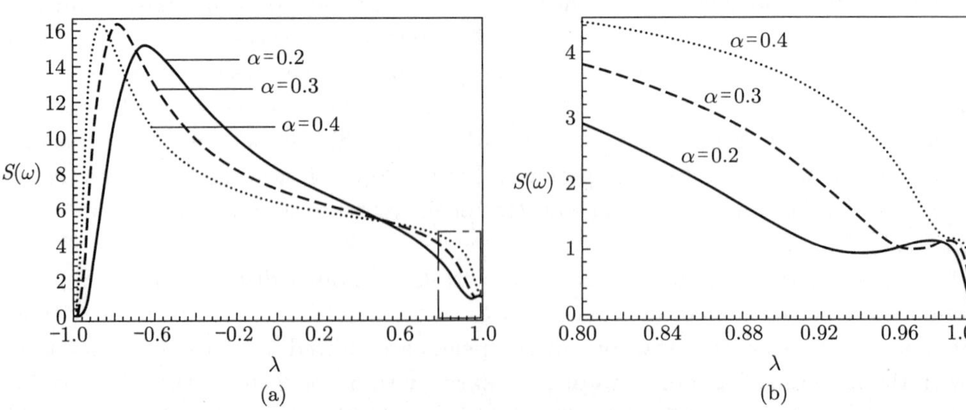

Fig. 21. Power spectrum of the delayed system (94) with correlated noises as a function in λ with $\varepsilon = 0.1$, $\tau = 250$, $\omega = 0.1$, $D = 0.5$. (a) for different values of additive noise intensity α; (b) drawing of partial enlargement inside the dashed-line frame.

4.2. *Stochastic Resonance*

This subsection focuses on the linear response of the delayed bistable system (94) subject to a harmonic excitation, i.e., $A_0 \neq 0$.

Assume that the signal amplitude is small. The transition rates in Eq.(104) are modulated with a frequency Ω according to the Arrhenius rate law

$$W'_{\pm}(t) = W_{\pm}(t)e^{\pm\gamma(t)}, \tag{112}$$

where $\gamma(t) = A_0 D^{-1}(1+\varepsilon/2)\cos\Omega t$. Let $\sigma(t) = n_+(t) - n_-(t)$, the governing equation of $\sigma(t)$ can be derived from Eqs.(105) and (112)

$$\begin{aligned}\frac{d\sigma(t)}{dt} = & (p_+ + p_-)(n_+ e^{\gamma(t)} - n_- e^{-\gamma(t)}) \\ & + (p_- - p_+)(n_+ e^{\gamma(t)} + n_- e^{-\gamma(t)})\sigma(t-\tau).\end{aligned} \tag{113}$$

Suppose that $\sigma(t) = \sigma_0(t) + \sigma_1(t)$ and $A_0 \ll 1$, we can check the following linearization of $\sigma(t)$ at solution $\sigma_0(t)$

$$\begin{aligned}\frac{d\sigma_1(t)}{dt} = & -(p_+ + p_-)\sigma_1(t) + (p_- - p_+)\sigma_1(t-\tau) \\ & + A_0 D^{-1}\cos\Omega t[(p_- - p_+)\sigma_0(t)\sigma_0(t-\tau) - (p_+ + p_-)],\end{aligned} \tag{114}$$

where $\sigma_0(t)$ is the solution of Eqs.(106) and (107), and $\sigma_1(t) = \bar{A}e^{i\Omega t}$ is the response of the system to the periodic signal $A_0 \cos\Omega t$. Here only the response at the frequency Ω for $t \to \infty$ contributes to the delta peak in the spectrum at Ω. So we give the expression of $\sigma_1(t)$ at $t \to \infty$

$$\sigma_1(t) = \mathrm{Re}\frac{(p_+ + p_-)A_0 D^{-1}(1+\varepsilon/2)e^{i\Omega t}}{(p_- - p_+)e^{-i\Omega\tau} - i\Omega - (p_+ + p_-)}. \tag{115}$$

Within the framework of the theory of the linear response approximation, the spectrum amplification is given in the following form

$$\eta = \frac{[(1+\varepsilon/2)(p_+ + p_-)]^2}{2D^2\left|(p_- - p_+)e^{-i\Omega\tau} - i\Omega - (p_+ + p_-)\right|^2}. \tag{116}$$

Figure 22 shows the effects of the multiplicative noise intensity D, the time delay τ, the excitation frequency Ω and the feedback gain ε on the spectrum amplitude for uncorrelated noises. In Fig.22(a), η is a function in D with different values of ε. For a positive feedback gain ($\varepsilon > 0$), η exhibits the typical SR with a pronounced peak with respect to D. The maxima become more pronounced with an increase in ε. For a negative feedback gain ($\varepsilon < 0$), η decreases monotonically with an increase in D, and the SR disappears. Thus, the positive feedback gain enhances the output of system, while the negative feedback gain suppresses the output of system. In Fig.22(b), η is a function in Ω with different values of τ. The curve of η demonstrates a resonance-like structure with an increase of Ω. This fact is different from the typical SR and is called the frequency SR.

Figure 23 shows the effects of the cross-correlation between noises and additive noise intensity α on the spectrum amplitude η. It is clear that the curve of η vs D exhibits a minimum firstly and a maximum later as shown in Fig.23(a). That is, η exists a suppression firstly and then a typical SR with a pronounced peak as a function of D when $\alpha = 0.05$. This typical phenomenon, i.e. resonant activation and giant suppression are both induced by λ. However, when $\alpha = 0.2$, the curve of η monotonically decreases with an increase of D and SR disappears (see Fig.23(b)). In other words, the suppression and the resonance disappear as α varies from 0.05 to 0.2. Furthermore, the curve of η decreases with an increase of λ.

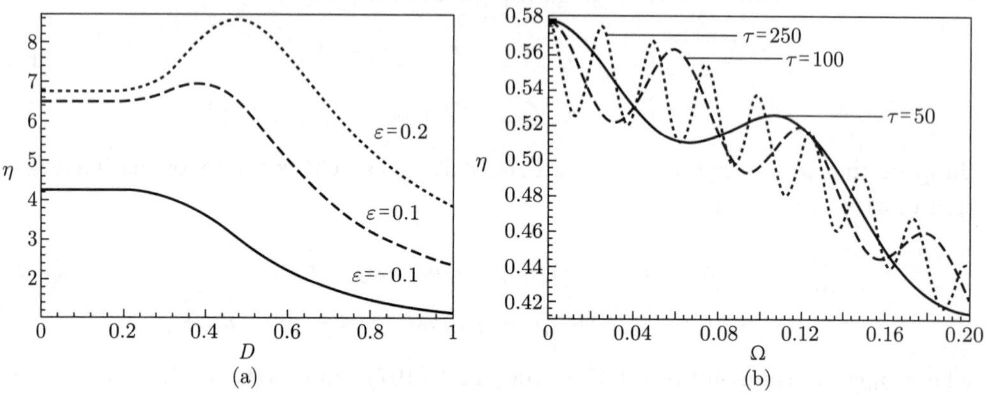

Fig. 22. The spectrum amplification η of the delayed system (94) with uncorrelated noises. (a) η versus D with different ε for $\tau = 250$, $\alpha = 0.3$; (b) η versus Ω with different τ for $\varepsilon = 0.1$, $\alpha = 1$, $D = 0.5$.

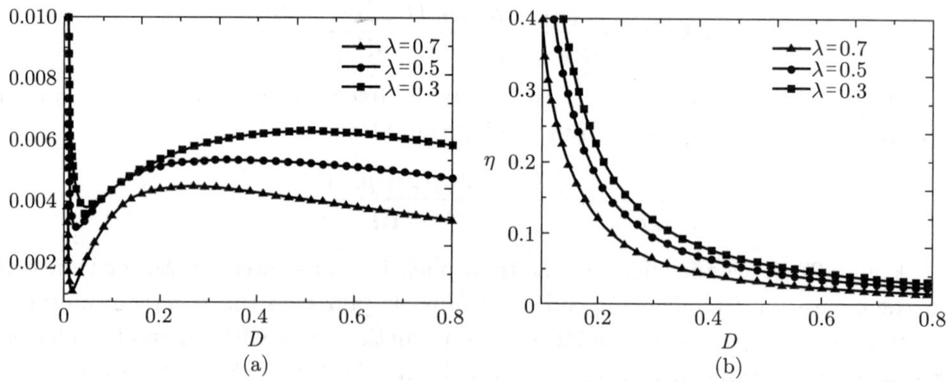

Fig. 23. The spectrum amplification η of the delayed system (94) with correlated noises versus D with $\tau = 250$, $\Omega = 1$, $\varepsilon = 0.05$. (a) $\alpha = 0.05$; (b) $\alpha = 0.2$.

5. Concluding Remarks

In this chapter we presented some interesting dynamical behaviors, such as stochastic jump, stochastic resonance and coherence resonance, existing in the DDE with added noises. The chapter shows that the method of multiple scales and the stochastic averaging method can be successfully applied to solving the problems in the principal response and the moment stability analysis of the delayed dynamical systems subject to random external and parametric excitations. Equipped with the theory of two-state model with delays, both CR and SR are explored in a delayed bistable system driven by additive and multiplicative noises, which can be uncorrelated or correlated. Different from the existing approaches, we also consider the multiplicative noise term that originally comes from the intrinsic fluctuations of system. Our work yields different results from the multiplicative noise model and the additive model. Moreover, our works focus on DDEs driven by Gaussian white noise which is a stochastic process with zero correlation time. When the correlation time of the fluctuation is negligible compared to the deterministic time scales of the system, we can choose the Gaussian white noise to model the random fluctuation. When the correlation time of the fluctuation is large, however, we must choose a stochastic process with non-zero correlation time, such as Ornstein-Uhlenbeck process, random telegraph process and non-Gaussian noise. Thus, our results may be extended to a class of stochastic systems with memory where the random fluctuations can be modeled by using these stochastic processes. Nevertheless, our results illustrate the dynamical behaviors in DDEs subject to both extrinsic and intrinsic random forces and may serve as a guide in many complex situations.

The study on the dynamics of a SDDE is obviously important for nonlinear dynamics and many interdisciplinary fields. However, the infinite dimensions of the solution space give rise to a tough problem to the theoretical treatment of DDE, especially to the SDDE. There are many challenges in the context of SDDE, for example, the derivation of FPK equations for SDDEs in general as for SDEs is still an open problem. Therefore, we expect more and more scientists in mathematics, physics and mechanics to develop new methods and concepts to deal with SDDE in near future.

Acknowledgments. Y. F. Jin thanks the support of the National Natural Science Foundation of China under Grant Nos. 10972032 and 11272051, the Excellent Young Scholars Research Fund of Beijing Institute of Technology under Grant No. 2010YS0101. H. Y. Hu thanks the support of the National Natural Science Foundation of China under Grant No. 51075032.

References

1. J. K. Hale, *Theory of Functional Differential Equations* (Springer-Verlag, New York, 1977).
2. R. D. Driver, *Ordinary and Delay Differential Equations* (Springer-Verlag, Berlin, 1977).
3. Y. X. Qin, Y. Q. Liu, L. Wang, and Z. X. Zheng, *Stability of Dynamic Systems with Delays* (Science Press, Beijing, 1989).

4. G. Stépán, *Retarded Dynamical Systems: Stability and Characteristic Functions* (Longman Scientific and Technical, Essex, 1989).
5. K. Gopalsamy, *Stability and Oscillations in Delay Differential Equations of Population Dynamics* (Kluwer Academic Publishers, Dordrecht, 1992).
6. Y. Kuang, *Delay Differential Equations with Applications to Population Dynamics* (Academic Press, New York, 1993).
7. J. K. Hale and S. M. V. Lunel, *Introduction to Functional Differential Equations* (Springer-Verlag, New York, 1993).
8. O. Diekmann, S. A. van Gils, S. M. Verduyn Lunel, and H. O. Walther, *Delay Equations, Functional-, Complex-, and Nonlinear Analysis* (Springer-Verlag, New York, 1995).
9. H. Y. Hu and Z. H. Wang, *Dynamics of Controlled Mechanical Systems with Delayed Feedback* (Springer, Heidelberg, 2002).
10. A. Einstein, Über die von der molekularkinetischen théorie der Wärme geforderte Bewegung von in ruhenden Flüssigkeiten suspendierten Teilchen, *Ann. Phys.*, **17**, 549–560 (1905).
11. M. Von Smoluchowski, Zur kinetischen théorie der Brownschen Molekularbewegung und der suspensionen, *Ann. Phys.*, **21**, 756–780 (1906).
12. P. Langevin, Sur la théorie de movement brownien, *C. R. Acad. Sci.*, **146**, 530–533 (1908).
13. N. Wiener, The mean of a functional of arbitrary elements, *Ann. Math.*, **22**, 66–72 (1920).
14. A. N. Kolmogorov, *Foundations of the Theory of Probability* (Chelsea, New York, 1950).
15. R. L. Stratonovich, *Introduction to the Theory of Random Noise* (Gordon and Breach, New York, 1963).
16. A. Papoulis, *Probability, Random Variables, and Stochastic Processes* (McGraw-Hill, New York, 1965).
17. W. Feller, *An Introduction to Probability Theory and Its Applications*, 2nd ed., Vol. II (Wiley, New York, 1974).
18. C. W. Gardiner, *Handbook of Stochastic Methods for Physics, Chemistry and the Natural Sciences*, 2nd ed. (Springer-Verlag, New York, 1997).
19. W. Horsthemke and R. Lefever, *Noise-induced Transitions: Theory and Applications in Physics, Chemistry, and Biology* (Springer-Verlag, New York, 1984).
20. L. Arnold, *Random Dynamical Systems* (Springer-Verlag, Berlin, 1998).
21. R. Kuske, Multiple-scales approximation of a coherence resonance route to chatter, *Computing in Science & Engineering*, **8**, 35–43 (2006).
22. A. Longtin, J. G. Milton, J. E. Bos, and M. C. Mackey, Noise and critical behavior of the pupil light reflex at oscillation onset, *Phys. Rev. A*, **41**, 6992–7005 (1990).
23. K. Vasilakos and A. Beuter, Effect of noise on a delayed visual feedback system, *J. Thero. Biol.*, **165**, 389–407 (1993).
24. I. Goychuk and P. Hänggi, Stochastic resonance in ion channels characterized by information theory, *Phys. Rev. E*, **61**, 4272–4280 (2000).
25. K. Patanarpeelert, T. D. Frank, R. Friedrich, P. J. Beek, and I. M. Tang, Theoretical analysis of destabilization resonances in time-delayed stochastic second-order dynamical systems and some implications for human motor control, *Phys. Rev. E*, **73**, 021901(1–17) (2006).
26. R. Kapral, E. Celarier, P. Mandel, and P. Nardone, Noisy delay-differential equations in optical bistability, N. B. Abraham and J. Chrostowski (Eds.), *Optical Chaos*, Vol. 667 of Proceedings of SPIE, International Society for Optical Engineering, Belling-

ham, WA, 175–182 (1986).
27. J. García-Ojalvo and R. Roy, Noise amplification in a stochastic Ikeda model, *Phys. Lett. A.*, **224**, 51–56 (1996).
28. T. Kreuz, S. Luccioli, and A. Torcini, Coherence resonance due to correlated noise in neuronal model, *Neurocomputing*, **70**, 1970–1976 (2007).
29. D. Wu and S. Q. Zhu, Stochastic resonance in FitzHugh-Nagumo system with time-delayed feedback, *Phys. Lett. A.*, **372**, 5299–5304 (2008).
30. Y. H. Hao, Y. B. Gong, and X. Lin, Multiple resonances with time delays and enhancement by non-Gaussian noise in Newman-Watts networks of Hodgkin-Huxley neurons, *Neurocomputing*, **74**, 1748–1753 (2011).
31. L. Stone, P. I. Saparin, A. Huppert, and C. Price, El Niño chaos: the role of noise and stochastic resonance on the ENSO cycle, *Geophys. Res. Lett.*, **25**, 175–178 (1998).
32. S.-E. A. Mohammed, *Stochastic Functional Differential Equations* (Longman Scientific and Technical, 1986).
33. X. Mao, *Stochastic Differential Equations and Their Applications* (Albion Publishing, 1997).
34. E. Buckwar, Introduction to the numerical analysis of stochastic delay differential equations, *J. Comput. Appl. Math*, **125**, 297–307 (2000).
35. U. Kuchler and E. Platen, Weak discrete time approximation of stochastic differential equations with time delay, *Math. Comput. Simul.*, **59**, 497–507 (2002).
36. S. Guillouzic, I. L'Heureux, and A. Longtin, Small delay approximation of stochastic delay differential equations, *Phys. Rev. E*, **59**, 3970–3982 (1999).
37. U. Küchler and B. Mensch, Langevins stochastic differential equation extended by a time-delayed term, *Stochastics Rep.*, **40**, 23–42 (1992).
38. D. R. Bell and S.-E. A. Mohammed, Smooth densities for degenerate stochastic delay equations with hereditary drift, *Annals Prob.*, **23**, 1875–1894 (1995).
39. J. Houlihan, D. Goulding, Th. Busch, C. Masoller, and G. Huyet, Experimental investigation of a bistable system in the presence of noise and delay, *Phys. Rev. Letts.*, **92**, 050601(1–4) (2004).
40. J. Losson and M. C. Mackey, Coupled map lattices as models of deterministic and stochastic differential delay equations, *Phys. Rev. E*, **52**, 115–128 (1995).
41. O. Elbeyli, J. Q. Sun, and G. Ünal, A semi-discretization method for delayed stochastic systems, *Communications in Nonlinear Science and Numerical Simulation*, **10**, 85–94 (2005).
42. A. H. Nayfeh, *Perturbation Methods* (John Wiley, New York, 1973).
43. A. H. Nayfeh and D. T. Mook, *Nonlinear Oscillations* (Wiley Interscinece, New York, 1979).
44. R. Khanin and M. P. Cartmell, Applying the perturbation method of multiple scales, *Math. Educ. Res.*, **8**, 19–26 (1999).
45. R. Khanin and M. P. Cartmell, Parallelization of perturbation analysis: application to large-scale engineering problems, *J. Symbol. Comput.*, **31**, 461–473 (2001).
46. A. H. Nayfeh, C.-M. Chin, and J. Pratt, Perturbation methods in nonlinear dynamics-application to machining dynamics, *ASME J. Manuf. Sci. Eng.*, **119**, 485–493 (1997).
47. C. L. Lee and C.-T. Lee, A higher order method of multiple scales, *J. Sound. Vib.*, **202**, 284–287 (1997).
48. Z. Ji and J. W. Zu, Method of multiple scales for vibration analysis of rotor-shaft systems with nonlinear bearing pedestal model, *J. Sound. Vib.*, **218**, 293–305 (1998).
49. T. J. Anderson, B. Balachandran, and A. H. Nayfeh, Nonlinear resonances in a flexible cantilever beam, *ASME J. Vibr. Acoust.*, **116**, 480–484 (1994).

50. T. J. Anderson, A. H. Nayfeh, and B. Balachandran, Experimental verification of the importance of the nonlinear curvature in the response of a cantilever beam, *ASME J. Vibr. Acoust.*, **118**, 21–27 (1996).
51. M. P. Cartmell, S. W. Ziegler, R. Khanin, and D. I. M. Forehand, Multiple scales analyses of the dynamics of weakly nonlinear mechanical systems, *Appl. Mech. Rev.*, **56**, 455–492 (2003).
52. S. Rajan and H. G. Davies, Multiple time scaling of the response of a duffing oscillatory to narrow band excitation, *J. Sound. Vib*, **123**, 497–506 (1988).
53. A. H. Nayfeh and S. J. Serhan, Response statistics of nonlinear systems to combined deterministic and random excitations, *Int. J. Nonlinear Mech.*, **25**, 493–509 (1990).
54. H. W. Rong, W. Xu, and T. Fang, Principal response of duffing oscillator to combined deterministic and narrow-band random parametric excitation, *J. Sound. Vib.*, **210**, 483–515 (1998).
55. R. H. Plaut and J. C. Hsieh, Nonlinear structural vibrations involving a time delay in damping, *J. Sound. Vib.*, **117**, 497–510 (1987).
56. H. Y. Hu, E. H. Dowell, and L. N. Virgin, Resonances of a harmonically forced Duffing oscillator with time delay state feedback, *Nonlinear Dyn.*, **15**, 311–327 (1998).
57. J. C. Ji and A. Y. T. Leung, Resonances of a nonlinear SDOF system with two time-delays on linear feedback control, *J. Sound. Vib.*, **253**, 985–1000 (2002).
58. N. N. Bogoliubov and Y. A. Mitropolsky, *Asymptotic Methods in the Theory of Nonlinear Oscillations* (Gordon and Breach, New York, 1961).
59. R. L. Stratonovich, *Topics in the Theory of Random Noise* (Gordon and Breach, New York, 1963).
60. R. Z. Khasminski, A limit theorem for the solution of differential equations with random right hand sides, *Theory Prob. Appl.*, **11**, 390–405 (1966).
61. G. C. Papanicolaou and W. Kohler, Asymptotic theory of mixing stochastic ordinary differential equations, *Commu. Pure Appl. Math.*, **27**, 641–668 (1974).
62. S. T. Ariaratnam, Dynamic stability of a column under random loading, *Proc. ICDSS.*, 255–265 (Pergamon Press, 1965).
63. J. B. Roberts and P. D. Spanos, Stochastic averaging: an approximate method of solving random vibration problems, *Int. J. Nonlinear Mech.*, **21**, 111–134 (1986).
64. J. R. Red-Horse and P. D. Spanos, A generalization to stochastic averaging in random vibration, *Int. J. Nonlinear Mech.*, **27**, 85–101 (1992).
65. Y. K. Lin, Some observations on stochastic averaging method, *Prob. Eng. Mech.*, **1**, 23–27 (1986).
66. Y. K. Lin and G. Q. Cai, *Probabilistic Structural Dynamics, Advanced Theory and Applications* (McGraw-Hill, New York, 1995).
67. W. Q. Zhu, Stochastic averaging methods in random vibration, *ASME Applied Mechanics Reviews*, **41**, 189–199 (1988).
68. W. Q. Zhu, Recent developments and applications of the stochastic averaging method in random vibration, *ASME Applied Mechanics Reviews*, **49**, 72–80 (1996).
69. W. Q. Zhu, Nonlinear stochastic dynamics and control in Hamiltonian formulation, *ASME Applied Mechanics Reviews*, **59**, 230–248 (2006).
70. Z. H. Liu and W. Q. Zhu, Stochastic averaging of quasi-integrable Hamiltonian systems with delayed feedback control, *J. Sound. Vib.*, **299**, 301–317 (2008).
71. Z. H. Liu and W. Q. Zhu, First-passage failure of quasi-integrable Hamiltonian systems under time-delayed feedback control, *J. Sound. Vib.*, **315**, 178–195 (2007).
72. Y. K. Lin and G. Q. Cai, Some thoughts on averaging techniques in stochastic dynamics, *Prob. Eng. Mech.*, **15**, 7–14 (2000).
73. S. Guillouzic, *Fokker-Planck Approach to Stochastic Delay Differential Equations*

(University of Ottawa, 2000).
74. S. Guillouzic, I. L'Heureux, and A. Longtin, Small delay approximation of stochastic delay differential equations, *Phys. Rev. E*, **59**, 3970–3982 (1999).
75. T. D. Frank, Delay Fokker-Planck equations, perturbation theory, and data analysis for nonlinear stochastic systems with time delays, *Phys. Rev. E*, **71**, 031106(1–14) (2005).
76. G. Giacomelli, R. Meucci, A. Lapucci, and F. T. Arecchi, Defects and space-like properties of delayed dynamical systems, *Phys. Rev. A*, **73**, 1099–1103 (1994).
77. T. D. Frank, Delay Fokker-Planck equations, Novikov's theorem, and Boltzmann distributions as small delay approximations, *Phys. Rev. E*, **72**, 011112(1–8) (2005).
78. T. D. Frank and P. J. Beek, Fokker-Planck equations for globally coupled many-body systems with time delays, *J. Stat. Mech.*, **10**, 10010–10030 (2005).
79. L. C. Du and D. C. Mei, Effects of time-delay in stationary properties of a logistic growth model with correlated noises, *Physica A*, **389**, 1189–1196 (2010).
80. C. H. Zeng, C. P. Zeng, A. L. Gong, and L. R. Nie, Effect of time delay in FitzHugh-Nagumo neural model with correlations between multiplicative and additive noises, *Physica A*, **389**, 5117–5127 (2010).
81. R. H. Shao and Y. Chen, Stochastic resonance in time-delayed bistable systems driven by weak periodic signal, *Physica A*, **388**, 977–983 (2009).
82. H. Q. Zhang, W. Xu, Y. Xu and D. X. Li, Stochastic time-delayed systems driven by correlated noises: steady-state analysis, *Physica A*, **388**, 3017–3023 (2009).
83. B. Mc Namara and K. Wiesenfeld, Theory of stochastic resonance, *Phys. Rev. A*, **39**, 4854–4869 (1989).
84. T. Ohira and Y. Sato, Resonance with noise and delay, *Phys. Rev. Letts.*, **82**, 2811–2815 (1999).
85. T. Ohira and Y. Sato, Delayed stochastic systems, *Phys. Rev. E*, **61**, 1247–1257 (2000).
86. L. S. Tsimring and A. Pikovsky, Noise-induced dynamics in bistable systems with delay, *Phys. Rev. Letts.*, **87**, 250602(1–4) (2001).
87. Y. Jiang, S. H. Dong, and M. Lozada-Cassou, Noise-induced coherence in bistable systems with multiple time delays, *Phys. Rev. E*, **69**, 056225(1–4) (2004).
88. M. Grigoriu, Control of time delay linear system with Gaussian white noise, *Prob. Eng. Mech.*, **12**, 89–96 (1997).
89. M. Di. Paola and A. Pirrotta, Time delay induced effects on control of linear systems under random excitation, *Prob. Eng. Mech.*, **16**, 43–51 (2001).
90. M. C. Mackey and I. G. Nechaeva, Solution moment stability in stochastic differential delay equations, *Phys. Rev. E*, **52**, 3366–3376 (1995).
91. J. Z. Lei and M. C. Michanel, Stochastic differential delay equation, moment stability, and application to hematopoietic stem cell regulation system, *SIAM. Appl. Math.*, **67**, 387–407 (2007).
92. O. Elbeyli, J. Q. Sun, and G. Ünal, A semi-discretization method for delayed stochastic systems, *Communications in Nonlinear Science and Numerical Simulation*, **10**, 85–94 (2005).
93. M. M. Klosek, Multiscale analysis of effects of additive and multiplicative noise on delay differential equations near a bifurcation point, *Acta Physica Polonica B*, **35**, 1387–1405 (2004).
94. M. M. Klosek and R. Kuske, Multiscale analysis of stochastic delay differential equations, *Multiscale Model. Simul.*, **3**, 706–729 (2005).
95. W. V. Wedig, Invariant measures and lyapunov exponents for generalized parameter fluctuations, *Structural Safety*, **8**, 13–25 (1990).

96. Z. K. Sun, W. Xu, and X. L. Yang, Response of nonlinear system to random narrow-band excitation with time delay state feedback, *Journal of Vibration Engineering*, **19**, 57–64 (2006).
97. Y. F. Jin and H. Y. Hu, Dynamics of a Duffing oscillator with two time delays in feedback control under narrow-band random excitation, *ASME Journal of Computational and Nonlinear Dynamics*, **3**, 021205(1–7) (2008).
98. T. T. Soong and M. Grigoriu, *Random Vibration of Mechanical and Structural Systems* (Prentice-Hall, Englewood Cliffs, NJ, 1993).
99. M. Shinozuka and C. -M. Jan, Digital simulation of random processes and its applications, *J. Sound. Vib.*, **25**, 111–128 (1972).
100. W. Q. Zhu, *Random Vibration* (Science Press, Beijing, 1992).
101. Y. F. Jin and H. Y. Hu, Principal resonance of a Duffing oscillator with delayed state feedback under narrow-band random parametric excitation, *Nonlinear Dyn.*, **50**, 213–227 (2007).
102. I. S. Gradshteyn and I. M. Ryzhik, *Tables of Integrals, Series and Products* (Academic Press, New York, 1980).
103. T. Fang and E. H. Dowell, Numerical simulations of jump phenomena in stable Duffing systems, *Int. J. Nonlinear Mech.*, **22**, 267–274 (1987).
104. Z. H. Wang and H. Y. Hu, Robust stabilization to nonlinear delayed systems via delayed state feedback: the averaging method, *J. Sound. Vib.*, **279**, 937–953 (2005).
105. L. Gammaitoni, P. Hänggi, P. Jung, and F. Marchesoni, Stochastic resonance, *Rev. Mod. Phys.*, **70**, 223–287 (1998).
106. R. Benzi, G. Parisi, and A. Vulpiani, The mechanism of stochastic resonance, *J. Phys. A*, **14**, L453–L457 (1981).
107. C. Nicolis and G. Nicolis, Stochastic aspects of climatic transitions-additive fluctuations, *Tellus*, **33**, 225–234.
108. S. Fauve and F. Heslot, Stochastic resonance in a bistable system, *Phys. Lett.*, **97A**, 5–7 (1983).
109. H. Gang, T. Ditzinger, C. Z. Ning, and H. Haken, Stochastic resonance without external periodic force, *Phys. Rev. Letts.*, **71**, 807–810 (1993).
110. A. Neiman, P. I. Saparin, and L. Stone, Coherence resonance at noisy precursors of bifurcations in nonlinear dynamical systems, *Phys. Rev. E*, **56**, 270–273 (1997).
111. A. Longtin, Autonomous stochastic resonance in bursting neurons, *Phys. Rev. E*, **55**, 868–876 (1997).
112. A. Pikovsky and J. Kurths, Coherence resonance in a noise-driven excitable system, *Phys. Rev. Letts.*, **78**, 775–778 (1997).
113. Y. F. Jin and H. Y. Hu, Coherence and stochastic resonance in a delayed bistable system, *Physica A*, **382**, 423–429 (2007).
114. Y. F. Jin, Noise-induced dynamics in a delayed bistable system with correlated noises, *Physica A*, **391**, 1928–1933 (2012).
115. A. J. R. Madureira, P. Hanggi, and H. S. Wio, Giant suppression of the activation rate in the presence of correlated white noise sources, *Phys. Lett. A*, **217**, 248–252 (1996).
116. Y. Jia, S. N. Yu, and J. R. Li, Stochastic resonance in a bistable system subject to multiplicative and additive noise, *Phys. Rev. E*, **62**, 1869–1878 (2000).
117. J. Wang, L. Cao, and D. J. Wu, Effect on the mean first passage time in symmetrical bistable systems by cross-correlation between noises, *Phys. Lett. A*, **308**, 23–30 (2003).
118. Y. F. Jin, W. Xu, M. Xu, and T. Fang, Stochastic resonance in linear system due to dichotomous noise modulated by bias signal, *J. Phys. A*, **38**, 3733–3742 (2005).

119. J. H. Li and Z. Q. Huang, Nonequilibrium phase transition in the case of correlated noises, *Phys. Rev. E*, **53**, 3315–3318 (2006).

Chapter 10

Some Studies on Delayed System Dynamics and Control

Guo-Ping Cai, Long-Xiang Chen, Kun Liu

Department of Engineering Mechanics, State Key Laboratory of Ocean Engineering
Shanghai Jiaotong University, Shanghai 200240, China
caigp@sjtu.edu.cn

In recent decades, the studies on delayed system dynamics have received more and more attention from researchers in automation, machinery, mechanics, and structural engineering, and many progresses have been made. This paper introduces studies on time delay problems by the authors in recent ten years, including: (1) a time delay identification using the particle swarm optimizer (PSO), (2) two time-delayed controllers for linear systems, (3) time-delayed controller for nonlinear systems, (4) parameter robustness of time-delayed controller, (5) robust H_∞ time-delayed controller based on the LMI technique, (6) delayed positive feedback technique, and (7) time delay experiments.

1. Introduction

Time delay exists inevitably in active control systems. Many factors, such as measurement of system variables, calculation of controller and processes for actuators to build up required control force, may result in non-synchronization of control force. For convenience of dynamical analysis and control design, time delay is often neglected before, but even small time delay may cause actuator to input energy into the control system when no energy is needed. This may cause degradation of control efficiency and instability of the system. Therefore, time delay should be compensated in phase of control design so as to eliminate its negative effect on control efficiency. On the other hand, some studies have shown that voluntary introduction of delay into control system can be beneficial to improving control performance and system stability. Delayed systems have complex dynamic behaviors and have been attracting more and more attention from researchers in mathematics, mechanics, structural engineering, automation and so on. Many progresses have been made.

The authors have done studies on delayed system dynamics for about ten years. This chapter summarizes the achievements the authors have made on this topic.

2. Time Delay Identification

Identification for time delay is a basic question in delayed system dynamics, but also is a difficult one. Most studies on time delay problems assume that the time delay in the control system is known, and parry the problem of how much the time delay is in control system. Although there exists inherent difficulty on the identification of time delay, especially with the experimental approach, it is nevertheless an important subject. We have studied the time delay identification using a flexible beam as an example. The approach is to change the identification into an optimization problem. The particle swarm optimizer (PSO) is used as the optimization method. Single and double time delays are both considered in the study.

Consider q PZT actuators to control the vibration of a cantilever beam. The discrete controlled modal equation can be written as

$$\mathbf{Z}(k+1) = \mathbf{FZ}(k) + \mathbf{GV}(k) \tag{1}$$

where $\mathbf{V}(k) = [V_1(k), \ldots, V_q(k)]^T$ is the control voltage of the PZT actuators.

By the classical discrete optimal control method, the controller can be obtained as

$$\mathbf{V}(k) = -\mathbf{LZ}(k) \tag{2}$$

where $\mathbf{L}$ is the feedback gain.

Consider that the control forces exerting on the beam by the actuators have the time delay τ_j, $j = 1, \ldots, q$, and assume that τ_j can be written as $\tau_j = l_j \overline{T}$, where $l_j > 1$ is a positive integer and $\overline{T}$ is the sampling period of the control system. Since the data sampling period in practice is often chosen to be very small, this assumption is reasonable. There are published results indicating that time delay has small effect on control efficiency and can be neglected in control design when it is smaller than one sampling period. It affects control efficiency only when it is larger than sampling period. Considering time delay in Eq.(1), we have

$$\mathbf{Z}(k+1) = \mathbf{FZ}(k) + \sum_{j=1}^{q} \mathbf{G}_j V_j(k - l_j). \tag{3}$$

Using the controller designed in the case of no time delay to control the system with time delay, namely substituting Eq.(2) into Eq.(3), we have

$$\mathbf{Z}(k+1) = \mathbf{FZ}(k) - \sum_{j=1}^{q} \mathbf{G}_j [\mathbf{L}_j \mathbf{Z}(k - l_j)]. \tag{4}$$

The beam response can be written in the following form using the assumed mode method

$$y(x,t) = \sum_{i=1}^{n} Y_i(x) \phi_i(t). \tag{5}$$

When an assumed time delay is taken, the response of the beam can be calculated using Eqs.(4) and (5).

Time delay may render the difference between the real and estimated responses of the beam. The real response is that observed by the sensor, the estimated one is that obtained numerically in the case that the assumed time delay is taken. So the sum of absolute value of the difference between the estimated and actual responses over a period of time can be taken as an objective function, given by

$$\min e = \int_{t_0}^{t_f} |y(t) - y^*(t)|\, dt \tag{6}$$

where t_0 and t_f are the start-stop time respectively, $y(t)$ is the estimated response of the beam in the case of assumed time delay, and $y^*(t)$ is the real response. The classical formula of the PSO are

$$\begin{cases} \bar{v}_{jk} = \omega v_{jk} + c_1 r_1 (p_{jk} - l_{jk}) + c_2 r_2 (p_{jg} - l_{jk}), \\ \bar{l}_{jk} = l_{jk} + \bar{v}_{jk}. \end{cases} \tag{7}$$

In Ref.1, we use the above identification method to identify the time delay in a beam system and the results prove the effectiveness of this method.

3. Two Time-delayed Controllers for Linear Structural Systems[2,3]

For linear structural system, some treating methods for time delay problem have been proposed, such as the expansion of Taylor series, the technique of phase shift, the advance estimation of state variables. Due to the inherent restrictions with each of the three methods, they can only deal effectively with some small time delay in control systems, but poorly with large one. The authors of this paper have recently proposed two new time-delayed controllers, one is in continuous form and the other is in discrete form[2,3]. These delay controllers are designed directly from time-delayed differential equation without any hypothesis in whole process of controllers, suitable for both small and large time delays. The effectiveness of the proposed controllers is verified by experiments in Refs.4–11. Here are brief descriptions of these two controllers.

3.1. *The Discrete Time-delayed Controller*[2]

Assume that q actuators are applied on the n-degree-of-freedom system, and those actuators have the delayed time $\tau_i (i = 1, \ldots, q)$. The motion equation of the system can be written as

$$\mathbf{M}\ddot{\mathbf{X}} + \mathbf{C}\dot{\mathbf{X}} + \mathbf{K}\mathbf{X} = \sum_{i=1}^{q} \mathbf{H}_i u_i(t - \tau_i) + \mathbf{D}(t) \tag{8}$$

where u_i represents the control force of the ith actuator and $\mathbf{D}(t)$ is the external excitation. In the state-space representation, equation (8) becomes

$$\dot{\mathbf{Z}}(t) = \mathbf{A}\mathbf{Z}(t) + \sum_{i=1}^{q} \mathbf{B}_i u_i(t - \tau_i) + \bar{\mathbf{D}}(t) \tag{9}$$

where $\mathbf{Z}(t) = [\mathbf{X}^T(t),\ \dot{\mathbf{X}}^T(t)]^T$. Zero-order holder is used in the control system, i.e., $u_i(t) = u_i(k)$, $k\overline{T} \leq t < (k+1)\overline{T}$, where k represents the k th step of control force and $u_i(k)$ denotes $u_i(k\overline{T})$. The zero-order holder represents that the actuators in the structure exert constant control forces on the structure during two adjacent sampling points, which makes sense because data sampling period is usually very short. The time delay τ_i can be written as $\tau_i = l_i\overline{T} - \bar{m}_i$, where $l_i > 1$ is a positive number and $0 \leq \bar{m}_i < \overline{T}$. Discretizing Eq.(9), one can obtain

$$\mathbf{Z}(k+1) = \mathbf{F}\mathbf{Z}(k) + \sum_{i=1}^{q} \mathbf{G}_i u_i(k - l_i) + \overline{\mathbf{D}}(k). \tag{10}$$

Augmenting the state variables in Eq.(10) as

$$\begin{cases} \mathbf{Z}_{2n+1}(k) = u_1(k - l_1), \\ \cdots\cdots\cdots\cdots \\ \mathbf{Z}_{2n+l_1}(k) = u_1(k-1), \\ \cdots\cdots\cdots\cdots \\ \mathbf{Z}_{2n+\sum\limits_{i=1}^{q-1} l_i + 1}(k) = u_q(k - l_q), \\ \cdots\cdots\cdots\cdots \\ \mathbf{Z}_{2n+\sum\limits_{i=1}^{q} l_i}(k) = u_q(k-1), \end{cases} \tag{11}$$

thus, equation (10) can be converted into the standard discrete form as

$$\overline{\mathbf{Z}}(k+1) = \overline{\mathbf{F}}\,\overline{\mathbf{Z}}(k) + \overline{\mathbf{G}}\mathbf{u}(k) + \widehat{\mathbf{D}}(k). \tag{12}$$

Equation (12) is a standard discrete state equation that contains no time delay and can be used to determine the system stability: the system is stable if all the eigenvalues of $\overline{\mathbf{F}}$ fall within the unit circle. Since equation (12) is in the standard form, all the control methods in discrete form could be used to design a controller for Eq.(12). It is observed from Eq.(11) that the controller designed contains not only the current step of state feedback term, but also the linear combination of some previous control terms.

3.2. *The Continuous Time-delayed Controller*[3]

In Section 3.1, the processing method for time delay is in a discrete form. The controller is designed by a particular augmentation of state variables and the dimension of the system is augmented. The delay controller can be also designed to be a continuous type. In this case, the dimension of the system keeps invariant.

This technology will be introduced below. To make the expression simple, assuming that only two actuators are applied in the control system and these two actuators have different time delay. Using an integral transformation for Eq.(9)

$$\begin{aligned}\mathbf{H}(t) &= \mathbf{Z}(t) + \mathbf{\Gamma}_1(t) + \mathbf{\Gamma}_2(t) \\ &= \mathbf{Z}(t) + \int_{-\tau_1}^{0} e^{-\mathbf{A}(\eta+\tau_1)}\mathbf{B}_1 u_1(t+\eta)d\eta + \int_{-\tau_2}^{0} e^{-\mathbf{A}(\eta+\tau_2)}\mathbf{B}_2 u_2(t+\eta)d\eta,\end{aligned} \tag{13}$$

thus a standard form without any explicit time delay can be obtained as

$$\dot{\mathbf{H}}(t) = \mathbf{A}\mathbf{H}(t) + \overline{\mathbf{B}}\mathbf{U}(t) + \mathbf{D}(t) \tag{14}$$

where $\mathbf{U}(t) = [u_1(t),\ u_2(t)]^T$, $\overline{\mathbf{B}} = [e^{-\mathbf{A}\tau_1}\mathbf{B}_1,\ e^{-\mathbf{A}\tau_2}\mathbf{B}_2]$, $\mathbf{\Gamma}_1(t)$ and $\mathbf{\Gamma}_2(t)$ represent two integral terms in Eq.(13), respectively. Equation (14) is in a standard continuous form, which means various active control methods could be used to design a controller for Eq.(14). However, the controller obtained is the function of state variable $\mathbf{H}(t)$ and $\mathbf{H}(t)$ contains two integral terms $\mathbf{\Gamma}_1(t)$ and $\mathbf{\Gamma}_2(t)$ which makes it not convenient to actually implement the controller. Therefore, something needs to be done with these two integral terms. These two integral terms are definite integral form, and may be discretized using the zero-order holder and in terms of the sampling period $\overline{T}$, for details please see the Ref.3. Like the discrete time-delayed controller, the designed controller contains not only the state feedback term of current step, but also the linear combination of control terms of certain previous steps.

4. Time-delayed Controller for Nonlinear Structural Systems[12]

Nonlinearity in engineering structures often is associated with the stiffness. Nonlinear stiffness can be described using the hysteretic model. The common model contains the Bouc-Wen model, bilinear model and trilinear model. Below we introduce our research work on time-delayed control design using a seismic-excited building structure.

Consider an n-story building with a rubber bearing isolation system. The rubber bearing isolation system is connected by an active tuned mass damper (ATMD). The structure undergoes a one-dimensional earthquake ground acceleration $\ddot{X}_g(t)$. Assume that the control force produced by the ATMD exists the time delay τ. The motion equation can be written as

$$\mathbf{M}\ddot{\mathbf{X}}(t) + \mathbf{C}\dot{\mathbf{X}}(t) + \mathbf{F}_k[\mathbf{X}(t)] = -\mathbf{M}_0\ddot{X}_g(t) + \mathbf{H}\overline{U}(t-\tau) \tag{15}$$

where $\mathbf{X} = [x_d, x_b, x_1, \ldots, x_n]^T$ is an $(n+2)\times 1$ vector of interstory drift, x_d and x_b are the interstory drift of ATMD and of rubber bearing isolation system, respectively, $x_1, \ldots, x_n$ are the interstory drift of each story unit of upper building structure, respectively; $\overline{U}(t-\tau)$ is the active control force. $\mathbf{F}_k[\mathbf{X}(t)]$ is the restoring force matrix, it can be written as

$$\mathbf{F}_k[\mathbf{X}(t)] = \mathbf{K}_1\mathbf{X}(t) + \mathbf{K}_2\mathbf{V}(t) \tag{16}$$

where $\mathbf{K}_1$ is an $(n+2)\times(n+2)$ elastic stiffness matrix; $\mathbf{K}_2$ is an $(n+2)\times(n+2)$ hysteretic stiffness matrix; $\mathbf{V} = [v_d, v_b, v_1, \ldots, v_n]^T$ is an $(n+2)\times 1$ vector denoting the hysteretic variable of each story unit. By the Bouc-Wen model, the ith element of $\dot{\mathbf{V}}$ can be written as

$$\dot{v}_i = f(\dot{x}_i, v_i) = D_i^{-1}(\bar{a}_i\dot{x}_i - \beta_i|\dot{x}_i||v_i|^{\bar{n}_i-1}v_i - \gamma_i\dot{x}_i|v_i|^{\bar{n}_i}) \tag{17}$$

in matrix representation, it is $\dot{\mathbf{V}} = \mathbf{f}(\dot{\mathbf{X}}, \mathbf{V})$. In the state-space representation, equation (15) becomes

$$\dot{\mathbf{Z}}(t) = \mathbf{A}\mathbf{Z}(t) + \mathbf{W}\ddot{X}_g(t) + \mathbf{B}\overline{U}(t-\tau) + \mathbf{B}_1\mathbf{V}(t) \tag{18}$$

where $\mathbf{Z}(t) = [\mathbf{X}^T(t),\ \dot{\mathbf{X}}^T(t)]^T$.

Using an integral transformation similar to Eq.(13) for Eq.(18), one can obtain

$$\dot{\overline{\mathbf{H}}}(t) = \mathbf{A}\overline{\mathbf{H}}(t) + \mathbf{W}\ddot{X}_g(t) + \overline{\mathbf{B}}\overline{U}(t) + \mathbf{B}_1\mathbf{V}(t) \tag{19}$$

where $\overline{\mathbf{B}} = e^{-\mathbf{A}\tau}\mathbf{B}$. Augmenting the state variables in Eq.(19) as $\widetilde{\mathbf{H}}(t) = [\overline{\mathbf{H}}^T,\ \mathbf{V}^T]^T$, so equation (19) can be written as

$$\dot{\widetilde{\mathbf{H}}}(t) = \mathbf{g}'\,[\widetilde{\mathbf{H}}(t)] + \widetilde{\mathbf{W}}\ddot{X}_g(t) + \widetilde{\mathbf{B}}\overline{U}(t) \tag{20}$$

where $\mathbf{g}'\,[\widetilde{\mathbf{H}}(t)]$ is given by

$$\mathbf{g}'\,[\widetilde{\mathbf{H}}(t)] = \begin{bmatrix} \mathbf{A}\overline{\mathbf{H}} + \mathbf{B}_1\mathbf{V} \\ \mathbf{f}(\dot{\mathbf{X}}, \mathbf{V}) \end{bmatrix}. \tag{21}$$

Thus, by the transformation and augmentation described above, equation (18) changes to Eq.(20) in the state space representation without any explicit time delay, which can be solved numerically step by step using the fourth-order Runge-Kutta method. In Ref.12, the instantaneous optimal control method is used to obtain the optimal controller.

When bilinear or trilinear model is used for description of nonlinear stiffness of structures, the stiffness term can be also divided into linear part and inelastic part. By transferring the inelastic part into the right hand of motion equation and regarding this term as external excitation, then using the integral transformation given in Section 3.2, the motion equation with time delay can be also transforming into a standard form without any explicit time delay, so various linear control theory can be used for controller design.

5. Parameter Robustness of Time-delayed Controller[9]

It is well-known that the controller designed using the robust control method is robust to the variance of structural intrinsic parameters and external disturbance, and the robust control method has been getting more and more attention in active control of structures. The time delay in practical engineering is mostly small, and is possibly time-varying and uncertain. One will be prompted to consider the following question — will the time-delayed controller designed using strong robust control method is robust for time delay? The authors of this paper have done some researches on this topic.

Assume that the q PZT actuators exist time delays. The state equation with considering the time delay τ_i and the external disturbance $e(t)$ can be written as

$$\dot{\mathbf{X}}(t) = \mathbf{A}\mathbf{X}(t) + \sum_{i=1}^{q} \mathbf{B}_i V_i(t-\tau_i) + \mathbf{B}_p e(t) \tag{22}$$

where $\mathbf{B}_p$ is a vector determined by the positions of external disturbance.

Making integral transformation similar to Eq.(13) for Eq.(22) and considering the controllable output equation, we have

$$\begin{cases} \dot{\mathbf{H}}(t) = \mathbf{A}\mathbf{H}(t) + \overline{\mathbf{B}}\mathbf{V}(t) + \mathbf{B}_p e(t), \\ \mathbf{z}(t) = \mathbf{C}_1\mathbf{H}(t) + \mathbf{D}_{12}\mathbf{V}(t) \end{cases} \tag{23}$$

where $\overline{\mathbf{B}} = [e^{-\mathbf{A}\tau_1}\mathbf{B}_1, \ldots, e^{-\mathbf{A}\tau_q}\mathbf{B}_q]$ and $\mathbf{V}(t) = [V_1(t), V_2(t), \ldots, V_q(t)]^T$. $\mathbf{z}(t)$ is controllable output of the system.

The task of H_∞ control is to design the state-feedback controller $\mathbf{V}(t) = \mathbf{K}\mathbf{H}(t)$ such that the following closed loop system is asymptotically stable

$$\begin{cases} \dot{\mathbf{H}}(t) = (\mathbf{A} + \overline{\mathbf{B}}\mathbf{K})\mathbf{H}(t) + \mathbf{B}_p e(t), \\ \mathbf{z}(t) = (\mathbf{C}_1 + \mathbf{D}_{12}\mathbf{K})\mathbf{H}(t), \end{cases} \tag{24}$$

and the norm of transfer function from $e(t)$ to $\mathbf{z}(t)$ satisfies

$$\|\mathbf{T}_{ze}(s)\|_\infty = \left\|(\mathbf{C}_1 + \mathbf{D}_{12}\mathbf{K})[s\mathbf{I} - (\mathbf{A} + \overline{\mathbf{B}}\mathbf{K})]^{-1}\mathbf{B}_p\right\| < \gamma \tag{25}$$

where $\gamma > 0$ is H_∞ performance index. From Ref.9 we know that, if there exist matrices $\overline{\mathbf{X}}^T = \overline{\mathbf{X}} > 0$ and $\mathbf{Y}$ satisfying the matrix inequality

$$\begin{bmatrix} \mathbf{A}\overline{\mathbf{X}} + \overline{\mathbf{B}}\mathbf{Y} + (\mathbf{A}\overline{\mathbf{X}} + \overline{\mathbf{B}}\mathbf{Y})^T & \mathbf{B}_p & (\mathbf{C}_1\overline{\mathbf{X}} + \mathbf{D}_{12}\mathbf{Y})^T \\ \mathbf{B}_p^T & -\mathbf{I} & \mathbf{0} \\ \mathbf{C}_1\overline{\mathbf{X}} + \mathbf{D}_{12}\mathbf{Y} & \mathbf{0} & -\gamma^2\mathbf{I} \end{bmatrix} < 0, \tag{26}$$

the closed-loop system (24) is asymptotically stable with the H_∞ performance index γ, and $\mathbf{K} = \mathbf{Y}\overline{\mathbf{X}}^{-1}$ is a H_∞ state-feedback controller of the system.

Using the matrix inequality (26) to solve the following optimization problem

$$\min \rho,$$
$$\text{s.t.} \begin{bmatrix} \mathbf{A}\overline{\mathbf{X}} + \overline{\mathbf{B}}\mathbf{Y} + (\mathbf{A}\overline{\mathbf{X}} + \overline{\mathbf{B}}\mathbf{Y})^T & \mathbf{B}_p & (\mathbf{C}_1\overline{\mathbf{X}} + \mathbf{D}_{12}\mathbf{Y})^T \\ \mathbf{B}_p^T & -\mathbf{I} & \mathbf{0} \\ \mathbf{C}_1\overline{\mathbf{X}} + \mathbf{D}_{12}\mathbf{Y} & \mathbf{0} & -\rho\mathbf{I} \end{bmatrix} < 0, \quad \overline{\mathbf{X}} > 0 \tag{27}$$

where $\rho = \gamma^2$, one can obtain the state feedback controller $\mathbf{V}(t) = \mathbf{KH}(t) = \mathbf{K}[\mathbf{X}(t) + \sum_{i=1}^{q} \mathbf{\Gamma}_i(t)]$. Equation (26) is a linear matrix inequality, so $\mathbf{K}$ can be obtained by solving this inequality using the MATLAB LMI Toolbox.

6. Robust $\mathbf{H}_\infty$ Time-delayed Controller Based on The LMI Technique[13]

Consider the linear time-delayed system

$$\begin{cases} \dot{\mathbf{x}}(t) = \mathbf{A}\mathbf{x}(t) + \mathbf{B}_1\mathbf{w}(t) + \mathbf{B}_2\mathbf{V}(t-\tau), \\ \mathbf{z}(t) = \mathbf{C}_1\mathbf{x}(t) + \mathbf{D}_{12}\mathbf{V}(t-\tau) + \mathbf{D}_{11}\mathbf{w}(t) \end{cases} \tag{28}$$

where $\mathbf{V}(t)$ is control force, and τ is time delay, $0 \le \tau \le \bar{\tau}$, where $\bar{\tau}$ is the upper bound of τ. The memoryless controller $\mathbf{V}(t) = \mathbf{Kx}(t)$ is used for the system (28), we have

$$\begin{cases} \dot{\mathbf{x}}(t) = \mathbf{A}\mathbf{x}(t) + \mathbf{B}_1\mathbf{w}(t) + \mathbf{B}_K\mathbf{x}(t-\tau), \\ \mathbf{z}(t) = \mathbf{C}_1\mathbf{x}(t) + \mathbf{D}_K\mathbf{x}(t-\tau) + \mathbf{D}_{11}\mathbf{w}(t). \end{cases} \tag{29}$$

Theorem 1. *Given $\bar{\tau}$, if there exist real matrices $\mathbf{P} > 0$, $\mathbf{R} > 0$, $\mathbf{Q} > 0$, and the two free-weighting matrices $\mathbf{N}_1$ and $\mathbf{N}_2$, satisfying the matrix inequality*

$$\Xi = \begin{bmatrix} \Xi_{11} & \Xi_{12} & \mathbf{PB}_1 & \bar{\tau}\mathbf{A}^T & \bar{\tau}\mathbf{N}_1 & \mathbf{C}_1^T \\ * & \Xi_{22} & \mathbf{0} & \bar{\tau}\mathbf{B}_K^T & \bar{\tau}\mathbf{N}_2 & \mathbf{D}_K^T \\ * & * & -\gamma^2\mathbf{I} & \bar{\tau}\mathbf{B}_1^T & \mathbf{0} & \mathbf{D}_{11}^T \\ * & * & * & -\bar{\tau}\mathbf{R}^{-1} & \mathbf{0} & \mathbf{0} \\ * & * & * & * & -\bar{\tau}\mathbf{R} & \mathbf{0} \\ * & * & * & * & * & -\mathbf{I} \end{bmatrix} < 0 \tag{30}$$

where $\Xi_{11} = \mathbf{PA} + \mathbf{A}^T\mathbf{P} + \mathbf{Q} + \mathbf{N}_1 + \mathbf{N}_1^T, \Xi_{12} = \mathbf{PB}_K - \mathbf{N}_1 + \mathbf{N}_2^T, \Xi_{22} = -\mathbf{Q} - \mathbf{N}_2 - \mathbf{N}_2^T$, the closed-loop system (29) is asymptotically stable with H_∞ performance index γ for any time delay τ satisfying $0 \le \tau \le \bar{\tau}$.

The proof of this Theorem can be found in Ref.13.

6.1. *Maximum Time Delay with a Known Controller*

For a control system, the task of control design is to obtain the feedback weighting **K**. The weighting **K** is different when different control design methods are used. We hope that the controller designed should have strong robustness to the variation of structural intrinsic parameters and external disturbance. As mentioned in the above, time delay exists inevitably in active control systems. Therefore, it is worth considering the following question: in what range of time delay the controller **K** designed in the case of no time delay is applicable? We certainly expect that this range to be as large as possible. In Ref.13, this question is solved by using the genetic algorithm and the feasp function of LMI Toolbox in MATLAB. The inequality (30) is a nonlinear matrix inequality, which cannot be solved directly using the LMI Toolbox. In Ref.13, the parameter-adjusting method is used to handle Eq.(30) by introducing two adjustable parameters λ and ρ to obtain the following new matrix inequality

$$\begin{bmatrix} \omega_{11} & \omega_{12} & \mathbf{B}_1 & \omega_{13} & \mathbf{0} & \overline{\mathbf{P}}\mathbf{C}_1^T & \overline{\mathbf{P}} \\ * & \omega_{22} & \mathbf{0} & \bar{\tau}\rho\mathbf{Y}^T\mathbf{B}_2^T & \bar{\tau}\overline{\mathbf{R}} & \rho\mathbf{Y}^T\mathbf{D}_{12}^T & \mathbf{0} \\ * & * & -\gamma^2\mathbf{I} & \bar{\tau}\mathbf{B}_1^T & \mathbf{0} & \mathbf{D}_{11}^T & \mathbf{0} \\ * & * & * & -\bar{\tau}\overline{\mathbf{R}} & \mathbf{0} & \mathbf{0} & \mathbf{0} \\ * & * & * & * & -\bar{\tau}\overline{\mathbf{R}} & \mathbf{0} & \mathbf{0} \\ * & * & * & * & * & -\mathbf{I} & \mathbf{0} \\ * & * & * & * & * & * & -\overline{\mathbf{Q}} \end{bmatrix} < 0. \tag{31}$$

If $\bar{\tau}$, λ, and ρ are all known, equation (31) becomes a linear matrix inequality, **K** can be obtained by solving this inequality using the MATLAB LMI Toolbox. Assume that the feedback weighting **K** has been obtained using a certain control design method such as the classical LQR method. Equation (31) is still nonlinear after the substitution of **K** into Eq.(31). Substituting a small delay τ into Eq.(31) and then using the feasp function of LMI Toolbox to verify if equation (31) is satisfied. If equation (31) holds, increasing τ, and then repeat the above process until equation (31) does not hold. In this way, the maximum time delay can be determined. The above process for determining the maximum time delay also involves the choosing of λ and ρ. Here we also use the genetic algorithm to optimize λ and ρ. The objective function used herein is the maximum time delay $\bar{\tau}$, given by

$$\max_{\lambda,\rho} \bar{\tau} \quad \text{subject to} \quad \text{LMI (31).} \tag{32}$$

The detailed calculation steps can be found in Ref.13. The maximum time delay for a known **K** can be determined using Eq.(32).

6.2. *Controller Design with Known Maximum Time Delay*

Next, we consider the solution of H_∞ controller with known $\bar{\tau}$. If $\bar{\tau}$, λ, and ρ are all known, equation (31) becomes a linear matrix inequality, $\mathbf{K}$ can be obtained by solving this inequality using the MATLAB LMI Toolbox. This controller can guarantee the closed-loop system (28) being asymptotically stable as long as the real time delay in the system is within $0 \leq \tau \leq \bar{\tau}$. Herein we consider the following question: when $\bar{\tau}$ is known, how to select λ and ρ so that $\mathbf{K}$ achieves better control effectiveness. In Ref.13, the genetic algorithm is used to optimize λ and ρ by minimizing the index γ in the LMI (31). The optimization problem is described as

$$\min_{\lambda,\rho} \gamma \quad \text{subject to} \quad \text{LMI (31)}. \tag{33}$$

After obtaining the optimal λ and ρ, substituting λ, ρ, and $\bar{\tau}$ into Eq.(31), the optimal $\mathbf{K}'$ can be obtained.

6.3. *The Largest Time Delay for System Stability with Unknown Controller*

In Section 6.1, the maximum time delay $\bar{\tau}$ corresponding to a known $\mathbf{K}$ is discussed. If $\mathbf{K}$ is unknown, how much will the maximum time delay be? For this case, how to obtain the feedback gain $\mathbf{K}$? These questions are discussed below.

From the above we know that system stability can be verified using the matrix inequality (31). The system is asymptotically stable if equation (31) is satisfied. Here we directly use the optimization problem given by Eq.(32) to determine the maximum time delay, where the objective function is $\bar{\tau}$, constraint condition is the matrix inequality (31), and the optimization variables are λ and ρ, given by

$$\max_{\lambda,\rho} \bar{\tau} \quad \text{subject to} \quad \text{LMI (31)}. \tag{34}$$

After the maximum time delay $\bar{\tau}$ is obtained, the corresponding feedback weighting $\overline{\mathbf{K}}$ can be solved using the method in Section 6.2. Since the feedback gain is not specified in advance in the solution of $\bar{\tau}$, this $\bar{\tau}$ is the biggest one of all the maximum time delays.

7. Delayed Positive Feedback Control Technique[14]

In active structural control, negative feedback is often used in control design, where positive feedback may only enlarge system responses and deteriorate system stability. This is the case when time delay is not involved in the control path. If time delay is appropriately introduced into the feedback control loop, positive feedback is not necessarily bad for control efficiency. On the contrary it may lead to a better control result. Zaihua Wang studied in detail the delayed positive control technique of linear vibration system with negative damping and using the Lambert W function to calculate the maximal real part of all the characteristic roots of delayed positive feedback control system. Next, we introduce the delayed positive feedback technique using a flexible beam as research object.

Consider the control of the first r_1 modes of the beam. The modal state equation can be written as

$$\dot{\mathbf{Z}}(t) = \overline{\mathbf{A}}\mathbf{Z}(t) + \overline{\mathbf{B}}\mathbf{V}(t) \tag{35}$$

where $\mathbf{Z}(t) = [\mathbf{\Phi}^T(t),\ \dot{\mathbf{\Phi}}^T(t)]^T$, $V_j(t)$ is the control term. If negative feedback technique is used, the controller can be written as $\mathbf{V}(t) = -\mathbf{LZ}(t)$, where $\mathbf{L}$ is the feedback gain, which can be obtained using the classical LQR method. Removing the negative sign, the positive feedback controller can be written as $\mathbf{V}(t) = \mathbf{LZ}(t)$.

Adding time delay τ_j in state parameter, the ith modal equation can be written as

$$\ddot{\phi}_i(t) + 2\zeta_i\omega_i\dot{\phi}_i(t) + \omega_i^2\phi_i(t) = \sum_{j=1}^{r_1}[L_{ji}\phi_i(t-\tau_j) + L_{j(r_1+i)}\dot{\phi}_i(t-\tau_j)] \tag{36}$$

where L_{ji} and $L_{j(r_1+i)}$ are the displacement and velocity feedback gains corresponding to the ith mode, respectively. The characteristic equation of Eq.(36) may be written as

$$P(s) = s^2 + 2\zeta_i\omega_i s + \omega_i^2 - \sum_{j=1}^{r_1}[L_{ji}e^{-s\tau_j} + sL_{j(r_1+i)}e^{-s\tau_j}] \tag{37}$$

where s is the Laplace variable. The characteristic roots are the solution of $P(s) = 0$. Equation (37) is a transcendental equation due to the existence of time delay, and the number of roots of $P(s) = 0$ is infinite. Define

$$\alpha_0 = \max\{\mathrm{Re}(s) : P(s) = 0\} \tag{38}$$

where $\mathrm{Re}(s)$ stands for the real part of s, α_0 represents the maximal real part of all the characteristic roots. For a given τ_j, if $\alpha_0 < 0$, all the characteristic roots will have negative real parts, so the delayed positive feedback controller can stabilize the system. The characteristic roots of Eq.(37) can be calculated numerically using the DDE-BIFTOOL Toolbox in MATLAB. By tracking changes of α_0 with time delay, the time delay available for positive feedback can be determined. Selecting proper time delay τ_j, the delayed positive feedback controller can be written as

$$\mathbf{V}(t) = \mathbf{LZ}(t-\tau_j). \tag{39}$$

8. Time Delay Experiments

Today many researches have been done on the time delay problem, the existing studies are on the theoretical basis and few efforts were ever made on the experiment. By working on a flexible beam, plate and manipulator, we have done some experimental studies on time delay. Below are brief introductions of these experiments.

8.1. *Continuous and Discrete Time-delayed Controllers*[5–8,11]

The experiment system is structured based on the DSP TMS320F2812. The detailed signal flow and process may consult Refs.5–11, 13, 14. The experiments of the discrete time-delayed controller can be found in Refs.6–8, 11 and the continuous time-delayed controller Refs.5, 7.

In Refs.7, 8, 11, the time-delayed experiment of a flexible aluminum alloy beam is presented, single and double delays are both considered, in which the discrete time-delayed controller is adopted. In the experiments, foil strain gauge is used as sensor and PZT patch actuator. The controller is designed using the discrete LQR method and VSC method, respectively. Free and forced vibrations of the beam are both considered. Experimental results indicate that the time-delayed controller can suppress the beam vibration effectively, and is available for both small and large time delay. Theoretical results agree well with experimental ones.

In Refs.6, 7, the time-delayed experiment of a flexible aluminum alloy plate is presented, single and double delays are both considered too, in which the discrete time-delayed controller is adopted. In the experiments, PZT patches are used as sensor and actuator, and their optimal positions on the plate are determined using the PSO. The controller is designed using the discrete LQR method. Experimental results verify the validity of the proposed time-delayed controller.

In Refs.5, 7, the time-delayed experiment of a flexible manipulator is done, where the continuous time-delayed controller is used and is designed using the optimal tracking control method. The designed controller is required to drive the system to reach 180° position within a second and the residual vibration of the manipulator after arriving at the end position must be suppressed by the controller. The joint torque is used to control the large motion of the system, and the PZT actuator stuck near the root of the manipulator is used for vibration control of the manipulator. In the experiment, three cases are considered: (i) The AC servo motor is only used for active control of the system and time delay is assumed for the motor; (ii) The AC servo motor and PZT actuator are both used, but only PZT actuator exists time delay; (iii) The AC servo motor and PZT actuator are both used and they exist different time delay. The feasibility and effectiveness of time-delayed feedback control is experimentally verified in Refs.5, 7.

8.2. *Parameter Robustness of Time-delayed Controller*[9]

In Ref.9, parameter robustness of time-delayed controller is experimentally studied using an epoxy resin plate as research object, where single and double delays are both considered, the treating process method is in continuous form and the controller is designed using the robust control method. In the experiment, PZT patches are used as sensor and actuator. The robustness of time-delayed controller to structural parameter and time delay are both studied.

8.3. *Robust H_∞ Time-delayed Controller*[10,13]

In Refs.10, 13, experiments of Robust H_∞ time-delayed control problem are presented for an epoxy resin plate, where PZT patches are used as sensor and actuator. The problem of maximum time delay for stability is experimentally studied firstly. By using the classical LQR method, the feedback weighting $\mathbf{K}$ without considering time delay is obtained. With the known $\mathbf{K}$, the maximum stable time delay $\bar{\tau}$ can be determined by using the method given in Refs.10, 13. Experimental results in Refs.10, 13 show that $\mathbf{K}$ is always available for suppressing the plate vibration when the real delay in the control system is within $[0, \bar{\tau}]$, and is unavailable when exceeds this range.

Then consider the case with known $\bar{\tau}$. By using the method given in Refs.10, 13, the optimal feedback weighting $\mathbf{K}'$ can be obtained. Experimental results in Refs.10, 13 indicate that the control effectiveness of $\mathbf{K}'$ is better than $\mathbf{K}$. When the delay in the system is larger than $\bar{\tau}$, $\mathbf{K}$ and $\mathbf{K}'$ are both unavailable.

Consider the case that $\bar{\tau}$ and $\mathbf{K}$ are all unknown in advance. The biggest time delay $\bar{\tau}$ for stability is solved at first using the optimization algorithm given in Section 6.3, and then H_∞ controller $\overline{\mathbf{K}}$ is determined using the method in Section 6.2. Experimental results in Refs.10, 13 show that when the real delay in the control system is smaller than $\bar{\tau}$, the plate vibration can be suppressed effectively, and the control effectiveness becomes worse when larger than $\bar{\tau}$.

8.4. *Delayed Positive Feedback Controller*[14]

In Ref.14, delayed positive feedback control experiment is presented on a flexible beam, where PZT patch is used as actuator and foil strain gauge sensor, and the LQR method is adopted to obtain feedback weighting. Three cases are considered in the experiment: (i) displacement feedback, (ii) velocity feedback, (iii) both displacement and velocity feedbacks. By numerical simulations, the changing curves of the maximal real-part of the characteristic roots varying with time delay for the above three cases are obtained to determine available time delay for delayed positive feedback. The experimental results indicate that, the delayed positive feedback controller can effectively reduce the beam vibration when the chosen time delay is within stable area.

9. Concluding Remarks

Time-delayed problem is a challenging research topic, where many aspects need further studies and more efforts. Time-delayed problems in the structural active control involve multiple disciplines such as structure dynamics, control theory and mathematics. A sound solution to these problems depends on an effective integration of all these fields and will make significant advances to engineering applications. This chapter summarizes our latest theoretical and experimental research on time-delayed problems. We hope that this research is useful to other scholars and also hopes for their suggestions to further advance the theory of delayed system dynamics.

Acknowledgments. This work was supported by the Key Project (Grant No. 11132001) and the General Projects (Grant Nos. 11072146 and 11002087) of the National Natural Science Foundation of China, and the Specialized Research Fund for the Doctoral Program of Higher Education of China (Grant No. 20110073110008).

References

1. L. X. Chen and G. P. Cai, Delay identification in time-delay dynamic system using particle swarm optimizer, *Chinese Journal of Applied Mechanics*, **27**, 433(2010). (in Chinese)
2. G. P. Cai and J. Z. Huang, Optimal control method for seismically excited building structures with time delay in control, *Journal of Engineering Mechanics-ASCE*, **128**, 602(2002).
3. G. P. Cai, J. Z. Huang, and S. X. Yang, An optimal control method for linear systems with time delay, *Computers and Structures*, **81**, 1539(2003).
4. G. P. Cai and L. X. Chen, Delayed feedback control experiments on some flexible structures, *Acta Mechanica Sinica*, **26**, 951(2010).
5. L. X. Chen and G. P. Cai, Experimental studies on active control of rotating flexible beam with time delay, *Chinese Journal of Theoretical and Applied Mechanics*, **40**, 520(2008). (in Chinese)
6. L. X. Chen, G. P. Cai, and J. Pan, Experimental study of delayed feedback control for a flexible plate, *Journal of Sound and Vibration*, **322**, 629(2009).
7. L. X. Chen, *Delayed Feedback Control and Experiment Study of Structural Vibration*, PhD Dissertation in Shanghai Jiaotong University (2009). (in Chinese)
8. L. X. Chen and G. P. Cai, Experimental study of time-delay variable structure control for a flexible beam under forced vibration, *Chinese Journal of Theoretical and Applied Mechanics*, **41**, 410(2009). (in Chinese)
9. T. Zhao, L. X. Chen, and G. P. Cai, Experimental study of H_∞ control for a flexible plate, *Journal of Vibration and Control* (in press, doi: 10.1177/1077546311419986).
10. T. Zhao, L. X. Chen, and G. P. Cai, Theoretical and experimental studies of H_∞ control for a flexible plate with time delay, *Chinese Journal of Theoretical and Applied Mechanics*, **43**, 1043(2011). (in Chinese)
11. J. Pan and G. P. Cai, Experiment study of a flexible cantilever beam with time delay in control, *Journal of Astronautics*, **29**, 206(2008). (in Chinese)
12. K. Liu, L. X. Chen, and G. P. Cai, Active control of a nonlinear and hysteretic building structure with time delay, *Structural Engineering and Mechanics*, **40**, 431(2011).
13. T. Zhao, L. X. Chen, and G. P. Cai, Experimental study of H_∞ control of flexible plate with time delay, *Shock and Vibration*, **20**, 227(2013).
14. K. Liu, L. X. Chen, and G. P. Cai, Experimental study of delayed positive feedback control for a flexible beam, *Theoretical & Applied Mechanics Letters*, **1**, 063003(2011).

Chapter 11

Switching Control of Uncertain Dynamical Systems with Time Delay

Jian-Qiao Sun
School of Engineering, University of California
Merced, CA 95344, USA
jqsun@ucmerced.edu

Xiao-Yan Zhang, Zhi-Chang Qin, Shun Zhong
Department of Mechanics, Tianjin University
Tianjin 300072, China

Uncertainties are common in dynamical systems. Switching controls are often used to deal with uncertainties in the system. This chapter discusses two cases of dynamical systems with uncertainties. The first case studies systems with uncertain time delays. Uncertain time delay is assumed to fall in a range with known upper and lower bounds. The supervisory control with hysteretic switching algorithm is proposed for the system. The second case deals with time-delayed systems with uncertain parameters and nonlinear behavior. Robust sliding mode controls are designed to regulate the system performance. In both cases, switching in the control plays an important role in stabilizing the system and maintaining proper performance.

1. Introduction

Whenever there is uncertainty, switching controls are used to maintain proper performance of dynamical systems. Uncertainty can be associated with the lack of the knowledge of the system leading to the model with uncertain parameters or functional structures. When there is time delay in the system, the delay is often uncertain as well. This chapter presents two case studies of analysis and control of dynamical systems with system uncertainties and uncertain time delay.

There are many studies of control systems with unknown and time-varying time delays. Chen et al. derived sufficient conditions for the existence of the guaranteed cost output-feedback controller in terms of matrix inequalities for uncertain dynamic systems with time delay[1]. A class of iterative learning controls with uncertain state delay and control delay has been studied[2]. The stability problem is treated in the integral quadratic constraint (IQC) framework. Kwon, Park, and Lee[3] investigated delay-dependent robust stability for neutral systems with the help of the Lyapunov method. Robust stability of systems with random time-varying delay is studied in Ref.4. Sufficient conditions for the exponential mean square stability of the system are derived by using the Lyapunov functional method and the linear matrix inequality (LMI) technique.

The Lyapunov method is also used for the stability analysis of systems with time-varying delay with known lower and upper bounds[5]. The Lyapunov function dependent on the known upper bound of uncertain state-delays is derived in the study of model predictive controls (MPC) for a constrained linear digital system with uncertain state-delays[6]. Jiang and Han[7] have studied robust stability of uncertain linear systems with interval time-varying delay. Stability of systems with uncertain time-varying bounded delays in the feedback loop is studied in Ref.8. The system is assumed to have time-varying structured uncertainties and interval time-varying delays. A compensation scheme consisting of a fuzzy-PID controller and a neural network compensator is proposed for real-time control over the network[9]. This scheme reduces the influence of time delays on stability while maintaining the system performance. According to Miller and Davison[10], given a finite-dimensional LTI plant and an upper bound on the admissible time delay, there is no general theory for designing a controller to handle an arbitrarily large uncertain delay. The authors further show that given a finite-dimensional LTI plant and an upper bound on the admissible time delay, there exists a linear periodic controller which robustly stabilizes the plant.

The supervisory control can be applied to the systems with uncertain time delay when the uncertain time delay is bounded with known lower and upper bounds[11]. The supervisory control proposes to use several estimates of uncertain parameters for the system model[12–15]. For each estimate of the parameter, a control is designed. A supervisor monitors the real-time response of the system, selects a plant model according to a switching criterion and implements the corresponding control.

Sliding mode control is one of the examples of variable structure controls and often contains a switching term to deal with system uncertainty. Reference 10 proposes a sliding surface in an augmented space formed by the system output and the reduced order estimator to stabilize nonlinear systems with mismatched nonlinear term and the nonlinear uncertainty subject to time-varying delay. The Lyapunov-Razumikhin approach is used to analyze the stability of the sliding motion. A virtual state concept for sliding surface design has been investigated[17,18]. Roh and Oh[19,20] study sliding mode control for the robust stabilization of uncertain linear input-delay systems with nonlinear parametric perturbations. The proposed sliding surface includes a predictor to compensate for the input delay of the system. The proposed sliding surface includes a predictor which consists of the current state and the past control during the period of delay. The predictor compensates for the input delay of the system. A state predictor with an adaptation scheme is used. References 21, 22 consider a delay-dependent sliding mode control for uncertain neutral delay systems. A delay-independent sliding mode control has been applied to a class of parabolic linear uncertain distributed parameter systems with time-varying delays[23]. A sliding mode control with passivity has been examined for a class of uncertain nonlinear singular time-delayed systems[24]. An integral-type switching surface function is designed by taking the singular matrix into account. A delay-dependent sufficient condition is proposed in terms of linear matrix inequality, which guarantees the sliding mode dynamics. The same integral sliding surface is used in the sliding mode control for uncertain stochastic systems with time-varying delay[25]. The method of linear matrix inequalities is used to derive conditions for the global stochastic stability. The work by Xu and Cao[26] makes use of time-delayed control during the sliding phase to reduce the switching gain of the sliding mode control. Song and Sun applied sliding mode control to a nonlinear system with parameter uncertainty and known time delay[27].

In Section 2, we present the supervisory control and switching rule. We review the mapping method and a high-order method for control design. Section 3 presents sliding mode controls for uncertain nonlinear dynamical systems with time delay.

2. Supervisory Control for Systems with Uncertain Time Delay

Consider a dynamical system with time delay given by

$$\dot{\mathbf{x}} = \mathbf{f}\left(\mathbf{x}\left(t\right), \mathbf{x}\left(t-\tau\right), t\right) + \mathbf{B}\mathbf{u}\left(t\right), \tag{1}$$

where $\mathbf{x} \in \mathbb{R}^n$, $\mathbf{u} \in \mathbb{R}^m$, $\mathbf{f}$ describes the system dynamics with time delay, and $\mathbf{B} = \{B_{ij}\}$ is the control influence matrix. The time delay τ is assumed to be slowly time-varying, and lies in an interval $[\tau_{\min}, \tau_{\max}]$ where the minimum and maximum time delays are assumed to be known. We discretize $[\tau_{\min}, \tau_{\max}]$ into $M_\tau - 1$ intervals so that $\tau_{\min} = \tau_1 < \tau_2 < \cdots < \tau_{M_\tau} = \tau_{\max}$. Consider M_τ models of the time-delayed system as

$$\dot{\mathbf{x}}_i = \mathbf{f}\left(\mathbf{x}_i\left(t\right), \mathbf{x}_i\left(t-\tau_i\right), t\right) + \mathbf{B}\mathbf{u}_i\left(t\right), \quad 1 \le i \le M_\tau. \tag{2}$$

Assume that we have obtained a set of M_τ optimal gains $\mathbf{K}_i$ $(1 \le i \le M_\tau)$ such that $\mathbf{u}_i\left(t\right) = -\mathbf{K}_i\mathbf{x}_i\left(t-\tau_i\right)$ for the set of time delays sampled from the interval $[\tau_{\min}, \tau_{\max}]$. The control $\mathbf{u}_i\left(t\right)$ must be stable for all the time delay $\tau \in [\tau_{\min}, \tau_{\max}]$.

Following the concept of the supervisory control[12–15], we define an estimation error as

$$\mathbf{e}_i = \mathbf{x}_i\left(t\right) - \mathbf{x}\left(t\right), \quad 1 \le i \le M_\tau, \tag{3}$$

where $\mathbf{x}\left(t\right)$ is the output of the system with unknown time delay. In the experiment, $\mathbf{x}\left(t\right)$ would be obtained from measurements. Consider a positive function of the estimation error $F_i(\mathbf{e}_i) > 0$. An example is $F_i(\mathbf{e}_i) = ||\mathbf{e}_i||^2$. Define a switching index $\pi_i(t)$ such that

$$\begin{gathered}\dot{\pi}_i(t) + \lambda_i\pi_i(t) = F_i(\mathbf{e}_i) \quad (\lambda_i > 0),\\ \pi_i(0) = 0,\end{gathered} \tag{4}$$

where the parameter λ_i defines the bandwidth of the low pass filter. The general solution for $\pi_i(t)$ can be obtained as

$$\pi_i(t) = e^{-\lambda_i t}\pi_i(0) + \int_0^t e^{-\lambda_i(t-\tau)}F_i(\mathbf{e}_i(\tau))d\tau. \tag{5}$$

The hysteretic switching algorithm is stated as follows[14,15]. Assume that the system is sampled at time interval Δt. At the kth time step, the system is under control with the gain $\mathbf{K}_j$ and the associated switching signal is $\pi_j(k)$. At the $(k+1)$th step, if there is an index i such that $\pi_i(k) < (1-\eta)\pi_j(k)$ where $\eta > 0$ is a small number, we switch to the gain $\mathbf{K}_i$. Otherwise, we continue with the gain $\mathbf{K}_j$. η is known as the hysteretic parameter and prevents the system from switching too frequently.

Many methods can be used to design controls in order to apply the hysteretic switching rule. In the following we introduce two methods.

2.1. *Optimal Feedback Gains via Mapping*

The mapping method to design optimal feedback control gains has been studied extensively[28–30]. The mapping of an extended state vector involving the delayed response can be constructed by the method of semi-discretization or continuous time approximation. For linear time-invariant as well as periodic systems[30–34], the mapping reads

$$\mathbf{y}(j+1) = \mathbf{\Phi}(\mathbf{K})\mathbf{y}(j), \quad j = 0, 1, 2, \ldots, \tag{6}$$

where $\mathbf{y}$ is a $n(N+1) \times 1$ vector, N is the discretization level such that $\Delta\tau = \tau/N$ is the sample time. The delayed feedback control $\mathbf{u}(t) = -\mathbf{K}\mathbf{x}(t-\tau)$ has been substituted to obtain the mapping. Therefore, the mapping $\mathbf{\Phi}$ is a function of the control gain $\mathbf{K}$. Note that the dimension of optimal gain $\mathbf{K}$ is $m \times n$ and is independent of the discretization level N.

Consider a compact and bounded region $\mathbf{\Omega} \subset \mathbb{R}^{m\times n}$ where $\mathbf{K} \in \mathbf{\Omega}$. We can find the domains of stability and optimal control gains in the region to minimize the largest magnitude of the eigenvalues of $\mathbf{\Phi}$. This leads to the following optimization problem

$$\min_{\mathbf{K}\in\mathbf{\Omega}} [\max|\lambda(\mathbf{\Phi})|] \quad \text{subject to} \quad |\lambda|_{\max} < 1. \tag{7}$$

The control performance criterion is the decay rate of the mapping $\mathbf{\Phi}$ over one iteration. In the frequency domain, we have found that the optimal feedback gains maximize the damping of the dominant closed-loop poles of the system that are closest to the imaginary axis of the s-plane[29].

2.2. *High-order Control*

In the above mapping method, the delayed response $\mathbf{x}(t-\tau)$ is discretized and becomes part of the extended state. In the following, we discretize the delayed control $\mathbf{u}(t-\tau)$ instead. This leads to a high-order control with better performance. Consider a linear system with a delayed control,

$$\dot{\mathbf{x}} = \mathbf{A}\mathbf{x}(t) + \mathbf{B}\mathbf{u}(t-\tau). \tag{8}$$

Let $\hat{t} = t/\tau$ be a scaled time. Denote $\hat{\mathbf{A}} = \tau\mathbf{A}$, $\hat{\mathbf{x}}(\hat{t}) = \mathbf{x}(\tau\hat{t})$, $\hat{\mathbf{B}} = \tau\mathbf{B}$, and $\hat{\mathbf{u}}(\hat{t}-1) = \mathbf{u}(\tau(\hat{t}-1))$. Equation (8) now reads

$$\frac{d\hat{\mathbf{x}}(\hat{t})}{d\hat{t}} = \hat{\mathbf{A}}\hat{\mathbf{x}}(\hat{t}) + \hat{\mathbf{B}}\hat{\mathbf{u}}(\hat{t}-1). \tag{9}$$

After scaling, the time delay is always one in the new time $\hat{t}$. Let $\Delta\tau = 1/N$. Following the concept of semi-discretization, we construct a mapping from Eq.(9) over the time interval $[\hat{t}_k, \hat{t}_{k+1}]$ where $\hat{t}_k = k\Delta\tau$.

$$\hat{\mathbf{x}}(k+1) = \tilde{\mathbf{A}}\hat{\mathbf{x}}(k) + \tilde{\mathbf{B}}\hat{\mathbf{u}}(k-N), \tag{10}$$

where

$$\tilde{\mathbf{A}} = e^{\hat{\mathbf{A}}\Delta\tau}, \quad \tilde{\mathbf{B}} = \int_0^{\Delta\tau} e^{\hat{\mathbf{A}}(\Delta\tau - s)}\hat{\mathbf{B}}ds, \tag{11}$$

and we denote $\hat{\mathbf{x}}((k+1)\Delta\tau)$ as $\hat{\mathbf{x}}(k+1)$, and $\hat{\mathbf{u}}(k\Delta\tau - N\Delta\tau)$ as $\hat{\mathbf{u}}(k-N)$. Recall that $N\Delta\tau = 1$. Introduce a $(n+Nm)\times 1$ extended state vector as

$$\mathbf{y}(k) = [\hat{\mathbf{x}}(k), \hat{\mathbf{u}}(k-N), \hat{\mathbf{u}}(k-N+1), \ldots, \hat{\mathbf{u}}(k-1)]^T. \tag{12}$$

Then, equation (10) can be written in terms of the extended vector without time delay as

$$\mathbf{y}(k+1) = \bar{\mathbf{A}}\mathbf{y}(k) + \bar{\mathbf{B}}\mathbf{u}(k), \tag{13}$$

where

$$\bar{\mathbf{A}} = \begin{bmatrix} \tilde{\mathbf{A}} & \tilde{\mathbf{B}} & \mathbf{0} & \cdots & \mathbf{0} \\ \mathbf{0} & \mathbf{0} & \mathbf{I} & \cdots & \mathbf{0} \\ \vdots & \vdots & \vdots & & \vdots \\ \mathbf{0} & \mathbf{0} & \mathbf{0} & \cdots & \mathbf{I} \\ \mathbf{0} & \mathbf{0} & \mathbf{0} & \cdots & \mathbf{0} \end{bmatrix}, \quad \bar{\mathbf{B}} = \begin{bmatrix} \mathbf{0} \\ \mathbf{0} \\ \vdots \\ \mathbf{0} \\ \mathbf{I} \end{bmatrix}. \tag{14}$$

Consider the full state feedback control[35]

$$\begin{aligned} \mathbf{u}(k) &= -\mathbf{K}\mathbf{y}(k) \\ &= -\mathbf{K}_1\hat{\mathbf{x}}(k) - \mathbf{K}_2\hat{\mathbf{u}}(k-N) \\ &\quad - \mathbf{K}_3\hat{\mathbf{u}}(k-N+1) - \cdots - \mathbf{K}_{N+1}\hat{\mathbf{u}}(k-1). \end{aligned} \tag{15}$$

We refer to this control as a high-order control because the gain matrix $\mathbf{K}$ is $m\times(n+Nm)$. Recall that the gain matrix for the control $\mathbf{u}(k) = -\mathbf{K}\mathbf{x}(k)$ is $m\times n$. Because of the much higher dimension of the gain matrix in the high-order approach, it is no longer practical to apply the optimization scheme in Eq.(7) to find optimal gains for the control $\mathbf{u}(k) = -\mathbf{K}\mathbf{y}(k)$.

The feedback gain of the high-order control can be more efficiently designed with the digital LQR optimal formulation to minimize a cost function[36]

$$J = \frac{1}{2}\sum_{k=0}^{\infty} \left[\mathbf{y}^T(k)\mathbf{Q}\mathbf{y}(k) + \mathbf{u}^T(k)\mathbf{R}\mathbf{u}(k)\right], \tag{16}$$

where $\mathbf{Q}$ is a $(n+Nm)\times(n+Nm)$ non-negative definite symmetric matrix and $\mathbf{R}$ is a $m \times m$ positive definite symmetric matrix. We obtain the optimal gain as

$$\mathbf{K} = [\mathbf{R} + \bar{\mathbf{B}}^T\mathbf{S}_\infty\bar{\mathbf{B}}]^{-1}\bar{\mathbf{B}}^T\mathbf{S}_\infty\bar{\mathbf{A}}, \tag{17}$$

where $\mathbf{S}_\infty$ satisfies the algebraic Riccati equation[37],

$$\mathbf{S}_\infty = \bar{\mathbf{A}}^T[\mathbf{S}_\infty - \mathbf{S}_\infty\bar{\mathbf{B}}[\mathbf{R} + \bar{\mathbf{B}}^T\mathbf{S}_\infty\bar{\mathbf{B}}]^{-1}\bar{\mathbf{B}}^T\mathbf{S}_\infty]\bar{\mathbf{A}} + \mathbf{Q}. \tag{18}$$

2.3. *Stability Requirements for Switching*

We apply both methods to a set of discrete time delays $\tau_i \in [\tau_{\min}, \tau_{\max}]$ $(1 \leq i \leq M_\tau)$ to create M_τ controls. Each control must be stable for any time delay $\tau_{\min} \leq \tau \leq \tau_{\max}$. With the mapping method, we can find the stability domains in the gain space and an optimal gain for the given time delay τ_i. Examples of the stability domains and optimal gains are shown in Fig.1. When the optimal gains for all the time delays τ_i fall in the intersection of all the stability domains, the feedback control designed for each time delay τ_i is stable for any time delay $\tau_{\min} \leq \tau \leq \tau_{\max}$. Consequently, it is possible to use the hysteretic algorithm to switch among the pre-designed controls and to keep the system stable all the time. When an optimal gain is out of the intersection, the control with that gain can destabilize the system with some time delay in the range $[\tau_{\min}, \tau_{\max}]$. The nature of conditional stability of the delayed control limits the size of the unknown time delay range $[\tau_{\min}, \tau_{\max}]$ because the stability domains change significantly with the time delay, particularly for periodic systems[29].

The LQR design for the high-order control does not yield stability domains explicitly. We have to numerically check the stability of the control designed for a given τ_i with respect to all the time delay τ in the range $[\tau_{\min}, \tau_{\max}]$. It should be pointed out that the size of the range $[\tau_{\min}, \tau_{\max}]$ in which the LQR optimal control designed for a given τ_i is stable depends on the robustness of the control with respect to the time delay τ.

2.4. *Example of LTI System*

2.4.1. *Low-order Feedback Control with Optimal Gains*

Consider the feedback control problem of the linear oscillator,

$$\dot{\mathbf{x}}(t) = \begin{bmatrix} 0 & 1 \\ -\omega^2 & -2\zeta\omega \end{bmatrix}\mathbf{x}(t) + \begin{bmatrix} 0 \\ 1 \end{bmatrix}u(t). \tag{19}$$

Take $\omega = 2$, $\zeta = 0.05$, and the discretization number of the time delayed response $N = 20$. Consider the state feedback control $u(t) = -[k_p, k_d]\mathbf{x}(t - \tau)$ with an uncertain transport delay τ. The nominal value of the time delay is $\tau_{\text{nom}} = \pi/30$. The lower and upper bounds of the time delay are given by $\tau_{\min} = 0.7\tau_{\text{nom}}$ and $\tau_{\max} = 1.6\tau_{\text{nom}}$. We sample four time delays, $\tau_1 = \tau_{\min}$, $\tau_2 = \tau_{\text{nom}}$, $\tau_3 = 1.3\tau_{\text{nom}}$, and $\tau_4 = \tau_{\max}$, to design optimal feedback gains according to the optimization scheme (7).

The optimal gains associated with the four time delays are listed in Table 1. The associated stability domains in the gain space are shown in Fig. 1. It is clear that the optimal gains of all the controls with different time delays fall in the intersection of the stability domains. Hence, it is possible to use the hysteretic algorithm to switch among the pre-designed controls and to keep the system stable all the time.

Table 1. Optimal feedback gains for the four certain time delay systems.

Time Delay	k_p	k_d
τ_1	13.970	6.3200
τ_2	5.3000	4.2800
τ_3	2.2400	3.2600
τ_4	1.2200	2.7500

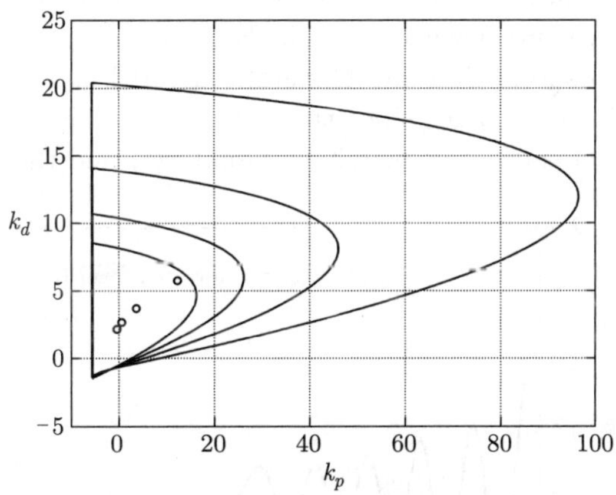

Fig. 1. Boundaries of stability domains (lines) in the gain space and the optimal feedback gains (o) for the LTI systems (29) with four sampled time delays τ_i $(i = 1, 2, 3, 4)$.

Figure 2 shows the closed-loop response of the system under the feedback control with all four sampled time delays when the system true time delay is taken to be τ_4, *which is assumed to be unknown and not considered in the control design.* As it can be seen from the figure, when the control designed for the time delay that is close to τ_4 is implemented, the performance is acceptable. Otherwise, the performance deteriorates as seen in the upper-left sub figure.

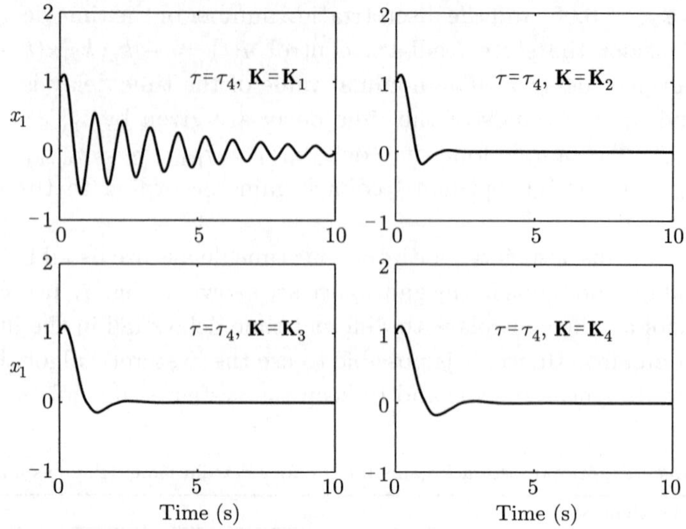

Fig. 2. The closed-loop response of the LTI system (19) under feedback controls designed for the sampled time delays when the system true time delay is τ_4 and is assumed to be unknown. When the feedback gain is $\mathbf{K}_2$, $\mathbf{K}_3$, or $\mathbf{K}_4$ for the time delay close to the actual time delay, the control performance is quite good. When the mismatch gap is large, or when $\mathbf{K}_1$ designed for τ_1 is implemented for the system with the true time delay τ_4, the performance deteriorates.

Next, we examine how well the hysteretic switching algorithm works. Assume that we start with a control with the gain $\mathbf{K}_1$ designed for τ_1 while the system delay is τ_4. Figure 3 shows that the hysteretic algorithm is able to switch to the control with thc gain $\mathbf{K}_3$ and $\mathbf{K}_4$ since the controls with these gains have a similar performance. Figure 4 shows the switching signal $\pi(t)$ and the control index.

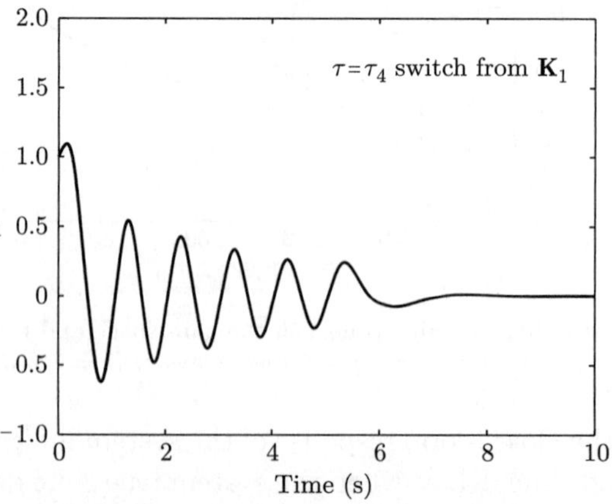

Fig. 3. The closed-loop response of the system (29) under the switching control when the initial gain of the control is $\mathbf{K}_1$ designed for τ_1 while the system true time delay is τ_4. This is to be compared with the upper-left sub figure in Fig.2.

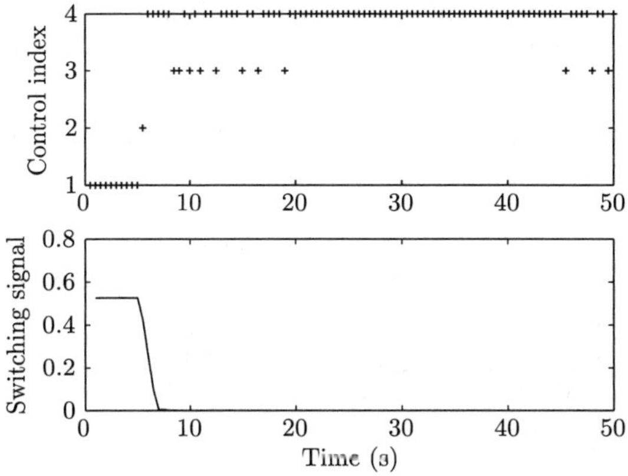

Fig. 4. The switching signal (bottom) and the control index (top) of the hysteretic switching algorithm for the closed-loop response of the LTI system (19) under the lower order control.

2.4.2. *High-order LQR Optimal Control*

Consider the system (19). The discretization number of the delayed control is set to be $N = 2^5$. We consider a feedback control $u(t-\tau)$ with an uncertain transport delay τ. The nominal value of the time delay is $\tau_{\text{nom}} = \pi/30$. $\tau_{\text{min}} = 0.1\tau_{\text{nom}}$ and $\tau_{\text{max}} = 9\tau_{\text{nom}}$. We sample ten time delays: $\tau_1 = \tau_{\text{min}}$, $\tau_2 = \tau_{\text{nom}}$, $\tau_3 = 2\tau_{\text{nom}}$, $\tau_4 = 3\tau_{\text{nom}}$, $\tau_5 = 4\tau_{\text{nom}}$, $\tau_6 = 5\tau_{\text{nom}}$, $\tau_7 = 6\tau_{\text{nom}}$, $\tau_8 = 7\tau_{\text{nom}}$, $\tau_9 = 8\tau_{\text{nom}}$, and $\tau_{10} = \tau_{\text{max}}$. The matrix $\mathbf{Q}$ for the LQR optimal control design is taken to be $Q_{ij} = 0$ for all i and j except for $Q_{11} = \omega^2$ and $Q_{22} = 1$, and $\mathbf{R}$, being a scalar in this case, is chosen to be 1.

Figures 5–7 show the closed-loop response of the system under the feedback

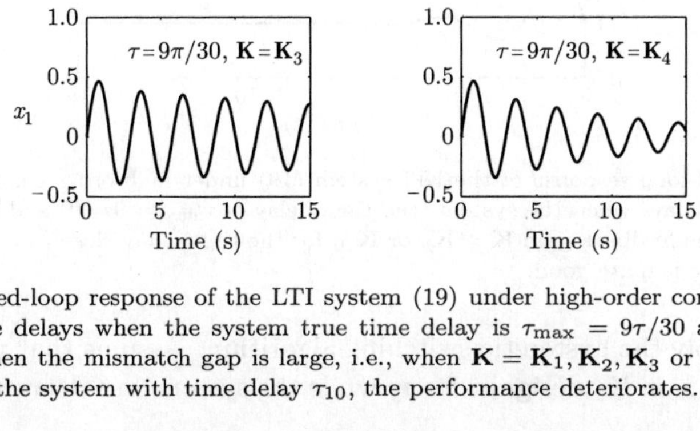

Fig. 5. The closed-loop response of the LTI system (19) under high-order controls designed for the sampled time delays when the system true time delay is $\tau_{\text{max}} = 9\tau/30$ and is assumed to be unknown. When the mismatch gap is large, i.e., when $\mathbf{K} = \mathbf{K}_1, \mathbf{K}_2, \mathbf{K}_3$ or $\mathbf{K}_4$ for τ_1 to τ_4 is implemented for the system with time delay τ_{10}, the performance deteriorates.

control designed for the ten sampled time delays. The true time delay of the system is taken to be $\tau_{10} = 9\pi/30$, *which is assumed to be unknown and not considered in the control design.* As it can be seen from the figures that when the control designed for the time delay that is close to τ_{10} is implemented, the performance is acceptable. Otherwise, the performance deteriorates.

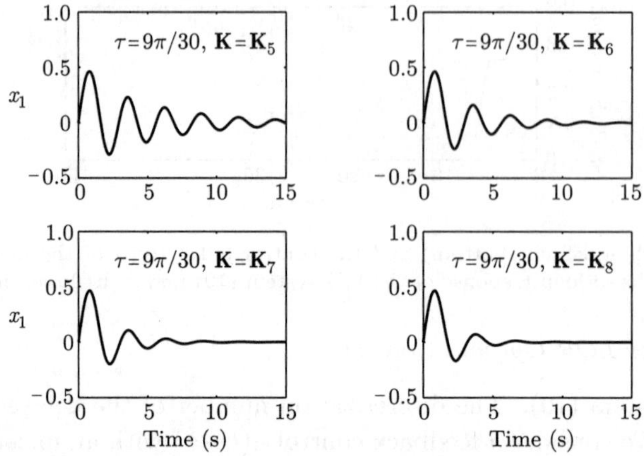

Fig. 6. The closed-loop response of the LTI system (19) under high-order controls designed for the sampled time delays when the system true time delay is $\tau_{\max} = 9\tau/30$ and is assumed to be unknown. When the feedback gain is $\mathbf{K}_7$ or $\mathbf{K}_8$ for the time delay close to the actual one, the control performance is quite good.

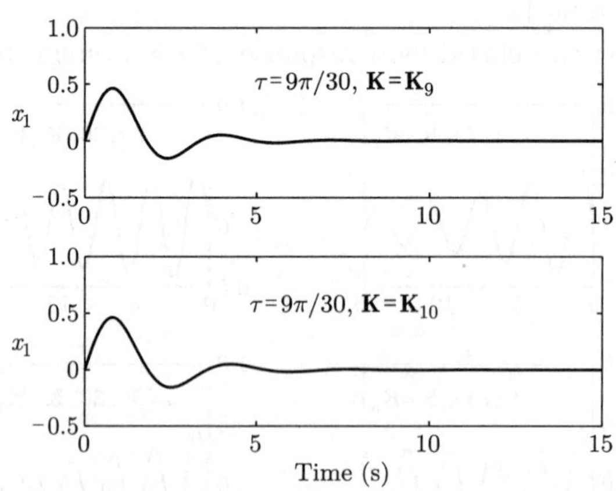

Fig. 7. The closed-loop response of the LTI system (19) under high-order controls designed for the sampled time delays when the system true time delay is $\tau_{\max} = 9\tau/30$ and is assumed to be unknown. When the feedback gain $\mathbf{K} = \mathbf{K}_9$ or $\mathbf{K}_{10}$ for the time delay close to the actual one, the control performance is quite good.

Next, we apply the hysteretic switching algorithm. Assume that we start with a control with the gain $\mathbf{K}_1$ designed for τ_1 while the system true delay is τ_{10}. Figure 8 shows that the hysteretic algorithm is able to switch to achieve the performance that can be achieved by the control with the gain $\mathbf{K}_{10}$. Figure 9 shows the switching signal $\pi(t)$ and the control index.

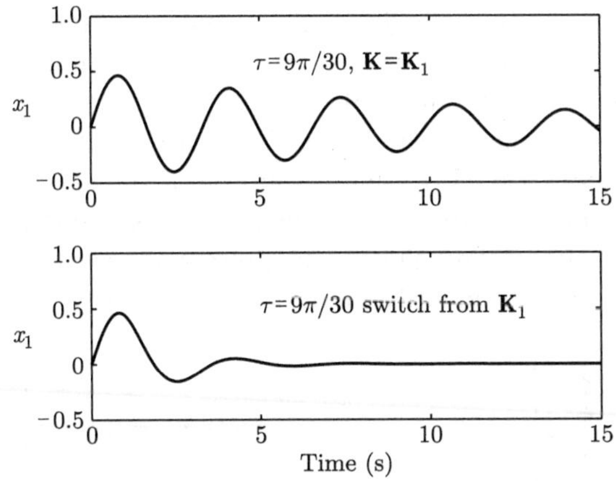

Fig. 8. The closed-loop response of the LTI system (19) under the switching high-order control when the initial gain of the control is $\mathbf{K}_1$ designed for τ_1 while the system true time delay is τ_{10} (bottom), as compared to the case when the gain is fixed at $\mathbf{K}_1$ (top).

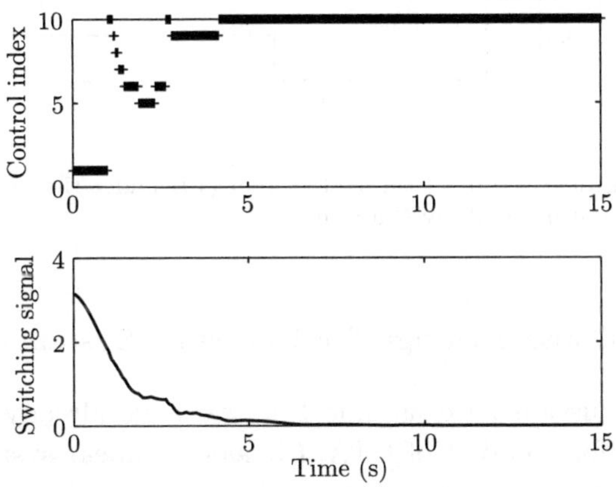

Fig. 9. The switching signal (bottom) and the control index (top) of the hysteretic switching algorithm for the closed-loop response of the LTI system (19) under the high-order control.

2.5. *Identification of Time Delay*

The switching index in Eq.(5) defines a criterion that can be used for identification of unknown time delay in dynamical systems. A similar index to that in Eq.(5) is used by Chen and Cai[38] and has provided an effective way to identify the time delay. It is natural to expect that as a consequence of switching controls, the switching index should point to the true underlining time delay in the system when the switching converges.

We consider the LTI system under the high-order control in the above example. We assume that the time delay of the control is time-varying as shown in the bottom part of Fig.10. The top part of the figure shows the switching index. The instances when the switching index changes match well with the moments when the time delay changes. When the switching index is flat, indicating the convergence of the supervisory control, it points to the index that matches the unknown time delay. Although we don't show the closed-loop response of the system, the performance of the control in response to the time-varying delay is comparable to that of the system with constant time delay. The results presented here demonstrate the ability of the supervisory control to identify unknown time delay.

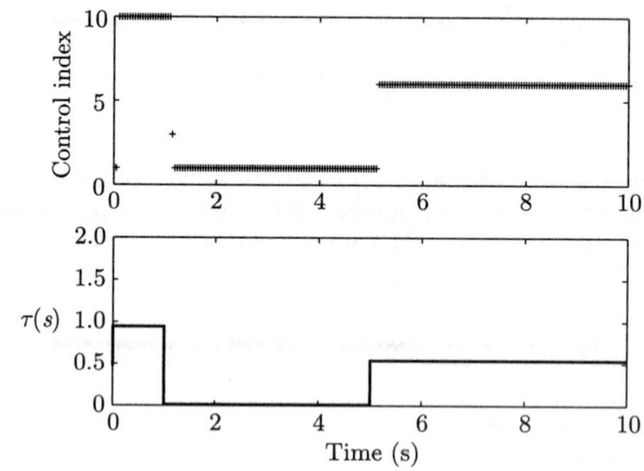

Fig. 10. The switching index (top) of the LTI system (19) under the high-order LQR optimal control and with the time-varying delay (bottom).

3. Sliding Mode Control Design for Uncertain Systems

In the following discussions, we assume that we have already taken care of the control delay in the formulation of CTA. Consider a linear system with uncertain nonlinear elements

$$\dot{\mathbf{y}} = \mathbf{F}\mathbf{y}(t) + \mathbf{G}\mathbf{u}(t) + \mathbf{g}(\mathbf{y},t), \tag{20}$$

where $\mathbf{y} \in \mathbb{R}^{n(N+1)}$, $\mathbf{F} \in \mathbb{R}^{n(N+1)\times n(N+1)}$, $\mathbf{G} \in \mathbb{R}^{n(N+1)\times m}$, $\mathbf{u} \in \mathbb{R}^m$, and $\mathbf{g} \in \mathbb{R}^{n(N+1)}$ $(n(N+1) \geq m)$. We assume that $\left|\mathbf{F} - \hat{\mathbf{F}}\right| \leq \tilde{\mathbf{F}}$ where $\hat{\mathbf{F}}$ is an estimate of $\mathbf{F}$, $\tilde{\mathbf{F}}$ is a known semi-positive definite matrix, and that $|\mathbf{g} - \hat{\mathbf{g}}| \leq \tilde{\mathbf{g}}$, where $\mathbf{g}$ is uncertain, $\hat{\mathbf{g}}$ is an estimate of $\mathbf{g}$, and $\tilde{\mathbf{g}} \in \mathbb{R}^{n(N+1)}$ is a known vector with non-negative elements.

Let $\mathbf{s}(\mathbf{y}) = \mathbf{S}\mathbf{y}$ be a sliding surface where $\mathbf{s} \in \mathbb{R}^m$ and $\mathbf{S} \in \mathbb{R}^{m\times n(N+1)}$. Assume that $\mathbf{SG}$ is invertible. When the system is sliding on $\mathbf{s}$, we have $\mathbf{s}(\mathbf{y}) = 0$ and

$\dot{\mathbf{s}}(\mathbf{y}) = \mathbf{S}\dot{\mathbf{y}} = \mathbf{0}$. From Eq.(20), we have

$$\mathbf{S}\dot{\mathbf{y}} = \mathbf{S}\mathbf{F}\mathbf{y}(t) + \mathbf{S}\mathbf{G}\mathbf{u}_{eq}(t) + \mathbf{S}\mathbf{g}(\mathbf{y}, t) = \mathbf{0}. \tag{21}$$

This leads to a nominal control $\mathbf{u}_{eq}$

$$\mathbf{u}_{eq}(t) = -(\mathbf{S}\mathbf{G})^{-1}(\mathbf{S}\mathbf{F}\mathbf{y}(t) + \mathbf{S}\mathbf{g}(\mathbf{y},t)). \tag{22}$$

Applying the estimates of the uncertain components and adding a switching term to it, we yield a control to be implemented

$$\mathbf{u}(t) = -(\mathbf{S}\mathbf{G})^{-1}(\mathbf{S}\hat{\mathbf{F}}\mathbf{y}(t) + \mathbf{S}\hat{\mathbf{g}}(\mathbf{y}, t)) - (\mathbf{S}\mathbf{G})^{-1}\mathbf{K}\mathrm{sgn}(\mathbf{s}), \tag{23}$$

where $\mathbf{K} \in \mathbb{R}^{m \times m}$ is a switching gain matrix to be determined. Consider a Lyapunov function

$$V = \frac{1}{2}\mathbf{s}^T\mathbf{s} \geq 0. \tag{24}$$

The time derivative of V reads

$$\begin{aligned} \dot{V} &= \mathbf{s}^T\mathbf{S}\dot{\mathbf{y}} \\ &= \mathbf{s}^T\mathbf{S}(\mathbf{F}\mathbf{y} + \mathbf{G}(-(\mathbf{S}\mathbf{G})^{-1}(\mathbf{S}\hat{\mathbf{F}}\mathbf{y} + \mathbf{S}\hat{\mathbf{g}}(t, \mathbf{y})) - (\mathbf{S}\mathbf{G})^{-1}\mathbf{K}\mathrm{sgn}(\mathbf{s})) + \mathbf{g}(t, \mathbf{y})) \\ &\leq |\mathbf{s}^T[\mathbf{S}(\mathbf{F} - \hat{\mathbf{F}})\mathbf{y} + \mathbf{S}(\mathbf{g} - \hat{\mathbf{g}})]| - \mathbf{s}^T\mathbf{K}\mathrm{sgn}(\mathbf{s}) \\ &\leq |\mathbf{s}^T\mathbf{S}\tilde{\mathbf{F}}\mathbf{y}| + |\mathbf{s}^T\mathbf{S}\tilde{\mathbf{g}}| - \mathbf{s}^T\mathbf{K}\mathrm{sgn}(\mathbf{s}). \end{aligned} \tag{25}$$

We choose a matrix $\mathbf{K}$ such that

$$\mathbf{s}^T\mathbf{K}\mathrm{sgn}(\mathbf{s}) = |\mathbf{s}^T\mathbf{S}\tilde{\mathbf{F}}\mathbf{y}| + |\mathbf{s}^T\mathbf{S}\tilde{\mathbf{g}}| + \gamma(\mathbf{s}), \tag{26}$$

where $\gamma(\mathbf{s}) > 0$ is a positive function of $\mathbf{s}$. Then, $\dot{V} \leq -\gamma(\mathbf{s})$ and the system is stable.

Consider a special case $\mathbf{K} = \kappa\mathbf{I}$ where κ is a scalar and $\mathbf{I}$ is a unit matrix. We have

$$\dot{V} \leq |\mathbf{s}^T\mathbf{S}\tilde{\mathbf{F}}\mathbf{y}| + |\mathbf{s}^T\mathbf{S}\tilde{\mathbf{g}}| - \kappa\sum_{i=1}^{m}|s_i| \equiv -\eta\sum_{i=1}^{m}|s_i|, \tag{27}$$

where

$$\kappa = \eta + \frac{|\mathbf{s}^T\mathbf{S}\tilde{\mathbf{F}}\mathbf{y}| + |\mathbf{s}^T\mathbf{S}\tilde{\mathbf{g}}|}{\sum_{i=1}^{m}|s_i|}, \qquad \eta > 0. \tag{28}$$

We point out that in the limit $\mathbf{s} \to 0$, the ratio in κ is finite.

3.1. *First Order System with Time Delay*

Consider a first order system

$$\dot{x}(t) = ax(t) + bx(t - \tau) + u(t) + g(x, t), \tag{29}$$

where $g(x, t) = \varepsilon x^3(t) + \delta \sin t$. The system constants are uncertain and fall in known ranges $-10 \leq a \leq -1$, $0 \leq b \leq 2$, $-0.1 \leq \varepsilon \leq 0.1$, and $-0.1 \leq \delta \leq 0.1$.

$\tau = 1$ is a constant time delay. In the simulation of the sliding mode control, we use the following as the unknown system:

$$\dot{x}(t) = -5x(t) + 0.5x(t-1) + 0.05x^3(t) + 0.05\sin t + u(t), \tag{30}$$

while the nominal system used in the sliding mode control design is taken to be

$$\dot{x}(t) = -2x(t) + x(t-1) + u(t). \tag{31}$$

This implies that the estimates $\hat{a} = -2$, $\hat{b} = 1$, and $\hat{g}(x,t) = 0$.

For the LPCTA method, we select the bandwidth ratio to be $r = 0.001$ and $N = 2^4$ based on our previous experience of computations. The extended vector is $\mathbf{y} \in \mathbb{R}^{(N+1)}$ and $N+1 = 17$. Consider a special index for designing the optimal sliding surface

$$J = \frac{1}{2}\int_{t_s}^{\infty} \mathbf{y}^T\mathbf{y}dt = \frac{1}{2}\int_{t_s}^{\infty} \mathbf{q}^T\mathbf{Q}\mathbf{q}dt. \tag{32}$$

Note that the selection of the matrix $\mathbf{Q}$ is done in Ref.11. We implement the sliding mode control for the special case when $\mathbf{K} = \kappa\mathbf{I}$. The control is given by Eq.(23) with $\eta = 1$.

Figure 11 shows the sliding surface response and sliding mode control of the system starting from the initial condition $x(0) = 1$. The sliding surface converges to the zero fast. It should be noted that the sliding mode control is designed based on the knowledge of upper and lower bounds of the parameter uncertainties and without using the knowledge of the parameters in the unknown system (30). The control performance is excellent because the control has no time delay. Next,we consider the effect of time delay of the control.

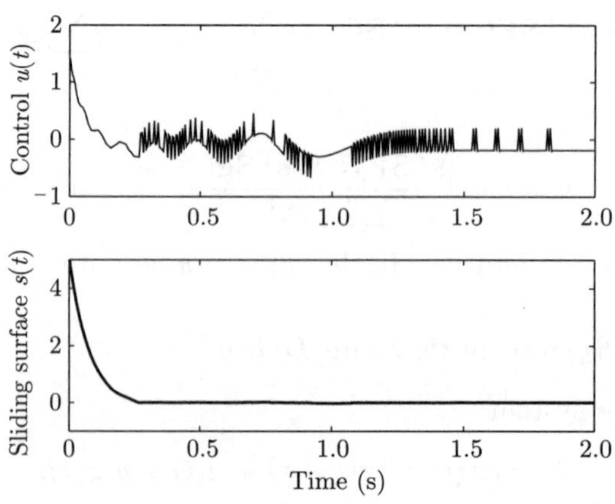

Fig. 11. The sliding control $u(t)$ and sliding function $s(t)$ of the first order nonlinear system (30). $x(0) = 1$.

3.2. *First Order System with Delayed Control*

Consider a first order system with delayed control

$$\dot{x}(t) = ax(t) + \varepsilon x^3(t) + \delta \sin t + bu(t - \tau_u), \tag{33}$$

where $g(x,t) = \varepsilon x^3(t) + \delta \sin t$. The system constants are uncertain and fall in the known ranges $-2 \leq a \leq 2$, $0.1 \leq b \leq 2.1$, $-0.1 \leq \varepsilon \leq 0.1$, and $-0.1 \leq \delta \leq 0.1$. $\tau_u = 1$ is a constant time delay. In the simulation of the sliding mode control, we use the following as the unknown system

$$\dot{x}(t) = x(t) + 0.005x^3(t) + 0.005 \sin t + u(t-1), \tag{34}$$

while the nominal system used in the sliding mode control design is taken to be

$$\dot{x}(t) = -x(t) + 0.5u(t-1). \tag{35}$$

This implies that the estimates $\hat{a} = -1$, $\hat{b} = 0.5$, and $\hat{g}(x,t) = 0$. Recall that the uncertainty in b becomes part of the uncertainty of the matrix $\mathbf{F}$ in Eq.(20) and the matrix $\mathbf{G}$ is fully determined. Hence, in the sliding mode control design, we only deal with the additive uncertainty in $\mathbf{F}$. We also note that the unknown open-loop system is unstable, while the nominal open-loop system is stable.

For the CTA method, we use the forward difference approximation for the derivative of the delayed control, and $M = 2^4$ based on our previous experience of computations. The extended vector is $\mathbf{y} \in \mathbb{R}^{(M+1)}$ and $M + 1 = 17$. To design the optimal sliding surface, we consider the same index as in Eq.(32), and follow the same steps as in the previous example. We implement the sliding mode control for the special case when $\mathbf{K} = \kappa\mathbf{I}$ and $\eta = 1$.

Figure 12 shows the response and the delayed sliding mode control of the system starting from the initial condition $x(0) = 1$. Before the control kicks in, the system is unstable and the response grows. The sliding mode control stabilizes the system.

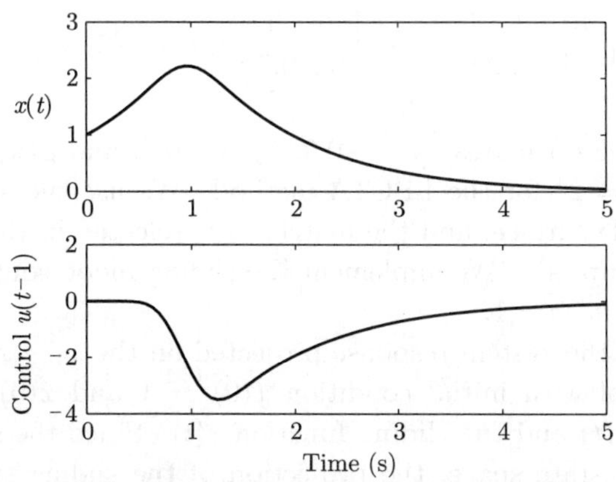

Fig. 12. The response $x(t)$ of the first order nonlinear system (34) and the delayed sliding mode control $u(t-1)$. $x(0) = 1$. The uncontrolled system is unstable.

Note that the delay of the control is shorter than $\tau_u = 1$. This is due to the fact that the numerical integration in the simulation uses the time step smaller than $\Delta\tau_u = 1/2^4$, which helps propagate the control $u(t)$ to $u(t - \tau_u)$ over a shorter time delay. In an experimental setting or a fully discretized simulation, we can establish a definite relation between the discretization time step $\Delta\tau_u$ and the sample time of the controller so that we can realize the exact time delay of the control.

3.3. *Second Order Uncertain System*

Consider a second-order system

$$\dot{\mathbf{x}}(t) = \begin{bmatrix} 0 & 1 \\ -k & -c \end{bmatrix} \mathbf{x}(t) + \begin{bmatrix} 0 & 0 \\ -k_p & -k_d \end{bmatrix} \mathbf{x}(t-\tau) + \mathbf{g}(\mathbf{x}, t) + \begin{bmatrix} 0 \\ 1 \end{bmatrix} u(t), \tag{36}$$

$$\mathbf{g}(\mathbf{x}, t) = \begin{bmatrix} 0 \\ \varepsilon x_1^3(t) + \delta \sin t \end{bmatrix}, \tag{37}$$

where $\mathbf{x} = [x_1(t), x_2(t)]^T$, $k = 4$, $c = 0.2$, and $\tau = \pi/2$. Other system parameters are assumed to be unknown, and to fall in a range with known upper and lower bounds: $-1 \le k_d \le 0$, $-1 \le k_p \le 0$, $-0.1 \le \varepsilon \le 0.1$, and $-0.1 \le \delta \le 0.1$. In the simulation of the sliding mode control, we use the following as the unknown system:

$$\dot{\mathbf{x}}(t) = \begin{bmatrix} 0 & 1 \\ -4 & -0.2 \end{bmatrix} \mathbf{x}(t) + \begin{bmatrix} 0 & 0 \\ 0.5 & 0.5 \end{bmatrix} \mathbf{x}(t-\pi/2) + \mathbf{g}(\mathbf{x}, t) + \begin{bmatrix} 0 \\ 1 \end{bmatrix} u(t), \tag{38}$$

$$\mathbf{g}(\mathbf{x}, t) = \begin{bmatrix} 0 \\ 0.05 x_1^3(t) + 0.05 \sin t \end{bmatrix}, \tag{39}$$

while the nominal system used in the sliding mode control design is taken to be

$$\dot{\mathbf{x}}(t) = \begin{bmatrix} 0 & 1 \\ -4 & -0.2 \end{bmatrix} \mathbf{x}(t) + \begin{bmatrix} 0 & 0 \\ 0.9 & 0.7 \end{bmatrix} \mathbf{x}(t-\pi/2) + \begin{bmatrix} 0 \\ 1 \end{bmatrix} u(t). \tag{40}$$

This implies that the estimates $\hat{k}_p = -0.9$, $\hat{k}_d = -0.7$, and $\hat{\mathbf{g}}(\mathbf{x}, t) = 0$. We select $r = 0.0001$ and $N = 2^4$ for the LPCTA method. We use the same index for the optimal sliding mode surface, and the matrix $\mathbf{Q}$ is selected in the same manner as in the previous examples[11]. We implement the sliding mode control for the special case when $\mathbf{K} = \kappa\mathbf{I}$ and $\eta = 1$.

Figure 13 shows the system response projected on the $x_1 - x_2$ plane under the sliding mode control with initial condition $x(0) = 1$ and $\dot{x}(0) = 1$. Figure 14 shows the control $u(t)$ and the sliding function $s(t)$. Since the system lives in an infinite dimensional state space, the projection of the sliding surface on the two dimensional $x_1 - x_2$ plane does not resemble the typical sliding surface of two dimensional dynamical systems.

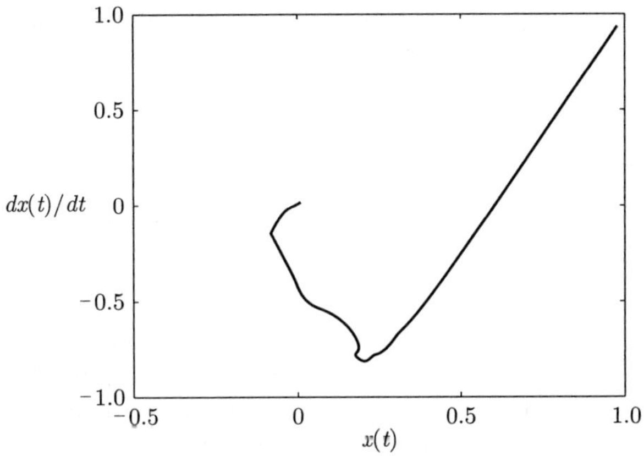

Fig. 13. The response trajectory in the $x_1 - x_2$ state space of the second order system (38).

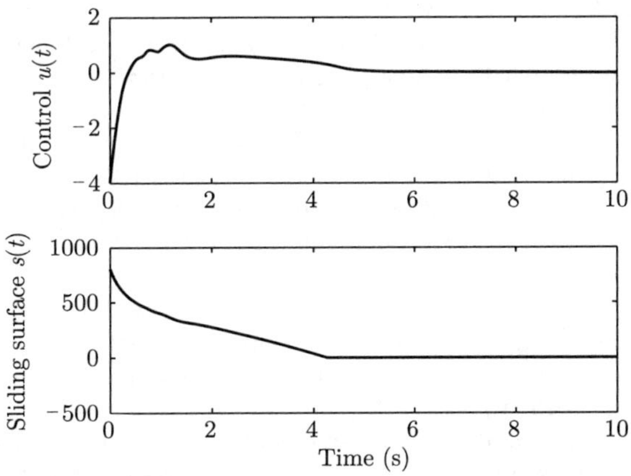

Fig. 14. The sliding control $u(t)$ and sliding function $s(t)$ of the second order system (38).

4. Concluding Remarks

We have demonstrated the need for switching controls when the time-delayed dynamical system has various uncertainties. The hysteretic switching algorithm can deal with feedback control with unknown time delay. When the upper and lower bounds of the unknown control delay are known, the set of feedback controls designed for the sampled time delays within the bounds can stabilize the system. The controls can be designed with the mapping method or the high-order design method. Numerical examples indicate that the performance of high-order controls tends to be better than the lower order controls designed by the mapping method. Sliding mode controls, on the other hand, can deal with system uncertainties. The sliding surface for the system with time delay is defined in the infinite dimensional state space. Its projection on the two dimensional plane spanned by the physical variables does not resemble the typical sliding surface of low dimensional dynamical systems without time delay. Furthermore, we have demonstrated that the switching index of the supervisory control serves as a criterion to identify unknown time delay in the system. It is thus possible to have a switching control to stabilize the system and to identify the unknown time delay at the same time.

References

1. W. H. Chen, Z. H. Guan, and X. M. Lu, Delay-dependent output feedback guaranteed cost control for uncertain time-delay systems, *Automatica*, **40** (7), 1263–1268 (2004).
2. G. Ji and Q. Luo, Iterative learning control for uncertain time-delay systems, *Dynamics of Continuous Discrete and Impulsive Systems-Series A-Mathematical Analysis*, **13**, 1300–1306 (2006).
3. O. M. Kwon, J. H. Park, and S. M. Lee, On delay-dependent robust stability of uncertain neutral systems with interval time-varying delays, *Applied Mathematics and Computation*, **203** (2), 843–853 (2008).
4. D. Yue, E. Tian, Y. Zhang, and C. Peng, Delay-distribution-dependent robust stability of uncertain systems with time-varying delay, *International Journal of Robust and Nonlinear Control*, **19** (4), 377–393 (2009).
5. Y. He, Q. G. Wang, C. Lin, and M. Wu, Delay-range-dependent stability for systems with time-varying delay, *Automatica*, **43** (2), 371–376 (2007).
6. X. B. Hu and W. H. Chen, Model predictive control for constrained systems with uncertain state-delays, *International Journal of Robust and Nonlinear Control*, **14** (17), 1421–1432 (2004).
7. X. F. Jiang and Q. L. Han, New stability criteria for linear systems with interval time-varying delay, *Automatica*, **44** (10), 2680–2685 (2008).
8. C. Y. Kao and A. Rantzer, Stability analysis of systems with uncertain time-varying delays, *Automatica*, **43** (6), 959–970 (2007).
9. C. L. Lin, C. H. Chen, and H. C. Huang, Stabilizing control of networks with uncertain time varying communication delays, *Control Engineering Practice*, **16** (1), 56–66 (2008).
10. D. E. Miller and D. E. Davison, Stabilization in the presence of an uncertain arbitrarily large delay, *IEEE Transactions on Automatic Control*, **50** (8), 1074–1089 (2005).
11. B. Song and J. Q. Sun, Supervisory control of dynamical systems with uncertain time delays, *Journal of Vibration and Acoustics*, **132** (6), (2010).
12. A. S. Morse, Supervisory control of families of linear set-point controllers – Part 1: Exact matching, *IEEE Transactions on Automatic Control*, **41** (10), 1413–1431 (1996).
13. A. S. Morse, Supervisory control of families of linear set-point controllers – Part 2: Robustness, *IEEE Transactions on Automatic Control*, **42** (11), 1500–1515 (1997).
14. J. P. Hespanha, D. Liberzon, and A. S. Morse, Logic-based switching control of a nonholomic system with parametric modeling uncertainty, *Systems & Control Letters*, **38** (3), 167–177 (1999).
15. J. P. Hespanha, D. Liberzon, and A. S. Morse, Hysteresis-based switching algorithms for supervisory control of uncertain systems, *Automatica*, **39** (2), 263–272 (2003).
16. X. G. Yan, S. K. Spurgeon, and C. Edwards, Sliding mode control for time-varying delayed systems based on a reduced-order observer, *Automatica*, doi:10.1016/j.automatica.2010.05.017 (2010).
17. Y. S. Song, S. H. Kwon, S. K. Park, D. C. Oh, and E. T. Jeung. Sliding mode con-

trol for time-delay system using a virtual state, *Proceedings of the American Control Conference*, **6**, 4669–4674 (2001).
18. S. Qu, M. Gong, X. Wang, and H. Qu. Sliding mode control for linear neutral time-delay systems. *Proceedings of Chinese Control and Decision Conference*, 4984–4987 (2008).
19. Y. H. Roh and J. H. Oh, Robust stabilization of uncertain input-delay systems by sliding mode control with delay compensation, *Automatica*, **35**, 1861–1865 (1999).
20. Y.-H. Roh and J. -H. Oh, Sliding mode control with uncertainty adaptation for uncertain input-delay systems, *International Journal of Control*, **73**(13), 1255–1260 (2000).
21. Y. Niu, J. Lam, and X. Wang. Sliding-mode control for uncertain neutral delay systems, *Proceedings of IEE Control Theory and Applications*, **151**, 38–44 (2004).
22. Y. He, J. Yi, Y. Liu, and Y. Jing. Delay-dependent sliding mode control for uncertain neutral delay systems, *Proceedings of Chinese Control and Decision Conference*, 5001–5005 (2008).
23. H. Xing, C. Gao, G. Tang, and D. Li, Variable structure sliding mode control for a class of uncertain distributed parameter systems with time-varying delays, *International Journal of Control*, **82**(2), 287–297 (2009).
24. L. Wu and W. X. Zheng, Passivity-based sliding mode control of uncertain singular time-delay systems, *Automatica*, **45**, 2120–2127 (2009).
25. Y. Niu, D. W. C. Ho, and J. Lam, Robust integral sliding mode control for uncertain stochastic systems with time-varying delay, *Automatica*, **41**, 873–880 (2005).
26. J. X. Xu and W. J. Cao, Synthesized sliding mode and time-delay control for a class of uncertain systems, *Automatica*, **36**, 1909–1914 (2000).
27. B. Song and J. Q. Sun, Sliding mode control of uncertain dynamical systems with time delay using the continuous time approximation method, *Journal of Vibration and Control*, doi: 10.1177/1077546311421795 (2011).
28. J. Sheng, O. Elbeyli, and J. Q. Sun, Stability and optimal feedback controls for time-delayed linear periodic systems, *AIAA Journal*, **42**(5), 908–911 (2004).
29. J. Sheng and J. Q. Sun, Feedback controls and optimal gain design of delayed periodic linear systems, *Journal of Vibration and Control*, **11**(2), 277–294 (2005).
30. J. Q. Sun, A method of continuous time approximation of delayed dynamical systems, *Communications in Nonlinear Science and Numerical Simulation.* **14**(4), 998–1007 (2008).
31. T. Insperger and G. Stepan. Semi-discretization of delayed dynamical systems. *Proceedings of ASME* 2001 *Design Engineering Technical Conferences and Computers and Information in Engineering Conference*, Pittsburgh, Pennsylvania (2001).
32. T. Insperger and G. Stepan, Semi-discretization method for delayed systems, *International Journal for Numerical Methods in Engineering*, **55**(5), 503–518 (2002).
33. T. Insperger and G. Stepan, *Semi-discretization for Time-delay Systems* (Springer, New York, 2011).
34. O. Elbeyli and J. Q. Sun, On the semi-discretization method for feedback control design of linear systems with time delay, *Journal of Sound and Vibration*, **273**, 429–440 (2004).
35. W. Kwon and A. Pearson, Feedback stabilization of linear systems with delayed control, *IEEE Transactions on Automatic Control*, **25**(2), 266–269 (1980).
36. G.-P. Cai, J.-Z. Huang, and S. X. Yang, An optimal control method for linear systems with time delay, *Computers & structures*, **81**, 1539–1546 (2003).
37. G. F. Franklin, J. D. Powell, and M. L. Workman, *Digital Control of Dynamic Systems* (Addison Wesley Longman, Berkeley, California, 1998).
38. L. X. Chen and G. P. Cai, Identification of time-delay in dynamical systems with the

method of particle swarm optimization, *Chinese Journal of Applied Mechanics*, **27**, 433–437 (2010).

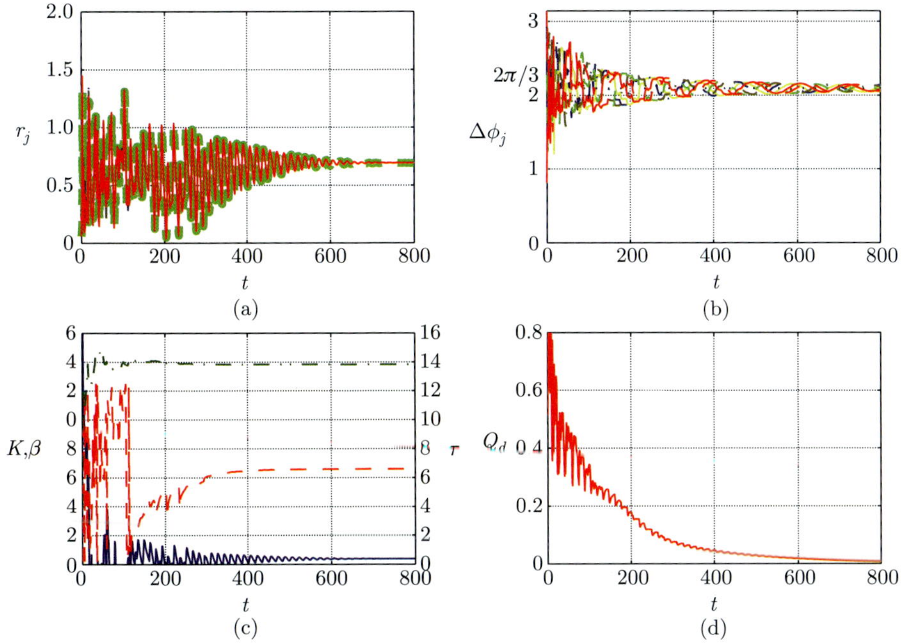

Fig. 7. Adaptive control of 3-cluster state in a network of 6 nodes by simultaneously tuning K, β, and τ according to Eqs. (23), (25), and (24), respectively. (a) absolute values $r_j = |z_j|$; (b) phase differences $\Delta\phi_j = \varphi_j - \varphi_{j+1}$; (c) temporal evolution of control parameters β (red dashed line), K (blue solid line), τ (green dashed-dotted line); (d) goal function. $\Gamma_\beta = \Gamma_K = 10$, $\Gamma_\tau = 0.05$. Other parameters as in Fig. 3.[58]

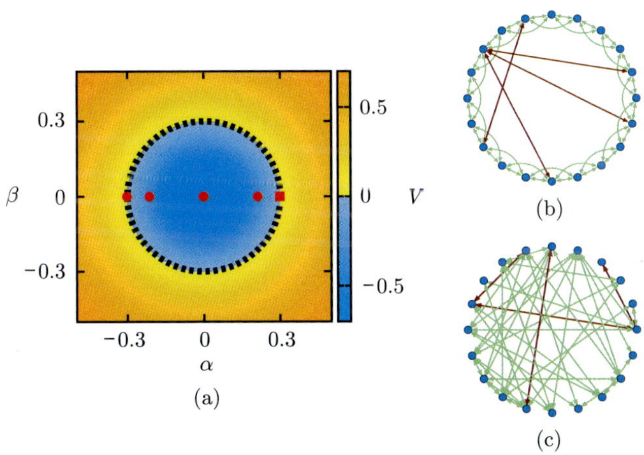

Fig. 8. (a) Master stability function for a network of FHN systems given by Eqs. (26). Dotted curve: $S((0,0),\sigma)$. Red circles (square): rescaled transversal (longitudinal) eigenvalues $\gamma_i = \sigma\nu_i$ of a bidirectionally coupled ring with $N = 8$ nodes. Parameters: $a = 1.3$, $\epsilon = 0.01$, $\sigma = 0.3$, $\tau = 1$. (b) Scheme of a bidirectional regular network ($N = 20$, $k = 2$), and (c) a random network ($N = 20$, fixed number of links kN) with excitatory coupling (green arrows) on which inhibitory links (red arrows) are superimposed.[32]

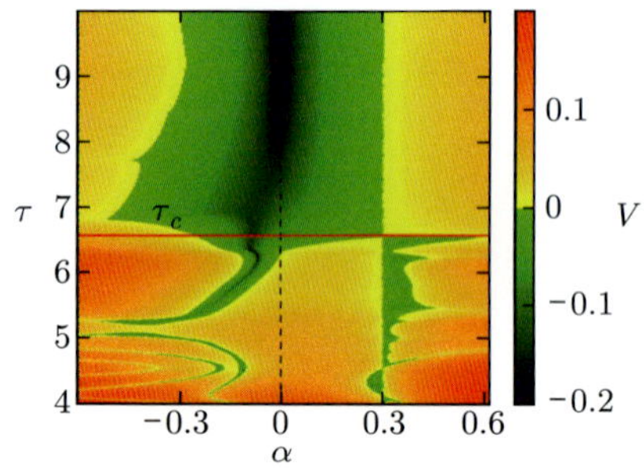

Fig. 9. Master stability function $\lambda_{\max}$ for a fixed coupling strength $\sigma = 0.3$ in the plane of the real part α ($\beta = 0$) and the delay time τ. The horizontal red line shows the position of the critical delay time τ_c. Parameter: $b = 0.95$.[33]

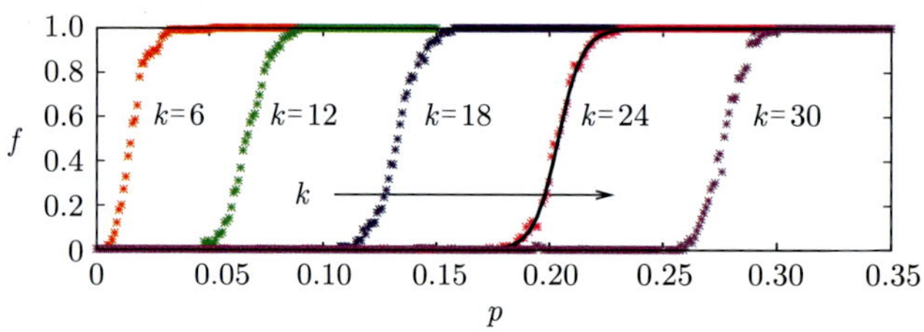

Fig. 10. Fraction of desynchronized networks f vs the probability of additional inhibitory links p for $N = 100$. k varies from 6 to 30. Number of realizations: 500 for each value of k. Parameters as in Fig. 8.[32]